Anti-SARS-CoV-2 Activity of Flavonoids

In the past years, COVID-19 caused millions of deaths and severely damaged not only public health but also the global economy. New variants continue to threaten human health. Thus, natural compounds for retarding infection are in high demand. This book summarizes the emerging research with flavonoids, whose application holds great promise of antivirus, anti-inflammation, antioxidative stress, and immunomodulatory. Topics include the role of flavonoids in preventing an inflammatory storm and how flavonoids interact with critical protein targets that are involved in the infection stages of COVID-19.

Key Features

- Highlights anti-SARS-CoV-2 drug discovery on natural products and medicinal plants.
- Includes all major subclasses of flavonoids that promisingly combat COVID-19.
- Proposes molecular mechanisms of flavonoids against protein targets of SARS-CoV-2.
- Contributions from an international team of leading researchers.
- Provides recommendations with respect to the future research.

Jen-Tsung Chen is a professor of cell biology at the National University of Kaohsiung in Taiwan. He also teaches genomics, proteomics, plant physiology, and plant biotechnology. Dr. Chen's research interests include bioactive compounds, chromatography techniques, plant molecular biology, plant biotechnology, bioinformatics, and systems pharmacology. He is an active editor of academic books and international journals to advance the exploration of multidisciplinary knowledge involving plant physiology, plant biotechnology, nanotechnology, materials science, ethnopharmacology, and systems biology. He serves as an associate editor, editorial board member, and guest editor in reputed journals. Dr. Chen published books in collaboration with international publishers and he is handling book projects on diverse topics such as drug discovery, herbal medicine, medicinal biotechnology, nanotechnology, bioengineering, plant functional genomics, plant speed breeding, epigenetics, functional RNAs, and CRISPR-based plant genome editing. Dr. Chen is a productive author in academic publications and has been included in the World's Top 2% Scientists 2023 by Stanford University.

Anti-SARS-CoV-2 Activity of Flavonoids

Edited by
Jen-Tsung Chen

CRC Press is an imprint of the
Taylor & Francis Group, an **informa** business

Designed cover image: Shutterstock

First edition published 2025
by CRC Press
2385 NW Executive Center Drive, Suite 320, Boca Raton FL 33431

and by CRC Press
4 Park Square, Milton Park, Abingdon, Oxon, OX14 4RN

CRC Press is an imprint of Taylor & Francis Group, LLC

ISBN: 978-1-032-55621-5 (hbk)
ISBN: 978-1-032-55976-6 (pbk)
ISBN: 978-1-003-43320-0 (ebk)

DOI: 10.1201/9781003433200

Typeset in Times
by codeMantra

Contents

Preface

The pandemic of COVID-19 is one of the most serious global public health crises of all time, and in the past years, it has caused over 6 million deaths and countless damage to economic and social activities all over the world. This disease is caused by an RNA virus, SARS-CoV-2, which tends to mutate quickly into variants, and therefore, continues to threaten human health. In the post-era of the COVID-19 pandemic, there is still a high demand to explore and develop effective treatments against SARS-CoV-2 variants with changing symptoms leading to a range of health problems including the tendency to develop into severe cases and probable death.

For decades, medicinal plants, traditional herbal medicines, and functional foods with natural bioactive compounds have been well proven to benefit human health at different levels such as antivirus, anti-inflammation, anti-oxidative stress, and immunomodulatory. Under the fundamentals of a paradigm shift, intensive research based on modern molecular tools was conducted in recent years to explore the role of bioactive compounds in COVID-19 management, particularly finding ways to combat SARS-CoV-2 variants. One of the most promising groups of bioactive compounds belongs to flavonoids that have been revealed for their potential in acting agents for COVID-19 management.

This book provides scientific proof based on a series of literature reviews organized by professors, researchers, and experts, focusing on the anti-SARS-CoV-2 activity of selected flavonoids highlighting molecular mechanisms that target the viral proteins to retard their life cycle and consequently against the disease. The role of medicinal plants and traditional herbal medicines in acting as anti-SARS-CoV-2 agents was comprehensively reviewed in this book, showing that their constituents of flavonoids including apigenin, baicalein, caflanone, catechins, epigallocatechin gallate, genistein, hesperidin/hesperetin, Icariin, luteolin, myricetin, naringin, naringenin, neohesperidin and quercetin can contribute to the inhibition of viral infection and subsequent replication.

Based on the advancements of bioinformatics tools, systems pharmacology including the studies using molecular docking and computer simulation gives significant contributions to enrich our understanding of the interaction between bioactive compounds with target proteins achieving a high resolution for paving ways to drug discovery. This book summarizes the current achievements in predicting the potential of selected flavonoids, such as subgroups of flavanones, flavones, and flavonols, for their anti-SARS-CoV-2 activity using the platforms of *in silico* technologies. In the subgroup of flavonols and their derivatives, the anti-SARS-CoV-2 activity came from the inhibitory effects in viral enzymes, and additionally, these compounds potentially control cytokine storms caused by the SARS-CoV-2 infection through regulating inflammatory mediators. For instance, one of the viral enzymes, RNA-dependent RNA polymerase (RdRp), plays a vital role in the life cycle of SARS-CoV-2 at the stages of replication and transcription, making it an ideal target for combating the disease by bioactive compounds. Selected flavonoids including anthocyanidin, chalcones, flavonols, flavone, and isoflavones were tested for their structural interactions with RdRp to evaluate the potential to inhibit its enzyme activity. Based on these studies, this book presents molecular insights into the potential ways to control COVID-19 by targeting crucial viral enzymes such as RdRp through flavonoids and derivatives. Comprehensive summaries for anti-SARS-CoV-2 studies using other types of protein targets, such as papain-like protease (PLpro) and 3-chymotrypsin-like protease (3CLpro), for testing the inhibitory activity of flavonoids were presented in this book. Interestingly, in addition to the inhibitory on protein targets, a detailed analysis of quercetin shows its considerable effects of antioxidant, anti-inflammatory, and immunomodulatory, which are capable of playing a crucial role in COVID-19 management.

During the infection process, one of the viral structural proteins, i.e., spike protein, has been identified as an essential component to interact with the host receptor, angiotensin-converting enzyme 2 (ACE2), and therefore, is undoubtedly a target for combating COVID-19. This book provides current findings using *in silico*, *in vitro*, and clinical studies highlighting the inhibitory activity of selected flavonoids for targeting the spike protein of SARS-CoV-2. In the meantime, some research also tests the potential of selected flavonoids on inhibitory targeting ACE2 for disrupting the entrance of SARS-CoV-2 into the host cell. Interestingly, in addition to the anti-SARS-CoV-2 activity by targeting proteins, some flavonoids, for instance, myricetin, quercetin, and rutin, were found to simultaneously possess anti-inflammatory and immunomodulatory effects which have the potential to mitigate the symptoms during the infection.

This book provides an updated literature review on the discovery of anti-SARS-CoV-2 agents derived from natural flavonoids. The underlying molecular mechanisms were illustrated to support the current findings chiefly based on coordinated *in silico*, *in vitro*, and clinical investigations. The knowledge provided by this book contributes to the management of global public health and thus supports the goal of establishing "Good Health and Well-Being", which is one of the "Sustainable Development Goals (SDGs) 2030" by the United Nations. This book is an ideal reference for students, teachers, researchers, professors, and experts in certain fields of life sciences, including bioactive compounds, biochemistry, bioinformatics, biomedical science, computational biology, drug discovery, ethnopharmacology, medicinal plants, molecular docking, natural products, systems/network pharmacology, pharmacology, traditional herbal medicines, and viral diseases. As the editor, I'd like to thank all the authors for their insightful chapters. The instruction and help from the publisher are so much appreciated.

List of Contributors

Jothi Dheivasikamani Abidharini
Department of Human Genetics and Molecular Genetics
Bharathiar University
India

M. Abirami
Department of Biochemistry
Shrimati Indira Gandhi College
India

Krishnendu Acharya
Department of Botany
University of Calcutta
India

Pouya Alipour
Department of Biology, Faculty of Basic Sciences
University of Maragheh
Maragheh, Iran.

Heba Alshater
Department of Forensic Medicine and Clinical Toxicology
Menoufia University
Shbien El-Kom, Egypt

Arumugam Vijaya Anand
Department of Human Genetics and Molecular Biology
Bharathiar University
India

Celia María Curieses Andrés
Hospital Clínico Universitario
Valladolid, Spain

Lin Ang
KM Science Research Division
Korea Institute of Oriental Medicine
Daejeon, Korea

Amna Hamad Abdallah Atia
Medicinal and Aromatic Plants and Traditional Medicine Research Institute
National Center for Research
Khartoum, Sudan

Seyed Abdulmajid Ayatollahi
Phytochemistry Research Center
Shahid Beheshti University of Medical Sciences
Tehran, Iran

Ali Azghar
Laboratory of Bioresources, Biotechnology, Ethnopharmacology and Health, Faculty of Sciences
University Mohammed First
Oujda, Morocco

Salah-eddine Azizi
Higher Institute of Nursing Professions and Health Techniques
Oujda, Morocco

Roodabeh Bahramsoltani
Department of Traditional Pharmacy, School of Persian Medicine
Tehran University of Medical Sciences
Tehran, Iran
And
Research Center for Clinical Virology
Tehran University of Medical Sciences
Tehran, Iran

B. Balamuralikrishnan
Department of Food Science and Biotechnology
Sejong University
Seoul, South Korea

Elmostapha Benaissa
Department of Bacteriology
Mohammed V Teaching Military Hospital
Rabat, Morocco

Atanu Bhattacharjee
Department of Biotechnology and Bioinformatics
North-Eastern Hill University
India

Arthi Boro
Department of Human Genetics and Molecular Genetics
Bharathiar University
India

Pallab Chakraborty
Department of Botany
Visva-Bharati University
India

William C. Cho
Department of Clinical Oncology
Queen Elizabeth Hospital
Hong Kong SAR, China

Maria Daglia
International Research Center for Food Nutrition and Safety
Jiangsu University
Zhenjiang, China

Mohammed Dalli
Higher Institute of Nursing Professions and Health Techniques
Oujda, Morocco

Nour Elhouda Daoudi
Higher Institute of Nursing Professions and Health Techniques
Oujda, Morocco

José Manuel Pérez de la Lastra
Institute of Natural Products and Agrobiology
CSIC-Spanish Research Council
La Laguna, Spain

Pascal D. Douanla
Department of Organic Chemistry
The University of Yaoundé-1
Yaoundé, Cameroon

Mostafa Elouennass
Department of Bacteriology
Mohammed V Teaching Military Hospital
Rabat, Morocco

Hesham R. El-Seedi
International Research Center for Food Nutrition and Safety
Jiangsu University
Zhenjiang, China

Niusha Esmaealzadeh
Department of Traditional Pharmacy, School of Persian Medicine
Tehran University of Medical Sciences
Tehran, Iran

Seyedeh Elham Faghih-Shirazi
Department of Physiology
Shiraz University of Medical Sciences
Shiraz, Iran

Burtram Fielding
Department of Medical Bioscience
University of the Western Cape
Cape Town, South Africa

P. Gajalakshmi
Department of Microbiology
Shrimati Indira Gandhi College
India

Neha Garg
Department of Medicinal Chemistry
Banaras Hindu University
India

Mojtaba Ghobadi
Department of Physiology
Shiraz University of Medical Sciences
Shiraz, Iran

Salar Hafez Ghoran
H.E.J. Research Institute of Chemistry
University of Karachi
Karachi, Pakistan

Zhiming Guo
School of Food and Biological Engineering
Jiangsu University
Zhenjiang, China

Puja Gupta
Department of Biotechnology
RIMT University
India

Arun Bahadur Gurung
Department of Biological Sciences
Indian Institute of Science Education and Research Kolkata
India

Solomon Habtemariam
Pharmacognosy Research & Herbal Analysis Services UK
Kent, United Kingdom

Baskaran Stephen Inbaraj
Department of Food Science
Fu Jen Catholic University
New Taipei City, Taiwan

S Jayashree
Department of Biotechnology
Stella Maris College (Autonomous)
India

Chetna Jhagta
Department of Medicinal Chemistry
Government College of Pharmacy
India

Celia Andrés Juan
Cinquima Institute and Department of Organic Chemistry
Valladolid University
Valladolid, Spain

Rhythm Kalsi
Department of Food Technology and Nutrition
Lovely Professional University
India

T. Karpagam
Department of Biochemistry
Shrimati Indira Gandhi College
India

Piyush Kashyap
Department of Food Technology and Nutrition
Lovely Professional University
India

Shaden A.M. Khalifa
Psychiatry and Psychology Department
Capio Saint Göran's Hospital
Stockholm, Sweden

Shiv Kumar
MMICT&BM (HM)
Maharishi Markandeshwar (Deemed to be) University
India

Ranjith Kumavath
Department of Biotechnology
Pondicherry University
India

Yassine Ben Lahlou
Department of Bacteriology
Mohammed V Teaching Military Hospital
Rabat, Morocco

Zisheng Luo
College of Biosystems Engineering and Food Science
Zhejiang University
Hangzhou, People's Republic of China

Adil Maleb
Laboratory of Microbiology, Faculty of Medicine and Pharmacy
University Mohammed First
Oujda, Morocco

R. Manikandan
Department of Biochemistry
Shrimati Indira Gandhi College
India

G Mayakkannan
Algaeliving Sdn Bhd
Kuala Lumpur, Malaysia

Vineet Mehta
Department of Pharmacology
Govt. College of Pharmacy
India

Ilyass Alami Merrouni
Higher Institute of Nursing Professions and Health Techniques
Oujda, Morocco

Alessandro Di Minno
Department of Pharmacy
University of Napoli Federico II
Naples, Italy

Amaresh Mishra
School of Biotechnology
Gautam Buddha University
India

Priyanka Nagu
Department of Pharmaceutics
Govt. College of Pharmacy
India

Deepak Nandi
Department of Biotechnology
School of Biosciences
RIMT University
India

Anand Kumar Pandey
Department of Biotechnology Engineering
Institute of Engineering and Technology
Bundelkhand University
India

Arun Parashar
School of Pharmaceutical Sciences
Shoolini University
India

Yamini Pathak
School of Biotechnology
Gautam Buddha University
India

Ayyadurai Pavithra
Department of Human Genetics and Molecular Genetics
Bharathiar University
India

Keenau Pearce
Department of Biotechnology
University of the Western Cape
Cape Town, South Africa

Eduardo Pérez-Lebeña
Sistemas de Biotecnología y Recursos Naturales
Valladolid, Spain

Francisco J. Plou
Institute of Catalysis and Petrochemistry
CSIC-Spanish Research Council
Madrid, Spain

Yixian Quah
Developmental and Reproductive Toxicology Research Group
Korea Institute of Toxicology
Daejeon, Korea

Roja Rahimi
Department of Traditional Pharmacy, School of Persian Medicine
Tehran University of Medical Sciences
Tehran, Iran
And
PhytoPharmacology Interest Group (PPIG)
Universal Scientific Education and Research Network (USERN)
Tehran, Iran

Naina Rajak
Department of Medicinal Chemistry
Banaras Hindu University
India

Sonu Ram
Department of Biotechnology
RIMT University
India

R. Ramya
Department of Biochemistry
Shrimati Indira Gandhi College
India

Shivani Rana
Department of Pharmacognosy
Govt. College of Pharmacy
India

Saravanan Renuka
Department of Chemistry and Biosciences
SASTRA Deemed to be University
India

Mohammed Roubi
Laboratory of Bioresources, Biotechnology, Ethnopharmacology and Health, Faculty of Sciences
University Mohammed the First
Oujda, Morocco

Abderrazak Saddari
Laboratory of Microbiology, Faculty of Medicine and Pharmacy
University Mohammed First
Oujda, Morocco

Palanisamy Sampathkumar
Department of Chemistry and Biosciences
SASTRA Deemed to be University
India

Joy Sarkar
Department of Botany
Dinabandhu Andrews College
India

A. Shanmugapriya
Department of Biochemistry
Shrimati Indira Gandhi College
India

Javad Sharifi-Rad
Facultad de Medicina
Universidad del Azuay
Cuenca, Ecuador

Dolly Sharma
Amity Institute of Biotechnology
Amity University
India

Minaxi Sharma
CARAH ASBL
Rue Paul Pastur. Ath, Belgium

Pankaj Sharma
Department of Pharmaceutics
Govt. College of Pharmacy
India

S Aruna Sharmili
Department of Biotechnology
Stella Maris College (Autonomous)
India

Km Shivangi
School of Biotechnology
Gautam Buddha University
India

S Shruthi
Department of Medical Biochemistry
Dr. A.L.M.PG Institute of Basic Medical Sciences
University of Madras
India

Vipendra Kumar Singh
School of Biosciences and Bioengineering
Indian Institute of Technology Mandi
India

Ramalingam Sivakumar
Department of Chemistry and Biosciences
SASTRA Deemed to be University
India

Eunhye Song
Global Cooperation Center
Korea Institute of Oriental Medicine
Daejeon, Korea

K Sonia
Department of Biochemistry
Ethiraj College for Women (Autonomous)
India

Kandi Sridhar
STLO, INRAe
Institut Agro Rennes Angers
Rennes, France

Gunna Sureshbabu Suruthi
Department of Human Genetics and Molecular Genetics
Bharathiar University
India

Fatemeh Taktaz
Phytochemistry Research Center
Shahid Beheshti University of Medical Sciences
Tehran, Iran

Eswar Rao Tatta
Department of Genomic Science
Central University of Kerala
India

Farah Khameis Farag Teia
Medicinal and Aromatic Plants and Traditional Medicine Research Institute
National Center for Research
Khartoum, Sudan

Mamta Thakur
College of Dairy and Food Technology
Rajasthan University of Veterinary & Animal Sciences
India

Vishwas Tripathi
School of Biotechnology
Gautam Buddha University
India

Jyoti Upadhyay
School of Biotechnology
Gautam Buddha University
India

Sekar Vijayakumar
Marine College
Shandong University
Weihai, PR China

Shalja Verma
Indian Institute of Technology, Roorkee
India

Viroj Wiwanitkit
Chandigarh University
India

Jianbo Xiao
University of Vigo
Vigo, Spain

Manju Yadav
Institute of Nuclear Medicine and Allied Sciences
India

Nermeen Yosri
Chemistry Department of Medicinal and Aromatic Plants
Beni-Suef University
Beni-Suef, Egypt

Sora Yasri
KM Center
Bangkok Thailand

Alireza Zali
Functional Neurosurgery Research Center
Shahid Beheshti University of Medical Sciences
Tehran, Iran

Chao Zhao
College of Marine Sciences
Fujian Agriculture and Forestry University
Fuzhou, China

Xiaobo Zou
School of Food and Biological Engineering
Jiangsu University
Zhenjiang, China

1 Flavonoids
Molecular Mechanisms of Anti-SARS-CoV-2 Activity and Their Role in COVID-19 Management

Priyanka Nagu, Pankaj Sharma, Arun Parashar, Minaxi Sharma, Baskaran Stephen Inbaraj, Vineet Mehta, and Kandi Sridhar

1.1 INTRODUCTION

The severe acute respiratory syndrome coronavirus 2 (SARS-CoV-2) is accountable for the extremely spreadable coronavirus disease 2019 (COVID-19) (Choe et al., 2021). The SARS-CoV-2 virus is a highly contagious virus that spreads globally at an alarming rate; afterward, the initial cases of this principally respiratory virus-related illness were formerly reported in Wuhan, Hubei Province, China, in late December 2019 (Ramírez-Mora et al., 2020; Sharma et al., 2021). SARS-CoV-2 has a high potential to mutate and various variants of SARS-CoV-2 have been reported, among which the Omicron variation has distinguished itself as the predominant circulating variant (Shrestha et al., 2022). More than 8 million individuals have lost their lives due to the SARS-CoV-2 virus, leaving a horrible ecological footprint (Toth and Szigeti, 2016; Ryan and Nanda, 2022).

The name coronaviruses refers to a class of wrapped viruses containing positive-sense, single-stranded RNA besides viral subdivisions that resemble a head-like crown (Machhi et al., 2020). They are members of the Nidovirales order, Coronaviridae family, and Orthocoronavirinae. They are able to communicate a disease to mammals, together with humans, triggering typically slight infectious conditions that sporadically result in severe outburst clusters, like those instigated by the Middle East Respiratory Syndrome (MERS) virus in 2012 and 2015 (Memish et al., 2020). The high mortality rate and lack of effective therapeutic intervention for the management of SARS-CoV-2 infection can be attributed to its high potential to mutate and evolve as a separate variant, which is equally contagious. So far, five different variants of the SARS-CoV-2 virus have been identified. In December 2020, the United Kingdom reported the identification of the variant named B.1.1.7, often denoted as the UK variant. This variant carried a specific set of mutations that raised concerns due to its potential for increased transmissibility (Walensky et al., 2021). Around the same time in December 2020, another variant labeled B.1.351 emerged in South Africa, commonly known as the South African variant. It contained a distinct combination of mutations, including the E484K mutation, which raised worries about its impact on vaccine efficacy (Bian et al., 2021). Early in January 2021, the P.1 variant, also known as the Brazilian variant, was detected in Brazil. This variant displayed a mutation similar to those found in the B.1.351 variant and led to concerns about its potential for increased transmissibility and immune escape (Boehm et al., 2021). In December 2020, the delta variant, designated as B.1.617.2, existed first in India. This variant brings significant attention due to its high transmissibility and potential association with increased disease severity (Araf et al., 2022). In November 2021, B.1.1.529 variant, universally mentioned as the Omicron variant, was revealed in South Africa. This variant quickly raised alarm due to its large number of mutations, including several in the spike protein region. There were concerns about its potential impact on transmissibility, vaccine effectiveness, and potential for immune evasion (Morshed, 2021).

Currently, there is not one legitimately permitted vaccine or antiviral treatment accessible for the preclusion or management of SARS-CoV-2 contamination and allied complications. However, certain medications have demonstrated some effectiveness against coronaviruses. Among them are protease inhibitors, such as Lopinavir and Ritonavir, which are typically used in the treatment of HIV (Pandya et al., 2022). Additionally, nucleoside analogs, neuraminidase inhibitors, and RNA synthesis inhibitors are being explored as potential medications (Frediansyah, 2021).

Recently, natural products are getting more consideration from researchers as an outcome of their significant biological properties. The most significant phytoconstituents are flavonoids, which are found in fruits, herbs, vegetables, and dried fruits. It has been discovered that flavonoids have a wide range of impacts on both microbes and animals with different structures and activities. Through the scavenging of free radicals, flavonoids show strong antioxidant action potential. Flavonoids offer a range of therapeutic benefits, such as cardiovascular protection, anti-diabetic, and antiviral activities. Recent studies show that herbal products including flavonoids might have the capabilities to normalize the fabrication and release of pro-inflammatory

DOI: 10.1201/9781003433200-1

cytokines, obstruct the advancement of the virus in host cells, and prevent the spread of viruses in host cells.

1.2 STRUCTURE AND LIFE CYCLE OF CORONAVIRUS AND ITS VARIANTS

SARS-CoV-2 is a single-stranded RNA virus enclosed within a protective capsule (Arikan et al., 2022). Its complete genome spans a length of 29,881 base pairs and codes for a sequence of 9,860 amino acids. This genetic information enables the virus to produce a variety of proteins, including both those involved in its structure and those with non-structural functions, through the expression of specific gene segments (Chaitanya, 2019). Within the Open Reading Frame (ORF) region, genetic coding occurs for a range of non-structural proteins, counting RNA-dependent polymerase, 3-chymotrypsin-like protease, and papain-like protease. Conversely, the genes denoted as spike glycoprotein (S), nucleocapsid protein (N), membrane protein (M), and envelope protein (E) are responsible for producing structural proteins in the case of SARS-CoV-2 (Amin et al., 2021). Specifically, the M and E in SARS-CoV-2 serve crucial roles in the process of virus budding, while the S spearheads the virus's effective entry into host cells (Zech, 2021). The n protein is crucial for the life cycle of SARS-CoV-2 and is known to regulate various vital activities after virus invasion (Wu et al., 2023). Figure 1.1 depicts the structure of the coronavirus, highlighting various associated proteins.

When SARS-CoV-2 comes in contact with the host cell, the S protein undergoes substantial structural rearrangement, which enables the virus to be assigned through the host cell membrane. The spike proteins are essential for attaching to host cells, which are pulmonary epithelial cells with a high expression of the receptor called angiotensin-converting enzyme 2 (ACE-2) receptors, which have been identified as a functional receptor for SARS-CoV-2 and allow it to enter the cell through endocytosis (Ahmad et al., 2021). Normally, S exists in a metastable, perfusion conformation; the S protein initially attaches to this receptor, starting the virus invasion into host cells. After binding to a receptor called ACE-2, the S protein is sliced by a protease twice: first at the outset of the S1/S2 cleavage site for priming and then near a fusion peptide in the S2 subunit for activation (Zhang et al., 2021). The primary breaking

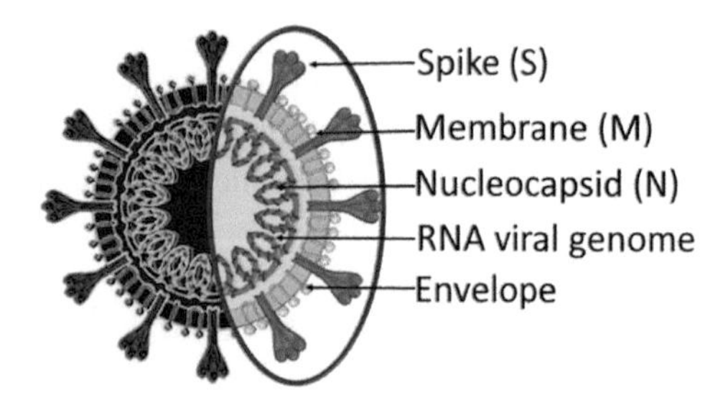

FIGURE 1.1 Structure of coronavirus. This figure is constructed by authors using BioRender, Toronto, Ontario, Canada.

seemingly stimulates the S protein, beginning modifications in conformation that allow the fusion of the viral and host cell membranes, whereas the second fragmentation stabilizes the S2 subunit at the attachment point (Pizzato et al., 2022). The virus enters pulmonary alveolar epithelial cells after membrane fusion, releasing the substances within the host cell. The virus replicates inside the host cell, creating a negative-strand RNA from the single-strand positive RNA that already exists through transcription (Parasher et al., 2021). Positive RNA strands are created by using the freshly generated negative-strand RNA as a template, and these positive RNA strands are subsequently used to translate new proteins of SARS-CoV-2 (Kovalev et al., 2014). The viral N protein fixes to the freshly formed genomic RNA, while the M protein aids in integration with the cellular endoplasmic reticulum (Zhou et al., 2014). The endoplasmic reticulum membrane encloses and transports this freshly produced nucleocapsid toward the lumen. After that, they are conveyed toward the cell membrane by the Golgi vesicles, before being subsequently expelled into the extracellular environment by exocytosis (Griffiths et al., 2012). The freshly generated virus particles can now invade surrounding epithelial cells and act as a source of infectious material for spreading the disease through respiratory droplets in the vicinity (Stadnytskyi et al., 2021). In addition to lung epithelial cells, the immune system, brain neuronal cells, and epithelial renal tubules are among the other key targets of SARS-CoV-2. The primary modes of SARS-CoV-2 transmission involve the exposure to respiratory droplets, which occurs through close contact with infected individuals or via droplet dispersion from individuals who may be asymptomatic, symptomatic, or in the pre-symptomatic phase (Masand et al., 2021). On the clinical level, COVID-19 is linked to a cough, an elevated temperature, difficulty in breathing, and finally respiratory system damage. Recent research suggests that COVID-19 is capable of attacking the CNS in addition to the above body systems. According to reports, 36% of COVID-19 patients experience neurological problems (Nagu et al., 2020).

The WHO has autonomously developed an ordering system for differentiating the developing variations of SARS-CoV-2 keen on variants of concern (VOCs) and variants of interest (VOIs). SARS-CoV-2 has several variants, some of which are considered important because of their potential to increase transmissibility or virulence. Five sub-genera or lineages make up the β-CoV genus (Preethi et al., 2022). Bats and rodents are the most likely gene bases of α and β CoVs, according to genomic analysis (Brussow, 2021). On the other hand, avian species appear to be the γ and δ CoV gene sources. The emergence of respiratory illness outbreaks can be attributed to a range of animal species, including camels, cattle, cats, and bats. These animals are vulnerable to a spectrum of respiratory, intestinal, hepatic, and neurological disorders caused by various members of this extensive virus family (Islam et al., 2021). These viruses have the ability to overcome species barriers for unidentified reasons and can infect humans with

TABLE 1.1
SARS-CoV-2 VOCs and VOIs

Variant	Variations	Lineage	Spike Mutations	Year Detected	Country
VOC	Alpha (α)	B.1.1.7	D614G, N501Y, P681H	September 2020	United Kingdom
	Beta (β)	B.1.351	A701V, N501Y, K417N, E484K, D614G	September 2020	South Africa
	Delta (δ)	B.1.617.2	P681R, T478K, L452R, D614G	December 2020	India
	Gamma (γ)	P.1	K417T, E484K, N501Y, D614G, H655Y	December 2020	Brazil
	Omicron	B.1.1.529	R346X, N460X, F490X	November 2021	South Africa and Botswana
VOI	Lambda (λ)	C.37	L452Q, F490S, D614G	December 2020	Peru
	Epsilon (ε)	B.1.427 and B.1.429	D614G, L452R	September 2020	United States of America
	Mu (μ)	B.1.621 and B.1.621.1	R346K, E484K, N501Y, D614G, P681H	January 2021	Colombia
	Zeta (ζ)	P.2	E484K, D614G	January 2021	Brazil
	Eta (η)	B.1.525	E484K, D614G, Q677H	December 2020	Nigeria
	Theta (θ)	P.3	E484K, N501Y, D614G, P681H	January 2021	The Philippines
	Iota (ι)	B.1.526	E484K, D614G, A701V	December 2020	United States of America
	Kappa (κ)	B.1.617.1	L452R, E484Q, D614G, P681R	December 2020	India

illnesses ranging from the common cold to more serious conditions. The overview of variants, linage, and spike mutations is given in Table 1.1.

1.3 PATHOPHYSIOLOGY OF SARS-COV-2

The most common means through which SARS-CoV-2 spreads in the population is by droplets created when the infected person coughs or sneezes (Figure 1.2). Within 6 feet of a person who is infected, the probability of transmission increases by several folds (Blocken et al., 2021). It has been suggested that the risk of acquiring SARS-CoV-2 infection from symptomatic subject persists if there is a short period of contact; however, there is a reduced chance of transmission during brief contact with asymptomatic subjects. Touching surfaces that have been exposed to the virus is another possible mechanism of SARS-CoV-2 transmission (Jones et al., 2020). Coughing, fever, and shortness of breath are common signs of COVID-19 infection, and reports of muscle soreness, anorexia, hyposmia, and taste loss are also becoming more common. The population most at risk for developing a serious illness is those over the age of 65, as well as those with existing medical conditions like high blood pressure, elevated blood sugar levels, and other long-term medical conditions affecting the heart, lungs, kidneys, and brain (Nagu et al., 2020).

The ACE-2 receptors for coronaviruses are present in the precursor cells, which become active upon the virus's entry into the body through respiratory or alternative pathways. The spikes on the virus attach to the ACE-2. As a result, ACE-2 receptors are subsequently downregulated. The downregulation of these receptors produces more angiotensin-2 (AT2). Further, increased AT2 production may lead to lung damage by raising pulmonary vascular permeability. SARS-CoV-2 can disrupt the normal functioning of the renin–angiotensin system (RAS) by targeting ACE-2 receptors. Also, in the brain, the virus gains entry to cells by binding to ACE-2 receptors, impairing the regulation of angiotensin II and leading to damage to ACE-2-producing tissues (Shirbhate et al., 2021). Additionally, the virus relies on the cofactor Transmembrane Serine Protease-2 (TMPRSS2), which eases viral entrance into cells, replication, and further infection (Heindl et al., 2024).

Figure 1.3 depicts the inflammatory response generated in the human body upon exposure to SARS-CoV-2 infection. To activate macrophages, the virus additionally secretes certain inflammatory mediators. IL-1, IL-6, and TNF are among the cytokines and chemokines that activate macrophages release into the bloodstream (Manik and Singh, 2022). This results in increased capillary permeability and vasodilation. Alveoli become compressed as a result of plasma leakage into the interstitial spaces of the alveoli cells. As a result, the levels of surfactant in alveolar epithelial cells drop, resulting in impeded gas exchange and alveolar collapse. The systemic inflammatory response is brought on in COVID-19 patients by a large proportion of CD14+ and CD16+ monocytes, which also generate IL-6 (Mudd et al., 2020). Therefore, it is evident that the cytokine storm that occurs in severe COVID-19 disease is primarily caused by a huge synthesis and release of pro-inflammatory cytokines. This may trigger a negative immunological response, which may ultimately result in breathing problems, cough, and acute respiratory distress syndrome (Middleton et al., 2020). Apart from this, the hypothalamus is impacted by the secreted IL-1, IL-6, and TNF-α. They induce the release of prostaglandin PGE-2, which raises the body temperature. Moreover, the hypoxic environment further leads to the development of tachycardia in infected individuals (Grossman, 2013). Septic shock and numerous organ failures are also possible outcomes from any of these bizarre inflammatory reactions (Weiss et al., 2020).

Antigen-presenting cells (APCs) show a fundamental role in commencing and coordinating immune responses against viral infections including COVID-19. When a viral infection occurs, APCs, such as dendritic cells and macrophages, recognize and engulf the virus particles.

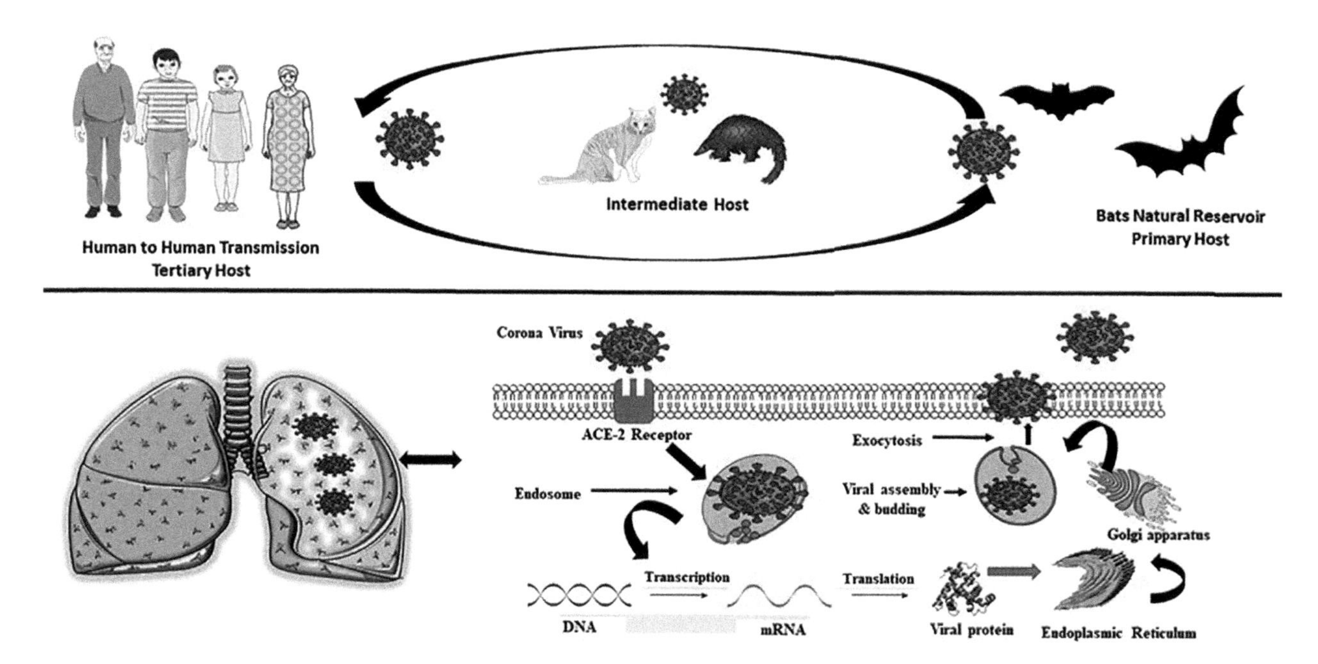

FIGURE 1.2 The life cycle of SARS-CoV-2. This figure is constructed by authors using BioRender, Toronto, Ontario, Canada.

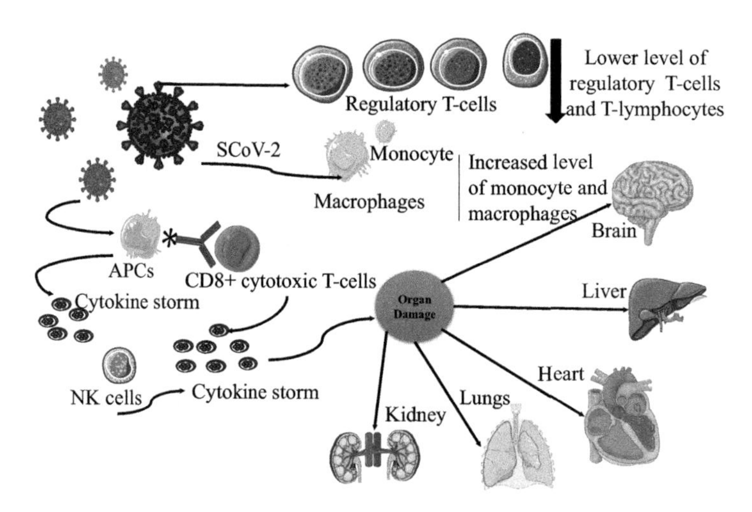

FIGURE 1.3 Inflammatory response of SARS-CoV-2. This figure is constructed by authors using BioRender, Toronto, Ontario, Canada.

Inside the APCs, the viral antigens are processed and presented on their cell surface using special molecules called major histocompatibility complex (MHC) molecules. These antigen–MHC complexes act as a "molecular fingerprint" that can be recognized by other immune cells (Stremersch et al., 2016). NK cells are fragments of the innate immune scheme and are responsible for detecting and eliminating virus-infected cells. When they encounter APCs presenting viral antigens, NK cells can recognize the abnormal MHC patterns and respond by releasing pro-inflammatory cytokines and exerting their cytotoxic activity against infected cells. This activation of NK cells provides immediate defense against the virus (Hammer et al., 2022).

On the other hand, CD8-positive cytotoxic cells are a key component of the adaptive immune system. They recognize viral antigens presented by APCs through their T-cell receptors. Once activated, CD8-positive cytotoxic cells multiply and differentiate into effector cells that directly target and kill virus-infected cells. This adaptive immune response is critical for eliminating the virus and providing long-term immunity (Moss, 2022). During this immune activation process, various pro-inflammatory cytokines are released by the activated immune cells, including both NK cells and CD8-positive cytotoxic cells. Cytokines are signaling molecules that help coordinate and regulate immune responses. However, in certain cases, the immune response can become overly intense, leading to an excessive and uncontrolled release of proinflammatory cytokines (Qudus et al., 2023). Considering COVID-19, a cytokine storm can occur when the immune system overreacts to the viral infection, causing an overwhelming production of cytokines. This can result in widespread inflammation and damage to various organs, including the brain, lungs, kidneys, liver, and heart. In severe cases of COVID-19, patients often exhibit lesser levels of NK cells, regulatory T-cells, and T-lymphocytes. These are important immune cell populations involved in controlling and regulating immune responses. The decreased levels of these cells suggest a dysregulated immune system in severe disease progression (Nusshag et al., 2023). Additionally, COVID-19 patients may show increased levels of monocytes and macrophages, which are types of immune cells that can produce pro-inflammatory cytokines. The elevated levels of these cells contribute to the heightened production of pro-inflammatory cytokines observed in severe cases of COVID-19 (Falck-Jones et al., 2023).

Overall, the immunologic features of COVID-19 involve the activation of APCs to current viral antigens to NK cells and CD8-positive cytotoxic cells, leading toward immune activation and the production of pro-inflammatory cytokines. In severe cases, an exaggerated immune response or cytokine storm can occur, resulting in organ damage. The observed alterations in immune cell populations reflect the dysregulation of the immune system in severe disease progression (Garduno et al., 2023).

1.4 TREATMENT AND THERAPIES FOR THE MANAGEMENT OF SARS-COV-2 INFECTION

The COVID-19 sickness has two phases in its pathology, the first of which occurs before or soon after the beginning of clinical symptoms and is characterized by the highest levels of SARS-CoV-2 replication (Sharma et al., 2021). During this phase of viral replication, antiviral drugs and antibody-based therapies are likely to be more efficient. The production of cytokines and the activation of the coagulation system, which results in a pro-thrombotic condition, are seen in the final stages of this phase (Choudhary et al., 2021). More than antiviral treatments, anti-inflammatory medications like corticosteroids, immune-modulating therapy, or a blend of these therapies are employed to reduce this hyperinflammatory condition (Bartoli et al., 2021). Convalescent plasma contains antibodies that develop in response to the infection in convalesced persons. These antibodies can potentially help those patients who are having SARS-CoV-2 infection by neutralizing the virus and reducing its severity. Monoclonal antibodies, such as bamlanivimab and imdevimab, have been used in high-risk individuals with mild to moderate symptoms to help neutralize the virus and prevent disease progression (Mengist et al., 2021). In some cases, corticosteroids like dexamethasone are prescribed to reduce excessive inflammation in the lungs, particularly in patients with severe disease or those requiring oxygen therapy (Singh et al., 2020).

All the above-mentioned therapeutic strategies are used for the symptomatic management of SARS-CoV-2 infection, but none of them can kill the SARS-CoV-2 virus. This could be the reason why this virus has continuously mutated and evolved as multiple variants since its discovery in late 2019. Recent time has witnessed significant research on herbal molecules in search of potential drug candidates for the management of SARS-CoV-2 infection (Jain et al., 2023). Although not much success has been achieved, this research has led to the identification of several crucial molecules that could be beneficial against SARS-CoV-2 infection. Flavonoids have demonstrated significant health benefits in humans in the past against a variety of ailments. They have been demonstrated as potential immune boosters, antiviral, antibacterial, anti-inflammatory, antioxidants, etc. (Noor et al., 2022). The flavonoid functional groups can interact with a variety of cellular targets and block a variety of pathways. They also have no systemic toxicity, have a strong history of working in synergy with conventional medications, and lack systemic toxicity (Ayilara et al., 2023). Flavonoids are candidates to be considered to obstruct the coronavirus life cycle and could help in the better management of SARS-CoV-2 infection, either directly or through adjuvant therapeutics.

1.5 FLAVONOIDS AS ANTIVIRAL AND SARS-COV-2 INHIBITOR

Single plant species have a tremendous degree of therapeutic potential when it comes to the study of herbal medicine. There have been questions concerning the pharmacologic usefulness of plants due to the possibility that a single plant could contain a large variety of biochemicals. Based on compelling evidence supporting their efficacy, plants are often categorized as antiviral, anti-inflammatory, immunomodulatory, and mixed-effects having various uses (Hemmami et al., 2023). The use of herbal medications has been demonstrated to have anti-inflammatory effects and may be essential in the management of SARS-CoV-2 infection (Asif et al., 2020). Increased levels of inflammatory markers like IL-6, erythrocyte sedimentation rate, and C-reactive protein, in addition to their direct antiviral effects, have also been linked to severe sickness and worse outcomes, most likely as a result of cytokine storm syndrome observed during SARS-CoV-2 infection (Henderson et al., 2021).

Various compounds called flavonoids are natural compounds produced in various fruits, vegetables, and plant-based foods. They are known for their diverse range of biological activities and have been extensively studied for their potential health benefits, which include antiviral activity. Numerous studies have demonstrated that they can obstruct different stages of the viral life cycle, including viral attachment, entry, replication, and release (Kausar et al., 2021). Flavonoid compounds have shown inhibitory effects against several important human pathogens, such as influenza viruses, human immunodeficiency viruses, herpes simplex viruses, hepatitis B and C viruses, and dengue viruses, among others (Zhang et al., 2018). However, it's important to note that the research on the specific effects of flavonoids against SARS-CoV-2, the virus that causes COVID-19, is limited, and no specific flavonoid has been proven to be an effective treatment for the virus so far (Kulandaisamy et al., 2022). Several studies have investigated the antiviral potential of flavonoids against other types of coronaviruses, such as SCoV and MERS-CoV, which are closely related to SARS-CoV-2 (Ahmad et al., 2021).

The fundamental structure of flavonoids consists of a C6-C3-C6 skeleton, forming a distinctive ring arrangement. They are characterized by their chemical structure, which consists of two-phenyl rings (A and B) and a three-carbon (C) bridge that makes up the 15-carbon skeleton that forms a heterocyclic ring (Santos et al., 2017). The A ring consists of carbons 2–3 and 6–8, the B ring consists of carbons 1, 4, and 5, and the C ring is formed by carbons 2, 3, and 4 (Figure 1.4). The basic structure of flavonoids is known as flavan, and various modifications can occur on this core structure, leading to a diverse range of flavonoid compounds (Samec et al., 2021). Flavonoids can undergo various modifications, including hydroxylation, methylation, glycosylation, and prenylation, among others. These modifications lead to the formation of different subclasses of flavonoids, such as flavones, flavonols, flavanones, isoflavones, anthocyanins, and proanthocyanidins (Table 1.2), each with its distinct structure and properties (George et al., 2017). These subclasses differ in the presence and position of functional groups such as hydroxyl (–OH), methoxy (–OCH3), or glycosyl (–O– sugar) groups attached to the flavan backbone.

These compounds possess remarkable biochemical properties, which have been linked to their potential in addressing various conditions such as cancer, Alzheimer's disease, and coronary artery diseases (Batiha et al., 2020). The presence and arrangement of these functional groups determine the chemical and biological properties of flavonoids. For example, hydroxyl groups contribute to the antioxidant activity of flavonoids, while glycosylic groups affect their solubility and bioavailability (Naso et al., 2023). The glycosylation of flavonoids is a common modification and plays a role in their transport, storage, and stability in plants (Vasudevan and Lee, 2020). Flavonoids also exhibit various chemical reactions, including oxidation, reduction, methylation, and conjugation. These reactions can further modify the structure and properties of flavonoids. Overall, the chemistry of flavonoids is diverse and complex, with their chemical structures and modifications contributing to their wide range of biological activities and potential health benefits (Batiha et al., 2020).

Studies have shown that some flavonoids, such as quercetin and epigallocatechin gallate (EGCG), have inhibitory effects on viral replication and may help reduce viral infectivity (Russo et al., 2020). It has been demonstrated that the flavonoid quercetin, which is present in a variety of fruits, vegetables, and herbs, has antiviral properties against numerous viruses. Some studies suggest that quercetin may constrain the entrance of the virus into host

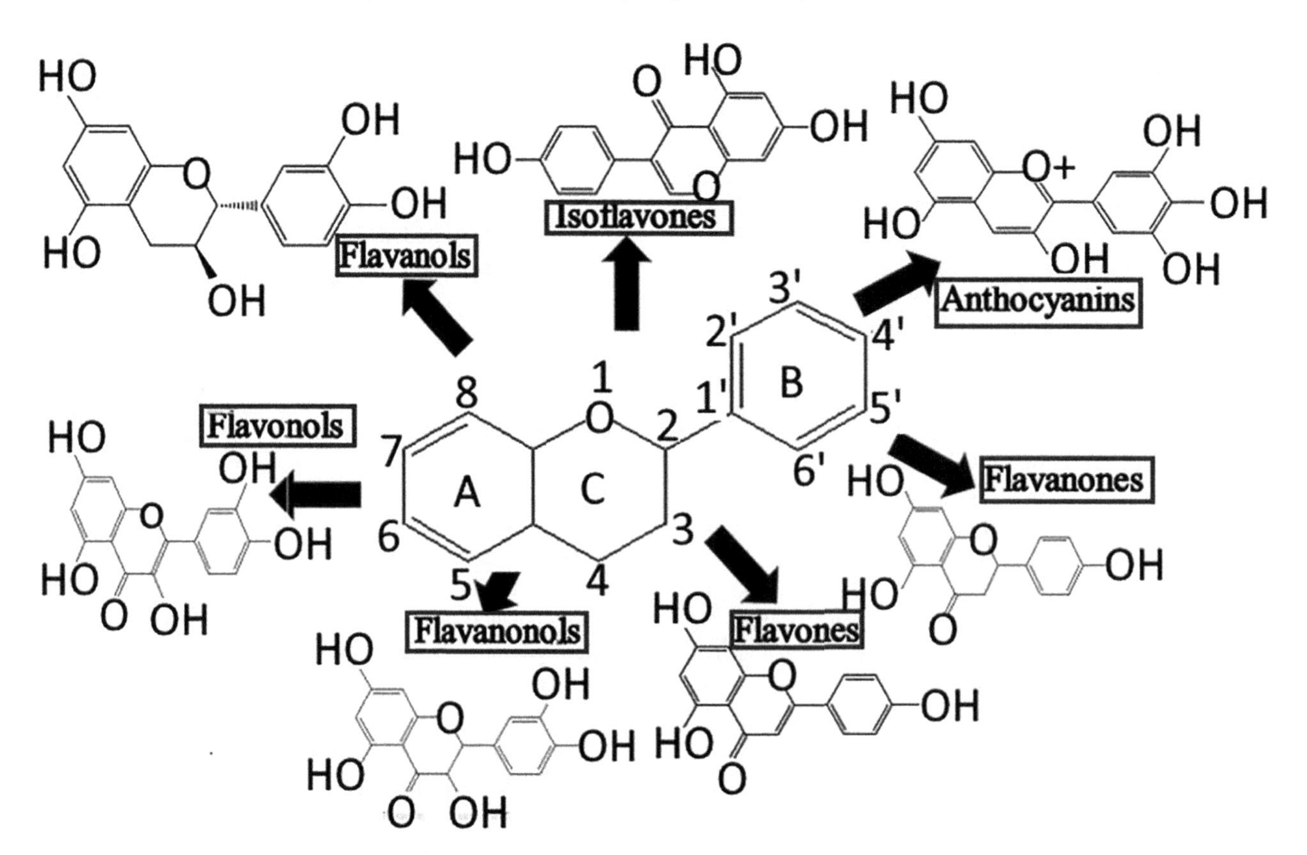

FIGURE 1.4 Structure of flavonoids. This figure is constructed by authors using BioRender, Toronto, Ontario, Canada.

TABLE 1.2
Flavonoids and Their Mechanism of Actions

Flavonoid	Mode of Action	References
Baicalein and baicalin	Inhibits SARS-CoV-2 RNA-dependent RNA polymerase activity	Zandi et al. (2021)
Baicalein	Inhibits SARS-CoV-2 replication and its 3C-like protease	Liu et al. (2021)
Luteolin	Inhibits SARS-CoV-2 RNA-dependent RNA polymerase activity	Munafo et al. (2022)
EGCG	Interferes with SARS-CoV-2 spike RBD–ACE-2 interaction	Ohishi et al. (2022)
Isorhamnetin	Binds to three residues involved in spike–ACE-2 interaction	Kaul et al. (2021)
Quercetin	Binds to three residues involved in spike receptor-binding domain and ACE-2 interaction	Williamson et al. (2020)
Neobaicalein	Binds to ACE-2 receptor	Hasan et al. (2021)
Oroxylin A	Binds to ACE-2 receptor	Zhang et al. (2023)
Scutellarin	Binds to ACE-2 receptor	Senthil et al. (2020)
Wogonin	Binds to ACE-2 receptor	Song et al. (2020)
Pelargonidin	Attenuates spike–ACE-2 interaction Reduces SARS-CoV-2 replication	Biagioli et al. (2021)
Epigallocatechin 3-O-gallate	HIV-1 RT-associated RNase	Mahboubi-Rabbani et al. (2021)
3-Deoxysappanchalcone	Decrease in viral genomic replication, DNA fragmentation	Hasan et al. (2021)
6-Hydroxy luteolin 7-O-β-d-glucoside	Decrease in neuraminidase	Bang et al. (2018)
Agathisflavone	Decrease in neuraminidase	Chaves et al. (2022)
Apigenin 7-O-β-d-(4-caffeoyl) glucuronide	Decrease HIV-1 integrase action	Lee et al. (2003)
Hesperidin	TNF-α, IL-6, and IFN-α	Guo et al. (2019)
Naringenin	Decrease the production of intracellular viral proteins	Tutunchi et al. (2020)

cells, reduce viral replication, and modulate the immune response (Milenkovic et al., 2021). However, it's imperative to note that this research was not specifically conducted on SARS-CoV-2. EGCG is a flavonoid abundant in green tea, and it has been deliberate for its antiviral properties against several viruses. Some research suggests that EGCG may prevent the virus from adhering to host cells and entering them and inhibit viral replication (Huseen et al., 2020). However, similar to quercetin, the studies on EGCG have not focused specifically on SARS-CoV-2 (Bashir et al., 2023). While these findings are promising, it's crucial to understand that the effectiveness of flavonoids as a treatment for SARS-CoV-2 is still speculative, and further research is needed.

Park et al. (2021) research has indicated that green tea polyphenols, in particular EGCG, prevent coronavirus enzyme activity and *in vitro* coronavirus propagation. This investigation looked at the coronavirus replication-inhibiting properties of green tea polyphenols in a mouse model. Jang et al. (2020) found that EGCG and theaflavin, the primary active components of green and black tea, respectively, are reported to have the potential to be beneficial in inhibiting SARS-CoV-2 activity. The 3CL-protease is necessary for the cleavage of the polyprotein of coronaviruses for the individual proteins to become functional. These findings imply that EGCG and theaflavin may be effective in treating SARS-CoV-2 infection.

The investigation of Xiao et al. (2021) on SARS-CoV-2 M^{pro} inhibitors reported myricetin to have strong inhibitory action with an IC50 of 3.684 M. Myricetin's chromone ring interacts with His41 via stacking in the binding pocket of SARS-CoV-2 M^{pro}, while its 3′-, 4′-, and 7-hydroxyl engage with Phe140, Asp187, and Glu166 through hydrogen bonds. These findings were significant in that they demonstrated that Myricetin inhibits the access of inflammatory cells and the release of inflammatory cytokines including IL-6, IL-1, TNF-α, and IFN-β, as well as the pulmonary inflammation caused by bleomycin.

A molecular docking study by Hadni et al. (2022) indicated that ligands including silibinin, tomentin a and b, amentoflavone, bilobetin, herbacetin, morin, , baicalein, and quercetin each displayed antagonistic ability for SARS-CoV-2 $3CL^{pro}$ and RBD. Furthermore, the molecular dynamics simulation studies predict that γ-glutamyl-S-allylcysteine shows the best results for M^{pro} of SARS-CoV-2 (Parashar et al., 2021).

A study by Joshi et al. (2021) focused on the screening of around 7,100 compounds, including active components utilized as antitussive drugs, antiretroviral phytochemicals, and synthesized antivirals, with SARS-CoV-2 M^{pro} as the major target. They discovered many naturally occurring compounds that firmly attach to the SARS-CoV-2 M^{pro}. These molecules include several flavonoids such as hesperidin, ε-viniferin, myricitrin, afzelin, phyllaemblicin B, and biorobin, suggesting a promising potential for flavonoids against SARS-CoV-2 infection. A separate study examined the possible inhibitory potentials of apigenin, fisetin, kolaflavanone, and remdesivir against SARS-CoV-2 main protease (6LU7). The findings of this investigation showed that all tested drugs had little or no toxicity and are plausible putative inhibitors of the SARS-CoV-2 main protease (Oladele et al., 2021). Conversely, additional investigational and experimental research is required to better understand their functions and confirm their effectiveness against COVID-19.

1.6 CONCLUSION

Flavonoids, present in a wide range of plants, could exhibit remarkable efficacy in both preventing and treating SARS-CoV-2 infection. Extensive studies have indisputably demonstrated the potent antiviral properties of flavonoids against SARS-CoV-2. These compounds effectively hinder the virus by targeting its key viral components, such as the ACE-2 receptor, TMPRSS2, M^{pro}, and receptor-binding domain. Additionally, flavonoids play a crucial role in mitigating inflammation caused by SARS-CoV-2 by blocking the production of various pro-inflammatory factors during the inflammatory response, which could help to counter various SARS-CoV-2-associated complications. To truly understand the efficacy and safety of these medications, there is a pressing need for more practical experimental research. Additionally, the low bioavailability of the ingested compounds requires consideration of methods to enhance their absorption in the body. Undeniably, flavonoids have demonstrated the ability to inhibit SARS-CoV-2 through direct and indirect mechanisms, making them a valuable group of anti-SARS-CoV-2 agents with significant therapeutic potential.

1.7 DECLARATION OF COMPETING INTEREST

The authors declare that they have no known competing financial interests or personal relationships.

REFERENCES

Ahmad I, Pawara R, Surana S, Patel H. The repurposed ACE2 inhibitors: SARS-CoV-2 entry blockers of Covid-19. *Topics in Current Chemistry*. 2021 Dec; 379:1–49.

Amin SA, Banerjee S, Ghosh K, Gayen S, Jha T. Protease targeted COVID-19 drug discovery and its challenges: Insight into viral main protease (Mpro) and papain-like protease (PLpro) inhibitors. *Bioorganic & Medicinal Chemistry*. 2021 Jan 1; 29:115860.

Araf Y, Akter F, Tang YD, Fatemi R, Parvez MS, Zheng C, Hossain MG. Omicron variant of SARS-CoV-2: Genomics, transmissibility, and responses to current COVID-19 vaccines. *Journal of Medical Virology*. 2022 May; 94(5):1825–32.

Arikan A, Sanlidag T, Sayan M, Uzun B, Uzun Ozsahin D. Fuzzy-based PROMETHEE method for performance ranking of SARS-CoV-2 IgM antibody tests. *Diagnostics*. 2022 Nov 17; 12(11):2830.

Asif M, Saleem M, Saadullah M, Yaseen HS, Al Zarzour R. COVID-19 and therapy with essential oils having antiviral, anti-inflammatory, and immunomodulatory properties. *Inflammopharmacology*. 2020 Oct; 28:1153–61.

Ayilara MS, Adeleke BS, Akinola SA, Fayose CA, Adeyemi UT, Gbadegesin LA, Omole RK, Johnson RM, Uthman QO, Babalola OO. Biopesticides as a promising alternative to synthetic pesticides: A case for microbial pesticides, phytopesticides, and nanobiopesticides. *Frontiers in Microbiology*. 2023 Feb 16; 14:1040901.

Bang S, Li W, Ha TK, Lee C, Oh WK, Shim SH. Anti-influenza effect of the major flavonoids from Salvia plebeia R. Br. via inhibition of influenza H1N1 virus neuraminidase. *Natural Product Research*. 2018 May 19; 32(10):1224–8.

Bartoli A, Gabrielli F, Alicandro T, Nascimbeni F, Andreone P. COVID-19 treatment options: A difficult journey between failed attempts and experimental drugs. *Internal and Emergency Medicine*. 2021 Mar; 16:281–308.

Bashir F, Komal A, Ibrahim A, Ain QT, Rehman B, Zaheer T. Exploring nature's invigorating power: Phytotherapy for SARS-CoV-2. *Phytopharmacological Communications*. 2023 Jun 30; 3(01):39–51.

Batiha GE, Beshbishy AM, Ikram M, Mulla ZS, El-Hack ME, Taha AE, Algammal AM, Elewa YH. The pharmacological activity, biochemical properties, and pharmacokinetics of the major natural polyphenolic flavonoid: Quercetin. *Foods*. 2020 Mar 23; 9(3):374.

Biagioli M, Marchianò S, Roselli R, Di Giorgio C, Bellini R, Bordoni M, Gidari A, Sabbatini S, Francisci D, Fiorillo B, Catalanotti B, Distrutti E, Carino A, Zampella A, Costantino G, Fiorucci S. Discovery of a AHR pelargonidin agonist that counter-regulates ACE2 expression and attenuates ACE2-SARS-CoV-2 interaction. *Biochemical Pharmacology*. 2021 Jun; 188:114564.

Bian L, Gao F, Zhang J, He Q, Mao Q, Xu M, Liang Z. Effects of SARS-CoV-2 variants on vaccine efficacy and response strategies. *Expert Review of Vaccines*. 2021 Apr 3; 20(4):365–73.

Blocken B, van Druenen T, Ricci A, Kang L, van Hooff T, Qin P, Xia L, Ruiz CA, Arts JH, Diepens JF, Maas GA. Ventilation and air cleaning to limit aerosol particle concentrations in a gym during the COVID-19 pandemic. *Building and Environment*. 2021 Apr 15; 193:107659.

Boehm E, Kronig I, Neher RA, Eckerle I, Vetter P, Kaiser L. Novel SARS-CoV-2 variants: The pandemics within the pandemic. *Clinical Microbiology and Infection*. 2021 Aug 1; 27(8):1109–17.

Brüssow H. COVID-19: Emergence and mutational diversification of SARS-CoV-2. *Microbial Biotechnology*. 2021 May 1; 14(3):756–68.

Chaitanya KV. Structure and organization of virus genomes. *Genome and Genomics: From Archaea to Eukaryotes*. 2019 Nov 18; 2019:1–30.

Chaves OA, Lima CR, Fintelman-Rodrigues N, Sacramento CQ, de Freitas CS, Vazquez L, Temerozo JR, Rocha ME, Dias SS, Carels N, Bozza PT. Agathisflavone, a natural biflavonoid that inhibits SARS-CoV-2 replication by targeting its proteases. *International Journal of Biological Macromolecules.* 2022 Dec 1; 222:1015–26.

Choe PG, Kim KH, Kang CK, Suh HJ, Kang E, Lee SY, Kim NJ, Yi J, Park WB, Oh MD. Antibody responses 8 months after asymptomatic or mild SARS-CoV-2 infection. *Emerging Infectious Diseases.* 2021 Mar; 27(3):928.

Choudhary S, Sharma K, Silakari O. The interplay between inflammatory pathways and COVID-19: A critical review on pathogenesis and therapeutic options. *Microbial Pathogenesis.* 2021 Jan 1; 150:104673.

Falck-Jones S, Österberg B, Smed-Sörensen A. Respiratory and systemic monocytes, dendritic cells, and myeloid-derived suppressor cells in COVID-19: Implications for disease severity. *Journal of Internal Medicine.* 2023 Feb; 293(2):130–43.

Frediansyah A, Tiwari R, Sharun K, Dhama K, Harapan H. Antivirals for COVID-19: A critical review. *Clinical Epidemiology and Global Health.* 2021 Jan 1; 9:90–8.

Garduno A, Martinez GS, Ostadgavahi AT, Kelvin D, Cusack R, Martin-Loeches I. Parallel dysregulated immune response in severe forms of COVID-19 and bacterial sepsis via single-cell transcriptome sequencing. *Biomedicines.* 2023 Mar 3; 11(3):778.

George VC, Dellaire G, Rupasinghe HV. Plant flavonoids in cancer chemoprevention: Role in genome stability. *The Journal of Nutritional Biochemistry.* 2017 Jul 1; 45:1–4.

Griffiths RE, Kupzig S, Cogan N, Mankelow TJ, Betin VM, Trakarnsanga K, Massey EJ, Lane JD, Parsons SF, Anstee DJ. Maturing reticulocytes internalize plasma membrane in glycophorin A-containing vesicles that fuse with autophagosomes before exocytosis. *Blood, The Journal of the American Society of Hematology.* 2012 Jun 28; 119(26):6296–306.

Grossman S. Body temperature regulation. *Porth's Pathophysiology: Concepts of Altered Health States*, edited by Shiela Grossman and Carol Mattson Porth. 2013 Aug 13, p. 216. Philadelphia, PA: Lippincott Williams & Wilkins.

Guo K, Ren J, Gu G, Wang G, Gong W, Wu X, Ren H, Hong Z, Li J. Hesperidin protects against intestinal inflammation by restoring intestinal barrier function and up-regulating Treg cells. *Molecular Nutrition & Food Research.* 2019 Jun; 63(11):1800975.

Hadni H, Fitri A, Benjelloun AT, Benzakour M, Mcharfi M. Evaluation of flavonoids as potential inhibitors of the SARS-CoV-2 main protease and spike RBD: Molecular docking, ADMET evaluation and molecular dynamics simulations. *Journal of the Indian Chemical Society.* 2022 Oct 1; 99(10):100697.

Hammer Q, Dunst J, Christ W, Picarazzi F, Wendorff M, Momayyezi P, Huhn O, Netskar HK, Maleki KT, García M, Sekine T. SARS-CoV-2 Nsp13 encodes for an HLA-E-stabilizing peptide that abrogates inhibition of NKG2A-expressing NK cells. *Cell Reports.* 2022 Mar 8; 38(10):110503.

Hasan MZ, Islam S, Matsumoto K, Kawai T. SARS-CoV-2 infection initiates interleukin-17-enriched transcriptional response in different cells from multiple organs. *Scientific Reports.* 2021 Aug 19; 11(1):16814.

Heindl MR, Rupp AL, Schwerdtner M, Bestle D, Harbig A, De Rocher A, Schmacke LC, Staker B, Steinmetzer T, Stein DA, Moulton HM. ACE2 acts as a novel regulator of TMPRSS2-catalyzed proteolytic activation of influenza A virus in airway cells. *Journal of Virology.* 2024 Mar; 12:e00102-24.

Hemmami H, Seghir BB, Zeghoud S, Ben Amor I, Kouadri I, Rebiai A, Zaater A, Messaoudi M, Benchikha N, Sawicka B, Atanassova M. Desert endemic plants in Algeria: A review on traditional uses, phytochemistry, polyphenolic compounds and pharmacological activities. *Molecules.* 2023 Feb 15; 28(4):1834.

Henderson LA, Canna SW, Friedman KG, Gorelik M, Lapidus SK, Bassiri H, Behrens EM, Ferris A, Kernan KF, Schulert GS, Seo P. American college of rheumatology clinical guidance for multisystem inflammatory syndrome in children associated with SARS-CoV-2 and hyperinflammation in pediatric COVID-19: Version 2. *Arthritis & Rheumatology.* 2021 Apr; 73(4):e13–29.

Huseen NH. Docking study of naringin binding with COVID-19 main protease enzyme. *Iraqi Journal of Pharmaceutical Sciences.* 2020 Dec 30; 29(2):231–8.

Islam A, Ferdous J, Islam S, Sayeed MA, Dutta Choudhury S, Saha O, Hassan MM, Shirin T. Evolutionary dynamics and epidemiology of endemic and emerging coronaviruses in humans, domestic animals, and wildlife. *Viruses.* 2021 Sep 23; 13(10):1908.

Jain S, Hussain A, Bhatt A, Nasa A, Navani NK, Mutreja R. Nanotechnology laying new foundations for combating COVID-19 pandemic. In *Oxides for Medical Applications*, edited by Piyush Kumar, Ganeshlenin Kandasamy, Jitendra Pal Singh, and Pawan Kumar Maurya. 2023 Jan 1, pp. 459–506. Cambridge: Woodhead Publishing.

Jang M, Park YI, Cha YE, Park R, Namkoong S, Lee JI, Park J. Tea polyphenols EGCG and theaflavin inhibit the activity of SARS-CoV-2 3CL-protease in vitro. *Evidence-Based Complementary and Alternative Medicine: eCAM.* 2020 Sep 17; 2020:5630838.

Jones NR, Qureshi ZU, Temple RJ, Larwood JP, Greenhalgh T, Bourouiba L. Two metres or one: What is the evidence for physical distancing in covid-19?. *BMJ.* 2020 Aug 25; 370:m3223.

Joshi RS, Jagdale SS, Bansode SB, Shankar SS, Tellis MB, Pandya VK, Chugh A, Giri AP, Kulkarni MJ. Discovery of potential multi-target-directed ligands by targeting host-specific SARS-CoV-2 structurally conserved main protease. *Journal of Biomolecular Structure and Dynamics.* 2021 Jun 13; 39(9):3099–114.

Kaul R, Paul P, Kumar S, Büsselberg D, Dwivedi VD, Chaari A. Promising antiviral activities of natural flavonoids against SARS-CoV-2 targets: Systematic review. *International Journal of Molecular Sciences.* 2021 Oct 14; 22(20):11069.

Kausar S, Said Khan F, Ishaq Mujeeb Ur Rehman M, Akram M, Riaz M, Rasool G, Hamid Khan A, Saleem I, Shamim S, Malik A. A review: Mechanism of action of antiviral drugs. *International Journal of Immunopathology and Pharmacology.* 2021 Mar; 35:20587384211002621.

Kovalev N, Pogany J, Nagy PD. Template role of double-stranded RNA in tombusvirus replication. *Journal of Virology.* 2014 May 15; 88(10):5638–51.

Kulandaisamy R, Kushwaha T, Dalal A, Kumar V, Singh D, Baswal K, Sharma P, Praneeth K, Jorwal P, Kayampeta SR, Sharma T. Repurposing of FDA approved drugs against SARS-CoV-2 papain-like protease: Computational, biochemical, and in vitro studies. *Frontiers in Microbiology.* 2022 May 10; 13:877813.

Lee JS, Kim HJ, Lee YS. A new anti-HIV flavonoid glucuronide from Chrysanthemum morifolium. *Planta Medica.* 2003 Sep; 69(09):859–61.

Liu H, Ye F, Sun Q, Liang H, Li C, Li S, Lu R, Huang B, Tan W, Lai L. *Scutellaria baicalensis* extract and baicalein inhibit replication of SARS-CoV-2 and its 3C-like protease *in vitro*. *Journal of Enzyme Inhibition and Medicinal Chemistry*. 2021 Dec; 36(1):497–503.

Machhi J, Herskovitz J, Senan AM, Dutta D, Nath B, Oleynikov MD, Blomberg WR, Meigs DD, Hasan M, Patel M, Kline P. The natural history, pathobiology, and clinical manifestations of SARS-CoV-2 infections. *Journal of Neuroimmune Pharmacology*. 2020 Sep; 15:359–86.

Mahboubi-Rabbani M, Abbasi M, Hajimahdi Z, Zarghi A. HIV-1 reverse transcriptase/integrase dual inhibitors: A review of recent advances and structure-activity relationship studies. *Iranian Journal of Pharmaceutical Research*. 2021 Spring; 20(2):333–69.

Manik M, Singh RK. Role of toll-like receptors in modulation of cytokine storm signaling in SARS-CoV-2-induced COVID-19. *Journal of Medical Virology*. 2022 Mar; 94(3):869–77.

Masand R, Jadhao A, Jadhav S. Review on modes of transmission of COVID-19. *International Journal of Current Microbiolo gy and Applied Sciences*. 2021 Feb; 10:1003–14.

Memish ZA, Perlman S, Van Kerkhove MD, Zumla A. Middle East respiratory syndrome. *The Lancet*. 2020 Mar 28; 395(10229):1063–77.

Mengist HM, Kombe AJ, Mekonnen D, Abebaw A, Getachew M, Jin T. Mutations of SARS-CoV-2 spike protein: Implications on immune evasion and vaccine-induced immunity. In *Seminars in Immunology*, edited by G. Kroemer. 2021 Jun 1, Vol. 55, p. 101533. New York: Academic Press.

Middleton EA, He XY, Denorme F, Campbell RA, Ng D, Salvatore SP, Mostyka M, Baxter-Stoltzfus A, Borczuk AC, Loda M, Cody MJ. Neutrophil extracellular traps contribute to immunothrombosis in COVID-19 acute respiratory distress syndrome. *Blood, The Journal of the American Society of Hematology*. 2020 Sep 3; 136(10):1169–79.

Milenkovic D, Ruskovska T, Rodriguez-Mateos A, Heiss C. Polyphenols could prevent SARS-CoV-2 infection by modulating the expression of miRNAs in the host cells. *Aging and Disease*. 2021 Aug; 12(5):1169.

Morshed MG. SARS-CoV-2: How science has advanced in the era of the COVID-19 pandemic. *IAHS Medical Journal*. 2021; 4(1):63–73.

Moss P. The T cell immune response against SARS-CoV-2. *Nature Immunology*. 2022 Feb; 23(2):186–93.

Mudd PA, Crawford JC, Turner JS, Souquette A, Reynolds D, Bender D, Bosanquet JP, Anand NJ, Striker DA, Martin RS, Boon AC. Targeted immunosuppression distinguishes COVID-19 from influenza in moderate and severe disease. *MedRxiv*. 2020 May 30; 2020:2020–05.

Munafò F, Donati E, Brindani N, Ottonello G, Armirotti A, De Vivo M. Quercetin and luteolin are single-digit micromolar inhibitors of the SARS-CoV-2 RNA-dependent RNA polymerase. *Scientific Reports*. 2022 Jun 22; 12(1):10571.

Nagu P, Parashar A, Behl T, Mehta V. CNS implications of COVID-19: A comprehensive review. *Reviews in the Neurosciences*. 2020 Dec 7; 32(2):219–34. https://doi.org/10.1515/revneuro-2020-0070.

Naso LG, Ferrer EG, Williams PA. Correlation of the anticancer and pro-oxidant behavior and the structure of flavonoid-oxidovanadium (IV) complexes. *Coordination Chemistry Reviews*. 2023 Oct 1; 492:215271.

Noor S, Mohammad T, Rub MA, Raza A, Azum N, Yadav DK, Hassan MI, Asiri AM. Biomedical features and therapeutic potential of rosmarinic acid. *Archives of Pharmacal Research*. 2022 Apr; 45(4):205–28.

Nusshag C, Wei C, Hahm E, Hayek SS, Li J, Samelko B, Rupp C, Szudarek R, Speer C, Kälble F, Schaier M. suPAR links a dysregulated immune response to tissue inflammation and sepsis-induced acute kidney injury. *JCI Insight*. 2023 Apr 4; 8(7):e165740.

Ohishi T, Hishiki T, Baig MS, Rajpoot S, Saqib U, Takasaki T, Hara Y. Epigallocatechin gallate (EGCG) attenuates severe acute respiratory coronavirus disease 2 (SARS-CoV-2) infection by blocking the interaction of SARS-CoV-2 spike protein receptor-binding domain to human angiotensin-converting enzyme 2. *PLoS One*. 2022 Jul 13; 17(7):e0271112.

Oladele JO, Oyeleke OM, Oladele OT, Oladiji AT. Covid-19 treatment: Investigation on the phytochemical constituents of Vernonia amygdalina as potential Coronavirus-2 inhibitors. *Computational Toxicology*. 2021 May 1; 18:100161.

Pandya M, Shah S, Dhanalakshmi M, Juneja T, Patel A, Gadnayak A, Dave S, Das K, Das J. Unravelling Vitamin B12 as a potential inhibitor against SARS-CoV-2: A computational approach. *Informatics in Medicine Unlocked*. 2022 Jan 1; 30:100951.

Parashar A, Shukla A, Sharma A, Behl T, Goswami D, Mehta V. Reckoning γ-Glutamyl-S-allylcysteine as a potential main protease (MPRO) inhibitor of novel SARS-CoV-2 virus identified using docking and molecular dynamics simulation. *Drug Development and Industrial Pharmacy*. 2021 May 4; 47(5):699–710.

Parasher A. COVID-19: Current understanding of its Pathophysiology, clinical presentation and treatment. *Postgraduate Medical Journal*. 2021 May 1; 97(1147):312–20.

Park J, Park R, Jang M, Park YI. Therapeutic potential of EGCG, a green tea polyphenol, for treatment of coronavirus diseases. *Life*. 2021 Mar 4; 11(3):197.

Pizzato M, Baraldi C, Boscato Sopetto G, Finozzi D, Gentile C, Gentile MD, Marconi R, Paladino D, Raoss A, Riedmiller I, Ur Rehman H. SARS-CoV-2 and the host cell: A tale of interactions. *Frontiers in Virology*. 2022 Jan 12; 364:1.

Preethi M, Roy L, Lahkar S, Borse V. Outlook of various diagnostics and nanodiagnostic techniques for COVID-19. *Biosensors and Bioelectronics*. 2022 Dec 1; 12:100276.

Qudus MS, Tian M, Sirajuddin S, Liu S, Afaq U, Wali M, Liu J, Pan P, Luo Z, Zhang Q, Yang G. The roles of critical pro-inflammatory cytokines in the drive of cytokine storm during SARS-CoV-2 infection. *Journal of Medical Virology*. 2023 Apr; 95(4):e28751.

Ramírez-Mora T, Retana-Lobo C, Reyes-Carmona J. COVID-19: Perspectives on the pandemic and its incidence in dentistry. *Odovtos International Journal of Dental Sciences*. 2020 Dec; 22(3):22–42.

Russo M, Moccia S, Spagnuolo C, Tedesco I, Russo GL. Roles of flavonoids against coronavirus infection. *Chemico-Biological Interactions*. 2020 Sep 1; 328:109211.

Ryan JM, Nanda S. *COVID-19: Social Inequalities and Human Possibilities*. 2022 Mar 13. London: Routledge.

Šamec D, Karalija E, Šola I, Vujčić Bok V, Salopek-Sondi B. The role of polyphenols in abiotic stress response: The influence of molecular structure. *Plants*. 2021 Jan 8; 10(1):118.

Santos EL, Maia BH, Ferriani AP, Teixeira SD. Flavonoids: Classification, biosynthesis and chemical ecology. *Flavonoids-From Biosynthesis to Human Health*. 2017 Aug 23; 13:78–94.

Senthil Kumar KJ, Gokila Vani M, Wang CS, Chen CC, Chen YC, Lu LP, Huang CH, Lai CS, Wang SY. Geranium and lemon essential oils and their active compounds downregulate angiotensin-converting enzyme 2 (ACE2), a SARS-CoV-2 spike receptor-binding domain, in epithelial cells. *Plants*. 2020 Jun; 9(6):770.

Sharma A, Balda S, Apreja M, Kataria K, Capalash N, Sharma P. COVID-19 diagnosis: Current and future techniques. *International Journal of Biological Macromolecules*. 2021 Dec 15; 193:1835–44.

Shirbhate E, Pandey J, Patel VK, Kamal M, Jawaid T, Gorain B, Kesharwani P, Rajak H. Understanding the role of ACE-2 receptor in pathogenesis of COVID-19 disease: A potential approach for therapeutic intervention. *Pharmacological Reports*. 2021 Dec 1; 73:1–2.

Shrestha LB, Foster C, Rawlinson W, Tedla N, Bull RA. Evolution of the SARS-CoV-2 omicron variants BA. 1 to BA. 5: Implications for immune escape and transmission. *Reviews in Medical Virology*. 2022 Sep; 32(5):e2381.

Singh AK, Majumdar S, Singh R, Misra A. Role of corticosteroid in the management of COVID-19: A systemic review and a Clinician's perspective. *Diabetes & Metabolic Syndrome: Clinical Research & Reviews*. 2020 Sep 1; 14(5):971–8.

Song HX, Wang H, Ma XR, Wang DF, Wang YL, Zou DX, Miao JX, Yang WP. Research on the potential mechanism of Huashi Baidu recipe against novel coronavirus pneumonia (COVID-19) by network pharmacology and molecular docking technology. *Journal of Hainan Medical College*. 2020 Dec 1; 26(23):1716–9.

Stadnytskyi V, Anfinrud P, Bax A. Breathing, speaking, coughing or sneezing: What drives transmission of SARS-CoV-2?. *Journal of Internal Medicine*. 2021 Nov; 290(5):1010–27.

Stremersch S, De Smedt SC, Raemdonck K. Therapeutic and diagnostic applications of extracellular vesicles. *Journal of Controlled Release*. 2016 Dec 28; 244:167–83.

Toth G, Szigeti C. The historical ecological footprint: From over-population to over-consumption. *Ecological Indicators*. 2016 Jan 1; 60:283–91.

Tutunchi H, Naeini F, Ostadrahimi A, Hosseinzadeh-Attar MJ. Naringenin, a flavanone with antiviral and anti-inflammatory effects: A promising treatment strategy against COVID-19. *Phytotherapy Research*. 2020 Dec; 34(12):3137–47.

Vasudevan UM, Lee EY. Flavonoids, terpenoids, and polyketide antibiotics: Role of glycosylation and biocatalytic tactics in engineering glycosylation. *Biotechnology Advances*. 2020 Jul 1; 41:107550.

Walensky RP, Walke HT, Fauci AS. SARS-CoV-2 variants of concern in the United States-challenges and opportunities. *JAMA*. 2021 Mar 16; 325(11):1037–8.

Weiss SL, Peters MJ, Alhazzani W, Agus MS, Flori HR, Inwald DP, Nadel S, Schlapbach LJ, Tasker RC, Argent AC, Brierley J. Surviving sepsis campaign international guidelines for the management of septic shock and sepsis-associated organ dysfunction in children. *Intensive Care Medicine*. 2020 Feb; 46:10–67.

Williamson G, Kerimi A. Testing of natural products in clinical trials targeting the SARS-CoV-2 (Covid-19) viral spike protein-angiotensin converting enzyme-2 (ACE2) interaction. *Biochemical Pharmacology*. 2020 Aug; 178:114123.

Wu W, Cheng Y, Zhou H, Sun C, Zhang S. The SARS-CoV-2 nucleocapsid protein: Its role in the viral life cycle, structure and functions, and use as a potential target in the development of vaccines and diagnostics. *Virology Journal*. 2023 Dec; 20(1):1–6.

Xiao T, Cui M, Zheng C, Wang M, Sun R, Gao D, Bao J, Ren S, Yang B, Lin J, Li X. Myricetin inhibits SARS-CoV-2 viral replication by targeting Mpro and ameliorates pulmonary inflammation. *Frontiers in Pharmacology*. 2021 Jun 17; 12:669642.

Zandi K, Musall K, Oo A, Cao D, Liang B, Hassandarvish P, Lan S, Slack RL, Kirby KA, Bassit L, Amblard F, Kim B, AbuBakar S, Sarafianos SG, Schinazi RF. Baicalein and baicalin inhibit SARS-CoV-2 RNA-dependent-RNA polymerase. *Microorganisms*. 2021 Apr 22; 9(5):893.

Zech F, Schniertshauer D, Jung C, Herrmann A, Cordsmeier A, Xie Q, Nchioua R, Prelli Bozzo C, Volcic M, Koepke L, Müller JA. Spike residue 403 affects binding of coronavirus spikes to human ACE2. *Nature Communications*. 2021 Nov 25; 12(1):6855.

Zhang B, Qi F. Herbal medicines exhibit a high affinity for ACE2 in treating COVID-19. *BioScience Trends*. 2023 Feb 28; 17(1):14–20.

Zhang Q, Xiang R, Huo S, Zhou Y, Jiang S, Wang Q, Yu F. Molecular mechanism of interaction between SARS-CoV-2 and host cells and interventional therapy. *Signal Transduction and Targeted Therapy*. 2021 Jun 11; 6(1):233.

Zhang XX, Wu QF, Yan YL, Zhang FL. Inhibitory effects and related molecular mechanisms of total flavonoids in Mosla chinensis Maxim against H1N1 influenza virus. *Inflammation Research*. 2018 Feb; 67:179–89.

Zhou T, Dang Y, Zheng YH. The mitochondrial translocator protein, TSPO, inhibits HIV-1 envelope glycoprotein biosynthesis via the endoplasmic reticulum-associated protein degradation pathway. *Journal of Virology*. 2014 Mar 15; 88(6):3474–84.

2 Traditional Herbal Medicines for COVID-19

The Impact of Flavonoids

Mohammed Dalli, Salah-eddine Azizi, Nour Elhouda Daoudi, Ali Azghar, Abderrazak Saddari, Ilyass Alami Merrouni, Mohammed Roubi, Elmostapha Benaissa, Yassine Ben Lahlou, Mostafa Elouennass, and Adil Maleb

2.1 INTRODUCTION

The novel coronavirus, termed COVID-19, emerged in December 2019 in Wuhan, a city in China. It rapidly spread from Wuhan to various other regions and nations. It is caused by the SARS-CoV-2 which infects primarily the respiratory tract [1]. Some works have suggested that this epidemic came from bats, which are known to carry various diseases [2]. In addition, there is a belief that the virus might have been transmitted to humans via a pangolin serving as an intermediary host. These pangolins were being sold at the Seafood Market in Wuhan, China [3]. However, until now, the exact origins of SARS-CoV-2 are still unknown and under investigation. Actually, SARS-CoV-2 affects the respiratory tract by infecting its cells, penetrating the body through different routes (the mouth, nose, or eyes), and using a specific protein named (S) protein. This protein binds to the angiotensin-converting enzyme 2 (ACE2) located on the surface of respiratory tract cells [4,5]. Once inside the respiratory system, the virus replicates and produces new viral particles using the host cell machinery. This replication process results in damage to the infected cells and the subsequent release of fresh viral particles into the body. This can trigger inflammation in the respiratory tract and the onset of symptoms like coughing, breathlessness, and fever [6]. In addition to these symptoms, individuals may also experience fatigue, muscle aches, sore throat, anosmia, ageusia, and headaches [7]. The virus can infect the brain, gastrointestinal tract, liver, heart, kidneys, and other organs [8]. As a result, several symptoms appear, ranging from mild to severe, depending on the individual's health situation and the level of the exposition to the virus. In fact, in severe cases, it affects the lungs and then causes pneumonia [6].

Scientists and public health around the world have been working, ever since the SARS-CoV-2 pandemic first appeared, to understand its transmission and find the best solutions to prevent and treat this illness. Moreover, there is an imperative need to research the efficiency and safety of certain medications for the treatment of COVID-19. Then, several drugs have been authorized by regulatory agencies for treating COVID-19 [9]. While it was reported that their effectiveness may change depending on the severity of the disease, it was signaled that each drug has its side effects [10]. Because of this, research on using bioproducts like flavonoids to treat SARS-CoV-2 infections is still being conducted. In this study, we gathered all available information on COVID-19 medication adverse effects, possible natural product inhibitory mechanisms, and the protective effects of flavonoids against SARS-CoV-2 infection from the databases Scopus, Elsevier, PubMed, and Web of Science.

2.1.1 Epidemiology

COVID-19 was identified for the first time in Wuhan, China, and quickly spread worldwide to become a global pandemic. Then, this virus has affected people of all ages and demographics [11]. However, severe and serious illness and death have been seen more in older persons and those with underlying medical issues [12]. The outbreak's epidemiology has varied across different countries, regions, and areas. According to WHO, the highest number of cases and deaths were observed in Europe, the Western Pacific, and the Americas [13].

The number of confirmed cases has grown annually since December 2019 (Figure 2.1). During the initial phases of the COVID-19 pandemic, there were only three confirmed cases and no deaths; however, as the virus spread globally, the number of infected individuals rose rapidly. More than 750 million positive cases were detected with a death rate of 6.84 by February 2023 [13] (Figure 2.1).

2.1.2 SARS-CoV-2 Structure

Coronaviridae is the family that the SARS-CoV-2 belongs to. Its genome is made up of single-stranded, positive-sense RNA. This RNA includes the genetic material required for viral transcription and replication. The genome, covering around 30,000 nucleotides, produces a range of structural and non-structural proteins. The virus has an envelope

DOI: 10.1201/9781003433200-2

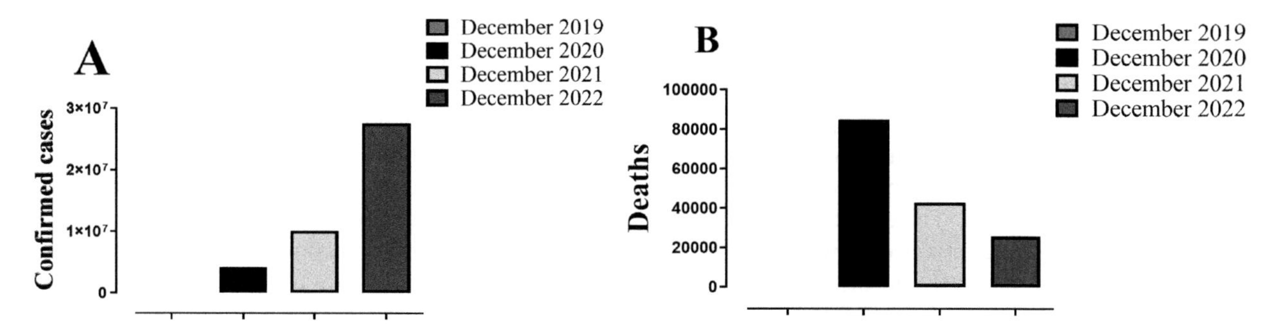

FIGURE 2.1 COVID-19 total number of deaths (b) and positive confirmed cases (a) as reported by the WHO [13].

made up of a lipid bilayer that originates from the host cell's membrane. The viral genetic material is encapsulated and safeguarded by this envelope. Additionally, the viral envelope is dotted with the presence of spike proteins (S proteins), giving the virus its distinctive crown-like appearance and giving rise to the term "coronavirus." These spike proteins are in charge of attaching to ACE2 receptors present in host cells and promoting the entrance of the virus. Inside the envelope, there is the nucleocapsid, housing the viral RNA genome and associated nucleocapsid (N) proteins. These proteins are integral to the encapsulation of RNA and the facilitation of viral replication. Additionally, the virus synthesizes multiple non-structural proteins, including RNA-dependent RNA polymerase (RdRp), and proteases, imperative for viral transcription and replication processes [14] (Figure 2.2).

2.1.3 Action Mechanism of SARS-CoV-2

COVID-19 is a contagious viral infection. The contact with an infected surface is considered among the main sources of infection with this virus where it can disseminate by inhalation, digestion, coughing, or sneezing [15]. Several studies were carried out in order to thoroughly comprehend the mechanism of action and mode of transmission of SARS-CoV-2. The SARS-CoV-2 action mechanism infection involves several steps that allow it to enter and replicate within human cells. Furthermore, the first step in the process starts with the linkage of the virus with the host cell via specific receptors. In particular, the virus's spike protein attaches itself to ACE2 receptors found on the outer membrane of the cells, including the cells of the respiratory tract. Following that, the virus enters the host cell via

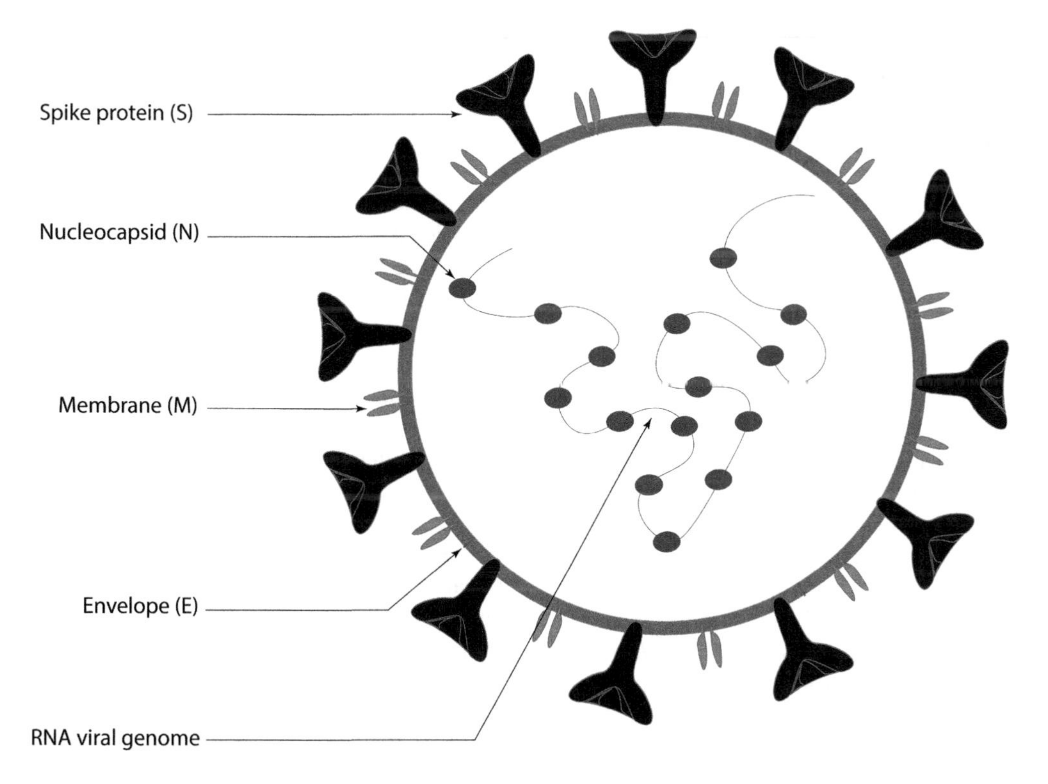

FIGURE 2.2 SARS-CoV-2 structure.

endocytosis, allowing its genetic material to penetrate the target cell. Once inside, the virus causes structural alterations in the Spike (S) glycoprotein. After this, cathepsin L proteolysis is carried out by intracellular proteases, and a membrane fusion mechanism within endosomes is activated. This fusion process opens up to release the virus into the host cell's cytoplasm. Thereafter, the RNA (single-stranded RNA molecule) produces new viral proteins by releasing its genetic material into the cell cytoplasm. Then, it is assembled into new virus particles, resulting in the generation of more viral RNA, proteins, and virus particles. The newly formed virus particles could then infect other cells in the same host or be transmitted to other individuals [16–18].

2.1.4 Herbal Medicines Containing Specific Flavonoids and Their Impact on COVID-19 Management

As natural polyphenols from medicinal and aromatic plants, flavonoids have shown antiviral effects against various respiratory viruses, especially COVID-19. Understanding the interactions between flavonoid-rich herbal remedies and the mechanisms of SARS-CoV-2 infection may provide valuable insights into novel therapeutic strategies for COVID-19. *Scutellaria baicalensis* is a well-known plant for its multifunctional potential such as anti-inflammatory, antioxidant, and antithrombotic agent [19]; meanwhile, it was reported that *S. baicalensis* ethanolic (IC_{50}=8.52 µg/mL) extract and its major compound baicalein (IC_{50}=0.39 µM) were able to inhibit SARS-CoV-2 3-CL protease. Furthermore, both the ethanolic extract and the primary flavonoid demonstrated the capability to impede the virus's replication within Vero cells, exhibiting EC50 values of 0.74 mg/mL and 2.9 mM, respectively [20]. Another study has reported that this plant has revealed its richness with baicalin and hesperetin, which were noted to exert highly effective activity in COVID-19 relief action [21] (Table 2.1). Furthermore, Shufeng Jiedu is a Chinese formulation formed of about eight plants that showed important action on respiratory tract infections. Thus, the evaluation of the anti-SARS-CoV-2 has significantly reduced viral loading in mice. Additionally, the formulation has exerted its anti-inflammatory potential by reducing inflammation factors in the lungs. Similarly, it was indicated that the bioactive compounds identified in Shufeng Jiedu formulation, such as quercetin, wogonin, and polydactin, bind strongly with COVID-19 protease (Table 2.1). Finally, it has been mentioned that the mixture has performed an improvement in the recovery period in infected patients [22]. Moreover, Capillarisin, a flavone present in *Artemisia capillaris* has demonstrated its anti-inflammatory and immunomodulatory potential by inducing a decrease in TNF-α, IL-1β, IL-1α, and IL-6 and by inducing antipyretic effect [23]. Also, the fruits of *Paulownia tomentosa*, a plant belonging to Scrophulariaceae family, were reported to be used to treat respiratory disorders. Also, this plant was found to be rich in geranylated flavonoids. The antiviral activity of these flavonoids named Tomentin showed different dose-dependent manner binding potential with SARS-CoV-2 main protease characterized with an IC_{50} value ranging between 5.0 and 14.4 µM [24] (Table 2.1). Another study has also suggested that *Allium sativum* could significantly contribute to the reduction of viral infection caused by SARS-CoV-2 and this is due to the presence of several compounds such as flavonoids. In fact, it was shown that garlic oil induced an inhibition of SARS-CoV-1 *in vivo* which suggests the blockage of structural viral proteins [25]. Also, kaempferol found in *Nigella sativa* seeds has demonstrated high binding affinity with SARS-CoV-2 protease C19MP [26]. Another compound from *Camellia sinensis* (green tea) named gallocatechin gallate was also reported to block Sars-CoV-19 replication [27]. Similarly, epigallocatechin gallate has also demonstrated its potency in inhibiting endoribonuclease (Nsp15) and blocking SARS-CoV-2 replication in cells with a concentration less than 1 µg/mL [28]. Finally, the computational study assessed by Chebaibi et al. [29] showed that several flavonoids from *Syzygium aromaticum* and *Citrus limon* were against COVID-19 6lu7 protease and 6y2e protease. While Ladanein, a flavonoid from *Marrubium peregrinum*, was also found to inhibit the entrance of the virus HCV into cells [30] (Table 2.1).

2.1.5 Flavonoids as Potential Inhibitors to SARS-CoV-2

Flavonoids or bioflavonoids are defined as natural substances and secondary metabolites that belong to the polyphenol family. It is found mainly in fruits, vegetables, nuts, seeds, and even medicinal plants [40]. These compounds possess a consistent chemical structure, consisting of a 15-carbon skeleton with two benzene rings (labeled A and B) connected by a heterocyclic pyran ring (named C). The flavonoid skeleton can be modified by various substitutions on the rings, resulting in a wide variety of flavonoid structures that give rise to a variety of molecules (flavones, flavonols, flavanones, anthocyanidins, and isoflavones) [41,42]. Flavonoids are endowed with high antioxidant, anti-inflammatory [43], anti-diabetic [44], antithrombogenic [45], anti-tumor [46], antiosteoporotic [47], antibacterial, antifungal, and antiviral properties [48], and could help in diminishing risk factor of diseases such as cardiovascular, atherosclerosis, and Alzheimer's disease [49–51]. Generally, humans have consumed flavonoids for thousands of years. Then, they are considered safe compounds when they are consumed in normal dietary amounts. However, it was signaled that some specific flavonoid compounds may interfere with the action of some therapeutic drugs [52]. Many studies have investigated the antiviral effect of flavonoids [53]. It was signaled that some flavonoids possess a potential antiviral effect by inhibiting viruses' replication, proteins translation, or viral entrance [54]. Indeed,

TABLE 2.1
Different Plants and Their Bioactive Compounds with Anti-SARS-CoV-2 Properties

Plant Name/Formulation	Bioactive Compounds	References
Scutellaria baicalensis	Baicalein	[19–21]
	hesperetin	
Shufeng Jiedu	Quercetin, wogonin, and polydactin	[22]
Artemisia capillaris	Capillarisin	[23]
Paulownia tomentosa	Tomentin	[24]
Allium sativum	Apigenin	[25]
	Quercetin	
	Kaempferol	
Camellia sinensis	Gallocatechin gallate	[27,28]
	Epigallocatechin gallate	
Citrus sp.	Naringenin	[31]
Green tea, black tea, coconuts, onions, grape seeds	Catechins	[32]
Syzygium aromaticum	Rhamnetin	[29]
*Citrus limo*n	Natsudaidain	
	Nobiletin	
	Sinensetin	
	Tangeretin	
	Neoeriocitrin	
Marrubium peregrinum	Ladanein	[30]
Tetradenia riparia	Luteolin	[33–35]
Epimedium sp.	Icariin	[36,37]
Nigella sativa	Kaempferol	[14,38,39]
	Apigenin	
	Quercetin	

flavonoids could play a major role in controlling viral infections. Consequently, scientists are keen on investigating the efficacy of these natural compounds against COVID-19.

2.2 QUERCETIN

Quercetin, a flavonoid, is renowned for its advantageous effects. It has demonstrated antiviral activity against various viruses, including the influenza virus, hepatitis B and C viruses, and HIV [55]. In addition, it was reported that this flavonoid exhibits both anti-infective and anti-replicative properties [56]. Quercetin could interact with proteases which prohibit the virus entry, inhibit and could significantly reduce inflammation caused by infection [57]. This phytochemical has been studied for its potential use as a preventive and curative approach to COVID-19. Furthermore, quercetin has demonstrated its ability to downregulate the expression of ACE2 and viral enzymes [58]. On the other hand, this bioactive compound showed anti-inflammatory action, which encourages its use in reducing clinical symptoms (fatigue, pro-appetite properties) [59] (Figure 2.3).

2.3 CATECHINS

Catechin is a type of flavonoid that belongs to the flavanols subclass. It is primarily present in green tea, black tea, coconuts, onions, grape seeds, and various medicinal plants [60]. It possesses several beneficial health effects including analgesic, antibacterial, antioxidant, and anti-inflammatory potentials [61]. Additionally, it displays antiviral activities against influenza A and B viruses by inhibiting viral receptor binding and sialidase functions. It also suppresses the viral fusion activity associated with the mumps virus [32,62,63]. Numerous conducted studies have demonstrated catechins' importance as an anti-SARS-CoV-2 by inhibition of the virus replication. These effects include boosting CD8^{+}-mediated adaptive immunity, reducing the occurrence of cytokine storms, and encouraging protective mechanisms that rely on autophagy to alleviate acute lung injury [64] (Figure 2.3).

2.4 EPIGALLOCATECHIN GALLATE

Epigallocatechin gallate or green tea catechin extracted from *Camellia sinensis* is another flavonoid that has been studied for its antiviral properties [65]. Research indicates its effectiveness in halting the replication of various viruses, including the influenza virus, herpes simplex virus, Zika virus, and hepatitis C virus. It does this by blocking virions from binding to host cells or their receptors [66]. Recently, it was reported that this flavonoid could attenuate SARS-CoV-2 infection by blocking the binding of the S protein to ACE2 [67] (Figure 2.3).

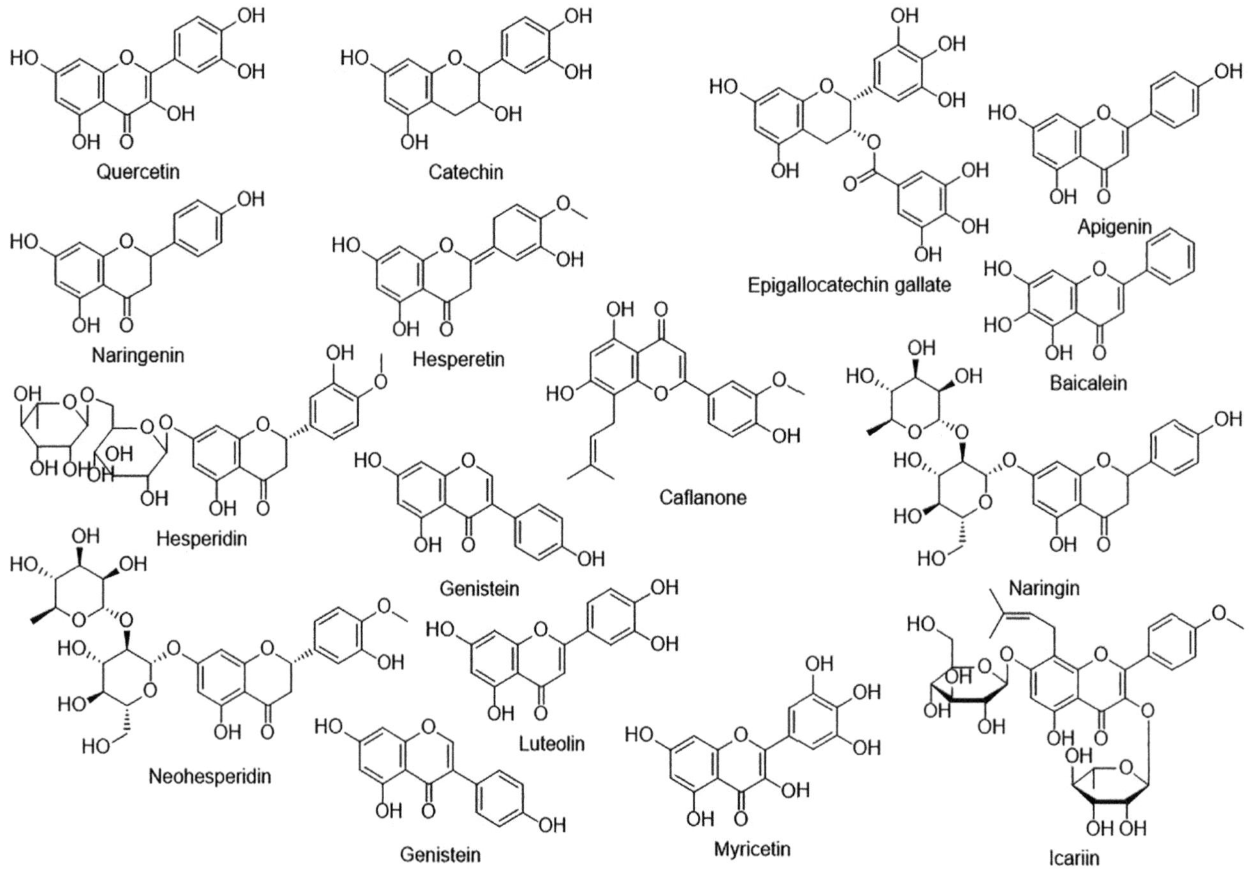

FIGURE 2.3 Flavonoids chemical structure responsible for the antiviral effect on SARS-CoV-2.

2.5 NARINGENIN

Naringenin is a flavonoid that belongs to the flavanones subclass. It is present widely in citrus fruits like lemons, oranges, bergamot, and tomatoes [68]. It possesses potential antioxidant and anti-inflammatory properties [69]. Moreover, it was reported that this flavanone possesses antiviral effects by reducing viral replication (Dengue virus, HCV, Zika virus, and Chikungunya virus) [70]. Also, naringenin has been investigated for its effect on COVID-19, and it was indicated that it might have an anti-COVID-19 effect by inhibiting protease, reducing inflammatory responses, and ACE2 receptor activity [70,71] (Figure 2.3).

2.6 HESPERIDIN/HESPERETIN

Hesperidin is a bioflavonoid that belongs to the flavanone subclass, and its aglycone metabolite is named hesperetin. These compounds are predominantly present in citrus and exhibit a large spectrum of effects, including the antimicrobial, anti-carcinogenic, and antioxidant properties [72,73]. Molecular docking investigation indicated that it can block the linkage of S proteins with ACE2 and also could reduce the expression of ACE2 and TMPRSS2 [74]. Furthermore, it was reported that the binding energy registered by hesperidin was lower than conventional drugs like lopinavir, ritonavir, and indinavir, which could suggest that this flavonoid is an effective antiviral source [75] (Figure 2.3).

2.7 NEOHESPERIDIN

Neohesperidin is a citrus flavonoid that belongs to the flavanone subclass. It possesses antioxidant, anti-inflammatory, anti-tumor, antiapoptotic, and antiviral effects [76,77]. Furthermore, according to *in silico* study, neohesperidin strongly binds to TMPRSS2 which suggests its high potential in reducing viral entry and could also be used as a supplement for the protection and treatment of viral infections [78] (Figure 2.3).

2.8 APIGENIN

Apigenin is one of the most common flavonoids widespread in plants, and it belongs to the flavones subclass. This phytochemical possesses several beneficial effects on amnesia, Alzheimer's disease, depression, and insomnia. Moreover, it is endowed with antioxidant, anti-diabetic, anti-inflammatory, anticancer, and antiviral effects against Epstein-Barr virus, buffalopox virus, picornaviruses, hepatitis C virus, African swine fever virus, and finally SARS-CoV-2 [79–81] (Figure 2.3).

2.9 NARINGIN

Naringin, frequently present in fruits and vegetables, is a flavonoid that possesses various health-promoting activities, including antiviral properties, immune-stimulating effects, and anti-inflammatory actions [82]. The assessed computational investigation indicated that naringin was able to interact with COVID-19 main protease with high binding energy [31]. A separate molecular docking investigation suggested that naringin exhibits a robust binding affinity to ACE2, implying its potential utility in thwarting coronavirus infection [83]. Naringin has also been identified for its ability to reduce the production of proinflammatory cytokines triggered by LPS in macrophage cells, hinting at its possible role in mitigating cytokine storms. Additionally, certain substances resulting from the breakdown of naringin by gut microbiota have demonstrated antiviral properties. For example, the primary metabolite of naringin broken down in this process is 3-(4'-hydroxyphenyl) propionic acid (HPPA) [84,85] (Figure 2.3).

2.10 LUTEOLIN

Luteolin, chemically denoted as 3,4,5,7-tetrahydroxy flavone, is a naturally occurring flavonoid present in numerous food sources such as apples, carrots, and cabbage, as well as specific medicinal herbs. Studies have indicated its efficacy against several viruses, including SARS-CoV-2 [86,87]. Many molecular docking investigations have revealed its robust ability to bind effectively to critical components of SARS-CoV-2, such as M^{pro}, PL^{pro}, RdRp, and the ACE2 receptor [88–90]. Additionally, a fluorescence resonance energy transfer (FRET) assay has provided evidence that luteolin is highly proficient at inhibiting M^{pro} activity, with IC_{50} values of 11.81 and 20.2 μM [91,92]. Additionally, research has demonstrated that luteolin can mitigate acute lung injury induced by sepsis in murine models through the inhibition of ICAM-1, NF-κB, oxidative stress, and the iNOS pathway [93]. Moreover, luteolin protects against lung injury in murine models caused by mercuric chloride, acting through the prevention of NF-κB activation and the stimulation of the Akt/Nrf2 pathway [94] (Figure 2.3).

2.11 CAFLANONE

Caflanone, known for its antiviral properties, was selected to assess its interaction with the ACE2 enzyme [95,96]. The promising results obtained from computational studies led to *in vitro* experiments that indicate caflanone's potential to hinder viral entry mechanisms, including cathepsin L, and modulate cytokine production [97]. *In vitro* investigations have demonstrated that caflanone may inhibit virus entry factors like Cathepsin, Belson Murine Leukemia Viral Oncogene Homolog 2 (ABL-2), and AXL Receptor Tyrosine Kinase (AXL-2), which aid in the transmission of the coronavirus from mother to fetus [98]. This phytochemical, derived from flavonoids, displays significant binding energy to the ACE2 receptor's protein S, helicase [99], and protease sites, which SARS-CoV-2 utilizes for cellular infection. Additionally, caflanone hinders the release of multiple cytokines, including Interleukin 1β, 6, 8, macrophage inflammatory protein 1α (Mip-1α), and Tumor Necrosis Factor-α (TNF-α) in *in vitro* studies [95] (Figure 2.3).

2.12 BAICALEIN

Baicalein demonstrates robust pharmacological potential, particularly in terms of its antiviral effectiveness [99]. Preclinical studies have shown that baicalein can effectively suppress cellular damage in Vero E6 cells induced by SARS-CoV-2. Furthermore, baicalein hinders the replication of the virus and lessens lung tissue deterioration in hACE2 transgenic mice infected with SARS-CoV-2. When administered, baicalein enhances respiratory function, reduces the accumulation of inflammatory cells in the pulmonary tissue, and lowers the serum concentrations of IL-1β and TNF-α in mice subjected to acute lung injury caused by LPS [97] (Figure 2.3).

2.13 GENISTEIN

Genistein, classified within the isoflavonoid category of flavonoids, is a multifunctional natural compound. Like other plant components, including lignans, which exhibit estrogenic properties, genistein serves as a prime example of a phytoestrogenic substance [100]. Genistein offers numerous therapeutic and pharmacological benefits, including its potential as an anti-carcinogenic agent [101], antimicrobial and antiviral properties [102], as well as antioxidant and anti-inflammatory activities [103]. These characteristics suggest that genistein may have beneficial effects in inhibiting the replication of SARS-CoV-2 [104] (Figure 2.3).

2.14 ICARIIN

Icariin, found in abundance as a secondary metabolite in Epimedium prenylflavonoids, an ancient Chinese medicinal plant, is well-regarded for its robust antioxidant and anti-inflammatory attributes. It exerts significant positive effects on various pathologies related to the nervous system, cardiovascular health, and the liver [36,37] (Figure 2.3).

2.15 MYRICETIN

Myricetin, known as 3,5,7,3′,4′,5′-hexahydroxyflavone, is a prevalent dietary flavonoid recognized for its contributions to health improvement through various pharmacological activities. These activities include analgesic, antioxidant, anti-inflammatory, antimicrobial, anti-tumor, anti-diabetic, and hepatoprotective properties; myricetin also offers substantial benefits in the regulation of immune processes [105]. Myricetin is recognized for its antiviral activity [95]. It offers a promising path for both therapeutic and preventive uses because of its combined effect in suppressing ATPase, helicase, polymerase, neuraminidase, and ICAM1 (Intercellular adhesion molecule 1), which together inhibit viral entry, transcription, replication, and budding.

These findings collectively emphasize the importance of further research into these phytochemicals in the battle against COVID-19 [95]. Another molecular docking study on *Phyllanthus emblica* and its derivate myricetin displayed favorable output against three target proteins (M^{pro}, receptor-binding domain (RBD), and nsp15 endoribonuclease) [106] (Figure 2.3).

2.16 CONCLUSION AND FUTURE PERSPECTIVES

In the end, this book chapter has shed light on the potential of flavonoids to inhibit viral infections, especially against SARS-CoV-2. Furthermore, this article has discussed different paths adopted by flavonoids to inhibit viral entry, replication, and transcription, as well as to alleviate inflammation. On the other hand, different flavonoids have demonstrated their high binding affinity toward target proteins as a promising key to mitigating acute lung injury, especially those related to COVID-19. In the coming years, it is crucial to carry out thorough clinical trials and investigations to confirm the effectiveness and safety of the tested compounds (*in vitro*, *in vivo*, and *in silico*) to face future threats of viral pandemics. Furthermore, it will be crucial to identify potential collaborative benefits and create standardized formulas to fully utilize the power of flavonoids and traditional herbal treatments.

REFERENCES

[1] Yuen E, Gudis DA, Rowan NR, et al. Viral infections of the upper airway in the setting of COVID-19: A primer for rhinologists. *Am J Rhinol Allergy* 2021; 35: 122–131.

[2] Rahman S, Ullah S, Shinwari ZK, et al. Bats-associated beta-coronavirus detection and characterization: First report from Pakistan. *Infect Genet Evol J Mol Epidemiol Evol Genet Infect Dis* 2023; 108: 105399.

[3] Zhang T, Wu Q, Zhang Z. Probable pangolin origin of SARS-CoV-2 associated with the COVID-19 outbreak. *Curr Biol* 2020; 30: 1346–1351.

[4] Hamming I, Timens W, Bulthuis MLC, et al. Tissue distribution of ACE2 protein, the functional receptor for SARS coronavirus. A first step in understanding SARS pathogenesis. *J Pathol* 2004; 203: 631–637.

[5] Li H, Wang Y, Ji M, et al. Transmission routes analysis of SARS-CoV-2: A systematic review and case report. *Front Cell Dev Biol* 2020; 8: 618.

[6] Subbarao K, Mahanty S. Respiratory virus infections: Understanding COVID-19. *Immunity* 2020; 52: 905–909.

[7] Weng L-M, Su X, Wang X-Q. Pain symptoms in patients with coronavirus disease (COVID-19): A literature review. *J Pain Res* 2021; 14: 147–159.

[8] Shah MD, Sumeh AS, Sheraz M, et al. A mini-review on the impact of COVID-19 on vital organs. *Biomed Pharmacother* 2021; 143: 112158.

[9] Tarighi P, Eftekhari S, Chizari M, et al. A review of potential suggested drugs for coronavirus disease (COVID-19) treatment. *Eur J Pharmacol* 2021; 895: 173890.

[10] Aygün İ, Kaya M, Alhajj R. Identifying side effects of commonly used drugs in the treatment of COVID-19. *Sci Rep* 2020; 10: 21508.

[11] Venkatesan P. The changing demographics of COVID-19. *The Lancet. Respiratory Medicine* 2020; 8: e95.

[12] Mahmoud M, Carmisciano L, Tagliafico L, et al. Patterns of comorbidity and in-hospital mortality in older patients with COVID-19 infection. *Front Med* 2021; 8: 726837.

[13] WHO. WHO Coronavirus (COVID-19) Dashboard | WHO Coronavirus (COVID-19) Dashboard With Vaccination Data, https://covid19.who.int/ (2023, accessed 20 February 2023).

[14] Dalli M, Azizi S, Azghar A, et al. Exploring the potential antiviral properties of *Nigella sativa* L. against SARS-CoV-2: Mechanisms and prospects. In: Chen JT (ed) *Ethnopharmacol Drug Discov COVID-19 Anti-SARS-CoV-2 Agents from Herbal Medicines and Natural Products*. Singapore: Springer, 2023, pp. 575–590.

[15] Boopathi S, Poma AB, Kolandaivel P. Novel 2019 coronavirus structure, mechanism of action, antiviral drug promises and rule out against its treatment. *J Biomol Struct Dyn* 2021; 39: 3409–3418.

[16] Harrison AG, Lin T, Wang P. Mechanisms of SARS-CoV-2 transmission and pathogenesis. *Trends Immunol* 2020; 41: 1100–1115.

[17] Kumar A, Narayan RK, Prasoon P, et al. COVID-19 mechanisms in the human body-what we know so far. *Front Immunol* 2021; 12: 693938.

[18] Jani R, Salavessa L, Delevoye C. Par ici la sortie ! Le SARS-CoV-2 utilise les lysosomes pour sortir de la cellule infectée. *Médecine/Sciences* 2021; 37: 716–719. https://doi.org/10.1051/medsci/2021100.

[19] Lin B. Chapter 12- integrating comprehensive and alternative medicine into stroke: Herbal treatment of ischemia. In: Watson RR (ed) *Complementary and Alternative Therapies and the Aging Population*. San Diego: Academic Press, 2009, pp. 229–274.

[20] Liu H, Ye F, Sun Q, et al. Scutellaria baicalensis extract and baicalein inhibit replication of SARS-CoV-2 and its 3C-like protease in vitro. *J Enzyme Inhib Med Chem* 2021; 36: 497–503.

[21] Al-Kuraishy HM, Al-Fakhrany OM, Elekhnawy E, et al. Traditional herbs against COVID-19: Back to old weapons to combat the new pandemic. *Eur J Med Res* 2022; 27: 1–11.

[22] Xia L, Shi Y, Su J, et al. Shufeng Jiedu, a promising herbal therapy for moderate COVID-19: Antiviral and anti-inflammatory properties, pathways of bioactive compounds, and a clinical real-world pragmatic study. *Phytomedicine* 2021; 85: 153390. https://doi.org/10.1016/j.phymed.2020.153390.

[23] Han S, Lee JH, Kim C, et al. Capillarisin inhibits iNOS, COX-2 expression, and proinflammatory cytokines in LPS-induced RAW 264.7 macrophages via the suppression of ERK, JNK, and NF-κB activation. *Immunopharmacol Immunotoxicol* 2013; 35: 34–42.

[24] Cho JK, Curtis-Long MJ, Lee KH, et al. Geranylated flavonoids displaying SARS-CoV papain-like protease inhibition from the fruits of Paulownia tomentosa. *Bioorganic Med Chem* 2013; 21: 3051–3057.

[25] Khubber S, Hashemifesharaki R, Mohammadi M, et al. Garlic (Allium sativum L.): A potential unique therapeutic food rich in organosulfur and flavonoid compounds to fight with COVID-19. *Nutr J* 2020; 19: 20–22.

[26] Sumaryada T, Pramudita CA. Molecular docking evaluation of some indonesian's popular herbals for a possible covid-19 treatment. *Biointerface Res Appl Chem* 2021; 11: 9827–9835.
[27] Zhao M, Yu Y, Sun L, et al. GCG inhibits SARS-CoV-2 replication by disrupting the liquid phase condensation of its nucleocapsid protein. *Nat Commun* 2019; 12: 1–14.
[28] Hong S, Seo SH, Woo SJ, et al. Epigallocatechin gallate inhibits the uridylate-specific endoribonuclease Nsp15 and efficiently neutralizes the SARS-CoV-2 strain. *J Agric Food Chem* 2021; 69: 5948–5954.
[29] Chebaibi M, Bousta D, Goncalves RFB, et al. Medicinal plants against coronavirus (SARS-COV-2) in morocco via computational virtual screening approach. *Res Sq* 2021; 1: 1–46.
[30] Badshah SL, Faisal S, Muhammad A, et al. Antiviral activities of flavonoids. *Biomed Pharmacother* 2021; 140: 111596.
[31] Amin Hussen NH. Docking study of naringin binding with COVID-19 main protease enzyme. *Iraqi J Pharm Sci* 2021; 29: 231–238.
[32] Fan F-Y, Sang L-X, Jiang M. Catechins and their therapeutic benefits to inflammatory bowel disease. *Molecules* 2017; 22: 484. https://doi.org/10.3390/molecules22030484.
[33] Shadrack DM, Deogratias G, Kiruri LW, et al. Luteolin: A blocker of SARS-CoV-2 cell entry based on relaxed complex scheme, molecular dynamics simulation, and metadynamics. *J Mol Model* 2021; 27: 221.
[34] Zhu J, Yan H, Shi M, et al. Luteolin inhibits spike protein of severe acute respiratory syndrome coronavirus-2 (SARS-CoV-2) binding to angiotensin-converting enzyme 2. *Phytother Res* 2023; 37: 3508–3521. https://doi.org/10.1002/ptr.7826.
[35] Alvarado W, Perez-Lemus GR, Menéndez CA, et al. Molecular characterization of COVID-19 therapeutics: Luteolin as an allosteric modulator of the spike protein of SARS-CoV-2. *Mol Syst Des Eng* 2022; 7: 58–66.
[36] Khezri MR, Nazari-Khanamiri F, Mohammadi T, et al. Potential effects of icariin, the epimedium-derived bioactive compound in the treatment of COVID-19: A hypothesis. *Naunyn Schmiedebergs Arch Pharmacol* 2022; 395: 1019–1027.
[37] Zeng Y, Xiong Y, Yang T, et al. Icariin and its metabolites as potential protective phytochemicals against cardiovascular disease: From effects to molecular mechanisms. *Biomed Pharmacother* 2022; 147: 112642.
[38] Dalli M, Bekkouch O, Azizi SE, et al. Nigella sativa l. Phytochemistry and pharmacological activities: A review (2019–2021). *Biomolecules* 2022; 12: 20. https://doi.org/10.3390/biom12010020.
[39] Dalli M, Daoudi NE, Azizi S, et al. Chemical composition analysis using HPLC-UV/GC-MS and inhibitory activity of different nigella sativa fractions on pancreatic α-amylase and intestinal glucose absorption. *Biomed Res Int* 2021; 2021: 12.
[40] Tapas DA, Sakarkar DM, Kakde R. Flavonoids as nutraceuticals: A review. *Trop J Pharm Res* 2008; 7(3): 7. https://doi.org/10.4314/tjpr.v7i3.14693.
[41] Beecher GR. Overview of dietary flavonoids: Nomenclature, occurrence and intake. *J Nutr* 2003; 133: 3248S–3254S.
[42] Geleijnse JM, Hollman PC. Flavonoids and cardiovascular health: Which compounds, what mechanisms? *Am J Clin Nutr* 2008; 88: 12–13.
[43] Maleki SJ, Crespo JF, Cabanillas B. Anti-inflammatory effects of flavonoids. *Food Chem* 2019; 299: 125124.
[44] Al-Ishaq RK, Abotaleb M, Kubatka P, et al. Flavonoids and their anti-diabetic effects: Cellular mechanisms and effects to improve blood sugar levels. *Biomolecules* 2019; 9: 430. https://doi.org/10.3390/biom9090430.
[45] Attaway JA, Buslig BS. Antithrombogenic and antiatherogenic effects of citrus flavonoids. Contributions of Ralph C. Robbins. *Adv Exp Med Biol* 1998; 439: 165–173.
[46] Fernández J, Silván B, Entrialgo-Cadierno R, et al. Antiproliferative and palliative activity of flavonoids in colorectal cancer. *Biomed Pharmacother* 2021; 143: 112241.
[47] Ma X-Q, Cheng Z, Zhang Y, et al. Antiosteoporotic flavonoids from Podocarpium podocarpum. *Phytochem Lett* 2013; 6: 118–122.
[48] Orhan DD, Ozçelik B, Ozgen S, et al. Antibacterial, antifungal, and antiviral activities of some flavonoids. *Microbiol Res* 2010; 165: 496–504.
[49] Bakhtiari M, Panahi Y, Ameli J, et al. Protective effects of flavonoids against Alzheimer's disease-related neural dysfunctions. *Biomed Pharmacother* 2017; 93: 218–229.
[50] Reed J. Cranberry flavonoids, atherosclerosis and cardiovascular health. *Crit Rev Food Sci Nutr* 2002; 42: 301–316.
[51] Ciumărnean L, Milaciu MV, Runcan O, et al. The effects of flavonoids in cardiovascular diseases. *Molecules* 2020; 25: 4320. https://doi.org/10.3390/molecules25184320.
[52] Corcoran MP, McKay DL, Blumberg JB. Flavonoid basics: Chemistry, sources, mechanisms of action, and safety. *J Nutr Gerontol Geriatr* 2012; 31: 176–189.
[53] Kaul TN, Middleton EJ, Ogra PL. Antiviral effect of flavonoids on human viruses. *J Med Virol* 1985; 15: 71–79.
[54] Badshah SL, Faisal S, Muhammad A, et al. Antiviral activities of flavonoids. *Biomed Pharmacother* 2021; 140: 111596.
[55] Wang HK, Xia Y, Yang ZY, et al. Recent advances in the discovery and development of flavonoids and their analogues as antitumor and anti-HIV agents. *Adv Exp Med Biol* 1998; 439: 191–225.
[56] Agrawal A. Pharmacological activities of flavonoids: A review. *Int J Pharm Sci Nanotechnol* 2011; 4: 1394–1398. https://doi.org/10.37285/ijpsn.2011.4.2.3.
[57] Di Petrillo A, Orrù G, Fais A, et al. Quercetin and its derivates as antiviral potentials: A comprehensive review. *Phytother Res* 2022; 36: 266–278.
[58] Imran M, Thabet HK, Alaqel SI, et al. The therapeutic and prophylactic potential of quercetin against COVID-19: An outlook on the clinical studies, inventive compositions, and patent literature. *Antioxidants (Basel, Switzerland)* 2022; 11: 876. https://doi.org/10.3390/antiox11050876.
[59] Di Pierro F, Derosa G, Maffioli P, et al. Possible therapeutic effects of adjuvant quercetin supplementation against early-stage COVID-19 infection: A prospective, randomized, controlled, and open-label Study. *Int J Gen Med* 2021; 14: 2359–2366.
[60] Ganeshpurkar A, Saluja AK. Protective effect of catechin on humoral and cell mediated immunity in rat model. *Int Immunopharmacol* 2018; 54: 261–266.
[61] Yimam M, Brownell L, Hodges M, et al. Analgesic effects of a standardized bioflavonoid composition from Scutellaria baicalensis and Acacia catechu. *J Diet Suppl* 2012; 9: 155–165.
[62] Song J-M, Lee K-H, Seong B-L. Antiviral effect of catechins in green tea on influenza virus. *Antiviral Res* 2005; 68: 66–74.

[63] Takahashi T, Kurebayashi Y, Tani K, et al. The antiviral effect of catechins on mumps virus infection. *J Funct Foods* 2021; 87: 104817.

[64] Yang C-C, Wu C-J, Chien C-Y, et al. Green tea polyphenol catechins inhibit coronavirus replication and potentiate the adaptive immunity and autophagy-dependent protective mechanism to improve acute lung injury in mice. *Antioxidants (Basel, Switzerland)* 2021; 11: 876.. https://doi.org/10.3390/antiox10060928.

[65] Narotzki B, Reznick AZ, Aizenbud D, et al. Green tea: A promising natural product in oral health. *Arch Oral Biol* 2012; 57: 429–435.

[66] Li J, Song D, Wang S, et al. Antiviral effect of epigallocatechin gallate via impairing porcine circovirus type 2 attachment to host cell receptor. *Viruses* 2020; 12: 176. https://doi.org/10.3390/v12020176.

[67] Ohishi T, Hishiki T, Baig MS, et al. Epigallocatechin gallate (EGCG) attenuates severe acute respiratory coronavirus disease 2 (SARS-CoV-2) infection by blocking the interaction of SARS-CoV-2 spike protein receptor-binding domain to human angiotensin-converting enzyme 2. *PLoS One* 2022; 17: e0271112.

[68] Salehi B, Fokou PVT, Sharifi-Rad M, et al. The therapeutic potential of naringenin: A review of clinical trials. *Pharmaceuticals (Basel)* 2019; 12: 11. https://doi.org/10.3390/ph12010011.

[69] Den Hartogh DJ, Tsiani E. Antidiabetic properties of naringenin: A citrus fruit polyphenol. *Biomolecules* 2019; 9: 99. https://doi.org/10.3390/biom9030099.

[70] Tutunchi H, Naeini F, Ostadrahimi A, et al. Naringenin, a flavanone with antiviral and anti-inflammatory effects: A promising treatment strategy against COVID-19. *Phytother Res* 2020; 34: 3137–3147.

[71] Alberca RW, Teixeira FME, Beserra DR, et al. Perspective: The potential effects of naringenin in COVID-19. *Immunol* 2020; 11: 570919.

[72] Parhiz H, Roohbakhsh A, Soltani F, et al. Antioxidant and anti-inflammatory properties of the citrus flavonoids hesperidin and hesperetin: An updated review of their molecular mechanisms and experimental models. *Phytother Res* 2015; 29: 323–331.

[73] Pyrzynska K. Hesperidin: A review on extraction methods, stability and biological activities. *Nutrients* 2022; 14: 2387. https://doi.org/10.3390/nu14122387.

[74] Cheng F-J, Huynh T-K, Yang C-S, et al. Hesperidin is a potential inhibitor against SARS-CoV-2 infection. *Nutrients* 2021; 13: 2800. https://doi.org/10.3390/nu13082800.

[75] Bellavite P, Donzelli A. Hesperidin and SARS-CoV-2: New light on the healthy function of citrus fruits. *Antioxidants (Basel, Switzerland)* 2020; 9: 742. https://doi.org/10.3390/antiox9080742.

[76] Ortiz A de C, Fideles SOM, Reis CHB, et al. Therapeutic effects of citrus flavonoids neohesperidin, hesperidin and its aglycone, hesperetin on bone health. *Biomolecules* 2022; 12: 626. https://doi.org/10.3390/biom12050626.

[77] Li A, Zhang X, Luo Q. Neohesperidin alleviated pathological damage and immunological imbalance in rat myocardial ischemia-reperfusion injury via inactivation of JNK and NF-κB p65. *Biosci Biotechnol Biochem* 2021; 85: 251–261.

[78] Alzaabi MM, Hamdy R, Ashmawy NS, et al. Flavonoids are promising safe therapy against COVID-19. *Phytochem Rev* 2022; 21: 291–312.

[79] Farhat A, Ben Hlima H, Khemakhem B, et al. Apigenin analogues as SARS-CoV-2 main protease inhibitors: In-silico screening approach. *Bioengineered* 2022; 13: 3350–3361.

[80] Salehi B, Venditti A, Sharifi-Rad M, et al. The therapeutic potential of apigenin. *Int J Mol Sci* 2019; 20: 1305. https://doi.org/10.3390/ijms20061305.

[81] Khandelwal N, Chander Y, Kumar R, et al. Antiviral activity of Apigenin against buffalopox: Novel mechanistic insights and drug-resistance considerations. *Antiviral Res* 2020; 181: 104870.

[82] Agrawal P, Agrawal C, Blunden G. Naringenin as a possible candidate against SARS-CoV-2 infection and in the pathogenesis of COVID-19. *Nat Prod Commun* 2021; 16: 12. https://doi.org/10.1177/1934578X211066723.

[83] Su W, Wang Y, Li P, et al. The potential application of the traditional chinese herb exocarpium citri grandis in the prevention and treatment of COVID-19. *Tradit Med Res* 2020; 5: 122–177.

[84] Chen T, Su W, Yan Z, et al. Identification of naringin metabolites mediated by human intestinal microbes with stable isotope-labeling method and UFLC-Q-TOF-MS/MS. *J Pharm Biomed Anal* 2018; 161: 262–272.

[85] Zeng X, Bai Y, Peng W, et al. Identification of naringin metabolites in human urine and feces. *Eur J Drug Metab Pharmacokinet* 2017; 42: 647–656.

[86] Chiou W-C, Lu H-F, Hsu N-Y, et al. Ugonin J acts as a SARS-CoV-2 3C-like protease inhibitor and exhibits anti-inflammatory properties. *Front Pharmacol* 2021; 12: 720018.

[87] Cui X-X, Yang X, Wang H-J, et al. Luteolin-7-O-glucoside present in lettuce extracts inhibits hepatitis B surface antigen production and viral replication by human hepatoma cells in vitro. *Front Microbiol* 2017; 8: 2425.

[88] Vincent S, Arokiyaraj S, Saravanan M, et al. Molecular docking studies on the anti-viral effects of compounds from kabasura kudineer on SARS-CoV-2 3CL(pro). *Front Mol Biosci* 2020; 7: 613401.

[89] Mohapatra PK, Chopdar KS, Dash GC, et al. In silico screening and covalent binding of phytochemicals of Ocimum sanctum against SARS-CoV-2 (COVID-19) main protease. *J Biomol Struct Dyn* 2023; 41: 435–444.

[90] Yalçın S, Yalçınkaya S, Ercan F. In silico detection of inhibitor potential of Passiflora compounds against SARS-Cov-2(Covid-19) main protease by using molecular docking and dynamic analyses. *J Mol Struct* 2021; 1240: 130556.

[91] Ryu YB, Jeong HJ, Kim JH, et al. Biflavonoids from Torreya nucifera displaying SARS-CoV 3CL(pro) inhibition. *Bioorg Med Chem* 2010; 18: 7940–7947.

[92] Shahhamzehei N, Abdelfatah S, Efferth T. In silico and in vitro identification of pan-coronaviral main protease inhibitors from a large natural product library. *Pharmaceuticals (Basel)* 2022; 15: 308. https://doi.org/10.3390/ph15030308.

[93] Rungsung S, Singh TU, Rabha DJ, et al. Luteolin attenuates acute lung injury in experimental mouse model of sepsis. *Cytokine* 2018; 110: 333–343.

[94] Liu B, Yu H, Baiyun R, et al. Protective effects of dietary luteolin against mercuric chloride-induced lung injury in mice: Involvement of AKT/Nrf2 and NF-κB pathways. *Food Chem Toxicol Int J Publ Br Ind Biol Res Assoc* 2018; 113: 296–302.

[95] Ngwa W, Kumar R, Thompson D, et al. Potential of flavonoid-inspired phytomedicines against COVID-19. *Molecules* 2020; 25: 2707. https://doi.org/10.3390/molecules25112707.
[96] Jo S, Kim S, Shin DH, et al. Inhibition of SARS-CoV 3CL protease by flavonoids. *J Enzyme Inhib Med Chem* 2020; 35: 145–151.
[97] Zloh M, Ilić M, Jojić N, et al. Narrative review of the progress in discovery of anti COVID-19 agents from traditional Chinese medicine using in silico approaches. *Longhua Chinese Med* 2020; 3: 19.
[98] Li M, Chen L, Zhang J, et al. The SARS-CoV-2 receptor ACE2 expression of maternal-fetal interface and fetal organs by single-cell transcriptome study. *PLoS One* 2020; 15: e0230295.
[99] Liskova A, Koklesova L, Samec M, et al. Targeting phytoprotection in the COVID-19-induced lung damage and associated systemic effects-the evidence-based 3PM proposition to mitigate individual risks. *EPMA J* 2021; 12: 325–347.
[100] Sharma A, Kaur R, Katnoria JK, et al. Family fabaceae: A boon for cancer therapy. In: Malik S (ed) *Biotechnology and Production of Anti-Cancer Compounds*. Cham: Springer International Publishing, pp. 157–175.
[101] Tuli HS, Tuorkey MJ, Thakral F, et al. Molecular mechanisms of action of genistein in cancer: Recent advances. *Front Pharmacol* 2019; 10: 1336.
[102] LeCher JC, Diep N, Krug PW, et al. Genistein has antiviral activity against herpes B virus and acts synergistically with antiviral treatments to reduce effective dose. *Viruses* 2019; 11: 499. https://doi.org/10.3390/v11060499.
[103] Ginwala R, Bhavsar R, Chigbu DI, et al. Potential role of flavonoids in treating chronic inflammatory diseases with a special focus on the anti-inflammatory activity of apigenin. *Antioxidants (Basel, Switzerland)* 2019; 8: 35. https://doi.org/10.3390/antiox8020035.
[104] Jafari A, Esmaeilzadeh Z, Khezri MR, et al. An overview of possible pivotal mechanisms of Genistein as a potential phytochemical against SARS-CoV-2 infection: A hypothesis. *J Food Biochem* 2022; 46: e14345.
[105] Agrawal PK, Agrawal C, Blunden G. Antiviral and possible prophylactic significance of myricetin for COVID-19. *Nat Prod Commun* 2023; 18: 1934578X231166283.
[106] Chikhale RV, Sinha SK, Khanal P, et al. Computational and network pharmacology studies of Phyllanthus emblica to tackle SARS-CoV-2. *Phytomedicine Plus Int J Phyther Phytopharm* 2021; 1: 100095.

3 Flavonoids as Potential Drugs for Combating SARS-CoV-2

Molecular Docking Studies

Celia María Curieses Andrés, José Manuel Pérez de la Lastra, Celia Andrés Juan, Francisco J. Plou, and Eduardo Pérez-Lebeña

3.1 INTRODUCTION

To date, a considerable amount of research has been devoted to investigating the potential therapeutic effect of polyphenols in alleviating the clinical symptoms caused by SARS-CoV-2 (Pérez de la Lastra et al., 2021). The Mediterranean diet has been shown to have beneficial effects on the prevention of various diseases such as diabetes, neurodegenerative diseases, cardiovascular diseases and metabolic syndrome due to its antioxidant components (Milton-Laskibar et al., 2023).

Polyphenols, organic chemicals found in plants, have a variety of properties that contribute to the promotion of human health (Pandey and Rizvi, 2009). Due to their antioxidant properties, these compounds can protect cells from damage caused by free radicals. DNA damage can be caused by unstable molecules known as free radicals (Andrés et al., 2023), proteins (Andrés et al., 2022) and lipids in cells (Juan et al., 2021).

Polyphenols can be found in a variety of foods. Fruits, vegetables, grains, legumes, nuts and seeds are all examples of foods that contain polyphenols. Other examples include legumes. At least one benzene ring containing two or more hydroxyl groups (OH) is a feature that distinguishes polyphenols from other types of compounds. They can act as donors of reducing species and as antioxidants (Andrés et al., 2023).

Flavonoids, a class of polyphenolic compounds, are widespread in the plant kingdom and are primarily responsible for the bright pigmentation of various fruits and vegetables. Flavonoids exhibit a variety of beneficial health properties, including their antioxidant, anti-inflammatory, cardiovascular and anti-carcinogenic effects. Research suggests that eating foods rich in polyphenols, particularly flavonoids, may be associated with a lower likelihood of developing chronic diseases such as heart disease, cancer, diabetes and Alzheimer's disease (Panche et al., 2016).

The extensive range of natural flavonoids exhibits considerable structural variability. Flavonoids comprise a class of phytochemical compounds derived from plants, fruits and vegetables. They have a characteristic structure known as C6-C3-C6, with the A-ring and B-ring consisting of phenolic components, while a pyran or flavan heterocycle, known as the C-ring, links them together (Dias et al., 2021). The observed structure is the result of the combination of two biosynthetic processes that take place in primary metabolism. First, it is produced from acetyl-CoA via the malonic acid metabolic pathway. Second, it is influenced by the shikimic acid metabolic pathway, which involves phenylalanine (Mouradov and Spangenberg, 2014) (Figure 3.1).

The compounds within a family have the same skeletal structure but show differences in the substitution pattern and the degree of oxidation of the heterocyclic C-ring. In addition, the individual compounds within the family show differences in the substitution pattern of the A and B rings. Figure 3.2 illustrates different families of flavonoids.

According to Panche et al. (2016), the compounds occur either in the form of aglycones or glycosides, and their biological effects are influenced by two factors: (i) structural variations and (ii) patterns of glycosylation. Polyphenols,

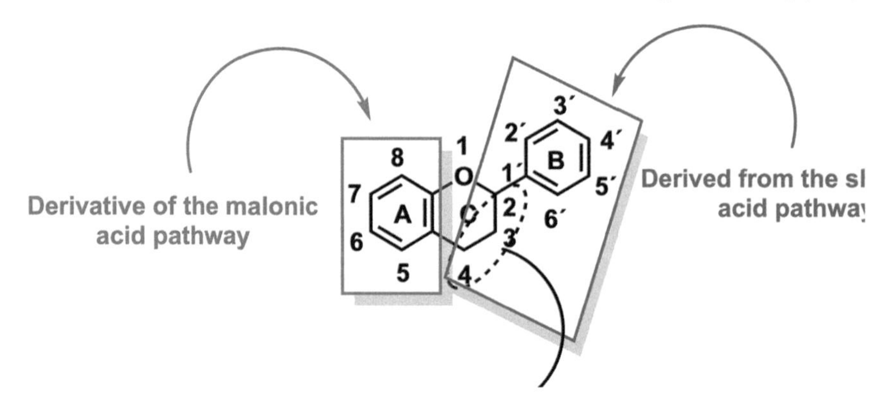

FIGURE 3.1 The basic skeleton of flavonoids.

DOI: 10.1201/9781003433200-3

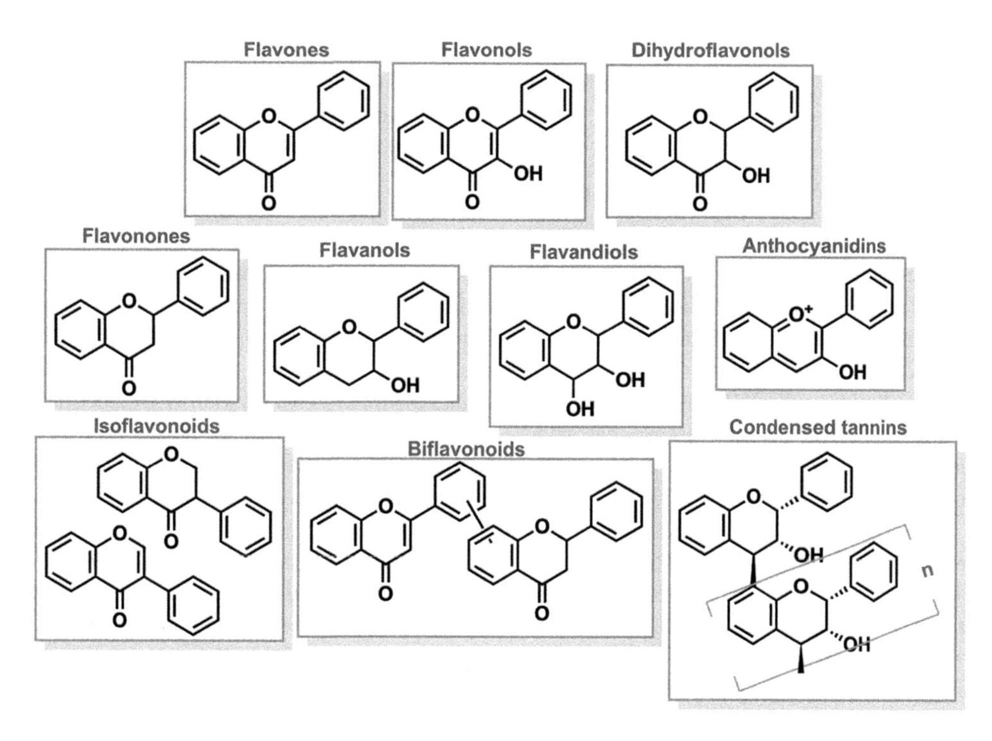

FIGURE 3.2 Classification of the main families of flavonoids.

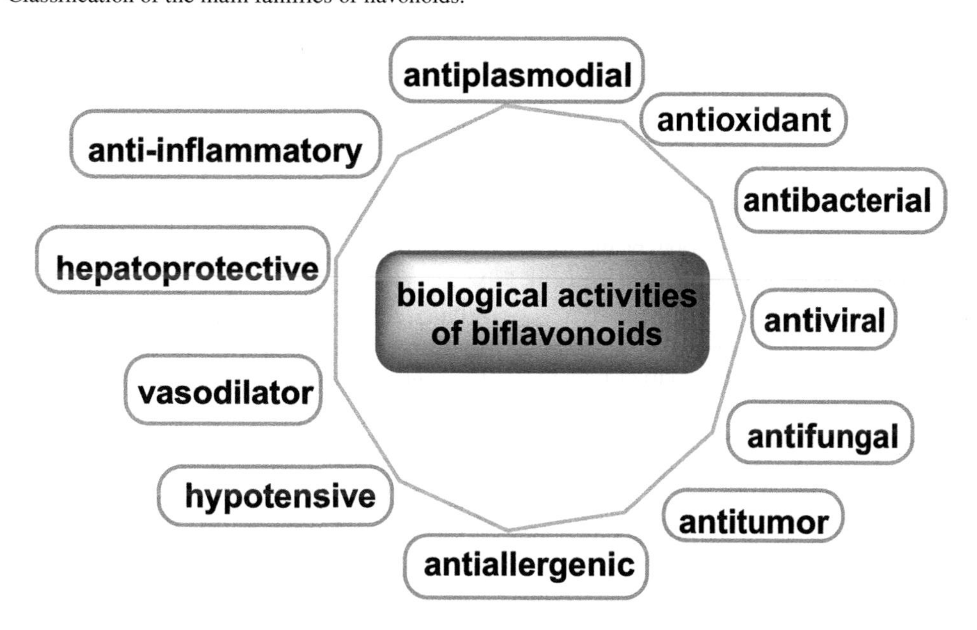

FIGURE 3.3 Pharmacological activities of biflavonoids.

which are found in many foods, are often glycosylated with monosaccharides such as galactose, glucose, arabinose, xylose and galacturonic acid, as well as other sugars. They also have the ability to form complexes with organic carboxylic acids, amines and lipids (Duthie et al., 2003).

These substances possess a variety of pharmacological properties, including antibacterial, antioxidant, anti-inflammatory, anticancer and antiviral effects (Ahmad et al., 2015; Badshah et al., 2021; Wang et al., 2018), In addition, they are generally considered safe, have low toxicity and have the potential to enhance the therapeutic effect of other drugs through synergistic interactions (Russo et al., 2020). Numerous flavonoids have been investigated for their potential antiviral activity, and a number of naturally occurring flavonoids have shown remarkable antiviral properties both in the laboratory and in living organisms. The biological activities of biflavonoids, which are flavonoid dimers containing a carbonyl group at C4 (such as chalcones, flavanones, flavones, flavanols, flavonols, aurones and isoflavones), are shown in Figure 3.3. These biflavonoids differ in the oxygenation pattern of their monomers, the degree of oxidation of the C3 fragment and the binding between the two flavonoid units. The bond in question may occur at several positions within the A-ring (in particular at positions 5, 6, 7 or 8), the B-ring (in particular at positions 2′, 3′, 4′, 5′ or 6′) or the C-ring (in particular at positions 2 or 3). This bond can be formed either by a C-C or a C-O-C bond. In exceptional cases, a methylene bridge connects the two rings (C-CH2-C). This process leads to many subtypes of biflavonoids (AA, BB, AB, CC, etc.), which can be linked by C-C or C-O-C bonds.

3.2 CORONAVIRUSES AND ANTIVIRAL AGENTS

The 21st century is characterized by several coronavirus pandemics. The SARS-CoV-1 virus first appeared in 2002, followed by MERS-CoV (Middle East Respiratory Syndrome Coronavirus) in 2012. Subsequently, in 2019, the betacoronavirus known as SARS-CoV-2 triggered the outbreak of the disease known as COVID-19 (Russo et al., 2020). The above-mentioned infectious disease has triggered a global health emergency that can be considered the deadliest in modern times (Cascella et al., 2023).

Coronaviruses, which belong to the nidoviruses, are a group of enveloped viruses with a single-stranded (+)RNA genome. Their discovery dates back to the 1960s in the United Kingdom and the United States, when researchers successfully isolated two viruses that cause the common cold in humans (Kapikian et al., 1969). Coronaviruses have a spherical or pleomorphic morphology characterized by a diameter of 80–120 nm. In 1968, electron microscope photos were published showing the coronal structures of the virus, which bore a striking resemblance to the "solar corona". Consequently, the Latin term "coronavirus" was chosen as the name for this virus family (Fung and Liu, 2019). Between 2003 and 2019, two different strains of extremely pathogenic viruses affecting humans emerged, namely the Severe Acute Respiratory Syndrome Coronavirus (SARS-CoV) in 2003 and the MERS-CoV in 2012. These virus strains were responsible for significant outbreaks that were classified as serious by the World Health Organization (WHO) and led to severe consequences (de Groot et al., 2013; Zhong et al., 2003).

At around 26,000–32,000 nucleotides, the size of the coronavirus RNA genome is comparatively larger than that of other RNA viruses (Hernandez Acosta et al., 2022). The coronavirus has four primary structural proteins, namely: (i) the trimeric glycoprotein spike (S), which is located on the surface of the viral envelope and plays a critical role in facilitating viral entry into host cells; (ii) the membrane or matrix protein (M); (iii) the envelope protein (E), which is essential for the assembly and release of virions; and (iv) the nucleocapsid protein (N), which is responsible for binding to the RNA genome, resulting in the formation of a helically symmetric nucleocapsid (Jackson et al., 2022).

The ongoing global impact of COVID-19 is due to the constant emergence of mutant strains of the novel coronavirus, which have varying degrees of virulence. According to the World Health Organization (WHO), the number of confirmed COVID-19 cases worldwide is 640,753,061 (as of October 25, 2023). Of these figures, a total of 533,593,551 people have successfully recovered, while 15,765,747 people have unfortunately died.

Significant progress has been made in the field of vaccine and antiviral drug research. However, the increasing frequency of infections, the uneven global distribution of vaccinations and the emergence of new variants of the original SARS-CoV-2 require an ongoing search for future antiviral treatments (Carabelli et al., 2023). The use of natural substances as antiviral drugs is a convincing approach amidst the complexity of drug development (Low et al., 2023). Since the emergence of SARS-CoV-2, a considerable number of articles, more than 15,000, have been devoted to the study of natural products. These studies have focused on the polyphenol family of natural products, which have been found to interrupt various stages of the coronavirus entry and replication cycle. Consequently, these natural products have shown promise as potential agents against SARS-CoV-2, as demonstrated by virtual *in silico*, *in vitro* and *in vivo* screening studies.

Antiviral agents, a major advance in medicine in recent decades, refer to substances that effectively inhibit the replication and spread of viral pathogens while minimizing the harmful effects on the host organism. Viral infections are characterized by a limited repertoire of antiviral drugs that are only effective against a narrow spectrum of pathogens (De Clercq, 2004). Most of these compounds are intended for the therapeutic treatment of human immunodeficiency virus (HIV) (Antonelli and Turriziani, 2012). The increase in drug-resistant strains is a significant problem, particularly in immunocompromised individuals, and it deserves considerable attention. Furthermore, for a large number of viral strains that cause persistent infections, such as HIV, there are no viable vaccines or effective therapeutic interventions that can eradicate the viral pathogen (Campos-Gonzalez et al., 2023).

3.3 MECHANISM OF ACTION OF ANTIVIRAL COMPOUNDS

Most approved drugs inhibit intracellular processes that influence the production and behavior of viral proteins and nucleic acids. The diagram in Figure 3.4 illustrates different levels at which antiviral drugs can exert their effect.

3.3.1 Adsorption and Entry Inhibitors

The primary phase of viral infection involves the process in which the virus attaches itself to the cell membrane. The process is made possible by the contact between the surface molecules of the virus and the receptors present on the surface of the cells (Ryu, 2017). Cells that express this particular receptor or these special receptors are called virus-sensitive. After the process of adsorption, the virus enters the cell, either by endocytosis or, only in the case of enveloped viruses, by fusion between the viral envelope and the cell membrane (Klasse et al., 1998). Approved inhibitors that specifically target these phases of the viral cycle have been authorized for the therapeutic treatment of HIV infection and the prophylaxis of RSV infection (Magden et al., 2005).

3.3.2 Capsid Uncoating Inhibitors

The process of viral uncoating is a crucial stage in the replication cycle of viruses, and researchers have successfully

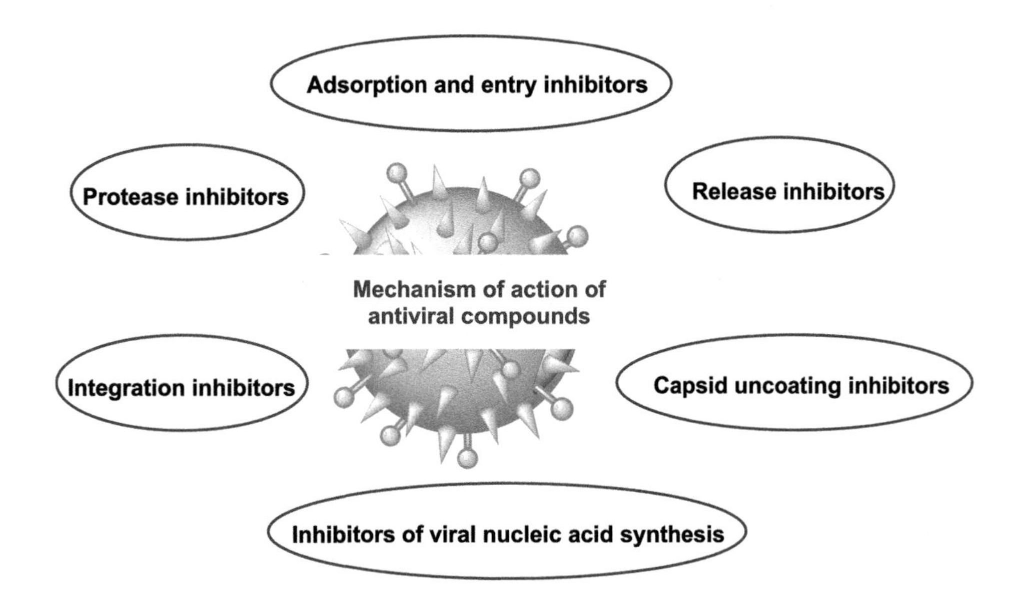

FIGURE 3.4 Levels at which antiviral compounds can act.

produced inhibitors that specifically target this phase to combat influenza viruses (Yamauchi and Greber, 2016).

3.3.3 Viral Nucleic Acid Synthesis Inhibitors

Viruses drive the process of nucleic acid synthesis through the use of viral enzymes, including DNA polymerase, RNA polymerase and reverse transcriptase (Choi, 2012). The importance of these polymerases in the viral life cycle has made them highly desirable targets for therapeutic intervention in various viral diseases. Nucleoside analogs (NAs) have emerged as the first class of polymerase inhibitors to show clinical efficacy. They are widely used for the treatment of various infections, including HBV, HSV-1, HSV-2 and HIV-1 (von Kleist et al., 2012). NAs are usually developed as prodrugs that require intracellular phosphorylation to generate a (deoxy)nucleoside triphosphate analog. This analog can then be integrated by the viral polymerase into a newly forming viral DNA or RNA strand (von Kleist et al., 2012). Most NAs are phosphorylated by intracellular kinases, with the exception of compounds used to treat herpes infections. In the case of these specific NAs, the first phosphorylation step is facilitated by the viral thymidine kinase, while the two subsequent phosphorylation steps are catalyzed by cellular enzymes (Piret and Boivin, 2011). After the incorporation process, the NAs effectively stop the polymerization machinery. In addition, the absence of the hydroxyl-3′-group on the deoxyribose residue prevents the binding of the subsequent incoming nucleotide (Kokoris and Black, 2002).

3.3.4 Integration Inhibitors

Antiretroviral agents that selectively inhibit integrase activity are approved for the therapeutic treatment of HIV infection (Zhao et al., 2022). HIV integrase is one of the three essential enzymes encoded by the viral gene. The enzyme known as integrase plays a crucial role in the replication process of HIV-1 by facilitating the integration of the viral genome into the DNA of the host cell. The enzymatic activity of this particular viral enzyme involves two distinct reactions. First, it facilitates the processing of the 3′ end of the viral DNA, leading to cleavage of the terminal GpT dinucleotide from each long terminal repeat. Second, it facilitates the process of strand transfer, in which the viral genome is integrated into the host chromosomal DNA to form a functional integrated proviral DNA (Craigie, 2012). Due to its essential role in the replication of HIV, integrase has been recognized as a validated target for the development of anti-HIV drugs.

3.3.5 Protease Inhibitors

Viral proteases are a promising starting point for the development of new antiviral treatments as they facilitate the processing of viral polyproteins and the maturation of precapsids. The catalytic activity of these proteases is essential for the production of infectious virions (Majerová and Novotný, 2021).

3.3.6 Release Inhibitors

The final stage of the viral life cycle involves the release of newly formed viral progeny from the cellular host. Certain drugs can prevent this stage by targeting viral proteins that play a role in this mechanism and are approved for the therapeutic treatment of influenza virus infections (Loregian et al., 2014).

3.4 CHARACTERISTICS OF SARS-COV-2

The use of genomic sequencing techniques in the study of SARS-CoV-2 has enabled the identification of important proteins and enzymes that are closely linked to the replication process (Haas et al., 2021), This analysis confirmed that these proteins and enzymes have considerable similarities (approx. 80%) with those of SARS-CoV (V'Kovski et al., 2021) and close similarities with the genomic sequences of other betacoronaviruses (Chen et al., 2021). The importance of three key proteins and enzymes involved in virus replication and the vaccination process is being elucidated using genome sequencing (Saravanan et al., 2022): PL^{pro} (papain-like protease), spike protein and $3CL^{pro}$ (3 chymotrypsin-like proteases).

ACE2, also known as angiotensin-converting enzyme 2, is a protein that acts as a receptor for SARS-CoV-2. It plays an important role in the renin–angiotensin system, a physiological system responsible for the regulation of blood pressure and various cardiovascular functions (Gheblawi et al., 2020), TMPRSS2, also known as serine transmembrane protease type 2, is a protein that is found on the cell surface of various organs in the human body, e.g., in the lungs, heart, kidneys, intestines and blood vessels. The virus uses TMPRSS2 to facilitate the cleavage of its spike protein. It is noteworthy that these molecules originate from the host cells (Zhang et al., 2022).

Both SARS-CoV and SARS-CoV-2 use the angiotensin-converting enzyme 2 (ACE2) receptor as a means to invade and infect host cells (Beyerstedt et al., 2021). ACE2 is an extracellular peptidase that is mainly found on the surface of airway epithelial cells and is responsible for facilitating the infection of human cells (Khan et al., 2022). The virus uses TMPRSS2 as an enzyme to cleave its spike protein, which serves as a key protein for cell entry, infection and replication. Consequently, these proteins primarily control the entry of the virus into the cells and the subsequent intracellular infection process (Glowacka et al., 2011). The spike protein and ACE 2 originate from the host organism, but $3CL^{pro}$ is associated with the virus and deserves great attention as a key target in the development of antiviral therapeutics (Huang et al., 2020).

The process of entry and invasion of the virus into the host cell, which is facilitated by the proteins ACE2 and TMPRSS2, shows similarities between MERS-CoV, SARS-CoV and SARS-CoV-2, which are three closely related viruses (Jackson et al., 2022). The key steps in this process are: (i) the attachment of the virus to the host cells, (ii) the binding of the virus to the ACE-2 receptor, (iii) the promotion of fusion between the virus and the infected cell, facilitated by the spike protein TMPRSS2, a serine-type protease, and (iv) the cleavage of the polypeptide responsible for replication by the enzyme $3CL^{pro}$ (Wu et al., 2020).

3.5 DRUGS AGAINST SARS-COV-2

The main aim of Covid-19 treatment is to prevent the spread of the virus and reduce lung damage in the early stages of the disease. This is achieved through the use of antiviral agents, which can be supplemented with antibiotics in cases where a bacterial superinfection is suspected (Sultana et al., 2020). There are currently no pharmaceutical interventions that provide a definitive cure for Covid-19. Therefore, clinical practice guidelines recommend the administration of medications that alleviate the symptoms associated with the infection (Jirjees et al., 2021). The most commonly used treatment options are antiviral drugs and inhibitors of the inflammatory response. The treatments used to date and their respective modes of action are briefly illustrated in Figure 3.5.

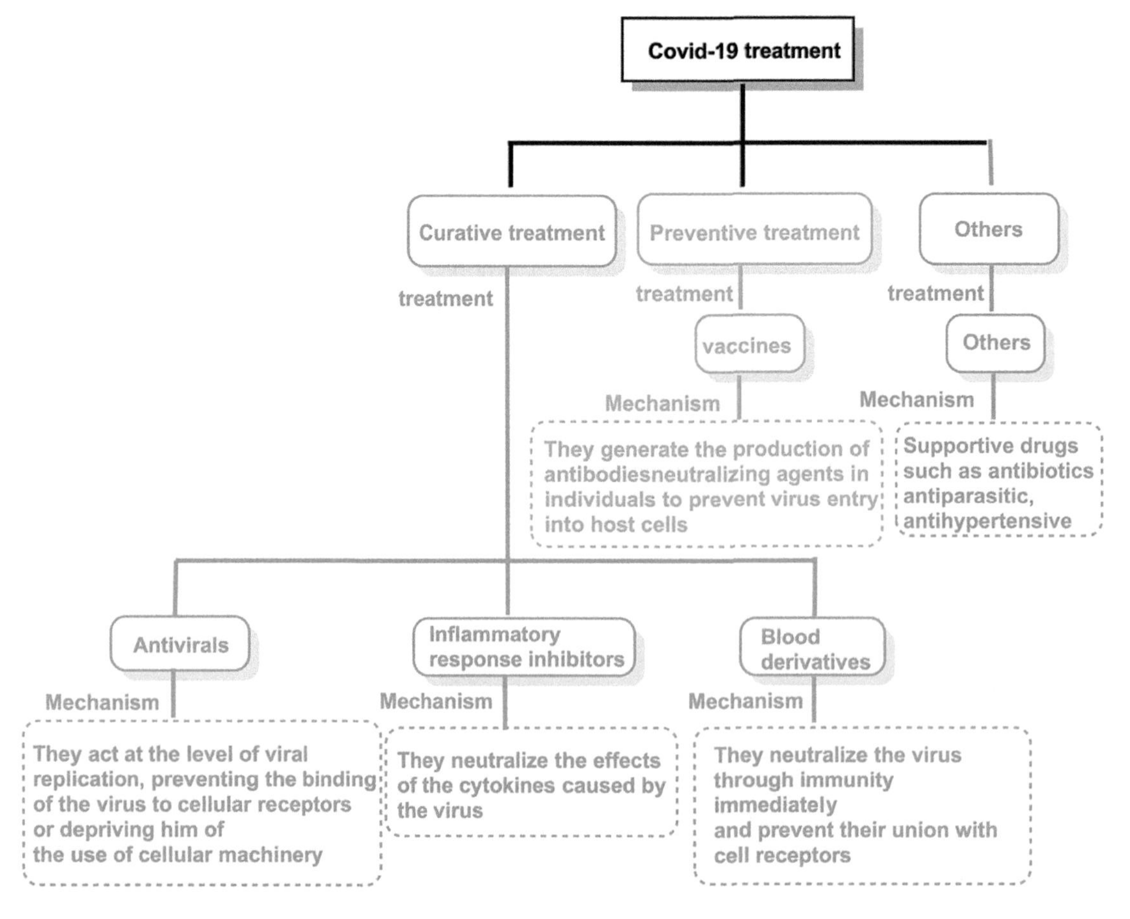

FIGURE 3.5 Treatments and mechanisms of action.

The categorization of anti-SARS-CoV-2 drugs can be based on their specific targets, which may include enzymes or proteins. The main categories are described below:

- **Protease inhibitors**: The main protease of SARS-CoV-2 is M^{pro}, which is necessary for virus assembly (Bulut, 2022). M^{pro} inhibitors, such as nafamostat and camostat, block the activity of this enzyme, preventing the virus from replicating (Farkaš et al., 2023). Nafamostat exhibits greater specificity in binding to the catalytic center compared to camostat. This is supported by the reported finding that nafamostat effectively inhibits SARS-CoV-2 infection at a lower concentration (Zhu et al., 2021). The potential repurposing of the two drugs camostat and nafamostat for the treatment of COVID-19 is based on their ability to inhibit the human enzyme TMPRSS2, which is essential for the proteolytic activation of the viral spike glycoprotein (S) (Zhu et al., 2021). The administration of lopinavir/ritonavir (LPV/r) in adult COVID-19 patients admitted to hospital has not shown significant benefits compared to conventional therapy. In the cohort of critically ill individuals diagnosed with COVID-19, the administration of lopinavir/ritonavir, hydroxychloroquine or a combination of both showed worse outcomes than the absence of antiviral therapy for the treatment of COVID-19 (Arabi et al., 2021). The importance of the major protease (M^{pro}) of the virus as a potential target for the development of inhibitors lies in its crucial role in the viral life cycle and its remarkable conservation in various coronaviruses.
- **Polymerase inhibitors**: The polymerase of SARS-CoV-2 is RdRp, which is responsible for viral RNA replication (Malone et al., 2022). RdRp inhibitors such as remdesivir and molnupiravir effectively inhibit the enzymatic function of RNA-dependent RNA polymerase (RdRp) and thus viral replication. Both molnupiravir and remdesivir have proven their efficacy as early interventions in COVID-19, leading to a reduction in the likelihood of hospitalization and mortality. Both pharmaceutical compounds show efficacy in combating the alpha, beta, delta and gamma forms of the coronavirus (Jayk Bernal et al., 2022). Remdesivir has proven to be a promising antiviral agent against a variety of RNA viruses, including certain coronaviruses such as SARS-CoV-2 (Wang et al., 2020). The prodrug in question is a NA, more precisely adenosine, which inhibits the viral polymerase by causing premature termination of the viral RNA strands being produced. This mechanism effectively inhibits viral replication. The antiviral efficacy of this active ingredient against SARS-CoV-2 has been demonstrated by experimental studies both in the laboratory (*in vitro*) and in living organisms (*in vivo*) (Malin et al., 2020; Razia et al., 2023). There are currently three antiviral treatments in the United States that specifically target SARS-CoV-2: Remdesivir (REM), molnupiravir (MOL) and the combination of nirmatrelvir and ritonavir (Paxlovid). These therapeutic measures work by inhibiting viral replication in the human body and thus slowing down the progression of the infection (Saravolatz et al., 2023). Paxlovid is an antiviral drug that is administered to adult patients diagnosed with COVID-19, especially those who do not require supplemental oxygen and who are at increased risk of developing severe manifestations of the disease (Phizackerley, 2022).
- **Entry inhibitors**: SARS-CoV-2 enters human cells through the interaction between the virus and ACE2. ACE2 inhibitors such as Camostat are able to prevent the virus from binding to ACE2 and thus impair the virus's ability to enter cells (Sharifkashani et al., 2020).
- **Immunomodulators**: Immunomodulatory agents such as monoclonal antibodies targeting the spike protein (S), receptor-binding domain (RBD) and ACE2 have been shown to be effective in enhancing the immune response against SARS-CoV-2 infection. In addition, pharmaceutical interventions such as tocilizumab and baricitinib have been identified as potential therapeutic options for the treatment of this viral infection (Pashaei and Rezaei, 2020). Tocilizumab (TCZ) is an FDA-approved recombinant humanized monoclonal antibody used for the therapeutic treatment of rheumatoid arthritis. The antibody has a high affinity for the receptors of interleukin-6 (IL-6R), resulting in inhibition of IL-6 signaling and subsequent suppression of the associated inflammatory response (Sheppard et al., 2017). The IL-6 signaling pathway plays an important role in the inflammatory immunological response observed in the alveoli of individuals with COVID-19 (Wang et al., 2022).

There are currently no drugs approved by the Food and Drug Administration (FDA) that can detect or prevent the entry and replication of the SARS-CoV-2 virus (Weston et al., 2020). Figure 3.6 shows a list of drugs used in the therapeutic treatment of SARS-CoV-2. In addition to the main categories mentioned above, there are alternative pharmaceutical compounds currently being researched that aim to selectively inhibit various enzymes or proteins associated with SARS-CoV-2. Research is currently underway on drugs that inhibit the spicule protein of the virus, a critical component for the virus to bind to ACE2 (Samavati and Uhal, 2020).

The choice of therapeutic measures for SARS-CoV-2 depends on various considerations, including the severity of the infection, the age of the patient and the presence

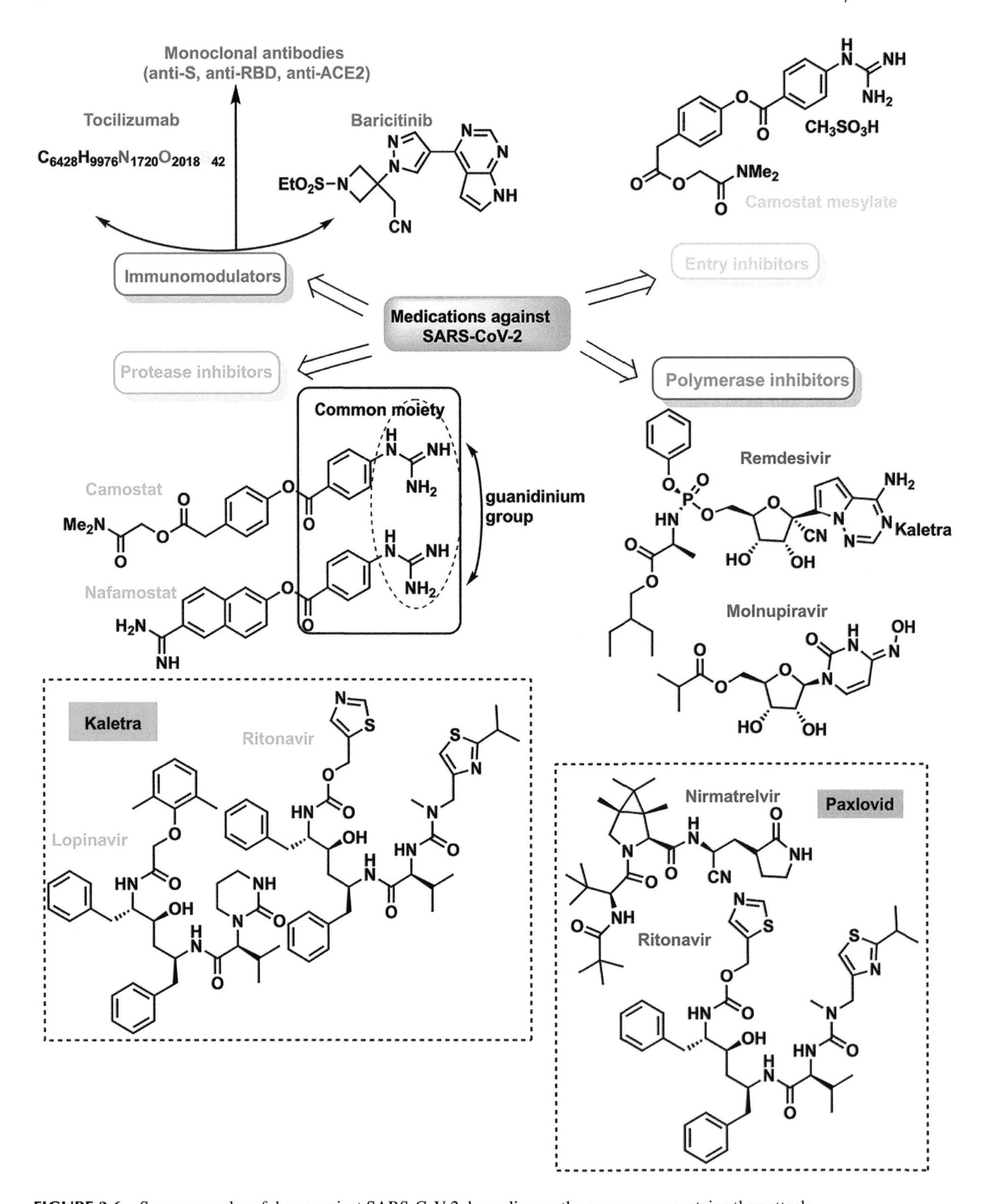

FIGURE 3.6 Some examples of drugs against SARS-CoV-2 depending on the enzymes or proteins they attack.

of concomitant diseases (Renu et al., 2020). The most important measure to combat the disease worldwide is the administration of SARS-CoV-2 vaccines. Administering vaccines to a large proportion of the population reduces the likelihood of infection and mitigates the possibility of experiencing the most severe manifestations of this disease (Suryawanshi and Biswas, 2023).

3.6 ACTIVITY OF FLAVONOIDS AGAINST SEVERAL VIRUSES AND SARS-COV-2

The World Health Organization (WHO) emphasizes the importance of traditional and complementary medicine (TCM) as an important resource, particularly in the area of primary healthcare (Basri et al., 2022). Medicinal plants have been shown to be a potential therapeutic option for the treatment of people suffering from COVID-19 due to their ability to improve overall health. There are not enough clinical studies to definitively prove the use of specific herbal resources. However, most authors tend to recommend the use of plant species that have shown antiviral properties in previous studies (Vaou et al., 2021).

The antiviral properties of flavonoids have been extensively researched and include a variety of DNA and RNA viruses (Lalani and Poh, 2020). Polyphenols exhibit broad-spectrum antiviral activity against several categories of viruses, including but not limited to influenza A (H1N1), hepatitis B and C (HBV/HCV), herpes simplex virus 1 (HSV-1) and HIV (Utomo et al., 2020), and on the virus that causes COVID-19 disease (SARS-CoV-2) (Paraiso et al., 2020).

Epigallocatechin gallate (EGCG) is a polyphenolic catechin that is mainly found in green tea. The antiviral activity of EGCG has been investigated against various virus strains (Chacko et al., 2010). It has numerous mechanisms of action against various viruses, including influenza A (especially H3N2 and B strains), HIV, calicivirus, HCV (hepatitis C virus) and dengue, among other pathogens (Liu et al., 2005; Song et al., 2005). Quercetin has significant antiviral properties and is able to inhibit the replication of various human viruses, including HSV-1, poliovirus type 1, parainfluenza virus type 3 (Pf-3), RSV a, HBV, HCV and influenza viruses (Cheng et al., 2015; Gonzalez et al., 2009; Gravina et al., 2011; Mucsi and Prágai, 1985; Wu et al., 2015).

Theaflavins (TFs) extracted from black tea are a special group of polyphenolic compounds that show direct antiviral effects on viral particles in many diseases, including HCV. These TFs effectively prevent the binding of viral particles to the receptor surface and thus hinder the progression of the infection (Chowdhury et al., 2018). Luteolin and quercetin show mechanisms that prevent the penetration of SARS-CoV into Vero E6 cells (Yi et al., 2004). The antiviral properties of certain naturally occurring flavones, including Apigenin, Baicalein, Chrysin, Luteolin, Scutellarein, Tangeretin, Wogonin and 6-hydroxyflavones, have been known since the 1990s. During this time, it was demonstrated that the combined administration of apigenin and acyclovir leads to an enhanced antiviral effect on herpes simplex virus types 1 and 2 (HSV-1 and HSV-2) in cell cultures (Mucsi et al., 1992).

The antiviral activity of flavonoids has been shown to be influenced by a significant structure–activity relationship. Since the discovery of HIV as the causative agent of acquired immunodeficiency syndrome (AIDS) (Zakaryan et al., 2017), several flavonoids possess inhibitory activity against HIV-1, e.g., flavones (Chrysin, Baicalein), flavonols (Quercetin, Kaempferol, Quercetagetin) and Acacetin and Kaempferol glycosides (Cushnie and Lamb, 2005).

There is a growing body of research focused on investigating the potential of flavonoids as a means of combating SARS-CoV-2 (Al-Karmalawy et al., 2021; Alzaabi et al., 2022; Bardelčíková et al., 2022; Zakaryan et al., 2017). Several studies have shown that flavonoids have an inhibitory effect on various targets and thus exhibit anti-SARS-CoV-2 activity. These targets include viral entry, viral mRNA, proteases and direct inhibition of viral replication (Al-Karmalawy et al., 2021; Alzaabi et al., 2022; Bardelčíková et al., 2022; Zakaryan et al., 2017), as well as indirectly affecting interferons and pro-inflammatory cytokines (Bardelčíková et al., 2022; Huang et al., 2020; Rizzuti et al., 2021; Xiong et al., 2021; Zakaryan et al., 2017; Zandi et al., 2021). The interaction between the spike RBD of SARS-CoV-2 and the human ACE2 receptor plays a crucial role in the prevention of SARS-CoV-2 infection (Bongini et al., 2020). Several studies have shown that some flavonoids have the ability to interfere with the interaction between the spike protein and ACE2 (Biagioli et al., 2021; Güler et al., 2021; Liu et al., 2020).

Baicalein, a type of flavonoid, has shown an inhibitory effect on the SARS-CoV-2 CL^{pro} enzyme in laboratory tests (*in vitro*). It has also shown the ability to inhibit the replication of SARS-CoV-2 *in vivo* (Su et al., 2020). Curcumin is one of the most studied molecules *in silico* and *in vitro* as a potent inhibitor of SARS-CoV-2 M^{pro} (Alici et al., 2022).

Several studies and review articles have documented the potential antiviral properties of Baicalein and Baicalin against various broad-spectrum DNA and RNA viruses. These viruses include herpes simplex virus type 1 (HSV1), influenza A virus (IAV), hepatitis B virus (HBV), dengue virus (DENV), Zika virus (ZIKV), chikungunya virus (CHIKV), respiratory syncytial virus (RSV), coxsackievirus B3 (CVB3) and coronaviruses such as the severe acute respiratory syndrome coronavirus (SARS-CoV), the MERS-CoV and the new coronavirus (SARS-CoV-2) (Huang et al., 2017; Moghaddam et al., 2014; Nayak et al., 2014; Pang et al., 2018; Shi et al., 2016; Xu et al., 2010; Zandi et al., 2012).

Several studies have demonstrated the antiviral efficacy of baicalin against SARS-CoV-2 *in vitro* (Chen et al., 2020; Jo et al., 2020; Russo et al., 2020; Su et al., 2020). Baicalein and baicalin were identified as the first examples of non-covalent, non-peptidomimetic inhibitors targeting the SARS-CoV-2 3CL protease. The potential efficacy of baicalein and baicalin in inhibiting SARS-CoV-2 infection *in vitro* did not lead to successful clinical trials with SARS-CoV-2 infected individuals, mainly due to their limited water solubility and low ability to be orally absorbed.

In silico computational studies have identified Rutin as an effective inhibitor of the main SARS-CoV-2 protease (M^{pro}) (Antonopoulou et al., 2022). In addition, the identification of its presence in many conventional antiviral drugs administered to Chinese patients with mild to moderate symptoms of COVID-19 suggests the potential for its use as a bioactive secondary metabolite in the fight against SARS-CoV-2 (Rizzuti et al., 2021).

Hesperetin is a good $3CL^{pro}$ inhibitor of SARS-CoV (Lin et al., 2005). It is important to recognize that the $3CL^{pro}$ of SARS-CoV-2 has a sequence similarity of 96% with SARS-CoV-2, with only 12 sequence sites showing variations. Therefore, hesperetin has promising inhibitory properties against the $3CL^{pro}$ of SARS-CoV-2 (Xu et al., 2020). Hesperidin is currently being investigated in phase II clinical trials for the therapeutic intervention of COVID-19 (Jahangirifard et al., 2022).

The results of the evaluation of 38 naturally occurring flavonoids against the active site of the SARS-CoV-2 M^{pro} protein by molecular docking and molecular dynamics simulations indicate a significant interaction between flavonoids, sugar residues and M^{pro} (Augustin et al., 2020).

The antiviral activity of various natural flavonoids was investigated and evaluated by Meng et al. using a SARS-CoV-2 spike pseudovirus system (Meng et al., 2023). The researchers investigated the relationship between the structure and activity of flavonoids as inhibitors of viral entry. Their results showed that the antiviral activity of flavonols depends on the presence of a planar C-ring comprising a double bond at C2-C3 and a carbonyl group at C-4. The presence of hydroxyl groups on the B-ring of flavonoids is an important factor in enhancing their antiviral activity. In addition, molecular docking analysis has shown that flavonols can form stable interactions with pocket 3, which represents the non-mutated regions of SARS-CoV-2 variants. This suggests that flavonols may also have an inhibitory effect against mutant strains (Meng et al., 2023).

The viral pathogen uses the spike protein as a means to attach to the host cells, while flavonoids can hinder the binding process between the spike protein and the cells. Consequently, the presence of flavonoids poses a challenge to the ability of the virus to successfully infect host cells. The spike protein consists of two different domains, namely the RBD and the fusion domain (Verma et al., 2022). The RBD of the virus has a specific affinity for the ACE2 enzyme receptor on human cells. Upon binding, the viral fusion domain facilitates the fusion of the viral membrane with the cell membrane, enabling the release of the viral genome into the host cell (Banerjee et al., 2022).

Flavonoids are able to prevent the attachment of the SARS-CoV-2 spike protein to human cells by two mechanisms: first, by binding to the RBD of the spike protein and hindering its interaction with ACE2, and second, by inducing structural changes in the RBD that impair its binding to ACE2 (Kashyap et al., 2022). The protein in question has attracted a great deal of attention in the scientific community due to its newly discovered status as the most important protein. As a result, extensive efforts have been made to explore methods to inhibit its function, with the ultimate goal of developing effective therapeutic approaches. The available evidence suggests that a wide range of flavonoids and polyphenols have inhibitory effects on ACE2 (Paraiso et al., 2020).

M^{pro}, an enzymatic protein used by the virus to replicate the genome, plays a crucial role in cleaving the viral genome into smaller fragments, facilitating their subsequent assembly into nascent viral particles (Citarella et al., 2021). The M^{pro} enzyme is characterized by its considerable size and complicated structure. It consists of two different subunits, each of which has an active site. The primary site of enzymatic activity within the larger subunit of M^{pro} is responsible for most of its cleavage functions (Lee et al., 2022). Flavonoids have the ability to inhibit the replication of the virus by inhibiting the M^{pro}. This inhibition is achieved by hindering the activity of the protease or altering its structure, thereby impairing the functionality of the virus (Lin et al., 2023).

The nucleoprotein is an important component that the virus uses to encapsulate its genetic material, enabling the formation of a condensed structure that can subsequently detach from the host cell. Flavonoids have the ability to interfere with the nucleoprotein and thus impair the ability of the virus to package itself and escape from the host cell (Artika et al., 2020). The nucleoprotein (N) of SARS-CoV-2 is a highly abundant protein that has a strong affinity for viral RNA, effectively wrapping it in a helical configuration known as a ribonucleocapsid (RNP). This RNP serves as the central component of the infectious virus and plays a critical role in facilitating viral replication (Bai et al., 2021). In addition to its involvement in viral replication and assembly, the SARS-CoV-2 nucleoprotein also has immunomodulatory properties that enable it to dampen the host's immune response and promote inflammation (Minkoff and tenOever, 2023).

The 3CL protease, sometimes referred to as papain-like protease, plays a crucial role in the replication of SARS-CoV-2. It is responsible for the cleavage of the precursor protein of the non-structural polypeptide (nsp) into its individual components (Moustaqil et al., 2021). The enzyme 3CL protease is relatively small compared to Mpro. It has a single active site, which makes it a promising candidate for the development of antiviral drugs (Báez-Santos et al., 2015). The enzyme in question is responsible for the hydrolysis of the viral polyprotein at 11 specific sites that are crucial for the formation of the structural and non-structural proteins of the virus. These proteins play a critical role in facilitating host cell infection, promoting viral replication and ultimately synthesizing new viral particles. The 3CL protease is encoded in the SARS-CoV-2 genome. It is a 306 amino acid long protein that is synthesized as a precursor polyprotein (Kneller et al., 2021).

3.7 MOLECULAR DOCKING STUDIES

The widespread use of computerized drug discovery is due to its effectiveness and cost-effectiveness. In the field of drug discovery, *in silico* methods are of great importance as they help in the identification of highly potent compounds, allow the exploration of relationships between molecular structure and biological activity, predict optimal structural groups, facilitate the efficient screening of numerous drug candidates and allow the assessment of the impact of possible mutations on the efficacy of these compounds. The scientific method known as molecular docking facilitates the rapid assessment of binding affinities between a particular protein of interest and a tiny molecule, such as a drug or potential drug candidate (Guedes et al., 2014). This approach is characterized by a high degree of efficiency and effectiveness. The main objective of this technique is to facilitate the binding of a small molecule to the protein structure, commonly referred to as pose generation, and to evaluate the degree of compatibility between the candidate

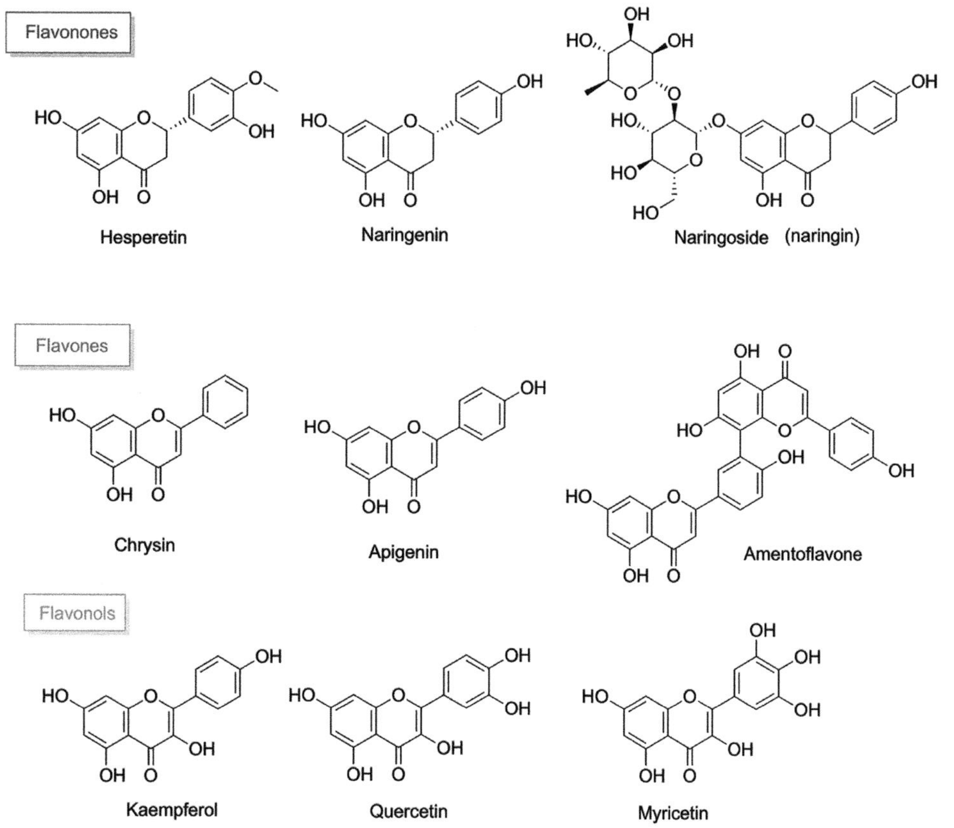

FIGURE 3.7 Molecular structures of the selected flavonoids.

molecule and the defined binding site, known as scoring (Ferreira et al., 2015). Numerous studies have investigated the efficiency and precision of various widely used molecular docking algorithms and have shown that they are capable of generating accurate molecular poses. This phenomenon results from the observation that a significant fraction of the formed binding modes exhibits conformational similarities with the co-crystallized ligands. Nevertheless, the scoring systems used by these programs still exhibit a considerable degree of inaccuracy, meaning that they cannot provide a reliable prediction (Muegge and Rarey, 2001). Recent computational studies have found that hesperidin, apigenin, luteolin, seselin, 6-gingerol, humulene epoxide, quercetin, kaempferol, curcumin and EGCG have inhibitory effects on various molecular targets involved in SARS-CoV-2 virus replication (Bachar et al., 2021). A number of flavonoids were selected for *in silico* studies, which are described in the following section: Figure 3.7 shows the presence of hesperetin, naringenin, chrysin, myricetin, quercetin, apigenin, kaempferol, amentoflavone and naringin.

We used the COVID-19 docking server (https://ncov.schanglab.org.cn/help.php) to study the possible interactions of some flavanones and flavones with the main protease and the Papain-like protease of the SARS-CoV-2. In the field of structural bioinformatics, the reliable estimation of the binding affinity of protein–ligand complexes is an important issue. To facilitate the generation of such a prediction, a scoring function (SF) is used. This function establishes a mathematical relationship between the X-ray crystal structures of protein–ligand complexes and their corresponding potential binding affinities. Numerous specialized functions (SFs) have been developed, which have been systematically divided into two main categories: classical SFs and machine-learning (ML) SFs. In contrast to traditional scoring functions (SFs), ML-based SFs have the ability to capture the intricate connection between protein–ligand binding affinity using a nonlinear method. These ML-based SFs have shown superior performance in terms of scoring accuracy compared to conventional SFs (Meli et al., 2022). The COVID-19 docking server uses the Autodock Vina as docking engine. In this study, we present (Figure 3.8) the expected binding energy for the highest ranked binding modes along with the Random Forest (RF) score value, which serves as a scoring tool for predicting structure-based binding affinity (Kong et al., 2020).

The flavanones Hesperetin, Naringenin and Naringin were selected as objects of investigation in order to study the influence of the hydroxyl position in the B-ring and the significance of the phenolic group at C-4′ and C-7. The structural composition of flavanones differs from other flavonoids by the presence of a chiral carbon at the C2 position and the absence of a substitution at the C3 position (Barreca et al., 2017; Memariani et al., 2021). Hesperetin is classified as a flavanone, a type of organic compound. Within this category, there are other glycosides, including hesperidin (also known as hesperetin-7-rutinoside). It is worth noting that hesperidin has a rather low water solubility, which is less than 5 µg/mL. Citrus fruits contain this compound, which is released as an aglycone when ingested. Hesperetin is a flavonoid compound that is frequently found in citrus fruits (Choi et al., 2022). Based on our docking studies performed on the SARS-CoV-2 main protease, the observed point values varied between –7.30 kcal/mol for Naringenin and –8.80 kcal/mol for Naringin (Figure 3.8). Naringin, a glycosylated flavanone, is extracted from the peel of selected citrus fruits and is responsible for their characteristic bitter taste. It is assumed that the concentration of naringin in fresh citrus peel is around 15–18 g/kg. The food sector uses the antioxidant properties of grapefruit extract to add flavour to confectionery, beverages and baked goods. It also serves as an oil stabilizer and contributes to the bitter taste found in commercial grapefruit juices. Naringenin is obtained from the natural compound naringin through

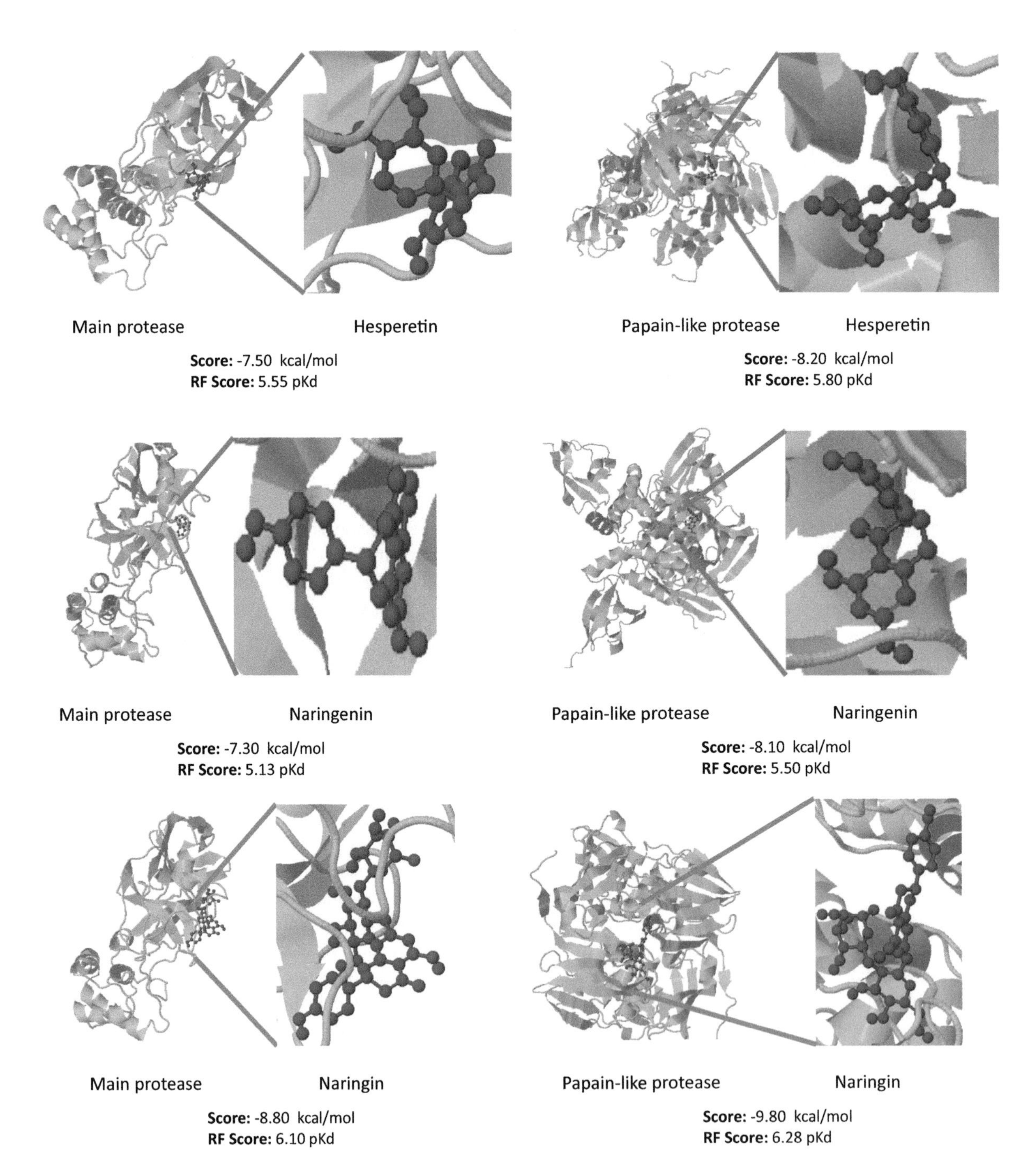

FIGURE 3.8 Molecular docking of the selected flavonoids with the main protease and the papain-like protease of the SARS-CoV-2 using the COVID-19 docking server (https://ncov.schanglab.org.cn/help.php).

(Continued)

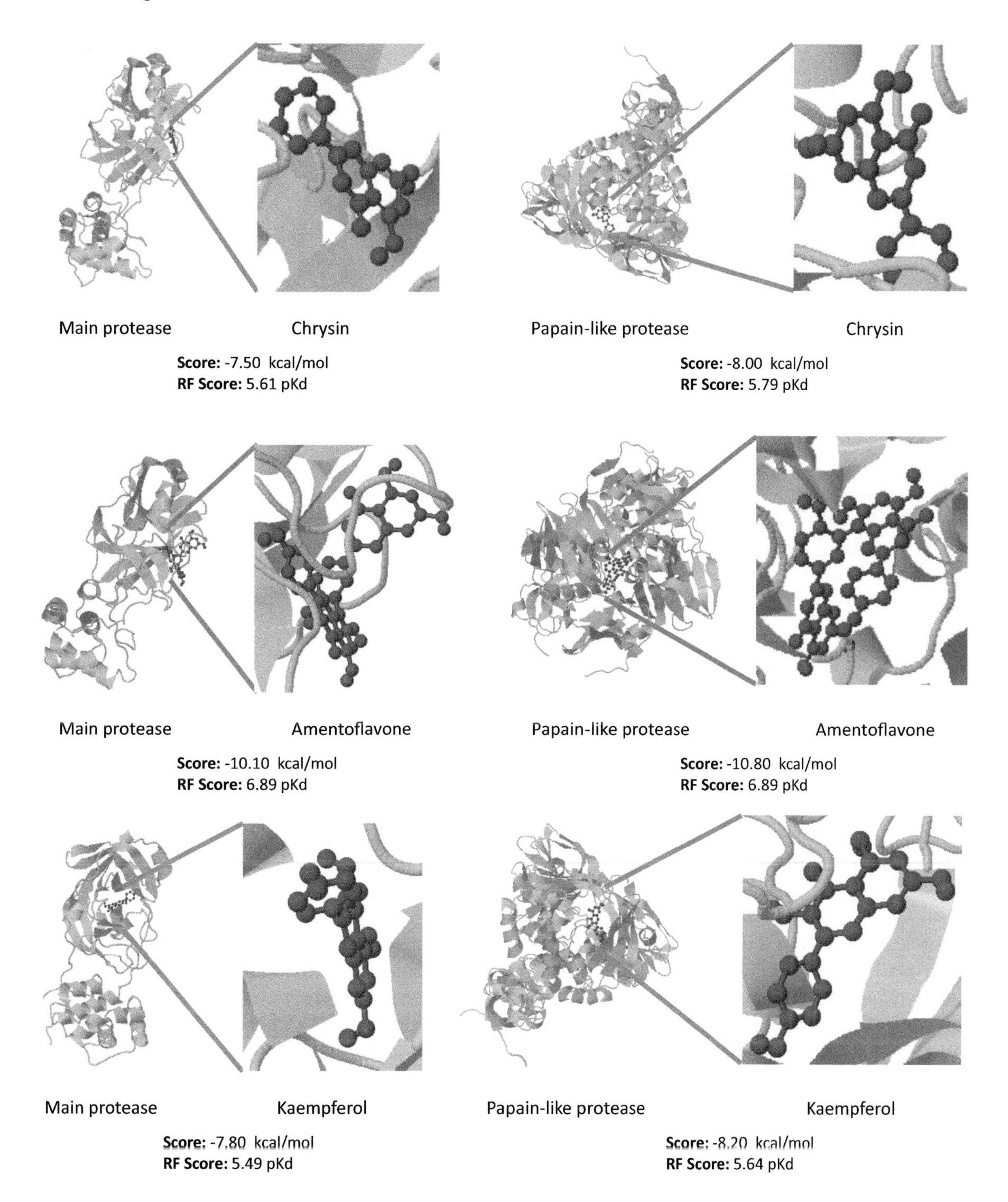

FIGURE 3.8 (*Continued*) Molecular docking of the selected flavonoids with the main protease and the papain-like protease of the SARS-CoV-2 using the COVID-19 docking server (https://ncov.schanglab.org.cn/help.php).

(*Continued*)

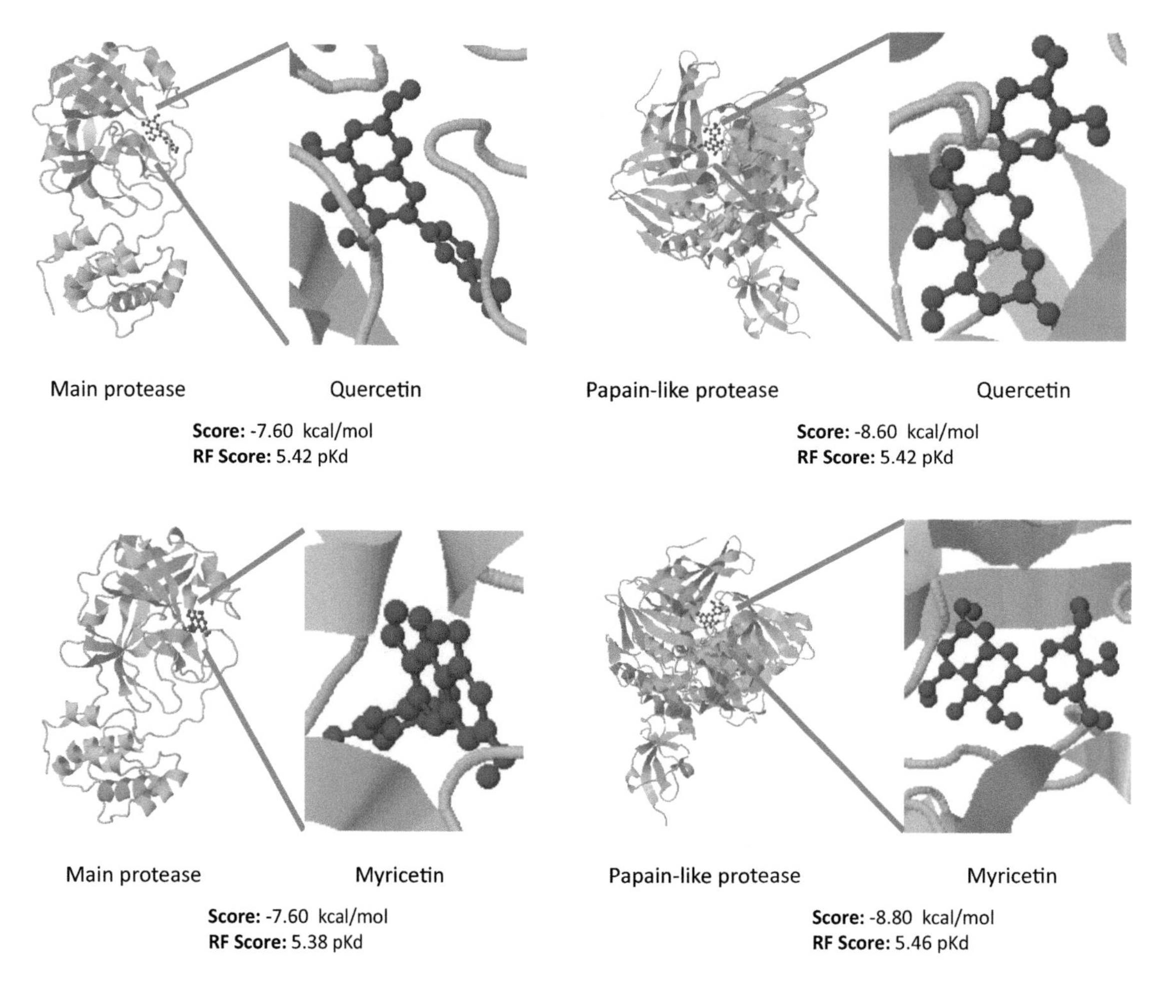

FIGURE 3.8 *(Continued)* Molecular docking of the selected flavonoids with the main protease and the papain-like protease of the SARS-CoV-2 using the COVID-19 docking server (https://ncov.schanglab.org.cn/help.php).

a process of acid hydrolysis (Shilpa et al., 2023). Naringin showed the most favorable binding affinity (–9.80 kcal/mol) in the docking analysis with the papain-like protease, while naringenin showed the least favorable binding affinity (–8.10 kcal/mol) among the three flavanones studied (Figure 3.8). Naringenin, the primary flavanone in grapefruit, has demonstrated the ability to inhibit the replication of hepatitis C viruses (Goldwasser et al., 2011).

The flavones Chrysin and Apigenin as well as the biflavone Amentoflanone were selected for this study. Chrysin, a flavone compound, has been found in various natural sources such as honey, propolis and certain plant species such as *Passiflora caerulea*, *Passiflora incarnata* and *Oroxylum indicum*. It is also worth noting that Chrysin is often used as an ingredient in dietary supplements. The safety of Chrysin consumption is considered acceptable within a range of 0.5–3 g/day (Campos et al., 2022).

Chrysin and Apigenin showed comparable docking values of –8.00 kcal/mol in the docking analysis with the papain-like protease of SARS-CoV-2 (Figure 3.8). Docking of the two molecules to the main protease showed that Apigenin achieved a higher score of –7.90 kcal/mol compared to Chrysin, which achieved a score of –7.50 kcal/mol. Apigenin, a type of flavone, occurs naturally in various plant sources, including parsley, celery and chamomile. Glycosides derived from Apigenin as an aglycone include a number of compounds, including Apiin (apigenin 7-O-apioglucoside), Apigetrin (apigenin 7-glucoside, identified in dandelion), Vitexin (apigenin 8-C-glucoside), Isovitexin (apigenin 6-C-glucoside), Rhoifolin (apigenin 7-O-neohesperidoside) and Shankoside (apigenin 6-C-glucoside 8-C-arabinoside) (Shankar et al., 2017).

With respect to Amentoflavone, our computational study showed a favorable interaction of this molecule with the main protease and the papain-like protease, which is reflected in lower values. The calculated values for these interactions are –10.10 and –10.80 kcal/mol, respectively (Figure 3.8). Nevertheless, it is essential to complement these theoretical interactions with *in vitro* and *in vivo* studies to validate the biological activities (Ekins et al., 2007).

Other flavonols used for the docking studies were Myricetin, Quercetin and Kaempferol. The interaction values of these flavonols showed a high similarity, ranging from –7.60 to –7.80 kcal/mol for their interaction with the main protease. Similarly, their interaction values ranged from –8.20 to –8.80 kcal/mol (Figure 3.8). Myricetin, a flavonol, is commonly found in a variety of plant sources such as vegetables, fruits, nuts, berries, tea and red wine. Myricetin has structural similarities to Fisetin, Luteolin and Quercetin and thus forms the basis for functional similarities with these other flavonols. Myricetin also occurs in its natural form as the glycoside Myricitrin (Semwal et al., 2016).

Quercetin, which was first identified by J. Rigaud in 1854, is a flavonol compound that is frequently found in fruits and vegetables. It is present in the form of glycosides and is one of the most abundant flavonoids in the human diet alongside naringenin and Rutin. Red onions have been found to contain around 10% quercetin by dry weight. Other food sources rich in Quercetin are apples, grapes, broccoli and tea. The hydrolysis of these glycosides takes place in the small intestine, whereby the Quercetin is effectively absorbed (Anand David et al., 2016).

Kaempferol, a flavonol compound, can be extracted from a variety of sources such as green tea, broccoli, grapefruit, grapes, Brussels sprouts, apples and other botanical specimens. It is partially soluble in aqueous solutions. Kaempferitrin and astragalin have been extracted from plant sources in the form of kaempferol glycosides (Kanakis et al., 2005). Amentoflavone, a biflavonoid compound consisting of bis-apigenin, is found in various plant species, including Ginkgo biloba, *Chamaecyparis obtusa*, *Hypericum perforatum* and *Xerophyta aplicata*. The substance exhibits antimalarial and anticancer properties, possibly due to its ability to inhibit fatty acid synthase. It also has an affinity for opioid receptors and acts as an antagonist (Bais and Abrol, 2015).

In summary, the investigation of naturally occurring plant chemicals, especially flavonoids such as Naringenin and Hesperetin, has shown promising prospects for combating SARS-CoV-2. The inherent antiviral properties of these compounds, particularly their ability to block key proteases essential for the functioning of the virus, highlight their potential as therapeutics for the treatment of COVID-19. The application of molecular docking methods in the study of interactions between flavonoids and key viral proteins has yielded significant results advocating the investigation of plant-derived chemicals as an important focus in the search for effective approaches for the prevention and treatment of COVID-19. The aforementioned research not only provides insights into the inhibitory properties of flavonoids against SARS-CoV-2, but also creates a basis for the discovery of new phytochemicals that have the potential to be effective therapeutic targets in combating the global pandemic. There is still a need for additional studies and in-depth investigations into the mechanisms of action and efficacy of flavonoids, with particular focus on their glucosides, to improve our understanding in this area. A comprehensive understanding of the intricacies of these interactions would not only improve our understanding of potential therapeutic interventions, but also open a promising avenue for the development of novel antiviral remedies from natural substances.

AUTHOR CONTRIBUTIONS

Conceptualization, C.M.C.A. and C.A.J.; investigation, C.M.C.A. and C.A.J.; writing—review and editing, C.M.C.A., C.A.J., J.M.P.d.l.L., E.P.L. and F.J.P.; supervision, C.M.C.A., C.A.J. and J.M.P.d.l.L.

FUNDING

This research was funded by Agencia Canaria de Investigación, Innovación y Sociedad de la Información (ACIISI) del Gobierno de Canarias, Project ProID2020010134, Caja Canarias, Project 2019SP43, the Spanish Ministry of Economy and Competitiveness (Grant PID2019–105838RB-C31) and the State Plan for Scientific, Technical Research and Innovation 2021–2023 from the Spanish Ministry of Science and Innovation (project PLEC2022–009507).

CONFLICTS OF INTEREST

The authors declare no conflict of interest. The funders had no role in the design of the study; in the collection, analyses or interpretation of data; in the writing of the manuscript or in the decision to publish the results.

REFERENCES

Ahmad, A., Kaleem, M., Ahmed, Z., & Shafiq, H. (2015). Therapeutic potential of flavonoids and their mechanism of action against microbial and viral infections-A review. *Food Res Int*, *77*, 221–235. https://doi.org/10.1016/j.foodres.2015.06.021

Alici, H., Tahtaci, H., & Demir, K. (2022). Design and various in silico studies of the novel curcumin derivatives as potential candidates against COVID-19-associated main enzymes. *Comput Biol Chem*, *98*, 107657. https://doi.org/10.1016/j.compbiolchem.2022.107657

Al-Karmalawy, A. A., Farid, M. M., Mostafa, A., Ragheb, A. Y., Sara, H. M., Shehata, M., Marzouk, M. M., et al. (2021). Naturally available flavonoid aglycones as potential antiviral drug candidates against SARS-CoV-2. *Molecules*, *26*(21), 6559. https://doi.org/10.3390/molecules26216559

Alzaabi, M. M., Hamdy, R., Ashmawy, N. S., Hamoda, A. M., Alkhayat, F., Khademi, N. N., Soliman, S. S. M., et al. (2022). Flavonoids are promising safe therapy against COVID-19. *Phytochem Rev*, *21*(1), 291–312. https://doi.org/10.1007/s11101-021-09759-z

Anand David, A. V., Arulmoli, R., & Parasuraman, S. (2016). Overviews of biological importance of quercetin: A bioactive flavonoid. *Pharmacogn Rev*, *10*(20), 84–89. https://doi.org/10.4103/0973-7847.194044

Andrés, C. M. C., Pérez de la Lastra, J. M., Andrés Juan, C., Plou, F. J., & Pérez-Lebeña, E. (2022). Impact of reactive species on amino acids-biological relevance in proteins and induced pathologies. *Int J Mol Sci*, *23*(22), 14049. https://doi.org/10.3390/ijms232214049

Andrés, C. M. C., Pérez de la Lastra, J. M., Andrés Juan, C., Plou, F. J., & Pérez-Lebeña, E. (2023). Chemical insights into oxidative and nitrative modifications of DNA. *Int J Mol Sci*, *24*(20), 15240. Retrieved from https://www.mdpi.com/1422-0067/24/20/15240

Andrés, C. M. C., Pérez de la Lastra, J. M., Andrés Juan, C., Plou, F. J., & Pérez-Lebeña, E. (2023). Polyphenols as antioxidant/pro-oxidant compounds and donors of reducing species: Relationship with human antioxidant metabolism. *Processes*, *11*(9), 2771. Retrieved from https://www.mdpi.com/2227-9717/11/9/2771

Antonelli, G., & Turriziani, O. (2012). Antiviral therapy: Old and current issues. *Int J Antimicrob Agents*, *40*(2), 95–102. https://doi.org/10.1016/j.ijantimicag.2012.04.005

Antonopoulou, I., Sapountzaki, E., Rova, U., & Christakopoulos, P. (2022). Inhibition of the main protease of SARS-CoV-2 (M(pro)) by repurposing/designing drug-like substances and utilizing nature's toolbox of bioactive compounds. *Comput Struct Biotechnol J*, *20*, 1306–1344. https://doi.org/10.1016/j.csbj.2022.03.009

Arabi, Y. M., Gordon, A. C., Derde, L. P. G., Nichol, A. D., Murthy, S., Beidh, F. A., et al. (2021). Lopinavir-ritonavir and hydroxychloroquine for critically ill patients with COVID-19: REMAP-CAP randomized controlled trial. *Intensive Care Med*, *47*(8), 867–886. https://doi.org/10.1007/s00134-021-06448-5

Artika, I. M., Dewantari, A. K., & Wiyatno, A. (2020). Molecular biology of coronaviruses: Current knowledge. *Heliyon*, *6*(8), e04743. https://doi.org/10.1016/j.heliyon.2020.e04743

Augustin, T. L., Hajbabaie, R., Harper, M. T., & Rahman, T. (2020). Novel small-molecule scaffolds as candidates against the SARS coronavirus 2 main protease: A fragment-guided in silico approach. *Molecules*, *25*(23), 5501. https://doi.org/10.3390/molecules25235501

Bachar, S. C., Mazumder, K., Bachar, R., Aktar, A., & Al Mahtab, M. (2021). A review of medicinal plants with antiviral activity available in Bangladesh and mechanistic insight into their bioactive metabolites on SARS-CoV-2, HIV and HBV. *Front Pharmacol*, *12*, 732891. https://doi.org/10.3389/fphar.2021.732891

Badshah, S. L., Faisal, S., Muhammad, A., Poulson, B. G., Emwas, A. H., & Jaremko, M. (2021). Antiviral activities of flavonoids. *Biomed Pharmacother*, *140*, 111596. https://doi.org/10.1016/j.biopha.2021.111596

Báez-Santos, Y. M., St John, S. E., & Mesecar, A. D. (2015). The SARS-coronavirus papain-like protease: Structure, function and inhibition by designed antiviral compounds. *Antiviral Res*, *115*, 21–38. https://doi.org/10.1016/j.antiviral.2014.12.015

Bai, Z., Cao, Y., Liu, W., & Li, J. (2021). The SARS-CoV-2 nucleocapsid protein and its role in viral structure, biological functions, and a potential target for drug or vaccine mitigation. *Viruses*, *13*(6), 1115. https://doi.org/10.3390/v13061115

Bais, S., & Abrol, N. (2015). Review on chemistry and pharmacological potential of amentoflavone. *Curr Res Neurosci*, *6*, 16–22.

Banerjee, S., Wang, X., Du, S., Zhu, C., Jia, Y., Wang, Y., & Cai, Q. (2022). Comprehensive role of SARS-CoV-2 spike glycoprotein in regulating host signaling pathway. *J Med Virol*, *94*(9), 4071–4087. https://doi.org/10.1002/jmv.27820

Bardelčíková, A., Miroššay, A., Šoltýs, J., & Mojžiš, J. (2022). Therapeutic and prophylactic effect of flavonoids in post-COVID-19 therapy. *Phytother Res*, *36*(5), 2042–2060. https://doi.org/10.1002/ptr.7436

Barreca, D., Gattuso, G., Bellocco, E., Calderaro, A., Trombetta, D., Smeriglio, A., Nabavi, S. M., et al. (2017). Flavanones: Citrus phytochemical with health-promoting properties. *Biofactors*, *43*(4), 495–506. https://doi.org/10.1002/biof.1363

Basri, N. F., Ramli, A. S., Mohamad, M., & Kamaruddin, K. N. (2022). Traditional and complementary medicine (TCM) usage and its association with patient assessment of chronic illness care (PACIC) among individuals with metabolic syndrome in primary care. *BMC Complement Med Ther*, *22*(1), 14. https://doi.org/10.1186/s12906-021-03493-x

Beyerstedt, S., Casaro, E. B., & Rangel, É. B. (2021). COVID-19: Angiotensin-converting enzyme 2 (ACE2) expression and tissue susceptibility to SARS-CoV-2 infection. *Eur J Clin Microbiol Infect Dis*, *40*(5), 905–919. https://doi.org/10.1007/s10096-020-04138-6

Biagioli, M., Marchianò, S., Roselli, R., Di Giorgio, C., Bellini, R., Bordoni, M., Fiorucci, S., et al. (2021). Discovery of a AHR pelargonidin agonist that counter-regulates ACE2 expression and attenuates ACE2-SARS-CoV-2 interaction. *Biochem Pharmacol*, *188*, 114564. https://doi.org/10.1016/j.bcp.2021.114564

Bongini, P., Trezza, A., Bianchini, M., Spiga, O., & Niccolai, N. (2020). A possible strategy to fight COVID-19: Interfering with spike glycoprotein trimerization. *Biochem Biophys Res Commun*, *528*(1), 35–38. https://doi.org/10.1016/j.bbrc.2020.04.007

Bulut, H. (2022). Drug development targeting SARS-CoV-2 main protease. *Glob Health Med*, *4*(6), 296–300. https://doi.org/10.35772/ghm.2022.01066

Campos, H. M., da Costa, M., da Silva Moreira, L. K., da Silva Neri, H. F., Branco da Silva, C. R., Pruccoli, L., Ghedini, P. C., et al. (2022). Protective effects of chrysin against the neurotoxicity induced by aluminium: In vitro and in vivo studies. *Toxicology*, *465*, 153033. https://doi.org/10.1016/j.tox.2021.153033

Campos-Gonzalez, G., Martinez-Picado, J., Velasco-Hernandez, T., & Salgado, M. (2023). Opportunities for CAR-T cell immunotherapy in HIV cure. *Viruses*, *15*(3), 789. https://doi.org/10.3390/v15030789

Carabelli, A. M., Peacock, T. P., Thorne, L. G., Harvey, W. T., Hughes, J., Peacock, S. J., Robertson, D. L., et al. (2023). SARS-CoV-2 variant biology: Immune escape, transmission and fitness. *Nat Rev Microbiol*, *21*(3), 162–177. https://doi.org/10.1038/s41579-022-00841-7

Cascella, M., Rajnik, M., Aleem, A., Dulebohn, S. C., & Di Napoli, R. (2023). Features, evaluation, and treatment of coronavirus (COVID-19). In *StatPearls*. Treasure Island, FL: StatPearls Publishing.

Chacko, S. M., Thambi, P. T., Kuttan, R., & Nishigaki, I. (2010). Beneficial effects of green tea: A literature review. *Chin Med*, *5*(1), 13. https://doi.org/10.1186/1749-8546-5-13

Chen, K. H., Wang, S. F., Wang, S. Y., Yang, Y. P., Wang, M. L., Chiou, S. H., & Chang, Y. L. (2020). Pharmacological development of the potential adjuvant therapeutic agents against coronavirus disease 2019. *J Chin Med Assoc*, *83*(9), 817–821. https://doi.org/10.1097/jcma.0000000000000375

Chen, Z., Boon, S. S., Wang, M. H., Chan, R. W. Y., & Chan, P. K. S. (2021). Genomic and evolutionary comparison between SARS-CoV-2 and other human coronaviruses. *J Virol Methods*, *289*, 114032. https://doi.org/10.1016/j.jviromet.2020.114032

Cheng, Z., Sun, G., Guo, W., Huang, Y., Sun, W., Zhao, F., & Hu, K. (2015). Inhibition of hepatitis B virus replication by quercetin in human hepatoma cell lines. *Virol Sin*, *30*(4), 261–268. https://doi.org/10.1007/s12250-015-3584-5

Choi, K. H. (2012). Viral polymerases. *Adv Exp Med Biol*, *726*, 267–304. https://doi.org/10.1007/978-1-4614-0980-9_12

Choi, S. S., Lee, S. H., & Lee, K. A. (2022). A comparative study of hesperetin, hesperidin and hesperidin glucoside: Antioxidant, anti-inflammatory, and antibacterial activities in vitro. *Antioxidants (Basel), 11*(8), 1618. https://doi.org/10.3390/antiox11081618

Chowdhury, P., Sahuc, M. E., Rouillé, Y., Rivière, C., Bonneau, N., Vandeputte, A., Séron, K., et al. (2018). Theaflavins, polyphenols of black tea, inhibit entry of hepatitis C virus in cell culture. *PLoS One, 13*(11), e0198226. https://doi.org/10.1371/journal.pone.0198226

Citarella, A., Scala, A., Piperno, A., & Micale, N. (2021). SARS-CoV-2 M(pro): A potential target for peptidomimetics and small-molecule inhibitors. *Biomolecules, 11*(4), 607. https://doi.org/10.3390/biom11040607

Craigie, R. (2012). The molecular biology of HIV integrase. *Future Virol, 7*(7), 679–686. https://doi.org/10.2217/fvl.12.56

Cushnie, T. P., & Lamb, A. J. (2005). Antimicrobial activity of flavonoids. *Int J Antimicrob Agents, 26*(5), 343–356. https://doi.org/10.1016/j.ijantimicag.2005.09.002

De Clercq, E. (2004). Antivirals and antiviral strategies. *Nat Rev Microbiol, 2*(9), 704–720. https://doi.org/10.1038/nrmicro975

de Groot, R. J., Baker, S. C., Baric, R. S., Brown, C. S., Drosten, C., Enjuanes, L., Ziebuhr, J., et al. (2013). Middle East respiratory syndrome coronavirus (MERS-CoV): Announcement of the coronavirus study group. *J Virol, 87*(14), 7790–7792. https://doi.org/10.1128/jvi.01244-13

Dias, M. C., Pinto, D. C. G. A., & Silva, A. M. S. (2021). Plant flavonoids: Chemical characteristics and biological activity. *Molecules, 26*(17), 5377. Retrieved from https://www.mdpi.com/1420-3049/26/17/5377

Duthie, G. G., Gardner, P. T., & Kyle, J. A. (2003). Plant polyphenols: Are they the new magic bullet? *Proc Nutr Soc, 62*(3), 599–603. https://doi.org/10.1079/pns2003275

Ekins, S., Mestres, J., & Testa, B. (2007). In silico pharmacology for drug discovery: Methods for virtual ligand screening and profiling. *Br J Pharmacol, 152*(1), 9–20. https://doi.org/10.1038/sj.bjp.0707305

Farkaš, B., Minneci, M., Misevicius, M., & Rozas, I. (2023). A tale of two proteases: MPro and TMPRSS2 as targets for COVID-19 therapies. *Pharmaceuticals, 16*(6), 834. Retrieved from https://www.mdpi.com/1424-8247/16/6/834

Ferreira, L. G., Dos Santos, R. N., Oliva, G., & Andricopulo, A. D. (2015). Molecular docking and structure-based drug design strategies. *Molecules, 20*(7), 13384–13421.

Fung, T. S., & Liu, D. X. (2019). Human coronavirus: Host-pathogen interaction. *Annu Rev Microbiol, 73*, 529–557. https://doi.org/10.1146/annurev-micro-020518-115759

Gheblawi, M., Wang, K., Viveiros, A., Nguyen, Q., Zhong, J. C., Turner, A. J., Oudit, G. Y., et al. (2020). Angiotensin-converting enzyme 2: SARS-CoV-2 receptor and regulator of the renin-angiotensin system: Celebrating the 20th anniversary of the discovery of ACE2. *Circ Res, 126*(10), 1456–1474. https://doi.org/10.1161/circresaha.120.317015

Glowacka, I., Bertram, S., Müller, M. A., Allen, P., Soilleux, E., Pfefferle, S., Pöhlmann, S., et al. (2011). Evidence that TMPRSS2 activates the severe acute respiratory syndrome coronavirus spike protein for membrane fusion and reduces viral control by the humoral immune response. *J Virol, 85*(9), 4122–4134. https://doi.org/10.1128/jvi.02232-10

Goldwasser, J., Cohen, P. Y., Lin, W., Kitsberg, D., Balaguer, P., Polyak, S. J., Nahmias, Y., et al. (2011). Naringenin inhibits the assembly and long-term production of infectious hepatitis C virus particles through a PPAR-mediated mechanism. *J Hepatol, 55*(5), 963–971. https://doi.org/10.1016/j.jhep.2011.02.011

Gonzalez, O., Fontanes, V., Raychaudhuri, S., Loo, R., Loo, J., Arumugaswami, V., French, S. W., et al. (2009). The heat shock protein inhibitor quercetin attenuates hepatitis C virus production. *Hepatology, 50*(6), 1756–1764. https://doi.org/10.1002/hep.23232

Gravina, H. D., Tafuri, N. F., Silva Júnior, A., Fietto, J. L., Oliveira, T. T., Diaz, M. A., & Almeida, M. R. (2011). In vitro assessment of the antiviral potential of trans-cinnamic acid, quercetin and morin against equid herpesvirus 1. *Res Vet Sci, 91*(3), e158–e162. https://doi.org/10.1016/j.rvsc.2010.11.010

Guedes, I. A., de Magalhães, C. S., & Dardenne, L. E. (2014). Receptor-ligand molecular docking. *Biophys Rev, 6*, 75–87.

Güler, H., Ay Şal, F., Can, Z., Kara, Y., Yildiz, O., Beldüz, A. O., Kolayli, S., et al. (2021). Targeting CoV-2 spike RBD and ACE-2 interaction with flavonoids of anatolian propolis by in silico and in vitro studies in terms of possible COVID-19 therapeutics. *Turk J Biol, 45*(4), 530–548. https://doi.org/10.3906/biy-2104-5

Haas, E. J., Angulo, F. J., McLaughlin, J. M., Anis, E., Singer, S. R., Khan, F., Alroy-Preis, S., et al. (2021). Impact and effectiveness of mRNA BNT162b2 vaccine against SARS-CoV-2 infections and COVID-19 cases, hospitalisations, and deaths following a nationwide vaccination campaign in Israel: An observational study using national surveillance data. *Lancet, 397*(10287), 1819–1829. https://doi.org/10.1016/s0140-6736(21)00947-8

Hernandez Acosta, R. A., Esquer Garrigos, Z., Marcelin, J. R., & Vijayvargiya, P. (2022). COVID-19 pathogenesis and clinical manifestations. *Infect Dis Clin North Am, 36*(2), 231–249. https://doi.org/10.1016/j.idc.2022.01.003

Huang, H., Zhou, W., Zhu, H., Zhou, P., & Shi, X. (2017). Baicalin benefits the anti-HBV therapy via inhibiting HBV viral RNAs. *Toxicol Appl Pharmacol, 323*, 36–43. https://doi.org/10.1016/j.taap.2017.03.016

Huang, S., Liu, Y., Zhang, Y., Zhang, R., Zhu, C., Fan, L., Shi, Y., et al. (2020). Baicalein inhibits SARS-CoV-2/VSV replication with interfering mitochondrial oxidative phosphorylation in a mPTP dependent manner. *Signal Transduct Target Ther, 5*(1), 266. https://doi.org/10.1038/s41392-020-00353-x

Huang, Y., Yang, C., Xu, X. F., Xu, W., & Liu, S. W. (2020). Structural and functional properties of SARS-CoV-2 spike protein: Potential antivirus drug development for COVID-19. *Acta Pharmacol Sin, 41*(9), 1141–1149. https://doi.org/10.1038/s41401-020-0485-4

Jackson, C. B., Farzan, M., Chen, B., & Choe, H. (2022). Mechanisms of SARS-CoV-2 entry into cells. *Nat Rev Mol Cell Biol, 23*(1), 3–20. https://doi.org/10.1038/s41580-021-00418-x

Jahangirifard, A., Mirtajani, S. B., Karimzadeh, M., Forouzmehr, M., Kiani, A., Moradi, M., Abedini, A., et al. (2022). The effect of hesperidin on laboratory parameters of patients with COVID 19: A preliminary report of a clinical trial study. *J Iran Med Council, 5*(1), 89–95.

Jayk Bernal, A., Gomes da Silva, M. M., Musungaie, D. B., Kovalchuk, E., Gonzalez, A., Delos Reyes, V., De Anda, C., et al. (2022). Molnupiravir for oral treatment of Covid-19 in nonhospitalized patients. *N Engl J Med, 386*(6), 509–520. https://doi.org/10.1056/NEJMoa2116044

Jirjees, F., Saad, A. K., Al Hano, Z., Hatahet, T., Al Obaidi, H., & Dallal Bashi, Y. H. (2021). COVID-19 treatment guidelines: Do they really reflect best medical practices to manage the pandemic? *Infect Dis Rep, 13*(2), 259–284. https://doi.org/10.3390/idr13020029

Jo, S., Kim, S., Kim, D. Y., Kim, M. S., & Shin, D. H. (2020). Flavonoids with inhibitory activity against SARS-CoV-2 3CLpro. *J Enzyme Inhib Med Chem, 35*(1), 1539–1544. https://doi.org/10.1080/14756366.2020.1801672

Juan, C. A., Pérez de la Lastra, J. M., Plou, F. J., & Pérez-Lebeña, E. (2021). The chemistry of reactive oxygen species (ROS) revisited: Outlining their role in biological macromolecules (DNA, lipids and proteins) and induced pathologies. *Int J Mol Sci, 22*(9), 4642. Retrieved from https://www.mdpi.com/1422-0067/22/9/4642

Kanakis, C. D., Tarantilis, P. A., Polissiou, M. G., Diamantoglou, S., & Tajmir-Riahi, H. A. (2005). DNA interaction with naturally occurring antioxidant flavonoids quercetin, kaempferol, and delphinidin. *J Biomol Struct Dyn, 22*(6), 719–724. https://doi.org/10.1080/07391102.2005.10507038

Kapikian, A. Z., James, H. D., Jr., Kelly, S. J., Dees, J. H., Turner, H. C., McIntosh, K., Chanock, R. M., et al. (1969). Isolation from man of "avian infectious bronchitis virus-like" viruses (Coronaviruses*) similar to 229E virus, with some epidemiological observations. *J Infect Dis, 119*(3), 282–290. https://doi.org/10.1093/infdis/119.3.282

Kashyap, P., Thakur, M., Singh, N., Shikha, D., Kumar, S., Baniwal, P., Inbaraj, B. S., et al. (2022). In silico evaluation of natural flavonoids as a potential inhibitor of coronavirus disease. *Molecules, 27*(19), 6374. https://doi.org/10.3390/molecules27196374

Khan, A. T.-A., Khalid, Z., Zahid, H., Yousaf, M. A., & Shakoori, A. R. (2022). A computational and bioinformatic analysis of ACE2: An elucidation of its dual role in COVID-19 pathology and finding its associated partners as potential therapeutic targets. *J Biomol Struct Dyn, 40*(4), 1813–1829. https://doi.org/10.1080/07391102.2020.1833760

Klasse, P. J., Bron, R., & Marsh, M. (1998). Mechanisms of enveloped virus entry into animal cells. *Adv Drug Deliv Rev, 34*(1), 65–91. https://doi.org/10.1016/s0169-409x(98)00002-7

Kneller, D. W., Li, H., Galanie, S., Phillips, G., Labbé, A., Weiss, K. L., Kovalevsky, A., et al. (2021). Structural, electronic, and electrostatic determinants for inhibitor binding to subsites S1 and S2 in SARS-CoV-2 main protease. *J Med Chem, 64*(23), 17366–17383. https://doi.org/10.1021/acs.jmedchem.1c01475

Kokoris, M. S., & Black, M. E. (2002). Characterization of herpes simplex virus type 1 thymidine kinase mutants engineered for improved ganciclovir or acyclovir activity. *Protein Sci, 11*(9), 2267–2272. https://doi.org/10.1110/ps.2460102

Kong, R., Yang, G., Xue, R., Liu, M., Wang, F., Hu, J., Chang, S., et al. (2020). COVID-19 docking server: A meta server for docking small molecules, peptides and antibodies against potential targets of COVID-19. *Bioinformatics, 36*(20), 5109–5111.

Lalani, S., & Poh, C. L. (2020). Flavonoids as antiviral agents for enterovirus A71 (EV-A71). *Viruses, 12*(2), 184. https://doi.org/10.3390/v12020184

Lee, J. T., Yang, Q., Gribenko, A., Perrin, B. S., Jr., Zhu, Y., Cardin, R., Hao, L., et al. (2022). Genetic surveillance of SARS-CoV-2 M(pro) reveals high sequence and structural conservation prior to the introduction of protease inhibitor paxlovid. *mBio, 13*(4), e0086922. https://doi.org/10.1128/mbio.00869-22

Lin, C. W., Tsai, F. J., Tsai, C. H., Lai, C. C., Wan, L., Ho, T. Y., Chao, P. D., et al. (2005). Anti-SARS coronavirus 3C-like protease effects of Isatis indigotica root and plant-derived phenolic compounds. *Antiviral Res, 68*(1), 36–42. https://doi.org/10.1016/j.antiviral.2005.07.002

Lin, L., Chen, D. Y., Scartelli, C., Xie, H., Merrill-Skoloff, G., Yang, M., Flaumenhaft, R., et al. (2023). Plant flavonoid inhibition of SARS-CoV-2 main protease and viral replication. *iScience, 26*(9), 107602. https://doi.org/10.1016/j.isci.2023.107602

Liu, S., Lu, H., Zhao, Q., He, Y., Niu, J., Debnath, A. K., Jiang, S., et al. (2005). Theaflavin derivatives in black tea and catechin derivatives in green tea inhibit HIV-1 entry by targeting gp41. *Biochim Biophys Acta, 1723*(1–3), 270–281. https://doi.org/10.1016/j.bbagen.2005.02.012

Liu, X., Raghuvanshi, R., Ceylan, F. D., & Bolling, B. W. (2020). Quercetin and its metabolites inhibit recombinant human angiotensin-converting enzyme 2 (ACE2) activity. *J Agric Food Chem, 68*(47), 13982–13989. https://doi.org/10.1021/acs.jafc.0c05064

Loregian, A., Mercorelli, B., Nannetti, G., Compagnin, C., & Palù, G. (2014). Antiviral strategies against influenza virus: Towards new therapeutic approaches. *Cell Mol Life Sci, 71*(19), 3659–3683. https://doi.org/10.1007/s00018-014-1615-2

Low, Z., Lani, R., Tiong, V., Poh, C., AbuBakar, S., & Hassandarvish, P. (2023). COVID-19 therapeutic potential of natural products. *Int J Mol Sci, 24*(11), 9589. https://doi.org/10.3390/ijms24119589

Magden, J., Kääriäinen, L., & Ahola, T. (2005). Inhibitors of virus replication: Recent developments and prospects. *Appl Microbiol Biotechnol, 66*(6), 612–621. https://doi.org/10.1007/s00253-004-1783-3

Majerová, T., & Novotný, P. (2021). Precursors of viral proteases as distinct drug targets. *Viruses, 13*(10), 1981. https://doi.org/10.3390/v13101981

Malin, J. J., Suárez, I., Priesner, V., Fätkenheuer, G., & Rybniker, J. (2020). Remdesivir against COVID-19 and other viral diseases. *Clin Microbiol Rev, 34*(1), e00162. https://doi.org/10.1128/cmr.00162-20

Malone, B., Urakova, N., Snijder, E. J., & Campbell, E. A. (2022). Structures and functions of coronavirus replication-transcription complexes and their relevance for SARS-CoV-2 drug design. *Nat Rev Mol Cell Biol, 23*(1), 21–39. https://doi.org/10.1038/s41580-021-00432-z

Meli, R., Morris, G. M., & Biggin, P. C. (2022). Scoring functions for protein-ligand binding affinity prediction using structure-based deep learning: A review. *Front Bioinform, 2*, 57.

Memariani, Z., Abbas, S. Q., Ul Hassan, S. S., Ahmadi, A., & Chabra, A. (2021). Naringin and naringenin as anticancer agents and adjuvants in cancer combination therapy: Efficacy and molecular mechanisms of action, a comprehensive narrative review. *Pharmacol Res, 171*, 105264. https://doi.org/10.1016/j.phrs.2020.105264

Meng, J. R., Liu, J., Fu, L., Shu, T., Yang, L., Zhang, X., Bai, L. P., et al. (2023). Anti-entry activity of natural flavonoids against SARS-CoV-2 by targeting spike RBD. *Viruses, 15*(1), 160. https://doi.org/10.3390/v15010160

Milton-Laskibar, I., Trepiana, J., Macarulla, M. T., Gómez-Zorita, S., Arellano-García, L., Fernández-Quintela, A., & Portillo, M. P. (2023). Potential usefulness of mediterranean diet polyphenols against COVID-19-induced inflammation: A review of the current knowledge. *J Physiol Biochem, 79*(2), 371–382. https://doi.org/10.1007/s13105-022-00926-0

Minkoff, J. M., & tenOever, B. (2023). Innate immune evasion strategies of SARS-CoV-2. *Nat Rev Microbiol, 21*(3), 178–194. https://doi.org/10.1038/s41579-022-00839-1

Moghaddam, E., Teoh, B. T., Sam, S. S., Lani, R., Hassandarvish, P., Chik, Z., Zandi, K., et al. (2014). Baicalin, a metabolite of baicalein with antiviral activity against dengue virus. *Sci Rep*, *4*, 5452. https://doi.org/10.1038/srep05452

Mouradov, A., & Spangenberg, G. (2014). Flavonoids: A metabolic network mediating plants adaptation to their real estate. *Front Plant Sci*, *5*. https://doi.org/10.3389/fpls.2014.00620

Moustaqil, M., Ollivier, E., Chiu, H. P., Van Tol, S., Rudolffi-Soto, P., Stevens, C., Gambin, Y., et al. (2021). SARS-CoV-2 proteases PLpro and 3CLpro cleave IRF3 and critical modulators of inflammatory pathways (NLRP12 and TAB1): implications for disease presentation across species. *Emerg Microbes Infect*, *10*(1), 178–195. https://doi.org/10.1080/22221751.2020.1870414

Mucsi, I., Gyulai, Z., & Béládi, I. (1992). Combined effects of flavonoids and acyclovir against herpesviruses in cell cultures. *Acta Microbiol Hung*, *39*(2), 137–147.

Mucsi, I., & Prágai, B. M. (1985). Inhibition of virus multiplication and alteration of cyclic AMP level in cell cultures by flavonoids. *Experientia*, *41*(7), 930–931. https://doi.org/10.1007/bf01970018

Muegge, I., & Rarey, M. (2001). Small molecule docking and scoring. *Rev Comput Chem*, *17*, 1–60.

Nayak, M. K., Agrawal, A. S., Bose, S., Naskar, S., Bhowmick, R., Chakrabarti, S., Chawla-Sarkar, M., et al. (2014). Antiviral activity of baicalin against influenza virus H1N1-pdm09 is due to modulation of NS1-mediated cellular innate immune responses. *J Antimicrob Chemother*, *69*(5), 1298–1310. https://doi.org/10.1093/jac/dkt534

Panche, A. N., Diwan, A. D., & Chandra, S. R. (2016). Flavonoids: An overview. *J Nutr Sci*, *5*, e47. https://doi.org/10.1017/jns.2016.41

Pandey, K. B., & Rizvi, S. I. (2009). Plant polyphenols as dietary antioxidants in human health and disease. *Oxid Med Cell Longev*, *2*(5), 270–278. https://doi.org/10.4161/oxim.2.5.9498

Pang, P., Zheng, K., Wu, S., Xu, H., Deng, L., Shi, Y., & Chen, X. (2018). Baicalin downregulates RLRs signaling pathway to control influenza A virus infection and improve the prognosis. *Evid Based Complement Alternat Med*, *2018*, 4923062. https://doi.org/10.1155/2018/4923062

Paraiso, I. L., Revel, J. S., & Stevens, J. F. (2020). Potential use of polyphenols in the battle against COVID-19. *Curr Opin Food Sci*, *32*, 149–155. https://doi.org/https://doi.org/10.1016/j.cofs.2020.08.004

Pashaei, M., & Rezaei, N. (2020). Immunotherapy for SARS-CoV-2: potential opportunities. *Expert Opin Biol Ther*, *20*(10), 1111–1116. https://doi.org/10.1080/14712598.2020.1807933

Pérez de la Lastra, J. M., Andrés-Juan, C., Plou, F. J., & Pérez-Lebeña, E. (2021). Impact of zinc, glutathione, and polyphenols as antioxidants in the immune response against SARS-CoV-2. *Processes*, *9*(3), 506. Retrieved from https://www.mdpi.com/2227-9717/9/3/506

Phizackerley, D. (2022). Three more points about paxlovid for covid-19. *BMJ*, *377*, o1397. https://doi.org/10.1136/bmj.o1397

Piret, J., & Boivin, G. (2011). Resistance of herpes simplex viruses to nucleoside analogues: Mechanisms, prevalence, and management. *Antimicrob Agents Chemother*, *55*(2), 459–472. https://doi.org/10.1128/aac.00615-10

Razia, D., Sindu, D., Grief, K., Cherrier, L., Omar, A., Walia, R., & Tokman, S. (2023). Molnupiravir vs remdesivir for treatment of covid-19 in lung transplant recipients. *J Heart Lung Transplant*, *42*(4), S165–S166.

Renu, K., Prasanna, P. L., & Valsala Gopalakrishnan, A. (2020). Coronaviruses pathogenesis, comorbidities and multi-organ damage - A review. *Life Sci*, *255*, 117839. https://doi.org/10.1016/j.lfs.2020.117839

Rizzuti, B., Grande, F., Conforti, F., Jimenez-Alesanco, A., Ceballos-Laita, L., Ortega-Alarcon, D., Velazquez-Campoy, A., et al. (2021). Rutin is a low micromolar inhibitor of SARS-CoV-2 main protease 3CLpro: Implications for drug design of quercetin analogs. *Biomedicines*, *9*(4), 375. https://doi.org/10.3390/biomedicines9040375

Russo, M., Moccia, S., Spagnuolo, C., Tedesco, I., & Russo, G. L. (2020). Roles of flavonoids against coronavirus infection. *Chem Biol Interact*, *328*, 109211. https://doi.org/10.1016/j.cbi.2020.109211

Ryu, W.-S. (2017). Virus life cycle. *Mole Virol Human Pathogenic Viruses*, *2017*, 31–45.

Samavati, L., & Uhal, B. D. (2020). ACE2, much more than just a receptor for SARS-COV-2. *Front Cell Infect Microbiol*, *10*, 317. https://doi.org/10.3389/fcimb.2020.00317

Saravanan, K. A., Panigrahi, M., Kumar, H., Rajawat, D., Nayak, S. S., Bhushan, B., & Dutt, T. (2022). Role of genomics in combating COVID-19 pandemic. *Gene*, *823*, 146387. https://doi.org/10.1016/j.gene.2022.146387

Saravolatz, L. D., Depcinski, S., & Sharma, M. (2023). Molnupiravir and nirmatrelvir-ritonavir: Oral coronavirus disease 2019 antiviral drugs. *Clin Infect Dis*, *76*(1), 165–171. https://doi.org/10.1093/cid/ciac180

Semwal, D. K., Semwal, R. B., Combrinck, S., & Viljoen, A. (2016). Myricetin: A dietary molecule with diverse biological activities. *Nutrients*, *8*(2), 90. https://doi.org/10.3390/nu8020090

Shankar, E., Goel, A., Gupta, K., & Gupta, S. (2017). Plant flavone apigenin: An emerging anticancer agent. *Curr Pharmacol Rep*, *3*(6), 423–446. https://doi.org/10.1007/s40495-017-0113-2

Sharifkashani, S., Bafrani, M. A., Khaboushan, A. S., Pirzadeh, M., Kheirandish, A., Yavarpour Bali, H., Rezaei, N., et al. (2020). Angiotensin-converting enzyme 2 (ACE2) receptor and SARS-CoV-2: Potential therapeutic targeting. *Eur J Pharmacol*, *884*, 173455. https://doi.org/10.1016/j.ejphar.2020.173455

Sheppard, M., Laskou, F., Stapleton, P. P., Hadavi, S., & Dasgupta, B. (2017). Tocilizumab (Actemra). *Hum Vaccin Immunother*, *13*(9), 1972–1988. https://doi.org/10.1080/21645515.2017.1316909

Shi, H., Ren, K., Lv, B., Zhang, W., Zhao, Y., Tan, R. X., & Li, E. (2016). Baicalin from Scutellaria baicalensis blocks respiratory syncytial virus (RSV) infection and reduces inflammatory cell infiltration and lung injury in mice. *Sci Rep*, *6*, 35851. https://doi.org/10.1038/srep35851

Shilpa, V. S., Shams, R., Dash, K. K., Pandey, V. K., Dar, A. H., Ayaz Mukarram, S., Kovács, B., et al. (2023). Phytochemical properties, extraction, and pharmacological benefits of naringin: A review. *Molecules*, *28*(15), 5623. https://doi.org/10.3390/molecules28155623

Song, J. M., Lee, K. H., & Seong, B. L. (2005). Antiviral effect of catechins in green tea on influenza virus. *Antiviral Res*, *68*(2), 66–74. https://doi.org/10.1016/j.antiviral.2005.06.010

Su, H., Yao, S., Zhao, W., Li, M., Liu, J., Shang, W., Xu, Y., et al.(2020). Discovery of baicalin and baicalein as novel, natural product inhibitors of SARS-CoV-2 3CL protease in vitro. *bioRxiv*. https://doi.org/10.1101/2020.04.13.038687

Su, H. X., Yao, S., Zhao, W. F., Li, M. J., Liu, J., Shang, W. J., Xu, Y. C., et al. (2020). Anti-SARS-CoV-2 activities in vitro of Shuanghuanglian preparations and bioactive ingredients. *Acta Pharmacol Sin*, *41*(9), 1167–1177. https://doi.org/10.1038/s41401-020-0483-6

Sultana, J., Cutroneo, P. M., Crisafulli, S., Puglisi, G., Caramori, G., & Trifirò, G. (2020). Azithromycin in COVID-19 patients: Pharmacological mechanism, clinical evidence and prescribing guidelines. *Drug Saf, 43*(8), 691–698. https://doi.org/10.1007/s40264-020-00976-7

Suryawanshi, Y. N., & Biswas, D. A. (2023). Herd immunity to fight against COVID-19: A narrative review. *Cureus, 15*(1), e33575. https://doi.org/10.7759/cureus.33575

Utomo, R. Y., Ikawati, M., & Meiyanto, E. (2020). Revealing the potency of citrus and galangal constituents to halt SARS-CoV-2 infection. *Preprints, 2020*, 2020030214.

V'Kovski, P., Kratzel, A., Steiner, S., Stalder, H., & Thiel, V. (2021). Coronavirus biology and replication: Implications for SARS-CoV-2. *Nat Rev Microbiol, 19*(3), 155–170. https://doi.org/10.1038/s41579-020-00468-6

Vaou, N., Stavropoulou, E., Voidarou, C., Tsigalou, C., & Bezirtzoglou, E. (2021). Towards advances in medicinal plant antimicrobial activity: A review study on challenges and future perspectives. *Microorganisms, 9*(10), 2041. https://doi.org/10.3390/microorganisms9102041

Verma, S., Patil, V. M., & Gupta, M. K. (2022). Mutation informatics: SARS-CoV-2 receptor-binding domain of the spike protein. *Drug Discov Today, 27*(10), 103312. https://doi.org/10.1016/j.drudis.2022.06.012

von Kleist, M., Metzner, P., Marquet, R., & Schütte, C. (2012). HIV-1 polymerase inhibition by nucleoside analogs: Cellular- and kinetic parameters of efficacy, susceptibility and resistance selection. *PLoS Comput Biol, 8*(1), e1002359. https://doi.org/10.1371/journal.pcbi.1002359

Wang, M., Cao, R., Zhang, L., Yang, X., Liu, J., Xu, M., Xiao, G., et al. (2020). Remdesivir and chloroquine effectively inhibit the recently emerged novel coronavirus (2019-nCoV) in vitro. *Cell Res, 30*(3), 269–271. https://doi.org/10.1038/s41422-020-0282-0

Wang, T. Y., Li, Q., & Bi, K. S. (2018). Bioactive flavonoids in medicinal plants: Structure, activity and biological fate. *Asian J Pharm Sci, 13*(1), 12–23. https://doi.org/10.1016/j.ajps.2017.08.004

Wang, X., Tang, G., Liu, Y., Zhang, L., Chen, B., Han, Y., Shao, S., et al. (2022). The role of IL-6 in coronavirus, especially in COVID-19. *Front Pharmacol, 13*, 1033674. https://doi.org/10.3389/fphar.2022.1033674

Weston, S., Coleman, C. M., Haupt, R., Logue, J., Matthews, K., Li, Y., . . . Frieman, M. B. (2020). Broad anti-coronavirus activity of food and drug administration-approved drugs against SARS-CoV-2 in vitro and SARS-CoV in vivo. *J Virol, 94*(21), e01218. https://doi.org/10.1128/jvi.01218-20

Wu, C., Liu, Y., Yang, Y., Zhang, P., Zhong, W., Wang, Y., Li, H., et al. (2020). Analysis of therapeutic targets for SARS-CoV-2 and discovery of potential drugs by computational methods. *Acta Pharm Sin B, 10*(5), 766–788. https://doi.org/10.1016/j.apsb.2020.02.008

Wu, W., Li, R., Li, X., He, J., Jiang, S., Liu, S., & Yang, J. (2015). Quercetin as an antiviral agent inhibits influenza A virus (IAV) entry. *Viruses, 8*(1), 6. https://doi.org/10.3390/v8010006

Xiong, Y., Zhu, G.-H., Zhang, Y.-N., Hu, Q., Wang, H.-N., Yu, H.-N., . . . Ge, G.-B. (2021). Flavonoids in Ampelopsis grossedentata as covalent inhibitors of SARS-CoV-2 3CLpro: Inhibition potentials, covalent binding sites and inhibitory mechanisms. *Int J Biol Macromol, 187*, 976–987. https://doi.org/https://doi.org/10.1016/j.ijbiomac.2021.07.167

Xu, G., Dou, J., Zhang, L., Guo, Q., & Zhou, C. (2010). Inhibitory effects of baicalein on the influenza virus in vivo is determined by baicalin in the serum. *Biol Pharm Bull, 33*(2), 238–243. https://doi.org/10.1248/bpb.33.238

Xu, J., Zhao, S., Teng, T., Abdalla, A. E., Zhu, W., Xie, L., Guo, X., et al. (2020). Systematic comparison of two animal-to-human transmitted human coronaviruses: SARS-CoV-2 and SARS-CoV. *Viruses, 12*(2), 244. https://doi.org/10.3390/v12020244

Yamauchi, Y., & Greber, U. F. (2016). Principles of virus uncoating: Cues and the snooker ball. *Traffic, 17*(6), 569–592. https://doi.org/10.1111/tra.12387

Yi, L., Li, Z., Yuan, K., Qu, X., Chen, J., Wang, G., . . . Xu, X. (2004). Small molecules blocking the entry of severe acute respiratory syndrome coronavirus into host cells. *J Virol, 78*(20), 11334–11339. https://doi.org/10.1128/jvi.78.20.11334-11339.2004

Zakaryan, H., Arabyan, E., Oo, A., & Zandi, K. (2017). Flavonoids: Promising natural compounds against viral infections. *Arch Virol, 162*(9), 2539–2551. https://doi.org/10.1007/s00705-017-3417-y

Zandi, K., Musall, K., Oo, A., Cao, D., Liang, B., Hassandarvish, P., . . . Schinazi, R. F. (2021). Baicalein and baicalin inhibit SARS-CoV-2 RNA-dependent-RNA polymerase. *Microorganisms, 9*(5), 893. https://doi.org/10.3390/microorganisms9050893

Zandi, K., Teoh, B. T., Sam, S. S., Wong, P. F., Mustafa, M. R., & Abubakar, S. (2012). Novel antiviral activity of baicalein against dengue virus. *BMC Complement Altern Med, 12*, 214. https://doi.org/10.1186/1472-6882-12-214

Zhang, Y., Sun, S., Du, C., Hu, K., Zhang, C., Liu, M., Dong, N., et al. (2022). Transmembrane serine protease TMPRSS2 implicated in SARS-CoV-2 infection is autoactivated intracellularly and requires N-glycosylation for regulation. *J Biol Chem, 298*(12), 102643. https://doi.org/10.1016/j.jbc.2022.102643

Zhao, A. V., Crutchley, R. D., Guduru, R. C., Ton, K., Lam, T., & Min, A. C. (2022). A clinical review of HIV integrase strand transfer inhibitors (INSTIs) for the prevention and treatment of HIV-1 infection. *Retrovirology, 19*(1), 22. https://doi.org/10.1186/s12977-022-00608-1

Zhong, N. S., Zheng, B. J., Li, Y. M., Poon, Xie, Z. H., Chan, K. H., Guan, Y., et al. (2003). Epidemiology and cause of severe acute respiratory syndrome (SARS) in guangdong, People's Republic of China, in February, 2003. *Lancet, 362*(9393), 1353–1358. https://doi.org/10.1016/s0140-6736(03)14630-2

Zhu, H., Du, W., Song, M., Liu, Q., Herrmann, A., & Huang, Q. (2021). Spontaneous binding of potential COVID-19 drugs (Camostat and Nafamostat) to human serine protease TMPRSS2. *Comput Struct Biotechnol J, 19*, 467–476. https://doi.org/10.1016/j.csbj.2020.12.035

4 Molecular Aspects on Anti-SARS-CoV-2 Activity of Flavonols

Inhibitory on Protein Targets, Immunomodulatory, and Anti-Inflammatory Effects

Piyush Kashyap, Mamta Thakur, Rhythm Kalsi, Kandi Sridhar, Minaxi Sharma, Shiv Kumar, and Baskaran Stephen Inbaraj

4.1 INTRODUCTION

COVID-19, a global health crisis of unprecedented magnitude, has been triggered by severe acute respiratory syndrome coronavirus 2 (SARS-CoV-2), which affects the lives of millions and exerts immense pressure on healthcare systems worldwide. According to Pang et al. (2023), the impact of an infection can range from being completely asymptomatic to fatal. SARS-CoV-2 primarily spreads through minute droplets coughed or exhaled by people who are infected. By August 14, 2023, there had been reported 6.9 million deaths and 769 million confirmed cases worldwide (WHO, 2023).

Amin and Jha (2020) reported that CoVs are RNA viruses having single strands and a positive polarity that are frequently found in both humans and wildlife. These viruses have been categorized by the International Committee on Taxonomy of Viruses under the family Coronaviridae, a subfamily Coronavirinae, and an order Nidovirales. The Coronavirinae includes four genera of CoVs: α-, β-, γ-, and δ- (Pang et al., 2023). Seven different CoVs can infect people. The α-CoVs include HCoV-NL63 and HCoV-229E, whereas the β-CoVs encompass SARS-CoV-1, HCoV-OC43, Middle East Respiratory Syndrome Coronavirus, HCoV-HKU1, and SARS-CoV-2 (Pillaiyar et al., 2020). According to Kashyap et al. (2022), the SARS-CoV-2 comprises the genomic RNA complexes and membrane that constitute the minimum of the following: (i) spike (S); (ii) transmembrane (M) and (iii) envelope (E) proteins (Figure 4.1). The ACE2 is the surface receptor of S proteins in the host cell. Once viral RNA enters host cells, both overlapping open-reading frames (ORF1a & b) undergo translation for the generation of bigger polyproteins – pp1a and pp1ab. To yield 16 non-structural proteins (NSPs), these polyproteins undergo further processing through the actions of the principal protease (M^{pro}) and papain-like protease (PL^{pro}) (Pang et al., 2023). These replicase genes, accountable for encoding the 16 NSPs, constitute nearly two-thirds of the genome. Subsequently, M^{pro}-mediated cleavage releases NSP4–NSP16 proteins, which are important for transcription and replication of viral genomes. Due to this, the efficient inhibition of M^{pro} may prevent viral RNA transcription and replication, hence limiting the spread of the virus (Lin et al., 2023).

This widespread problem has highlighted the urgent requirement for efficient therapeutic measures to reduce the transmission of the virus and related effects. To treat COVID-19, a variety of therapeutic modalities have been used, with antiviral medications receiving a lot of attention. Notably, regulatory bodies have authorized the use of some antiviral medications, such as Remdesivir, to treat COVID-19, which works primarily by preventing viral replication (Lim et al., 2020). Currently, numerous biotic molecules have emerged as probable therapeutic aims for SARS-CoV-2. Among these candidates, extensive research has been conducted on the receptor-binding domain (RBD) of S protein and 3-chymotrypsin-like protease ($3CL^{pro}$) as possible targets for coronaviruses (Premkumar et al., 2020; Zhang et al., 2020).

Besides these, there is a growing emphasis in the scientific literature on the investigation and identification of new chemicals that target SARS-CoV-2. The naturally occurring polyphenolic chemicals known as flavonols – a subclass of flavonoids – exist in diverse plant-based sources. Antioxidant, anti-inflammatory, and immunomodulatory capabilities of these substances have long been recognized for their numerous health benefits (Kothari et al., 2020; Ullah et al., 2020; Barreca et al., 2021). Recent work suggested that flavonols have a promising role to play in the complex strategy used to fight coronavirus, which includes direct virus suppression, modulation of host immunological responses, and mitigation of excessive inflammation (Zhuo et al., 2023). Shortly after the X-ray crystallographic structure of SARS-CoV-2 M^{pro} became accessible, various molecular docking studies were published, forecasting strong binding interactions between numerous flavonols and the S1 binding pocket of SARS-CoV-2 M^{pro} (Fink et al., 2022;

DOI: 10.1201/9781003433200-4

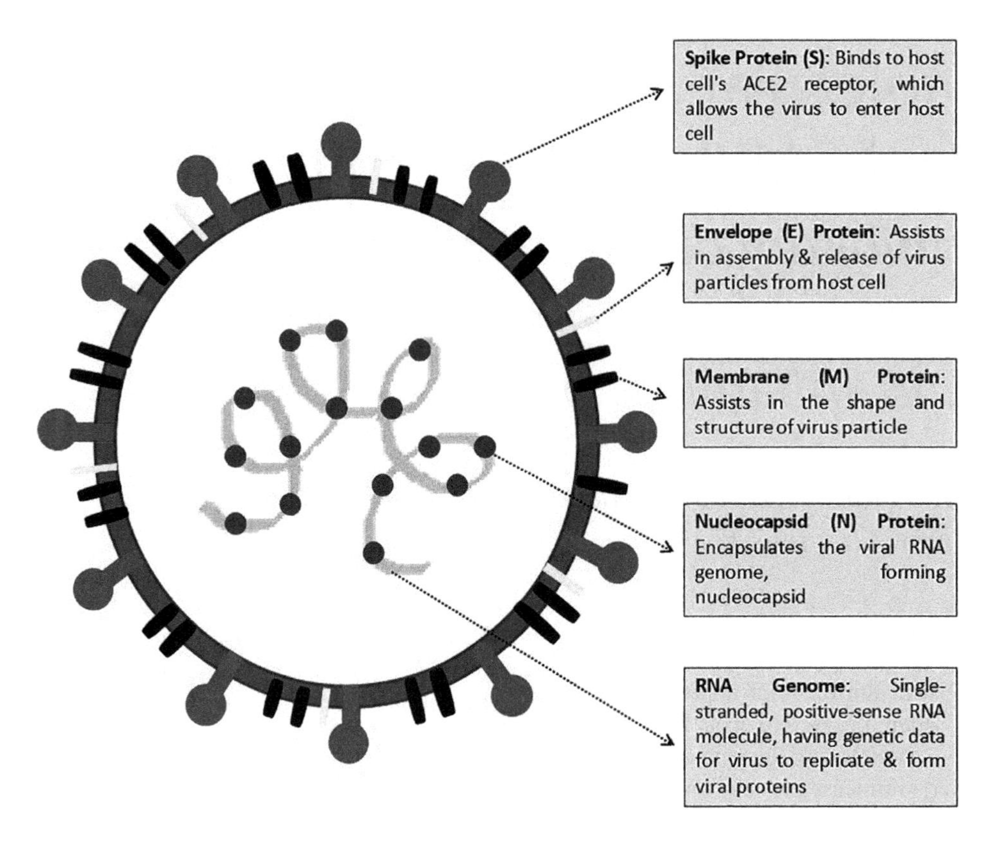

FIGURE 4.1 Structure of SARS-CoV-2 virus. This figure is constructed by authors using Microsoft 365® Apps, Microsoft Cooperation, Redmond, Washington, United States.

Hadni et al., 2022). The hunt for new methods to stop viral replication, lessen immunological deregulation, and reduce inflammation has been more serious as the globe struggles to deal with the problems caused by COVID-19.

Natural substances like flavonols have drawn a lot of attention in this area due to their ability to provide multiple modes of action against the virus and its consequences. This chapter aims to present a thorough overview of molecular mechanisms underlying flavonols' anti-SARS-CoV-2 activity, focusing on their inhibition of key viral protein targets, their ability to control host immune responses, and their function in reducing the inflammatory effects frequently observed in case of COVID-19. The budding physiological advantages of these natural substances in the context of COVID-19 can be better understood by exploring the complex interplay between flavonols and both viral and host variables. The current chapter aims to incorporate the possible COVID-19 mitigation and SARS-CoV-2 prevention techniques by a methodical analysis of the existing corpus of literature.

4.2 FLAVONOLS' INHIBITORY EFFECTS AND SARS-COV-2 PROTEINS

The flavonols – an essential subgroup of flavonoids – are identified by the hydroxyl group which is present in the C-ring at the C-3 position. The most researched chemicals against coronavirus, according to data analysis, are flavonols – mainly quercetin, kaempferol, fisetin, myricetin, and their derivatives (Russo et al., 2020; Gorla et al., 2021; Mouffouk et al., 2021). Additionally, these compounds possess distinct biological properties that improve human immunity against oxidative stress, HIV, inflammation, and cardiovascular diseases (Sharma et al., 2018; Nakanishi et al., 2020; Zaragozá et al., 2020). Flavonols could be noticed as promoter substances that can cause the suppression of viral proteins and serve as a probable source for COVID-19-fighting medications and vaccines. The details of the inhibitory effects of flavonols on viral proteins are explained in Table 4.1.

CoVs produce three distinct viral proteases – $3CL^{pro}$, PL^{pro}, and M^{pro}. Among these, the life cycle, replication, and functional protein growth of the virus depend on $3CL^{pro}$, PL^{pro}, and M^{pro}, respectively (Verma et al., 2021; Pang et al., 2023). Consequently, these proteases have emerged as the most favorable for making antiviral medications, as discussed by Tabari et al. (2021). Black garlic extract-derived quercetin and its derivative, with an IC_{50} value of 137 μg/mL, have displayed notable efficacy in inhibiting $3CL^{pro}$ by engaging with a protease substrate-binding site, as reported by Du et al. (2021), Nguyen et al. (2021), and Su et al. (2021). Abian et al. (2020) have shown that quercetin functions as an antagonist at the active site of $3CL^{pro}$ and induces a destabilizing effect on the protease's thermal stability, which is dose-dependent. Conversely, rutin and myricetin display restricted preventive capabilities when interacting with the naphthalene inhibitor binding region of PL^{pro}.

TABLE 4.1
Summary of Inhibitory Impacts of Flavonols against SARS-CoV-2 Proteins

Flavonol(s)	Source or Extract	Method Used	Inhibitory Effects against SARS-CoV-2 Proteins	References
Kaempferol	Shanghai Yuanye Bio-Technology Co., Ltd. (Shanghai, China) (purity ≥ 98%)	Dual split protein assays	Interaction of kaempferol with heptad repeat (HR) parts of SARS-CoV-2 S2 subunits, deformation of HR1 and direct reaction with lysine units of HR2 area, preventing SARS-CoV-2 infection by inhibiting membrane fusion, having a wide-spectrum anti-fusion probability	Gao et al. (2023)
Quercetin/isorhamnetin glucosides	–	Absorption, Distribution, Metabolism, Excretion, and Hepatotoxicity (ADMET) analysis	Binding affinities: −10.0 and −10.1 kcal/mol, Inhibition constant: 0.51–7.27×10^{-7} μM. Interactions involving 4 and 8 hydrogen bonds, with glucosides acting as allosteric inhibitors of SARS-CoV-2-Mpro	Adegbola et al. (2023)
Myricetin	Macklin Inc. (Shanghai, China) (Assay 97%)	Molecular docking, bilayer interferometry assays, immunocytochemistry, and pseudoviruses assays	Inhibition of HCoV-229E and SARS-CoV-2 replication (EC_{50} 55.18 μM) in vitro, blocking of SARS-CoV-2 virus entry facilitators and relieving the serine/threonine protein kinase 1 (RIPK1)-driven inflammation through the RIPK1/NF-κB pathway	Pan et al. (2023)
Kaempferol and quercetin	PubChem Compound Database as SMILES strings	Fluorescence resonance energy transfer (FRET) peptide substrate MCA-AVLQSGFR–Lys(Dnp)-Lys-NH_2 (Isca Biochemicals, UK)	Binding affinity of kaempferol: −54.0 kcal/mol, quercetin: −50.9 kcal/mol, kaempferol-3-O-rutinoside: −44.0 kcal/mol; kaempferol showed 45% $3CL^{pro}$ inhibition than kaempferol-3-O-rutinoside (18%), while quercetin showed > 50% inhibition, IC_{50} value for quercetin – 23.4 μM	Bahun et al. (2022)
Rutin	*Saussurea costus* and *Saussurea involucrata*	Absorption, Distribution, Metabolism, Excretion, and Hepatotoxicity (ADMET) analysis	Binding energy with M^{pro}: −9.14 kcal/mol; rutin formed seven H-bonds, three at Cys 145 and four at Thr 26, Gly 143, His 163, Thr 26, and one pi-H bond at Gln 189 which played an important role in inhibiting M^{pro}	Houchi and Messasma (2022)
Quercetin, Kaempferol, Myricetin, Fisetin, Morin, Galangin	Ligand Library	AutoDock Vina interlinked with AutoDock MGL tools	The binding energies of morin, fisetin, and quercetin were −8.0, −8.0, and −7.2 kcal/mol, respectively; exhibition of pi-S interaction, pi-alkyl, and pi-sigma with Met165, Met49, and His41 by morin; fisetin inhibitor showed four H-bonds with His41, His163, Leu141, and Phe140, as well as pi-alkyl interactions with Met165, Met49, and Cys145 with M^{pro} of COVID-19	Imran et al. (2022)
Kaempferol, quercetin, myricetin, quercetin-3-O-glycoside, and rutin. Dihydroflavonols used were (+)-dihydroquercetin, (+)-dihydrokaempferol, and (+)-dihydromyricetin	Sigma-Aldrich	SARS-CoV-2 Assay Kit (BPS bioscience)	Binding of kaempferol, quercetin, myricetin, isoquercitrin, (+)-dihydrokaempferol, rutin, (+)-dihydroquercetin, and (+)-dihydromyricetin to minimum two sub-sites (S1, S1′, S2, and S4) in binding pocket and inhibition of SARS-CoV-2 Mpro; affinity scores varied from −8.8 to −7.4 (kcal/mol)	Zhu et al. (2022)

(Continued)

TABLE 4.1 (*Continued*)
Summary of Inhibitory Impacts of Flavonols against SARS-CoV-2 Proteins

Flavonol(s)	Source or Extract	Method Used	Inhibitory Effects against SARS-CoV-2 Proteins	References
Dihydromyricetin	Topscience Co. Ltd (Shanghai, China)	FRET-based enzymatic assay	IC_{50} of dihydromyricetin was 1.716±0.419 μM; interaction of dihydrochromone ring of dihydromyricetin with imidazole side chain of His163 through π–π stacking and formation of H-bond by 1-oxygen of dihydromyricetin with backbone nitrogen of Glu166; 3-, 7-, 3′-, and 4′-hydroxyl of dihydromyricetin interaction with Gln189, Leu141, Arg188, and Thr190 through H- bonds	Xiao et al. (2021)
Herbacetin	Library	Tryptophan-based assay	IC_{50} values –53.90 μM, phenyl moiety of herbacetin takes the S1 site using Glu166, while chromen-4-one scaffold is located in the S2 site using H-bonds with His41 and Gln189	Jo et al. (2020)

According to Zhu et al. (2022), the flavonols like quercetin, kaempferol, myricetin, rutin, and derivatives can reduce the M^{pro} activity by interacting to a minimum of two sub-sites: S1, S1′, S2, and S4. Their ratings for affinity varied from 8.8 to 7.4. Such flavonols also decreased the HCoV-229E M^{pro} potential and affinities were 7.1–7.8. These antioxidative flavonols and dihydroflavonols have the potential to stop the virus, according to *in vitro* inhibition studies. In another investigation, Owis et al. (2020) described the molecular docking as well as various interactions between the main protease's N3 binding site and *Salvadora persica* L. derived flavonol glycosides of kaempferol and isorhamnetin. These examined flavonols had a noticeable difference in binding stability than the standard medicine darunavir. This indicates that the flavonol skeleton itself has some sort of function. The comparison of these data showed that the B-ring lacks surplus methoxyl group at C-3′ and shows rutinose moiety in the C-ring of flavonol structure which could boost the binding strength. According to ul-Qamar et al. (2020), two flavonols – myricitrine and myricetin-3-O-D-glucopyranoside – have been widely employed in Chinese medicine from historic times against 3CL^{pro} of SARS-CoV-2. Nearly 32,297 molecules were subjected to docking analysis to assess their efficacy. The findings were compared to positive controls such as nelfinavir, prulifloxacin, and colistin. These compounds exhibited a strong affinity and the ability to effectively bind to both the catalytic dyad (composed of Cys-145 and His-41) and the receptor-binding site of 3CL^{pro}. They were also observed to establish close interactions with the conserved catalytic dyad residues. Importantly, these compounds are known to be physiologically active, safe for use, and naturally occurring in a diverse range of therapeutic plants and species.

Several flavonols, including astragalin, quercetin and derivatives, kaempferol, and quercitrin, were examined for their propensity for interacting with 3CL^{pro} and PL^{pro}. The results showed that all the examined compounds had constant interactions with higher binding affinities to proteins in positive control remdesivir. Quercetin-3-O-glucoside achieved the highest docking score in the study conducted by Hiremath et al. (2021). This compound established eight hydrogen bonds with specific amino acid components, including His74, Arg83, Tyr155, Asn157, and His176, within the binding pocket of the PL^{pro} protein. Other investigations have also revealed that flavonols possess the capability to bind ACE2 receptors and S protein. The chemical composition of these compounds plays a pivotal role in deciding their potential for reaction to these proteins. Strong hydrogen bonds are formed between the ortho di-hydroxyl groups of flavonols' B-ring and the S protein, contributing to the stability of this complex. Pandey et al. (2021a) conducted docking studies involving isorhamnetin, fisetin, kaempferol, and quercetin against the spike protein. The results indicated that these compounds exhibited significant binding affinities to the S2 domain of S protein than hydroxychloroquine. The steric hindrance due to the methoxyl group at the C-3′ position of the B-ring in isorhamnetin reduced the synthesis of H-bonds and changed the makeup of interacting residues rather than quercetin and its derivative. Moreover, the potential of H-bond formation is minimized due to the lack of the –OH group in the C-3′ position. Galanin, morin, and myricetin demonstrated favorable docking results for spike glycoprotein as opposed to the traditional drugs abacavir and hydroxychloroquine (Pandey et al., 2021b). Additionally, the herbacetin, kaempferol, and morin showed an excellent capacity to bind spike protein (Tallei et al., 2020). Due to its strong binding potential with spike proteins, quercetin exhibited a better potential for connecting the ACE2 receptor in comparison to hydroxychloroquine and other antiviral medications. It had also strong binding potentials with spike proteins (Gu et al., 2021; Gasmi et al., 2022).

RdRp is an important constituent of SARS-CoV-2 replication and transcription machinery. It is also known as nsp12 and facilitates the viral RNA generation due to cofactors nsp7 and nsp8. RdRp is frequently targeted in docking studies, and compounds capable of inhibiting its function are being explored as possible medications against COVID-19 infections, as discussed by Gasmi et al. (2022). Compounds such as quercetin, kaempferol, and their derivatives have been investigated as budding RdRp protein inhibitors. When comparing binding energy values, it was observed that, except for quercetin-3-O-sulfate, most quercetin derivatives exhibited higher binding scores in comparison to native compounds. Considering various quercetin derivatives, the rutin was potent, demonstrating the significance of the rutinoside group in boosting hydrogen bonding and electrostatic reactions (Rahman et al., 2021). Further, the compounds with glucuronic unit substitutions have reduced the RdRp binding energies. The existence of –OH and –COOH groups may be responsible for the glucuronic moiety's reactivity. The location of the glucuronic unit's attachment to the quercetin had an impact on its responsiveness. These findings show that quercetin's reactivity is decreased when a glucuronic unit is substituted at the C-3 position as opposed to the C-7 and C-3′ positions (Mouffouk et al., 2021). This showed that kaempferol's glucuronide derivatives are more potent than its sulfate derivatives. The most active component of this examination was kaempferol-3-O-rutinose, which had the highest activity of any other molecule. Additionally, the reactivity of aglycones – kaempferol and quercetin – is decreased by the addition of a sulfate unit at the C-3 position (da Silva et al., 2020). However, the flavonols' inhibitory effects against SARS-CoV-2 must be validated by clinical studies to develop suitable medicines at the earliest. In a particular investigation (NCT04401202) exploring the impact of *Nigella sativa* seed oil, which is abundant in quercetin and kaempferol, it was revealed that during its phase 2 evaluation, 62.1% of individuals in the intervention group, who received 500mg soft gel capsules of this oil twice a day, experienced recovery within a 2-week timeframe. In contrast, only 36% of individuals in the untreated group showed signs of improvement (Topcagic et al., 2017).

These findings suggest that the association of S protein and ACE2 receptor can be destabilized by flavonols, thereby inhibiting viral entry. Additionally, they can interfere with the activities of various replication enzymes involving $3CL^{pro}$, PL^{pro}, and RdRp. These compounds may reduce the risk of COVID-19 symptoms. Flavonols attributed to various biological properties lead to modulation of the immune response. However, for a systematic understanding of mechanisms underlying the interactions between flavonols and virus proteins, as well as to assess their safety, efficacy, and bioavailability in clinical settings, further extensive research is warranted.

4.3 COMBINATION THERAPIES AGAINST SARS-COV-2

Research is continuing to determine whether flavonols can be used in conjunction with other treatments to treat COVID-19. Flavonols may be used in combination therapy with other antiviral medications, monoclonal antibodies, or conventional treatments to increase the efficacy of those regimens. Combining these substances might increase their antiviral impacts, but clinical studies are required to prove their safety and effectiveness.

According to Swain et al. (2022), the blend of darunavir and quercetin-3-rhamnoside emerged as the least toxic yet highly effective pharmaceutical combination against SARS-CoV-2-M^{pro}, as indicated by various parameters such as binding affinity, toxicity, drug-likeness score, and advanced molecular docking-simulation metrics including RMSD, root mean square fluctuation, protein's radius of gyration, and H-bond integration analyses. Specifically, darunavir exhibited a docking and drug-likeness score of 10.25 kcal/mol and 0.60, respectively, while the quercetin derivative displayed values of 10.90 kcal/mol and 0.82, respectively. This combination improved the potential of darunavir and flavonols against SARS-CoV-2, contributing to the stability of pharmacokinetic profiles during treatment, encompassing factors like absorption, distribution, excretion, metabolism, and toxicity. Saakre et al. (2021) suggested the use of quercetin and zinc in tandem as a budding method against COVID-19, demonstrating heightened antiviral activity with minimal cytotoxicity. Quercetin can play a dual function, directly blocking the viral $3CL^{pro}$ and indirectly promoting RdRp inhibition through its zinc ionophore properties, which assist in zinc uptake. The antiviral attributes of zinc might encompass its capacity to inhibit the RNA virus RdRp, subsequently averting viral invasion of the body's cells and tissues. Although the risk-to-reward ratio appears to favor zinc supplementation in COVID-19, as suggested by Pal et al. (2021), the use of zinc supplementation as a therapeutic and preventive measure for viral diseases like COVID-19 is however in initial phases of scientific and experimental research, with ongoing investigations into its antiviral mechanisms, clinical benefits, and appropriate dosage. Furthermore, due to their shared antiviral and immunomodulatory effects, quercetin and other nutrients like vitamins C and D may synergistically combat viruses, as discussed by Agrawal et al. (2020). The presence of ascorbates may enhance the potency of quercetin, which can be recycled. Utilizing a combination of quercetin and vitamin C in a multidrug approach could enhance the immune system's defense mechanisms. This can be achieved by impeding virus entry, replication, enzyme activities, and assembly. Additionally, it may stimulate early interferon production, control interleukin levels, and foster the development of T cells and phagocytic function (Biancatelli et al., 2020).

In addition to this, the anti-SARS-COVID-2 effects of rutin, when combined with acetylsalicylic acid, vitamin C, and vitamin D3, along with the supplementation of calcium and magnesium, were investigated by Mazik et al. (2022). The capacity of rutin to hinder different phases of the viral life cycle stands as a notable advantage. Rutin has also been associated with inhibiting key components of SARS-CoV-2, including the papain PL^{pro}, RdRp, and helicase, as potential inhibitors, as noted by Wu et al. (2020), Agrawal et al. (2021), and Rahman et al. (2021). Besides, vitamin C has demonstrated no apparent adverse effects while affecting various biological mechanisms linked to COVID-19 infections. The usage of acetylsalicylic acid was considered favorable for COVID-19 patients, as claimed by Merzon et al. (2021). Empirical studies indicate that individuals who had been prescribed a low dose of acetylsalicylic acid for cardiometabolic issues before contracting SARS-CoV-2 exhibited a considerably less severe course of COVID-19 (Chow et al., 2021; Liu et al., 2021). Various investigations have offered significant proof of an association between diminished vitamin D levels and increased severity of COVID-19. These studies have suggested that supplementing with vitamin D3 might enhance the progression of SARS-CoV-2 infections, as highlighted by Borsche et al. (2021), Liu et al. (2021), and Dror et al. (2022). Given these findings, there is potential to combine various therapeutic components such as ascorbic acid and acetylsalicylic acid. This approach offers the opportunity to mitigate the progression of a severe illness and may be of interest for both medical treatment and prophylactic purposes. Therefore, this strategy must be thoroughly validated through well-designed clinical investigations.

In addition to being used as medications, antibiotics are also used in antigen and immunoglobulin testing for SARS-CoV-2 identification. Casirivimab and imdevimab, two monoclonal antibodies, have received emergency use authorization for combating COVID-19. In spite of the fact that flavonols and other compounds have shown promise in preclinical and laboratory studies, combining them with monoclonal antibodies may increase their effectiveness by targeting various disease-related aspects (Hwang et al., 2022). There is a clear justification to shift the focus to efficient combination medicines because there is no gold standard treatment and numerous repurposed monotherapy strategies have failed to provide a meaningful advantage. Future randomized controlled studies concentrating on safe and efficient combinations may improve overall therapy efficacy against COVID-19.

4.4 MODULATION OF INFLAMMATORY PATHWAYS BY FLAVONOLS

The pathways influencing inflammation include the regulation of the NLRP3 inflammasome, TLRs or BRD4, the stimulation of Nrf2, or the impacts on ACE2. TLRs or NLRP3 inflammasomes are thought to mediate the antiviral and immunomodulatory effects of flavonols (Figure 4.2) (Liskova et al., 2021). The NLRP3 inflammasome is a cluster of several proteins found within the cell's cytoplasm, and its formation can be prompted by a range of internal or external warning signals. Excessive activation of the NLRP3 inflammasome has been associated with the onset of various inflammatory ailments in humans, suggesting that it may offer a potential avenue for treating these diseases (Jiang et al., 2020). One fundamental element of the inflammatory reaction involves the regulation of the NLRP3 inflammasome, which initiates the conversion of pro-IL-1β and pro-IL-18 into active cytokines. This process is initiated in response to AngII stimulation (Ratajczak and Kucia, 2020).

Toll-like receptors (TLRs) have a significant function in identifying viral particles and initiating the innate immune system's response. The activation of TLRs is another inflammatory pathway, which leads to the release of pro-inflammatory cytokines (Khanmohammadi and Rezaei, 2021). Activation of TLR in macrophages and monocytes leads to the production of IL-6. TLRs are recognized as crucial stimulators of IL-6 that have a pivotal function in stimulating T-cells and have the ability to influence the function of Th17 cells, turning them into pro-inflammatory self-reactive T-cells. Furthermore, IL-6 can stimulate the generation of acute-phase proteins like CRP (Roshanravan et al., 2020).

Another pathway influenced by flavonols is the modulation and activation of BRD4. NF-κB recruits BRD4, resulting in the initiation of proinflammatory signaling mediated by NF-κB. However, BRD4 inhibitors reduce macrophage recruitment as well as T-cell infiltration. Nrf2's function is linked to the regulation of inflammation, affecting both its initiation and resolution by suppressing proinflammatory signals like IL-6 and IL-1β (Cuadrado et al., 2020). Although ACE2 plays a critical role in viral entry, it paradoxically offers protective effects by converting AngII into Ang1–7 (Behl et al., 2020). As a result of ACE2 cleavage, Ang1–7 binds to the Mas receptor, thereby decreasing the generation of anti-inflammatory cytokines (Mery et al., 2020). The spike protein of SARS-CoV-2 attaches to ACE2 and downregulates it via intracellular binding, resulting in a rise in AngII concentration and the stimulation of the AngII/AT1R axis. This increase in AngII production and AT1R activation is associated with pro-inflammatory responses (Liskova et al., 2021).

4.5 IMMUNOMODULATORY AND ANTI-INFLAMMATORY EFFECTS OF FLAVONOLS

A potentially fatal complication linked with acute respiratory distress syndrome (ARDS) due to cytokine storm has been seen in fierce cases of COVID-19

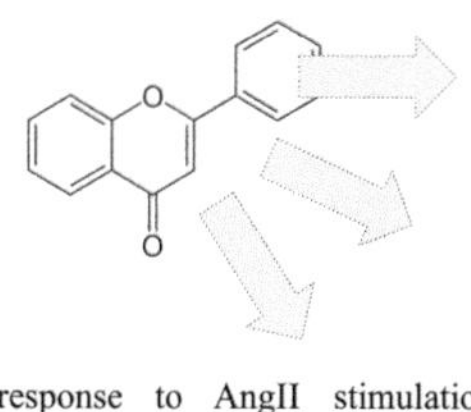

1 In response to AngII stimulation, activation of the NLRP3 inflammasome occurs. The NLRP3 inflammasome triggers the cleavage and release of pro-IL-1β and pro-IL-18 into biologically active cytokines

1) Modulation of NLPR3 inflammasome

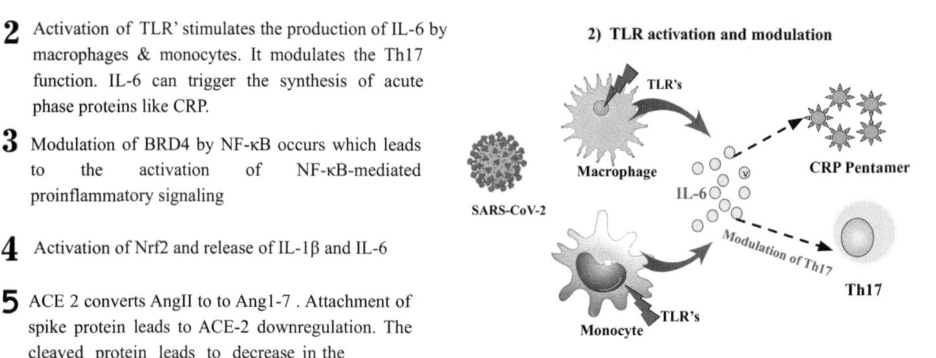

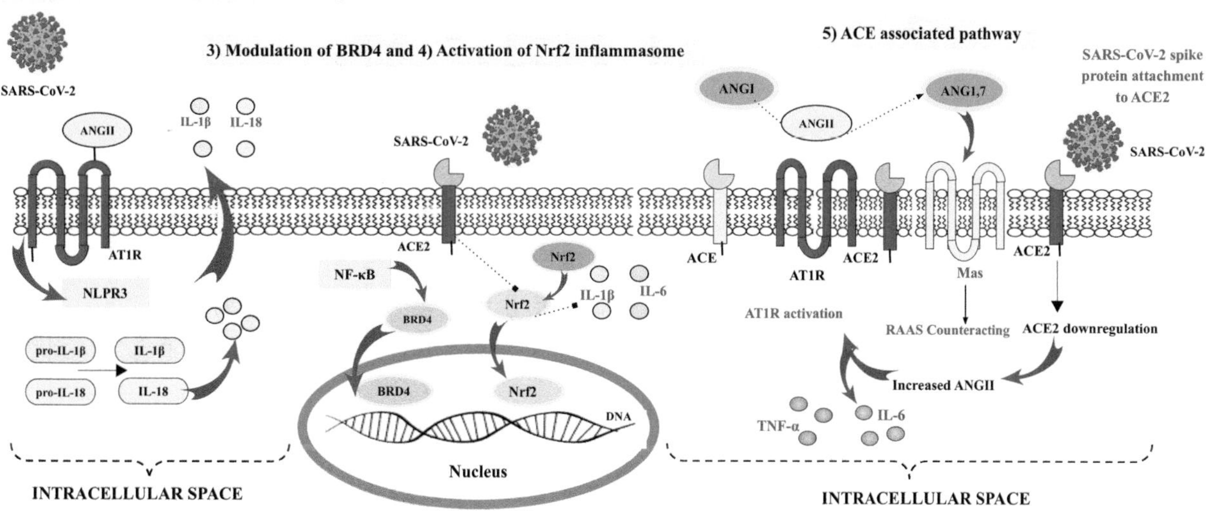

FIGURE 4.2 Different inflammatory pathways associated with the effect of flavonoids targeted on SARS-CoV-2. This figure is adapted from Liskova et al. (2021) and is an open-access article (copyright © 2021 the authors) distributed under the terms and conditions of the Creative Commons Attribution (CC BY) license.

(Mahmudpour et al., 2020). Approximately 33% of individuals affected by COVID-19 who are hospitalized may be affected by these cases, with a mortality rate of 40% (Tzotzos et al., 2020). In severe cases of COVID-19, patients display elevated levels of pro-inflammatory cytokines in their bloodstream compared to individuals with milder manifestations of the illness (Qin et al., 2020). It has also been demonstrated that PAK1 (RAC/CDC42-activated kinase) plays an important role in the surge of cytokines release, referred to as cytokine storm. This hyperactive immune response results in the death of hospitalized patients who have been infested with SARS-CoV-2. Various clinical trials of synthetic medicines have been done in patients that target the specific mechanisms associated with this immune procedure, including ciclesonide, hydroxychloroquine, ivermectin, and ketorolac, which are inhibitors of PAK1 (Maruta and He, 2020). Additionally, herbal immunomodulators are also used, which are capable of stimulating or inhibiting the immune's system innate and adaptive responses. Certain plants, with various components, have the ability to regulate the immune system, helping to shield the body from viral infections by strengthening its defenses instead of directly targeting the infection itself. Numerous phytochemicals are present in these plants having immunomodulatory properties. These compounds include anthraquinones, flavonoids, terpenoids, alkaloids, polyphenols, polysaccharides, proteins, fatty acids, and compounds containing sulfur. Flavonoids and their glycosides have been documented to have the most prominent compounds among all (Table 4.2). Flavonoids like quercetin, naringenin, and naringin were shown to induce metabolic shifts in cultured human macrophages that ran counter to the proinflammatory metabolic changes influenced by lipopolysaccharide (LPS) and IFN-γ stimulation (Mendes et al., 2019). Moreover, flavonoids regulate the functions of immune cells by enhancing the efficiency of NK cells and cytotoxic T lymphocytes. They also govern macrophage activity by influencing lysosomal functions and releasing nitric oxide (Sassi et al., 2017). Hesperidin exerted potent immune-modulating properties in rats subjected to rigorous training and exhaustive exercise, demonstrating enhanced NK cell cytotoxicity, an increased proportion of phagocytic monocytes, a modulation of macrophage cytokine secretion, and an increase in T helper cells (Ruiz-Iglesias et al., 2020). A recent investigation conducted by Bellavite and Donzelli (2020) focused primarily on the nutraceutical properties of citrus fruits and

TABLE 4.2
Inflammatory and Immunomodulatory Effects of Flavonols

Flavonols	Type of Study	Samples	Study Design	Effects	References
Myricetin	*In vivo*	Primary peritoneal macrophages from Male mice (6–8 weeks old) of a C57BL/6 background	Treatment with nigericin (10 μM) and myricetin (75 μM) for 35 min	Preventing the formation of the NLRP3 inflammasome	Chen et al. (2019)
Apigenin	*In vivo*	MSU-induced peritonitis in mice	Oral administration of apigenin (100 mg/kg)	Decreased the number of neutrophils and monocytes Inhibition of IL-1β	Lim et al. (2018)
Naringenin	*In vitro*	Human A549 cells were subjected to ZIKV infection, while ZIKV infection was induced in monocyte-derived dendritic cells	The dosage used was 15.6, 31.25, 62.5, and 125 μM	Inhibit the multiplication of viruses or the formation of virus particles	Cataneo et al. (2019)
Flavone -c- glycoside	*In vitro*	Isolated neutrophils from blood samples of humans	Treated with pure methanolic extract of *Lophatherum gracile*	Reduce the occurrence of respiratory bursts and the production of neutrophils extracellular traps in activated human neutrophils. Inhibit COVID-19 infection through ACE2 binding	Chen et al. (2023)
Rosmarinic acid and Luteolin	*In vitro*	Human lung epithelial cell line (type II pneumocytes), A549 cells (CCL-185™)	Rosmarinic acid and Luteolin (0–20 μg/mL) were extracted from the treatment for 24 hours	The exposure of cells to Spike S1 showed dose-dependent inflammation resulting in decreases in the gene expression associated with IL-6, IL-1β, IL-18, and NLRP3 Reductions in the release of cytokines Luteolin exhibits better inflammatory properties in comparison to Rosmarinic acid	Dissook et al. (2023)
Multiple compounds	*In vivo*	Mouse angiotensin-converting enzyme (ACE2) microinjected female hACE2 mice Calu-3 cells	Oral administration of *Huashi baidu* extract at 4 mg/kg for 5 consecutive days	Licochalcone B, glycyrrhisoflavone, and echinatin demonstrated reduced SARS-CoV-2-induced IL-1β P17 release in a dose-dependent manner	Xu et al. (2023)
Naringin	*In vitro* *In vivo*	RAW 264.7 macrophage cell line Myocardial ischemia/reperfusion injury model of Sprague–Dawley rats	Treated with different concentrations (10, 20, and 40 μg/mL) for 18 hours. Treated with naringin at 4 mg/kg body weight (BW) 10 minutes before LAD occlusion	Naringin reduced the production of cytokines (COX 2, iNOS, IL-1β, and IL-6) that are induced by LPS in macrophage cell lines Limit cytokine activity by inhibiting HMGB1 expression in a mouse model	Liu et al. (2022)

showed that hesperidin, a flavonoid component of citrus fruits, could act as a potential antiviral, antioxidant, and inflammation-modulating substance against SARS-CoV-2. Hesperidin improved the disrupted levels of inflammatory mediators in rats experiencing renal damage caused by ischemia/reperfusion (Meng et al., 2020). Moreover, in mice subjected to simultaneous treatment with hesperidin and LPS, hesperidin prompted suppressing inflammatory reactions, resulting in decreased levels of IL-33 and TNF-α (Al-Rakabi et al., 2020). Nonetheless, it is imperative that future preclinical, epidemiological, and clinical studies verify the hypothesized role of citrus flavonoid hesperidin in the prevention of COVID-19 (Bellavite and Donzelli, 2020). Studies have demonstrated that kaempferol, a compound present in propolis, can reduce the levels of proinflammatory mediators, including IL-6, TNF-alpha, and VEGF (vascular endothelial growth factor), through an ERK-NFkB-cMyc-p21 pathway (Machado et al., 2012; Da et al., 2019). As a result, increased levels of IL-10, a regulatory cytokine, were observed. Based on docking studies, it appeared that kaempferol inhibited SARS-CoV-2's 3A ion channel (Ren et al., 2019). Lutein inhibited inflammation induced by IL-1β in chondrocytes of rats (Fei et al., 2019). A similar outcome was observed in a study involving apigetrin, which is a glucoside form of apigenin, in mice suffering from acute otitis media (Hadrich and Sayadi, 2018; Guo et al., 2019). The water-based extract from Smilax campestris, mainly consisting of catechin and quercetin glycosylated derivatives, exhibited a decrease in the secretion of proinflammatory cytokines, particularly TNF-α, IL-1β, IL-6, IL-8, and MCP-1, in LPS-stimulated macrophages derived from the THP-1 monocytic cell line (Salaverry et al., 2020). Additionally, apigenin was found to mitigate inflammation by lowering the levels of plasma IL-6, TNF-α, and interferon-γ (IFN-γ) *in vivo* (Jung et al., 2020).

Quercetin notably inhibits the progression of ALI in a mouse model infected with IAV (Liu et al., 2016). It also curtails the activation of the NLRP3 inflammasome and the secretion of IL-1a and IL-18 (Jiang et al., 2016). Furthermore, quercetin pre-treatment hinders the IL-6 release induced by LPS in neutrophils and dendritic cells, thereby enhancing IL-6-induced STAT3 signaling (Manjunath and Thimmulappa, 2022). Elevated concentrations of mucin proteins MUC1 and MUC5AC have been seen in COVID patients, but the administration of quercetin has been shown to reduce mucin synthesis and decrease mucus production, thus potentially alleviating breathing difficulties associated with the disease (Lu et al., 2021). Quercetin's block of tyrosine phosphorylation in both the EGFR and NF-κB pathways led to a decrease in mucin synthesis. Additionally, in human airway epithelial cells NCI-H292, the presence of quercetin resulted in a reduction of MUC5AC expression. Consequently, this reduced mucus production and alleviated respiratory challenges (Yang et al., 2012). The co-administration of quercetin and ascorbic acid was also recommended for preventing and treating respiratory tract infections (Biancatelli et al., 2020). Furthermore, quercetin has shown its ability to block the ACE2 receptor of SARS-CoV-2, as confirmed by a FRET assay utilizing recombinant ACE2 and Mca-APK as substrates (Rahman et al., 2021). Quercetin has also been revealed to be an immunomodulatory agent that can be used to treat and prevent COVID-19. Quercetin's ability to modulate the immune response may be beneficial in mitigating the excessive immune activation and inflammation evident in critical cases of COVID-19. Furthermore, quercetin has been found to synergize with vitamin C, another popular immune-boosting nutrient. This combination has shown promise in both prophylaxis and therapy, particularly in high-risk populations. Quercetin has shown promising results in the treatment of COVID-19. According to Di Pierro et al. (2023), quercetin consumption resulted in faster recovery from COVID-19 than the control group. Furthermore, within the treatment group, quercetin reduced the duration of SARS-CoV-2 presence by 68%, whereas in the control group, it was reduced by 24%.

Genistein also inhibits the signal transduction pathways activated by cytokines in cells of the immune system. It exerts immunoregulatory effects by suppressing the generation of cytokines and chemokines. In addition to this, apigenin has also been shown to have immunity-modulating properties both *in vitro* and *in vivo*, inhibiting IL-6, IL-10, INF-1, and STAT-3 and stimulating TNF-1. In addition, it impedes the function of Th1 (T-helper 1) and Th17 cells, and it also disrupts the NF-κB activation pathway. Additionally, it obstructs the translocation of NF-κB into the cell nucleus and interferes with the phosphorylation and degradation of I-κBα when reactivated peripheral blood CD4 T-cells and leukemic Jurkat T-cell lines are stimulated via the TCR. This subsequently results in a decrease in eosinophil count (Akter et al., 2023). Given these findings, it can be conferred that flavonols might possess anti-inflammatory properties against COVID-19.

A flavone, scutellarein, gives Mexican oregano and sweet orange their characteristic bitter taste and is also widely distributed in various fern species, such as *Asplenium belangeri* and *Pseudolysimachion longifolium*. In LPS-stimulated RAW264.7 cells, scutellarein exhibited a significant decrease in nitric oxide (NO) generation and led to a decline in the mRNA expression levels of both inducible nitric oxide synthase (iNOS) and TNF-α (Sung et al., 2015). Additionally, it reduced the phosphorylation levels of NF-β activating enzymes upstream. It also inhibited the activity of Src kinase and Src autophosphorylation. Scientists also utilized a FRET assay to investigate the inhibitory impacts of scutellarein and its derivatives on M^{pro}. Their findings revealed that one of its analogs, 4′-O-methylscutellarein, exhibited a significantly lower 50% inhibition concentration (IC50 value of 0.40±0.03 μM) compared to Scutellarein, which achieved 50% inhibition at 5.68±0.48 μM (Sassi et al., 2017). Moreover, it was noted that the ATPase activity of SARS-CoV NSP13 was hindered by 4′-O-methylscutellarein, as determined by a colorimetric-based ATP hydrolysis assay (Montone et al., 2021).

Nonetheless, it is essential to carry out *in vitro* and clinical investigations to ascertain the function of Scutellarein in COVID-19 infection.

4.6 CONCLUSION

COVID-19 has caused a substantial loss of lives and various forms of harm, making it imperative that we develop treatments that can prevent or reduce the viral infection. The flavonols are capable of inhibiting several critical viral targets, which facilitate the entry and replication of the virus, for example, Mpro, RBD of the S protein, RdRp, as well as the human ACE2 receptor and TMPRSS2. Moreover, these substances have been demonstrated to suppress the synthesis of numerous pro-inflammatory cytokines and pathways associated with inflammation, thus exhibiting anti-inflammatory and immunoregulatory properties. Despite the promising results of studies conducted thus far using flavonoids against COVID-19, the literature on this topic remains incomplete, and there is insufficient evidence to support its application to COVID-19 patients. Therefore, precise and detailed research studies must be conducted to determine the exact mechanisms underlying the action of flavonoids as well as to identify the population eligible to receive such treatment and to determine whether the compounds are bioavailable or if they can be enhanced in efficacy as single compounds or when combined with other agents targeting the COVID-19 virus. The effectiveness of flavonol-rich foods and nutraceuticals in preventing and treating viral infection requires further investigation.

DECLARATION OF COMPETING INTEREST

The authors declare that they have no known competing financial interests or personal relationships.

REFERENCES

Abian, O., Ortega-Alarcon, D., Jimenez-Alesanco, A., Ceballos-Laita, L., Vega, S., Reyburn, H. T., Rizzuti, B., & Velazquez-Campoy, A. (2020). Structural stability of SARS-CoV-2 3CLpro and identification of quercetin as an inhibitor by experimental screening. *International Journal of Biological Macromolecules*, *164*, 1693–1703.

Adegbola, P. I., Fadahunsi, O. S., Ogunjinmi, O. E., Adegbola, A. E., Ojeniyi, F. D., Adesanya, A., Olagoke, E., Adisa, A. D., Ehigie, A.F., Adetutu, A., & Semire, B. (2023). Potential inhibitory properties of structurally modified quercetin/isohamnetin glucosides against SARS-CoV-2 Mpro; molecular docking and dynamics simulation strategies. *Informatics in Medicine Unlocked*, *37*, 101167.

Agrawal, P. K., Agrawal, C., & Blunden, G. (2020). Quercetin: Antiviral significance and possible COVID-19 integrative considerations. *Natural Product Communications*, *15*(12), 1934578X20976293.

Agrawal, P. K., Agrawal, C., & Blunden, G. (2021). Rutin: A potential antiviral for repurposing as a SARS-CoV-2 main protease (Mpro) inhibitor. *Natural Product Communications*, *16*(4), 1934578X21991723.

Akter, R., Rahman, M. R., Ahmed, Z. S., & Afrose, A. (2023). Plausibility of natural immunomodulators in the treatment of COVID-19-A comprehensive analysis and future recommendations. *Heliyon*, 9, e17478.

Al-Rikabi, R., Al-Shmgani, H., Dewir, Y. H., & El-Hendawy, S. (2020). In vivo and in vitro evaluation of the protective effects of hesperidin in lipopolysaccharide-induced inflammation and cytotoxicity of cell. *Molecules*, *25*(3), 478.

Amin, S. A., & Jha, T. (2020). Fight against novel coronavirus: A perspective of medicinal chemists. *European Journal of Medicinal Chemistry*, *201*, 112559.

Bahun, M., Jukić, M., Oblak, D., Kranjc, L., Bajc, G., Butala, M., & Ulrih, N. P. (2022). Inhibition of the SARS-CoV-2 3CLpro main protease by plant polyphenols. *Food Chemistry*, *373*, 131594.

Barreca, D., Trombetta, D., Smeriglio, A., Mandalari, G., Romeo, O., Felice, M. R., & Nabavi, S. M. (2021). Food flavonols: Nutraceuticals with complex health benefits and functionalities. *Trends in Food Science & Technology*, *117*, 194–204.

Behl, T., Kaur, I., Bungau, S., Kumar, A., Uddin, M. S., Kumar, C., Pal, G., Sahil, Shrivastava, K., Zengin, G., & Arora, S. (2020). The dual impact of ACE2 in COVID-19 and ironical actions in geriatrics and pediatrics with possible therapeutic solutions. *Life Sciences*, 257, 118075.

Bellavite, P., & Donzelli, A. (2020). Hesperidin and SARS-CoV-2: New light on the healthy function of citrus fruits. *Antioxidants*, *9*(8), 742.

Biancatelli, R. M. L.C., Berrill, M., Catravas, J. D., & Marik, P. E. (2020). Quercetin and vitamin C: An experimental, synergistic therapy for the prevention and treatment of SARS-CoV-2 related disease (COVID–19). *Frontiers in Immunology*, *11*, 1451.

Borsche, L., Glauner, B., & von Mendel, J. (2021). COVID-19 mortality risk correlates inversely with vitamin D3 status, and a mortality rate close to zero could theoretically be achieved at 50 ng/mL 25 (OH) D3: Results of a systematic review and meta-analysis. *Nutrients*, *13*(10), 3596.

Cataneo, A. H. D., Kuczera, D., Koishi, A. C., Zanluca, C., Silveira, G. F., Arruda, T. B. D., & Bordignon, J. (2019). The citrus flavonoid naringenin impairs the in vitro infection of human cells by Zika virus. *Scientific Reports*, *9*(1), 16348.

Chen, H., Lin, H., Xie, S., Huang, B., Qian, Y., Chen, K., Niu, Y., Shen, H. M., Cai, J., Li, P., & Wu, Y. (2019). Myricetin inhibits NLRP3 inflammasome activation via reduction of ROS-dependent ubiquitination of ASC and promotion of ROS-independent NLRP3 ubiquitination. *Toxicology and Applied Pharmacology*, *365*, 19–29.

Chen, Y. L., Chen, C. Y., Lai, K. H., Chang, Y. C., & Hwang, T. L. (2023). Anti-inflammatory and antiviral activities of flavone C-glycosides of Lophatherum gracile for COVID-19. *Journal of Functional Foods*, *101*, 105407.

Chow, J. H., Khanna, A. K., Kethireddy, S., Yamane, D., Levine, A., Jackson, A. M., & Mazzeffi, M. A. (2021). Aspirin use is associated with decreased mechanical ventilation, intensive care unit admission, and in-hospital mortality in hospitalized patients with coronavirus disease 2019. *Anesthesia & Analgesia*, *132*(4), 930–941.

Cuadrado, A., Pajares, M., Benito, C., Jiménez-Villegas, J., Escoll, M., Fernández-Ginés, R., yague, A. J., Lastra, D., Manda, G., Rojo, A. I., & Dinkova-Kostova, A. T. (2020). Can activation of NRF2 be a strategy against COVID-19?. *Trends in Pharmacological Sciences*, *41*(9), 598–610.

Da, J., Xu, M., Wang, Y., Li, W., Lu, M., & Wang, Z. (2019). Kaempferol promotes apoptosis while inhibiting cell proliferation via androgen-dependent pathway and suppressing vasculogenic mimicry and invasion in prostate cancer. *Analytical Cellular Pathology*, *2019*, 1907698.

da Silva, F. M. A., da Silva, K. P. A., de Oliveira, L. P. M., Costa, E. V., Koolen, H. H., Pinheiro, M. L. B., & de Souza, A. D. L. (2020). Flavonoid glycosides and their putative human metabolites as potential inhibitors of the SARS-CoV-2 main protease (Mpro) and RNA-dependent RNA polymerase (RdRp). *Memórias do Instituto Oswaldo Cruz*, *115*, e200207.

Di Pierro, F., Khan, A., Iqtadar, S., Mumtaz, S. U., Chaudhry, M. N. A., Bertuccioli, A., & Zerbinati, N. (2023). Quercetin as a possible complementary agent for early-stage COVID-19: Concluding results of a randomized clinical trial. *Frontiers in Pharmacology*, *13*, 1096853.

Dissook, S., Umsumarng, S., Mapoung, S., Semmarath, W., Arjsri, P., Srisawad, K., & Dejkriengkraikul, P. (2023). Luteolin-rich fraction from Perilla frutescens seed meal inhibits spike glycoprotein S1 of SARS-CoV-2-induced NLRP3 inflammasome lung cell inflammation via regulation of JAK1/STAT3 pathway: A potential anti-inflammatory compound against inflammation-induced long-COVID. *Frontiers in Medicine*, *9*, 1072056.

Dror, A. A., Morozov, N., Daoud, A., Namir, Y., Yakir, O., Shachar, Y., Lifshitz, M., Segal, E., Fisher, L., Mizrachi, M., & Sela, E. (2022). Pre-infection 25-hydroxyvitamin D3 levels and association with severity of COVID-19 illness. *PLoS One*, *17*(2), e0263069.

Du, A., Zheng, R., Disoma, C., Li, S., Chen, Z., Li, S., Liu, P., Zhou, Y., Shen, Y., Liu, S., & Xia, Z. (2021). Epigallocatechin-3-gallate, an active ingredient of traditional Chinese medicines, inhibits the 3CLpro activity of SARS-CoV-2. *International Journal of Biological Macromolecules*, *176*, 1–12.

Fei, J., Liang, B., Jiang, C., Ni, H., & Wang, L. (2019). Luteolin inhibits IL-1β-induced inflammation in rat chondrocytes and attenuates osteoarthritis progression in a rat model. *Biomedicine & Pharmacotherapy*, *109*, 1586–1592.

Fink, E. A., Bardine, C., Gahbauer, S., Singh, I., White, K., Gu, S., Wan, X., Ary, B., Glenn, I., O'Connell, J., et al. (2022). Large library docking for novel SARS-CoV-2 main protease noncovalent and covalent inhibitors. *Protein Science*, *32*, e4712. https://doi.org/10.1101/2022.07.05.498881.

Gao, J., Cao, C., Shi, M., Hong, S., Guo, S., Li, J., Liang, T., Song, P., Xu, R., & Li, N. (2023). Kaempferol inhibits SARS-CoV-2 invasion by impairing heptad repeats-mediated viral fusion. *Phytomedicine*, *118*, 154942.

Gasmi, A., Mujawdiya, P. K., Lysiuk, R., Shanaida, M., Peana, M., Gasmi Benahmed, A., & Bjørklund, G. (2022). Quercetin in the prevention and treatment of coronavirus infections: A focus on SARS-CoV-2. *Pharmaceuticals*, *15*(9), 1049.

Gorla, U. S., Rao, K., Kulandaivelu, U. S., Alavala, R. R., & Panda, S. P. (2021). Lead finding from selected flavonoids with antiviral (SARS-CoV-2) potentials against COVID-19: An in-silico evaluation. *Combinatorial Chemistry & High Throughput Screening*, *24*(6), 879–890.

Gu, Y. Y., Zhang, M., Cen, H., Wu, Y. F., Lu, Z., Lu, F., Liu, X. S., & Lan, H. Y. (2021). Quercetin as a potential treatment for COVID-19-induced acute kidney injury: Based on network pharmacology and molecular docking study. *PloS One*, *16*(1), e0245209.

Guo, H., Li, M., & Xu, L. J. (2019). Apigetrin treatment attenuates LPS-induced acute otitis media though suppressing inflammation and oxidative stress. *Biomedicine & Pharmacotherapy*, *109*, 1978–1987.

Hadni, H., Fitri, A., Benjelloun, A. T., Benzakour, M., & Mcharfi, M. (2022). Evaluation of flavonoids as potential inhibitors of the SARS-CoV-2 main protease and spike RBD: Molecular docking, ADMET evaluation and molecular dynamics simulations. *Journal of the Indian Chemical Society*, *99*(10), 100697.

Hadrich, F., & Sayadi, S. (2018). Apigetrin inhibits adipogenesis in 3T3-L1 cells by downregulating PPARγ and CEBP-α. *Lipids in Health and Disease*, *17*, 1–8.

Hiremath, S., Kumar, H. V., Nandan, M., Mantesh, M., Shankarappa, K. S., Venkataravanappa, V., Basha, C. J., & Reddy, C. L. (2021). In silico docking analysis revealed the potential of phytochemicals present in Phyllanthus amarus and Andrographis paniculata, used in Ayurveda medicine in inhibiting SARS-CoV-2. *3 Biotech*, *11*, 1–18.

Houchi, S., & Messasma, Z. (2022). Exploring the inhibitory potential of Saussurea costus and Saussurea involucrata phytoconstituents against the Spike glycoprotein receptor binding domain of SARS-CoV-2 Delta (B. 1.617. 2) variant and the main protease (Mpro) as therapeutic candidates, using Molecular docking, DFT, and ADME/Tox studies. *Journal of Molecular Structure*, *1263*, 133032.

Hwang, Y. C., Lu, R. M., Su, S. C., Chiang, P. Y., Ko, S. H., Ke, F. Y., & Wu, H. C. (2022). Monoclonal antibodies for COVID-19 therapy and SARS-CoV-2 detection. *Journal of Biomedical Science*, *29*(1), 1–50.

Imran, M., Iqbal, S., Hussain, A., Uddin, J., Shahzad, M., Khaliq, T., Ahmad, A. R., Mushtaq, L., Kashif, M., & Mahmood, K. (2022). In silico screening, SAR and kinetic studies of naturally occurring flavonoids against SARS CoV-2 main protease. *Arabian Journal of Chemistry*, *15*(1), 103473.

Jiang, H., Gong, T., & Zhou, R. (2020). The strategies of targeting the NLRP3 inflammasome to treat inflammatory diseases. *Advances in Immunology*, *145*, 55–93.

Jiang, W., Huang, Y., Han, N., He, F., Li, M., Bian, Z., & Zhu, L. (2016). Quercetin suppresses NLRP3 inflammasome activation and attenuates histopathology in a rat model of spinal cord injury. *Spinal Cord*, *54*(8), 592–596.

Jo, S., Kim, S., Kim, D. Y., Kim, M. S., & Shin, D. H. (2020). Flavonoids with inhibitory activity against SARS-CoV-2 3CLpro. *Journal of Enzyme Inhibition and Medicinal Chemistry*, *35*(1), 1539–1544.

Jung, U. J., Cho, Y. Y., & Choi, M. S. (2016). Apigenin ameliorates dyslipidemia, hepatic steatosis and insulin resistance by modulating metabolic and transcriptional profiles in the liver of high-fat diet-induced obese mice. *Nutrients*, *8*(5), 305.

Kashyap, P., Thakur, M., Singh, N., Shikha, D., Kumar, S., Baniwal, P., & Inbaraj, B. S. (2022). In silico evaluation of natural flavonoids as a potential inhibitor of coronavirus disease. *Molecules*, *27*(19), 6374.

Khanmohammadi, S., & Rezaei, N. (2021). Role of Toll-like receptors in the pathogenesis of COVID-19. *Journal of Medical Virology*, 93(5), 2735–2739.

Kothari, D., Lee, W. D., & Kim, S. K. (2020). Allium flavonols: Health benefits, molecular targets, and bioavailability. *Antioxidants*, *9*(9), 888.

Lim, H., Min, D. S., Park, H., & Kim, H. P. (2018). Flavonoids interfere with NLRP3 inflammasome activation. *Toxicology and Applied Pharmacology*, *355*, 93–102.

Lim, J., Jeon, S., Shin, H. Y., Kim, M. J., Seong, Y. M., Lee, W. J., Choe, K. W., Kang, Y. M., Lee, B., & Park, S. J. (2020). Case of the index patient who caused tertiary transmission of coronavirus disease 2019 in Korea: The application of lopinavir/ritonavir for the treatment of COVID-19 pneumonia monitored by quantitative RT-PCR. *Journal of Korean Medical Science*, *35*(6), e79.

Lin, B., Cheng, L., Zhang, J., Yang, M., Zhang, Y., Liu, J., & Qin, X. (2023). Immunology of SARS-CoV-2 infection and vaccination. *Clinica Chimica Acta*, *540*, 117390.

Liskova, A., Samec, M., Koklesova, L., Samuel, S. M., Zhai, K., Al-Ishaq, R. K., Abotaleb, M., Nosal, V., Kajo, K., & Kubatka, P. (2021). Flavonoids against the SARS-CoV-2 induced inflammatory storm. *Biomedicine & Pharmacotherapy*, *138*, 111430.

Liu, Z., Zhao, J., Li, W., Shen, L., Huang, S., Tang, J., Duan, J., Fang, F., Huang, Y., Chang, H., & Zhang, R. (2016). Computational screen and experimental validation of anti-influenza effects of quercetin and chlorogenic acid from traditional Chinese medicine. *Scientific Reports*, *6*(1), 19095.

Liu, N., Sun, J., Wang, X., Zhang, T., Zhao, M., & Li, H. (2021). Low vitamin D status is associated with coronavirus disease 2019 outcomes: A systematic review and meta-analysis. *International Journal of Infectious Diseases*, *104*, 58–64.

Liu, Q., Huang, N., Li, A., Zhou, Y., Liang, L., Song, X., Yang, Z., & Zhou, X. (2021). Effect of low-dose aspirin on mortality and viral duration of the hospitalized adults with COVID-19. *Medicine*, *100*(6), e24544.

Liu, W., Zheng, W., Cheng, L., Li, M., Huang, J., Bao, S., Xu, Q., & Ma, Z. (2022). Citrus fruits are rich in flavonoids for immunoregulation and potential targeting ACE2. *Natural Products and Bioprospecting*, *12*(1), 4.

Lu, W., Liu, X., Wang, T., Liu, F., Zhu, A., Lin, Y., Luo, J., Ye, F., He, J., & Zhong, N. (2021). Elevated MUC1 and MUC5AC mucin protein levels in airway mucus of critical ill COVID-19 patients. *Journal of Medical Virology*, *93*(2), 582.

Machado, J. L., Assunçao, A. K. M., da Silva, M. C. P., Reis, A. S. D., Costa, G. C., Arruda, D. D. S., & Nascimento, F. R. F. D. (2012). Brazilian green propolis: anti-inflammatory property by an immunomodulatory activity. *Evidence-Based Complementary and Alternative Medicine*, *2012*, 157652.

Mahmudpour, M., Roozbeh, J., Keshavarz, M., Farrokhi, S., & Nabipour, I. (2020). COVID-19 cytokine storm: The anger of inflammation. *Cytokine*, *133*, 155151.

Manjunath, S. H., & Thimmulappa, R. K. (2022). Antiviral, immunomodulatory, and anticoagulant effects of quercetin and its derivatives: Potential role in prevention and management of COVID-19. *Journal of Pharmaceutical Analysis*, *12*(1), 29–34.

Maruta, H., & He, H. (2020). PAK1-blockers: Potential therapeutics against COVID-19. *Medicine in Drug Discovery*, *6*, 100039.

Mazik, M. (2022). Promising therapeutic approach for SARS-CoV-2 infections by using a rutin-based combination therapy. *ChemMedChem*, *17*(11), e202200157.

Mendes, L. F., Gaspar, V. M., Conde, T. A., Mano, J. F., & Duarte, I. F. (2019). Flavonoid-mediated immunomodulation of human macrophages involves key metabolites and metabolic pathways. *Scientific Reports*, *9*(1), 14906.

Meng, X., Wei, M., Wang, D., Qu, X., Zhang, K., Zhang, N., & Li, X. (2020). The protective effect of hesperidin against renal ischemia-reperfusion injury involves the TLR-4/NF-κB/iNOS pathway in rats. *Physiology International*, *107*(1), 82–91.

Mery, G., Epaulard, O., Borel, A.-L., & Toussaint, B. (2020). COVID-19: Underlying adipokine storm and angiotensin 1–7 umbrella. *Frontiers in* Immunology, *11*, 1714.

Merzon, E., Green, I., Vinker, S., Golan-Cohen, A., Gorohovski, A., Avramovich, E., & Magen, E. (2021). The use of aspirin for primary prevention of cardiovascular disease is associated with a lower likelihood of COVID-19 infection. *The FEBS Journal*, *288*(17), 5179–5189.

Montone, C. M., Aita, S. E., Arnoldi, A., Capriotti, A. L., Cavaliere, C., Cerrato, A., Lammi, C., Pioversana, S., Ranaldi, G., & Laganà, A. (2021). Characterization of the trans-epithelial transport of green tea (C. sinensis) catechin extracts with in vitro inhibitory effect against the SARS-CoV-2 papain-like protease activity. *Molecules*, *26*(21), 6744.

Mouffouk, C., Mouffouk, S., Mouffouk, S., Hambaba, L., & Haba, H. (2021). Flavonols as potential antiviral drugs targeting SARS-CoV-2 proteases (3CLpro and PLpro), spike protein, RNA-dependent RNA polymerase (RdRp) and angiotensin-converting enzyme II receptor (ACE2). *European Journal of Pharmacology*, *891*, 173759.

Nakanishi, I., Ohkubo, K., Shoji, Y., Fujitaka, Y., Shimoda, K., Matsumoto, K. I., Mukuhara, K., & Hamada, H. (2020). Relationship between the radical-scavenging activity of selected flavonols and thermodynamic parameters calculated by density functional theory. *Free Radical Research*, *54*(7), 535–539.

Nguyen, T. T. H., Jung, J. H., Kim, M. K., Lim, S., Choi, J. M., Chung, B., Kim, D. W., & Kim, D. (2021). The inhibitory effects of plant derivate polyphenols on the main protease of SARS coronavirus 2 and their structure-activity relationship. *Molecules*, *26*(7), 1924.

Owis, A. I., El-Hawary, M. S., El Amir, D., Aly, O. M., Abdelmohsen, U. R., & Kamel, M. S. (2020). Molecular docking reveals the potential of Salvadora persica flavonoids to inhibit COVID-19 virus main protease. *RSC Advances*, *10*(33), 19570–19575.

Pal, A., Squitti, R., Picozza, M., Pawar, A., Rongioletti, M., Dutta, A. K., Sahoo, S., Goswami, K., & Prasad, R. (2021). Zinc and COVID-19: Basis of current clinical trials. *Biological Trace Element Research*, *199*, 2882–2892.

Pan, H., He, J., Yang, Z., Yao, X., Zhang, H., Li, R., Xiao, Y., Zhao, C., Jiang, H., Liu, Y., & Liu, L. (2023). Myricetin possesses the potency against SARS-CoV-2 infection through blocking viral-entry facilitators and suppressing inflammation in rats and mice. *Phytomedicine*, *116*, 154858.

Pandey, P., Rane, J. S., Chatterjee, A., Kumar, A., Khan, R., Prakash, A., & Ray, S. (2021a). Targeting SARS-CoV-2 spike protein of COVID-19 with naturally occurring phytochemicals: An in silico study for drug development. *Journal of Biomolecular Structure and Dynamics*, *39*(16), 6306–6316.

Pandey, P., Khan, F., Rana, A. K., Srivastava, Y., Jha, S. K., & Jha, N. K. (2021b). A drug repurposing approach towards elucidating the potential of flavonoids as COVID-19 spike protein inhibitors. *Biointerface Research in Applied Chemistry*, *11*(1), 8482–8501.

Pang, X., Xu, W., Liu, Y., Li, H., & Chen, L. (2023). The research progress of SARS-CoV-2 main protease inhibitors from 2020 to 2022. *European Journal of Medicinal Chemistry*, *257*, 115491.

Pillaiyar, T., Meenakshisundaram, S., & Manickam, M. (2020). Recent discovery and development of inhibitors targeting coronaviruses. *Drug Discovery Today*, *25*(4), 668–688.

Premkumar, L., Segovia-Chumbez, B., Jadi, R., Martinez, D. R., Raut, R., Markmann, A. J., Cornaby, C., Bartelt, L., Weiss, S., & de Silva, A. M. (2020). The receptor-binding domain of the viral spike protein is an immunodominant and highly specific target of antibodies in SARS-CoV-2 patients. *Science Immunology*, *5*(48), eabc8413.

Qin, C., Ziwei, M. P. L. Z. M., Tao, S. Y. M. Y., Ke, P. C. X. M. P., & Shang, M. M. P. K. (2020). Dysregulation of immune response in patients with COVID-19 in Wuhan, China; clinical infectious diseases; Oxford academic. *Clinical Infectious Diseases*. 71(15), 762–768.

Rahman, F., Tabrez, S., Ali, R., Alqahtani, A. S., Ahmed, M. Z., & Rub, A. (2021). Molecular docking analysis of rutin reveals possible inhibition of SARS-CoV-2 vital proteins. *Journal of Traditional and Complementary Medicine*, *11*(2), 173–179.

Ratajczak, M. Z., & Kucia, M.(2020). SARS-CoV-2 infection and overactivation of Nlrp3 inflammasome as a trigger of cytokine "storm" and risk factor for damage of hematopoietic stem cells. *Leukemia*, *34*, 1–4.

Ren, J., Lu, Y., Qian, Y., Chen, B., Wu, T., & Ji, G. (2019). Recent progress regarding kaempferol for the treatment of various diseases. *Experimental and Therapeutic Medicine*, *18*(4), 2759–2776.

Roshanravan, N., Seif, F., Ostadrahimi, A., Pouraghaei, M., & Ghaffari, S. (2020). Targeting cytokine storm to manage patients with COVID-19: A mini-review. *Archives of Medical Research*, *51*(7), 608–612.

Ruiz-Iglesias, P., Estruel-Amades, S., Camps-Bossacoma, M., Massot-Cladera, M., Franch, À., Pérez-Cano, F. J., & Castell, M. (2020). Influence of hesperidin on systemic immunity of rats following an intensive training and exhausting exercise. *Nutrients*, *12*(5), 1291.

Russo, M., Moccia, S., Spagnuolo, C., Tedesco, I., & Russo, G. L. (2020). Roles of flavonoids against coronavirus infection. *Chemico-Biological Interactions*, *328*, 109211.

Saakre, M., Mathew, D., & Ravisankar, V. (2021). Perspectives on plant flavonoid quercetin-based drugs for novel SARS-CoV-2. *Beni-Suef University Journal of Basic and Applied Sciences*, *10*(1), 1–13.

Salaverry, L. S., Parrado, A. C., Mangone, F. M., Dobrecky, C. B., Flor, S. A., Lombardo, T., & Rey-Roldán, E. B. (2020). In vitro anti-inflammatory properties of Smilax campestris aqueous extract in human macrophages, and characterization of its flavonoid profile. *Journal of Ethnopharmacology*, *247*, 112282.

Sassi, A., Bzéouich, I. M., Mustapha, N., Maatouk, M., Ghedira, K., & Chekir-Ghedira, L. (2017). Immunomodulatory potential of hesperetin and chrysin through the cellular and humoral response. *European Journal of Pharmacology*, *812*, 91–96.

Sharma, A., Sharma, P., Tuli, H. S., & Sharma, A. K. (2018). Phytochemical and pharmacological properties of flavonols. In: *eLS: Encyclopedia for Life Science*, John Wiley & Sons, Ltd, Chichester. https://doi.org/10.1002/9780470015902.a0027666.

Su, H., Yao, S., Zhao, W., Zhang, Y., Liu, J., Shao, Q., Wang, Q., Li, M., Xie, H., Shang, H., & Xu, Y. (2021). Identification of pyrogallol as a warhead in design of covalent inhibitors for the SARS-CoV-2 3CL protease. *Nature Communications*, *12*(1), 3623.

Sung, N. Y., Kim, M. Y., & Cho, J. Y. (2015). Scutellarein reduces inflammatory responses by inhibiting Src kinase activity. *The Korean Journal of Physiology & Pharmacology: Official Journal of the Korean Physiological Society and the Korean Society of Pharmacology*, *19*(5), 441.

Swain, S. S., Singh, S. R., Sahoo, A., Hussain, T., & Pati, S. (2022). Anti-HIV-drug and phyto-flavonoid combination against SARS-CoV-2: A molecular docking-simulation base assessment. *Journal of Biomolecular Structure and Dynamics*, *40*(14), 6463–6476.

Tabari, M. A.K., Iranpanah, A., Bahramsoltani, R., & Rahimi, R. (2021). Flavonoids as promising antiviral agents against SARS-CoV-2 infection: A mechanistic review. *Molecules*, *26*(13), 3900.

Tallei, T. E., Tumilaar, S. G., Niode, N. J., Kepel, B. J., Idroes, R., Effendi, Y., Sakib, S A., & Emran, T. B. (2020). Potential of plant bioactive compounds as SARS-CoV-2 main protease (M pro) and spike (S) glycoprotein inhibitors: A molecular docking study. *Scientifica*, *2020*, 6307457.

Topcagic, A., Zeljkovic, S. C., Karalija, E., Galijasevic, S., & Sofic, E. (2017). Evaluation of phenolic profile, enzyme inhibitory and antimicrobial activities of Nigella sativa L. seed extracts. *Bosnian Journal of Basic Medical Sciences*, *17*(4), 286.

Tzotzos, S. J., Fischer, B., Fischer, H., & Zeitlinger, M. (2020). Incidence of ARDS and outcomes in hospitalized patients with COVID-19: A global literature survey. *Critical Care*, *24*(1), 1–4.

ul Qamar, M. T., Alqahtani, S. M., Alamri, M. A., & Chen, L. L. (2020). Structural basis of SARS-CoV-2 3CLpro and anti-COVID-19 drug discovery from medicinal plants. *Journal of Pharmaceutical Analysis*, *10*(4), 313–319.

Ullah, A., Munir, S., Badshah, S. L., Khan, N., Ghani, L., Poulson, B. G., & Jaremko, M. (2020). Important flavonoids and their role as a therapeutic agent. *Molecules*, *25*(22), 5243.

Verma, D., Mitra, D., Paul, M., Chaudhary, P., Kamboj, A., Thatoi, H., Janmeda, P., Jain, D., & Mohapatra, P. K. D. (2021). Potential inhibitors of SARS-CoV-2 (COVID 19) proteases PLpro and Mpro/3CLpro: molecular docking and simulation studies of three pertinent medicinal plant natural components. *Current Research in Pharmacology and Drug Discovery*, *2*, 100038.

WHO (2023). Coronavirus Disease (COVID-19) Pandemic World Health Organization. https://www.who.int/emergencies/diseases/novel-coronavirus-2019

Wu, C., Liu, Y., Yang, Y., Zhang, P., Zhong, W., Wang, Y., Wang, Q., Xu, Y., Li, M., Li, X., & Li, H. (2020). Analysis of therapeutic targets for SARS-CoV-2 and discovery of potential drugs by computational methods. *Acta Pharmaceutica Sinica B*, *10*(5), 766–788.

Xiao, T., Wei, Y., Cui, M., Li, X., Ruan, H., Zhang, L., Bao, J., Ren, S., & Zhou, H. (2021). Effect of dihydromyricetin on SARS-CoV-2 viral replication and pulmonary inflammation and fibrosis. *Phytomedicine*, *91*, 153704.

Xu, H., Li, S., Liu, J., Cheng, J., Kang, L., Li, W., Zhong, Y., Wei, C., & Huang, L. (2023). Bioactive compounds from Huashi Baidu decoction possess both antiviral and anti-inflammatory effects against COVID-19. *Proceedings of the National Academy of Sciences*, *120*(18), e2301775120.

Yang, T., Luo, F., Shen, Y., An, J., Li, X., Liu, X., Ying, B., Liao, Z., & Wen, F. (2012). Quercetin attenuates airway inflammation and mucus production induced by cigarette smoke in rats. *International Immunopharmacology*, *13*(1), 73–81.

Zaragozá, C., Villaescusa, L., Monserrat, J., Zaragozá, F., & Álvarez-Mon, M. (2020). Potential therapeutic anti-inflammatory and immunomodulatory effects of dihydroflavones, flavones, and flavonols. *Molecules*, *25*(4), 1017.

Zhang, H., Penninger, J. M., Li, Y., Zhong, N., & Slutsky, A. S. (2020). Angiotensin-converting enzyme 2 (ACE2) as a SARS-CoV-2 receptor: Molecular mechanisms and potential therapeutic target. *Intensive Care Medicine*, *46*, 586–590.

Zhu, Y., Scholle, F., Kisthardt, S. C., & Xie, D. Y. (2022). Flavonols and dihydroflavonols inhibit the main protease activity of SARS-CoV-2 and the replication of human coronavirus 229E. *Virology*, *571*, 21–33.

Zhuo, L. I., Hang, X. I. E., Chunping, T. A. N. G., Lu, F. E. N. G., Changqiang, K. E., Yechun, X. U., & Yang, Y. E. (2023). Flavonoids from the roots and rhizomes of Sophora tonkinensis and their in vitro anti-SARS-CoV-2 activity. *Chinese Journal of Natural Medicines*, *21*(1), 65–80.

5 A Comprehensive on Flavonoids Targeting RNA-Dependent RNA Polymerase of SARS-CoV-2

Anand Kumar Pandey and Shalja Verma

5.1 INTRODUCTION

Coronaviruses, due to their high mutability and infectivity, have often trembled the society. Severe Acute Respiratory Syndrome Coronavirus 2 (SARS-CoV-2), the cause of the current COVID-19 pandemic due to its high morbidity and mortality rate, posed a strong challenge to health security in the near past thereby encouraging the scientific community to evaluate effective ways to prevent and cure COVID-19 [1]. Numerous antivirals were identified of which some could land up as effective drugs for clinical use, but the synthetic nature of these drugs resulted in drastic side effects. Investigation of natural compounds having antiviral properties against SARS-CoV-2 thus became the need of the scenario [2]. Flavonoids constitute a large class of phytochemicals containing phenolic groups and have been identified to have diverse bioactive properties, including antioxidant, anti-inflammatory, antibacterial, antifungal, antiaging, and antiviral effects, in treating diverse diseases ranging from microbial infections and common lifestyle diseases to complicated neurodegenerative diseases and cancers [3]. These phytochemicals have been classified into different classes namely flavone, flavanol, flavonol, flavanone, isoflavone, and anthocyanidin, and have been evaluated to structurally inhibit the effective enzymes of SARS-CoV-2 which mediate viral essential mechanisms [4].

RNA-dependent RNA polymerase (RdRp) is a multi-subunit transcription replication complex of SARS-CoV-2 which plays an essential role in viral replication machinery and consists of non-structural proteins (nsp12, nsp8, and nsp7) [5]. Nsp12 is the core component of polymerase, where nsp8 and nsp7 act as accessory factors to enhance its activity by increasing processivity and template binding. The essentiality of RdRp proposed a target of antivirals to investigate treatment against SARS-CoV-2 [6]. Nucleotide analogs are the most commonly identified inhibitors of RdRp to obstruct viral replication and transcription. Remdesivir, a well-known prodrug, gets converted into remdesivir triphosphate inside the cell and acts as a nucleotide analog. It inhibits the functioning of RdRp by getting integrated into the extending RNA chain and thus terminates the process of RNA extension [7,8]. The amino acid residues of RdRp playing a role in RNA binding and catalysis are highly conserved, thus highlighting the conserved replication mechanism of RdRp in a wide range of viruses [9]. Several other nucleotide analogs such as ribavirin, favipiravir, EIDD-2801, and galidesivir in turn efficiently inhibit the replication machinery of SARS-CoV-2 by targeting RdRp. EIDD-2801 shows manyfold higher inhibition activity against RdRp of SARS-CoV-2 compared to remdesivir due to extra hydrogen bonding of cytidine ring N4 hydroxyl group of EIDD-2801 with Lys 545 of RdRp and an additional hydrogen bond of cytidine base of drug with the guanine base of the template RNA strand [10,11]. Apart from nucleotide analogs, other synthetic inhibitors targeting other sites of RdRp essential for RNA template binding, nucleotide entry, and RNA exit have also been explored [12]. Suramin, a synthetic polysulfonate, has a 20-fold higher potency to inhibit RdRp than remdesivir and acts by blocking the binding of RNA into the RNA binding site of RdRp [13]. Structural determination of suramin with RdRp revealed its binding at two sites to inhibit RdRp, one of which directly inhibited the binding of RNA template and another site clashed with primer RNA strand lying in close vicinity of the catalytic site of RdRp [14]. Though the effectiveness of these nucleotide analogs and RNA template blockers is immense, the side effects associated with these synthetic drugs encourage the investigation of natural counterparts utilizing these known molecules as basic structural templates [15].

Abundant studies have been conducted to investigate natural flavonoids to inhibit RdRp that can exhibit comparable inhibition to that of the known synthetic inhibitors. Several studies have established the effectiveness of the investigated flavonoids in *in vitro* and *in vivo* conditions, but a large part of studies have done preliminary *in silico* analysis. Thus, this chapter comprises comprehensive structural details of RdRp targetable sites, different classes of flavonoids, and studies conducted to characterize natural flavonoid molecules as RdRp inhibitors to provide natural treatment against SARS-CoV-2. This piece of work will further encourage the scientific community to look for a new scope of investigation in this concern that would probably lead to the development of natural high-potential inhibitor drugs against the dreaded disease of COVID-19.

DOI: 10.1201/9781003433200-5

5.2 FLAVONOIDS AND THEIR CLASSIFICATION

Flavonoids comprise a group of phenolic compounds synthesized by the phenylpropanoid pathway in plants. They are low-molecular-weight compounds containing a benzo-γ-pyrone structure and are universally present in plants [16]. Majorly these natural polyphenols are present in the nucleus of mesophylls and inside the ROS generation centers. They are eminent regulators of growth factors like auxin in plants [17]. The chemical structure of flavonoids contains a 15-C skeleton consisting of two rings of benzene A and B which are linked through a pyrane heterocyclic C ring. Flavonoids are classified into different classes based on their oxidation level and substitution patterns of the C ring. The class of flavonoids where the B ring is attached to the third position of the C ring is known as isoflavones, while the compounds where the B ring associates with the fourth position of the C ring are classified as neoflavonoids [18]. Moreover, those compounds in which ring B is associated at the second position can be classified into different subgroups including flavonols, flavones, flavanones, catechins, chalcones, and anthocyanins, based on features of the C ring. Within a class, individual compounds vary based on the substitutions on rings A and B. The activity and chemical nature of flavonoids are structure-dependent as well as depend on the hydroxylation degree, conjugations, substitutions, and polymerization degree [16].

Flavonols are ketone-containing flavonoids and are the building blocks of proanthocyanins. They contain a hydroxyl group at the third position of the C ring which may be glycosylated. They also have diverse hydroxylation and methylation patterns and comprise the largest flavonoid subgroup. They are abundantly found in vegetables and fruits including kale, tomatoes, berries, grapes, onion, and lettuce. Red wine and tea also contain flavonols. The most common flavonols are kaempferol, myricetin, quercetin, and fisetin [19–21].

Flavones contain a double bond between the second and third carbon of the C ring and a ketone at the fourth position. A hydroxy group at the fifth position of the A ring is common in flavones of fruits and vegetables, while other positions also contain hydroxylation mostly at the seventh position of the A ring and the third and fourth positions of the B ring, which may vary as per the taxonomic classification of fruits and vegetables [20,21]. Flavones are widely distributed in flowers, leaves, and fruits in the form of glycosides. Red peppers, celery, parsley, mint, chamomile, and ginkgo biloba are major flavone sources, while apigenin, luteolin, and tangeretin are common flavones. Further, polymethoxylated flavones, nobiletin, tangeretin, and sinensetin are found in citrus peels [20].

Flavanones are usually found in fruits of the citrus family and convey the bitter taste of citrus peel and juice. They are known as dihydroflavones because they have a saturated C ring making the marginal difference between flavones and flavanones. Naringenin and hesperetin are common flavanones and are responsible for numerous health benefits because of their antioxidant effect [22].

Catechins or flavanols are 3-hydroxy flavanone derivatives that have a hydroxyl group bound to the third position of the C ring. They do not contain a double bond between the third and second position of the C ring and are abundant in apples, banana, blueberries, pears, and peaches [23].

Anthocyanins are color pigments found in flowers, plants, and fruits, for example, delphinidin, cyanidin, malvidin, peonidin, and pelargonidin. They are usually present in layers of outer cells of fruits like red grapes, cranberries, black currants, raspberries, blackberries, etc. The color of anthocyanins is pH dependent and also varies by acylation or methylation on A and B ring hydroxyl groups [23,24].

Chalcones are a subclass of flavonoids that is characterized by the absence of a C ring found in the basic structure of the flavonoid skeleton. They are thus known as open-chain flavonoids. Mainly chalcones consist of arbutin, phloridzin, chalconaringenin, and phloretin. These chalcones occur in sufficient amounts in pears, tomatoes, strawberries, wheat products, and bearberries [25].

Isoflavonoids comprise a distinctive group of flavonoids that have limited distribution in the plant kingdom and are majorly found in leguminous plants and soybeans. They are also found in microbes. Isoflavonoids play a lead role as precursors for phytoalexins development in plant-to-microbe interactions. Daidzein and genistein isoflavones are common phytoestrogens due to their estrogenic activity in animals. This class of flavonoids by inducing metabolic and hormonal changes can influence different pathways of diseases [26,27].

Neoflavonoids are a polyphenolic class of compounds and have a 4-phenylchromen backbone with no substitution of the hydroxyl group at the second position compared to 2-phenylchromen-4-one backbone of flavonoids. Neoflavonoid namely calophyllolide was the first of this class isolated from seeds of *Calophyllum inophyllum*. This class of flavonoids is found in the timber and bark of the *Mesua thwaitesii* plant [28].

5.3 BIOACTIVE PROPERTIES OF FLAVONOIDS

Flavonoids are known to convey diverse beneficial health effects and can treat numerous diseases. The antioxidant, anti-inflammatory, antimicrobial, hepatoprotective, anticancer, and neuroprotective properties of flavonoids prove them outstanding natural treatments that can be procured via dietary sources. The exceptional chemistry of flavonoids specifies their health benefits [28].

Oxidative stress is a common pathology of a diverse variety of diseases including cancers. The antioxidant properties of flavonoids due to the presence of phenolic groups encourage their utilization in treating conditions associated with oxidative stress [29]. The substitution, configuration, and number of functional hydroxyl groups in flavonoids display antioxidant effects by scavenging free radicals and chelating metal ions thus preventing the free radical generation [18]. The configuration of the hydroxyl group in the B ring is the lead determinant of RNS (Reactive Nitrogen Species) and ROS (Reactive Oxygen Species) scavenging as it gives an electron and hydrogen to peroxyl, hydroxyl, and peroxynitrite radicals, thus stabilizing them leading to the formation of stable radicals of flavonoids. Different mechanisms of antioxidant effects have been reported for different flavonoids including reduction of ROS generation by either inhibiting responsible enzymes or chelating elements leading to the generation of free radicals, protection, or upregulation of antioxidant defenses and ROS scavenging [29,30]. Flavonoids often utilize a combination of these mechanisms to mediate antioxidant effects such as ROS scavenging and inhibition of ROS-generating enzymes. The lead enzymes playing a role in the generation of ROS are glutathione-S-transferase, microsomal monooxygenase, mitochondrial succinoxidase, NADH oxidase, etc. Peroxidation of lipids is common in oxidative stress conditions [31]. Flavonoids convey lipid protection by several mechanisms. Free ions of metals encourage the formation of ROS by reducing hydrogen peroxide with the generation of highly reactive radicals of hydroxyl. Because of their low potential of redox, flavonoids thermodynamically reduce the highly oxidizing free radicals by donating hydrogen atoms [32]. Also, the ability of flavonoids to chelate metal ions like copper, iron, etc. inhibits the generation of free radicals, e.g., quercetin has an exceptional ability to stabilize and chelate iron [33]. The major inhibition of peroxidation of lipids is mediated by the 3,4-catechol structure of the B ring of flavonoids which is effective in scavenging peroxynitrite, peroxyl, and superoxide radicals. The oxidation of B-ring containing catechol produces a stable radical of orthosemiquinone, which acts as an effective radical scavenger [34]. Rutin and epicatechin are lipid peroxidation inhibitors and strong scavengers of free radicals in *in vitro* conditions. The flavones that do not possess a catechol system result in unstable radical formation and thus exhibit a weak potential for radical scavenging [35]. Further, the flavonoids with unsaturated bond between 2 and 3 carbon in the C ring along with a 4-oxo functional group are more effective antioxidants than those lacking these two features. The heterocycle of flavonoids mediates the antioxidant activity by allowing conjugation between the aromatic rings and the incidence of the free 3-hydroxyl group. It is reported that the OH group of the B ring binds to 3-OH via hydrogen bonding, thus aligning the B ring with the heterocycle and the A ring. Because of this intramolecular hydrogen bonding, the effect of 3-OH is increased by the presence of 3,4-catechol, thereby enhancing the antioxidant effect of flavonoids that contain such bonding [36]. Furthermore, the glycoside derivatives of flavonoids are less potent antioxidants compared to the aglycones, and their antioxidant potential decreases with the increase in glycosidic moieties [37]. Hence, the effective chemical features of flavonoids convey their efficient antioxidant effects and thus can treat oxidative stress conditions in different diseases.

The antimicrobial effect of flavonoids is evident in plants itself which produce them as they are synthesized in response to infections of microbes in plants. Several flavonoids including galangin, apigenin, flavonol glycosides, flavone, flavanones, isoflavones, and chalcones have been evaluated to possess antibacterial effects [38]. These flavonoids do not have a specific site of action rather they possess multiple cellular targets. Inhibition of bacterial proteins, such as adhesins, transport proteins of the bacterial cell envelope, and other enzymes, by flavonoids via structural interaction and formation of hydrogen, covalent, and non-covalent bonds conveys the antibacterial effect. Several lipophilic flavonoids have the potential to disrupt the membranes of microbes [39]. Catechins due to their reduced C3 unit convey an antibacterial effect and have been investigated to inhibit *Streptococcus mutans*, *Vibrio cholerae*, *Shigella*, and some other pathogenic bacteria. The catechins can inactivate the cholera toxin produced by *V. cholera* and inhibit the glycosyltransferase of *S. mutans* due to their complexing effect [40]. A past study reported that flavonoids by forming hydrogen bonds or intercalating with the stacking nucleic acid bases can inhibit the synthesis of DNA and RNA and thus can convey an antibacterial effect [41]. Inhibition of DNA gyrase of *Escherichia coli* by flavonoids (e.g., apigenin and quercetin) has also been reported, thus providing a different mechanism to demonstrate antimicrobial property [39]. Moreover, sophoraflavanone G and naringenin have shown antibacterial effects against the streptococci and methicillin-resistant strain of *Staphylococcus aureus* (MRSA) via alteration or reduction in the fluidity of inner and outer layers of the membrane in the hydrophobic and hydrophilic regions. The basis of such an effect lies in the 5,7-dihydroxylation of ring A and the 2,6- or 2,4-dihydroxylation of ring B in flavanone. The OH group at position 5 in flavones and flavanones is of importance for their anti-MRSA activity [42,43]. Moreover, the antistaphylococcal activity was mediated by the substitution of C10 and C8 chains of flavan-3-ol class of flavonoids. Antibacterial activity of 5-hydroxy isoflavanones and 5-hydroxy flavanones containing additional three, two, or one OH groups at 4′, 2′, and 7 positions was evident against *Streptococcus sobrinus* and *S. mutans* [18]. Licochalcones A and C extracted from roots of *Glycyrrhiza inflata* inhibited *Micrococcus luteus* and *S. aureus* thus displaying their antibacterial effect by a similar action mode as used by respiratory chain inhibiting antibiotics [44]. Further analysis reported that the site of inhibition of these antibiotic flavonoids lies between CoQ and cytochrome c of the electron transport chain (ETC) of bacteria [43]. Hence, all these facts reveal the antibacterial effect of flavonoids proving them potential drug candidates for various infectious diseases.

Inflammation is a common process that comes into play when any tissue damage, microbial infection, and chemical irritation occur. It is initiated by immune cells' migration from vessels of blood to the damage site along with the release of mediators. Following this, inflammatory cells are recruited, and proinflammatory cytokines, ROS, and RNS are released to repair the damaged tissue and to eliminate foreign pathogens. Though normal inflammatory conditions are self-limiting and rapid, prolonged inflammation often causes serious disorders [45]. This system of inflammation can be modified by pharmacologic agents, diet, pollution in the environment, and the presence of chemicals in food. Several flavonoid family members are effective in combating the function of inflammatory cells and the immune system [46]. Many flavonoids like apigenin, hesperidin, quercetin, and luteolin pose analgesic and anti-inflammatory effects by modulating the functions of enzyme systems that mediate the inflammatory processes like serine-threonine protein kinase and tyrosine-protein kinase [47]. Leadingly these flavonoids competitively bind to these kinases at the active site and inhibit them, thereby inhibiting the cell activation and signal transduction processes associated with the immune system [48]. Flavonoids also inhibit the isoforms expression of inducible cyclooxygenase, nitric oxide synthase, and lipoxygenase that produce large amounts of prostanoids, nitric oxide, leukotrienes, and several mediators of inflammatory systems like chemokines, cytokines, or adhesion molecules [49]. Phosphodiesterases which play a role in cell activation are also inhibited by flavonoids [50]. Moreover, the lead target of the anti-inflammatory effect of flavonoids is cytokines biosynthesis, as cytokines mediates circulating leukocyte adhesion to injury sites. Some flavonoids inhibit prostaglandin production, which are chemicals responsible for proinflammatory signaling [47].

Treatment with silymarin has the potential to reverse the inflammatory changes induced by carrageenan [51]. Quercetin has also been observed to inhibit the secretion of immunoglobulins IgM, IgG, and IgA, stimulated by mitogens, in *in vitro* studies. Many flavonoids inhibit the adhesion of platelets, their secretion, and aggregation at a concentration of 1–10 mM. These effects of flavonoids are associated with the inhibition of the metabolism of arachidonic acid (AA) by carbon monoxide. Several flavonoids by acting as cyclic AMP phosphodiesterase inhibitors inhibit the function of platelets [52].

Flavonoids have also been investigated for their anticancer effects [53]. Onions and apples are major quercetin sources, which prevent cancers of the lungs, prostate, breast, and stomach. Moreover, intake of fruits, vegetables, and wine has been documented to reduce cancers including endometrium, lungs, esophagus, colon, and stomach [54]. The lead mechanisms reported to downregulate cancers on flavonoid administration are arrest of the cell cycle, inhibition of tyrosine kinase, downregulation of p53 mutant protein, inhibition of heat shock protein, inhibition of Ras protein expression, etc. [55]. Fatty acid synthase has been reported to be overexpressed in cancer conditions. Inhibition of fatty acid synthase and hence lipogenesis has been evaluated by flavanol epigallochatechin-3-gallate in prostate cancer cells. This inhibition of lipogenesis strongly arrests the growth of cancer cells and hence leads to cell death [56]. Quercetin has also been reported to provide relief in breast cancers as it interacts with and inhibits type II nuclear estrogen [57]. Past study on chemically induced cancer of the mammary gland determined that genistein suppresses cancer development without posing any toxic effect on the reproductive or endocrine system [58]. Hesperidin, a glycoside flavanone, has been investigated to inhibit cancer of the mammary gland and colon induced by azoxymethanol. Several flavonoids have been reported for their anti-mutagenic property [59]. A carbonyl functional group of flavone nuclei at C-4 has been found to convey the anticancer properties of these compounds. Moreover, flavone-8-acetic acid has been evaluated for its anti-tumor effect. Robinetin, ellagic acid, myricetin, and quercetin show inhibition against tumorigenicity on mouse skin. High flavonoids and isoflavones consumption thus show protective effects against prostate cancer. Also, it is a well-known fact that oxidative stress can initiate cancer, and the use of potent antioxidants can combat cancer progression, thus diet containing flavonoids can be beneficial in preventing or curing cancers [18].

The antiviral effects of flavonoids have also been documented in numerous researches though the mechanism of antiviral activity differs with the chemistry of flavonoid and the viral infection. Various flavonoids have been identified that can structurally inhibit the viral enzymes, thus developing a structure–function association between the enzyme and the flavonoid. A comparative antiviral study revealed that flavan-3-ol posed better inhibition against HIV-2, HIV-1, and immunodeficiency viral infection compared to flavanones and flavones. Baicalin has been reported to inhibit HIV-1 replication and infection. Baicalein along with other flavonoids including hinokiflavone and robustaflavone inhibited the reverse transcriptase of HIV-1. Further study on HIV-1 revealed the obstructed entry of HIV-1 in the chemokine co-receptor and CD4 expressing cells, along with reverse transcriptase antagonism by flavone-O-glycoside [60]. Catechins inhibit HIV-1 DNA polymerase, thereby conveying their antiviral effects [61]. Robinetin and demethylated gardenin A flavonoid inhibit the proteinase of HIV-1 thus preventing its infection [62]. Further synergistic effects of flavonols and flavones have been studied. Luteolin and kaempferol display synergistic inhibitory effects against the herpes simplex virus [63]. Synergism between other antivirals and flavonoids gave combinations of quercetin and 5-ethyl-2-dioxyuridine and quercetin and acyclovir as potent treatment against herpes simplex virus and pseudorabies infection, respectively [64]. Concludingly, flavonols were found to be more effective than flavones to treat type I herpes simplex virus. Further, quercetin, naringin, hesperetin, and daidzein display anti-dengue effects against different DENV-2 infection stages and cycles of replications. Quercetin showed the most effective antiviral effect against the DENV-2 viral strain in the Vero cell line [65]. Several flavonoids including dihydrofisetin, dihydroquercetin, pelargonidin chloride, leucocyanidin, and catechin have been investigated to show antiviral effects against the polio virus, respiratory syncytial virus, herpes simplex virus, and sindbis virus. The most prevalent mechanisms followed by flavonoids to convey antiviral effects are inhibition of polymerase enzyme, binding of viral nucleic acid, or binding of capsid proteins [66].

The subsequent section illustrates the structure of RdRp, the lead drug target of SARS-CoV-2, and its effective sites that have been utilized for the development of natural antivirals including flavonoids.

5.4 RDRP AS A DRUG TARGET AND ITS STRUCTURAL INSIGHTS

RdRp is a key enzyme required for genome replication and gene transcription in coronaviruses. The key catalytic unit for replication and transcription comprises non-structural protein-12 (nsp12), while the nsp7 and nsp8 together act as cofactors and support the catalysis. The lead role of nsp12 makes it a key drug target for designing drugs analogous to nucleotides as inhibitors of RdRp. RdRp has been utilized as a drug target against several viruses including coronaviruses, Zika virus, and hepatitis C virus. The active site architecture of RdRp is conserved in a widespread range of positive-sense RNA viruses and provides a common structural template for the development of broad-spectrum drugs against RNA viruses and also favors the repurposing of drugs [12,14,67].

RdRp of SARS-CoV-2 consists of an nsp12 subunit of 932 amino acids, an nsp8 subunit of 198 amino acids, and an nsp7 subunit of 83 amino acids. The nsp12 possesses a RdRp right-hand domain ranging from residue Ser367 to Phe920, and nidovirus-specific domain at N-terminal from Asp60 to Arg249 which is similar to nidovirus RdRp-associated nucleotidyl transferase (NiRAN) domain. Additionally, residue Ala4 to Thr28 also contributes to the NiRAN domain. The interface domain extending from Ala250 to Arg365 connects the RdRp domain with the NiRAN domain. The conserved RdRp domain architecture contains three subdomains including finger (Leu366 to Ala581 and Lys621 to Gly679), palm (Thr582 to Pro620 and Thr680 to Gln815), and a thumb (His816 to Glu920) subdomain. Further, an N-terminal hairpin ranging from Asp29 to Lys50 inserts in the groove between the palm of the RdRp domain and the NiRAN domain. The pair of nsp7–nsp8 shows a similar structure as that of the nsp7–nsp8 pair of SARS-CoV. Moreover, 13 extra residues at the nsp8 N-terminal show the bent in the long golf club shape shaft (Figure 5.1) [68].

The catalytic site of RdRp is present in the conserved polymerase domain which is made up of 7 motifs A–G lying leadingly in the palm domain. The residue Asp618 that binds with the divalent cation lies in an A motif ranging from 611 to 626 residues and is conserved in almost all polymerases of viruses including poliovirus and hepatitis C virus [12]. The catalytic triad of Ser759–Asp760–Asp761 lies between two strands of the C motif which ranges from 753 to 767 residues. Similar to the conventional architecture of RNA polymerases for RNA synthesis, the entry position of primer-template, nucleotide triphosphates, and the exit path of nascent RNA strand have a positive charge, are accessible to solvent, and converge into a central cavity where template-dependent RNA synthesis is mediated [14, 68]. Further, the entry channel of nucleotide triphosphates is made up of Lys545, Arg553, and Arg555 hydrophilic residues present in the F motif. The RNA template enters the active site (present in the A and C motifs) via a groove clamped by F and G motifs, while the primer strand is supported by the E motif and the thumb subdomain. Eventually, the hybrid of template and product gets released via the exit tunnel of RNA lying on the front side of the polymerase [12,14,68]. All these motifs together communicate to mediate genome replication and gene transcription by RdRp.

Such in-depth structural insights reveal the effective regions of RdRp which can be targeted for drug designing and have been exploited by numerous researches. The subsequent section illustrates different studies that have characterized RdRp inhibition potential of flavonoids as natural antiviral compounds and can be utilized in drug development.

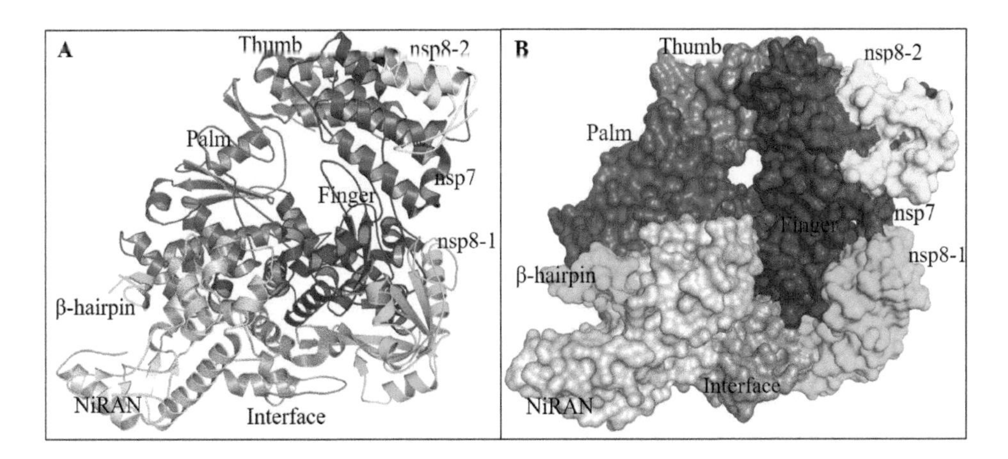

FIGURE 5.1 The cartoon and surface representation of structure of RNA-dependent RNA polymerase of SARS-CoV-2 displaying its different domains [68].

5.5 FLAVONOIDS AS INHIBITORS OF RDRP AGAINST SARS-COV-2

Flavonoids being the natural components present in plants pose least or no side effects and have exceptional bioactive properties. Numerous studies have characterized the antiviral effects of flavonoids in different diseases including COVID-19. Previous investigations have proved the potential of flavonoids and their derivatives to inhibit the essential proteins of SARS-CoV-2. RdRp being the key enzyme playing a role in the replication and transcription of virus has been targeted in several studies and effective flavonoids possessing effective RdRp inhibition potential have been revealed. Some of the major investigations including *in silico* and *in vitro* studies have been illustrated as follows (Figure 5.2).

Munafò et al. [69] investigated potential flavonoids against essential motifs of RdRp using *in silico* and *in vitro* analysis. The RNA cavity lying in the G motif and N terminus of the B motif containing the key residues Lys577, Gln573, Arg569, Lys500, and Asn497, and the nucleotide triphosphate entry site lying near the active site containing lead residues Asp865, Arg836, Arg555, Arg553, and Lys551 were screened. Their analysis revealed high negative binding energies of −7.62 kcal/mol for quercetin and −5.23 kcal/mol for luteolin at the nucleotide triphosphate entry site, while the binding energies at the RNA cavity were −7.69 kcal/mol for quercetin and −6.18 kcal/mol for luteolin. Further, analysis of interactions revealed interactions of the two flavonoids with Arg836, His816, Ser814, Arg555, Lys551, Ser549, and His439 residues at the nucleoside triphosphate entry site and Tyr689, Lys577, Gln573, Arg569, Lys500, Asn497, and Asn496 at the RNA cavity. The *in vitro* analysis gave significant IC50 values of 6.9 μM and 4.6 μM for quercetin and luteolin, respectively, proving the antiviral effect of the two flavonoids. The high structural similarity of the two flavonoids provides a template that fits and effectively inhibits the RdRp at respective sites [69].

Rameshkumar et al., in their study targeting RdRp as a drug target, identified specific flavonoids binding at the interface between the polymerase domain and NiRAN domain extending to the finger subdomain of the polymerase domain. Albireodelphin gave the highest negative binding score of −9.8 kcal/mol and formed effective hydrogen bonds with Arg514, Glu402, His401, Glu398, Asn397, Asn394, Phe390, Asp382, His378, Asp350, and Arg348. Further, delphinidin 3-O-beta-D-glucoside 5-O-(6-coumaroyl-beta-D-glucoside; (−)-maackiain-3-O-glucosyl-6″-O-malonate; cyanidin-3-(p-coumaroyl)-rutinoside-5-glucoside; cyanidin-3-(6-p-caffeoyl)glucoside; api genin-7-(6″-malonylglucoside); agathisflavone; cupressuflavone, dracorubin, and amentoflavone gave the effective negative binding energies of −9.7 to −6 kcal/mol [70].

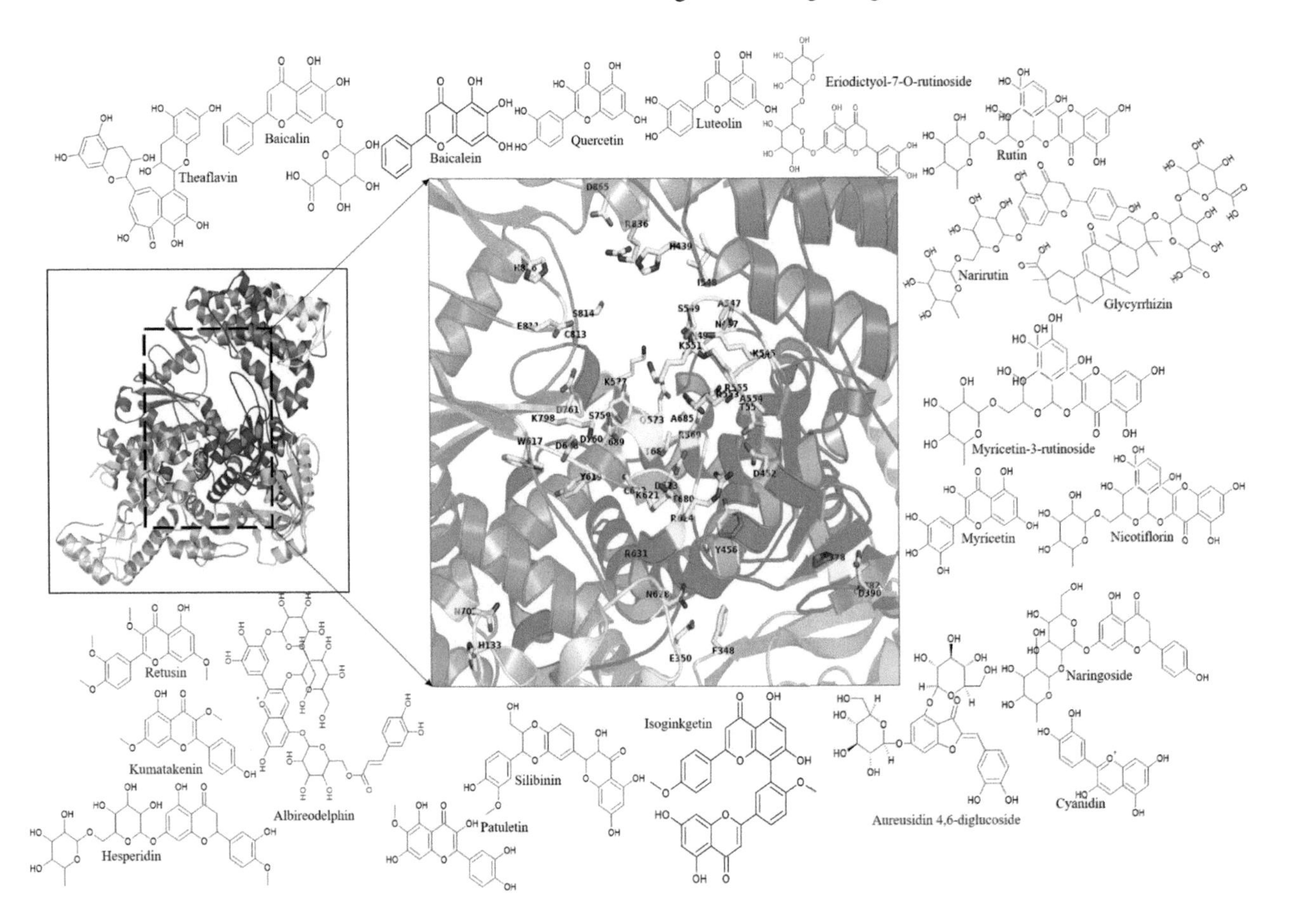

FIGURE 5.2 The cartoon representation of major sites targeted to inhibit RNA-dependent RNA polymerase of SARS-CoV-2 in different studies by different natural flavonoids [69–82].

Lung et al., in their flavonoid screening study against the catalytic site of RdRp, revealed that theaflavin was the most potent candidate with a binding energy of –9.11 kcal/mol. Interaction analysis of the theaflavin–RdRp complex revealed significant interactions and hydrogen bonding with Arg624, Arg553, and Asp452 residues [71].

Leal et al. analyzed the flavonoid composition of *Siparuna* species, a Brazilian folk medicinal plant, against the RdRp of SARS-CoV-2. They extracted retusin and kumatakenin and analyzed their antiviral effect against SARS-CoV-2 in Calu-3 and VeroE6 cell lines. Their analysis revealed a higher selective index compared to control lopinavir and chloroquine [72].

Ogunyemi et al. evaluated that 12α-epi-millettosin and usararotenoid A are effective flavonoids having high binding affinity and interactions with lead amino acids of RdRp active site, divalent cation binding site, and nucleotide triphosphate entry site [73].

da Silva et al. evaluated that nicotiflorin, rutin, and their sulfate and glucuronide derivatives are potential inhibitors of RdRp. Nicotiflorin was found to be the most potential candidate and interacted with the key residues of RdRp including Asp760 of the C motif and Trp617 and Asp623 of the A motif via hydrogen bonds and π–anion interactions. Rutin also formed effective interactions with lead residues of the A motif (Arg624 and Asp623) and with catalytic residues of the C motif (Asp760 and Ser759). Further, considered sulfate and glucuronide derivates including quercetin-3-O-glucuronide, quercetin-7-O-glucuronoid, quercetin-3-O-sulfate, kaempferol-4′-O-sulfate, and kaempferol-3-O-glucuronide, in turn, formed effective interactions with the key catalytic residues of C motifs of RdRp thus proving the considered compound as potential candidates for RdRp inhibition [74].

Vijayakumar et al., in their docking analysis of 23 flavonoids and 25 indole chalcones, found that cyanidin was the most effective flavonoid for RdRp inhibition with a binding energy of –7.7 kcal/mol forming lead hydrogen bond interaction with Asp761 of the RNA binding site. Moreover, their analysis revealed that other flavonoids with effective binding energy ranging from –6.4 to –8.7 kcal/mol formed interactions with the β-hairpin residues, but the role of these residues in RdRp action mechanisms is still unknown [75].

Metwaly et al. considered patuletin, a rare flavonoid from *Tagetes patula* flowers, for RdRp inhibition analysis based on its structural similarity with F86, a ligand for RdRp of SARS-CoV-2. Docking analysis of patuletin with RdRp resulted in binding energy of –20 kcal/mol at the active pocket of RdRp which was comparable to the binding energy of –23 kcal/mol of F86. The lead interactions mediating the binding included interactions of the pyrocatechol moiety of patuletin with Thr680 and Cys622 residues via hydrogen binding, electrostatic interactions with Asp623 and Cys622, and interaction of 3,5,7-Trihydroxy-6-methoxy-4H-chromen-4-one moiety with Arg555, Urd20, and Urd10 residues of RdRp via five hydrogen bonds. In addition, MD simulation analysis of patuletin with RdRp further confirmed the stable interaction of patuletin with lead residues of RdRp disclosing its exclusive RdRp inhibition potential [76].

Faisal et al. targeted an allosteric site of RdRp with 100 flavonoids and identified myricetin, naringoside, and aureusidin-4,6-diglucoside as lead potential candidates. Further, MD simulation-based analysis proved stable interactions of the considered compounds at the target site of RdRp. Along with this, the toxicity analysis and human intestinal absorption of all three compounds gave promising outcomes proving their potential for further investigations [77].

Raj et al. performed *in vitro* and *in silico* analysis of flavonoids containing 4H-chromen-4-one scaffold to evaluate their antiviral potential against SARS-CoV-2. *In vitro* antiviral analysis on infected Vero cell lines displayed the effective antiviral potential of afzelin and isoginkgetin. The IC50 value of isoginkgetin was 22.81 μM compared to 7.18, 11.49, and 11.63 μM for remdesivir, lopinavir, and chloroquine revealing the effectiveness of the flavonoid. Docking analysis of the considered molecules with RdRp of SARS-CoV-2 revealed a binding energy of –8.3 kcal/mol for isoginkgetin with significant interaction with Tyr689, Ala688, Ala685, Ser682, Arg569, Lys500, and Asn496 forming five hydrogen bonds. Further, DFT and MD simulations disclosed stable interaction of isoginkgetin with RdRp [78].

Zandi et al., in their *in vitro* biochemical and cell culture-based studies, exhibited the antiviral potential of baicalin and baicalein against the SARS-CoV-2. Their results made it evident that both the flavonoids targeted RdRp to inhibit the proliferation of SARS-CoV-2, whereas baicalein showed high inhibition potential. Further *in silico* analysis revealed high binding energies of –8.7 and –7.8 kcal/mol for baicalein and baicalin, respectively, with significant interaction at the His133 and Asn705 of nucleotidyl transferase domain and polymerase palm subdomain, respectively. As the considered flavonoids are not analogous to nucleosides, their inhibition mechanism by binding at a site apart from the catalytic site seems probable [79].

Hamdy et al. identified 10 flavonoids having antiviral effects. Of all considered flavonoids, silibinin displayed a significant selectivity index for SARS-CoV-2 with an excellent IC50 value of 0.042 μM. Toxicity analysis proved that silibinin was safe even at a 7-fold higher concentration than the IC50 value of SARS-CoV-2. In addition, silibinin at a concentration of 0.031 μM showed more than 90% virucidal activity. Molecular docking analysis exhibited the effective binding energy of –7.15 kcal/mol with three hydrogen bond interactions with Ile548, Asp618, and Urd-20 along with two bond interactions with Mg1004 and Mg1005. The interaction of silibinin with Asp-618 was aromatic hydrogen bond interaction, while Lys545, Ala547, Ile548, Lys551, Asp618, Asp760, Glu811, Ser814, and Arg836 formed hydrophobic interactions [80].

Puttaswamy et al., in their study on plant secondary metabolites, showed that several flavonol glycosides along with tannins, triterpenes, and anthocyanins were found to have effective interaction with RdRp. The flavanol glycosides including erodictyol-7-O-glycoside and narirutin showed the highest negative binding energy. Most of the considered plant metabolites interacted with the catalytic pocket of RdRp, thus reducing the activity of RdRp by sterically hindering substrate interaction. Eridictyol-7-O-rutinoside, a flavanone glycoside leadingly found in lemon, revealed –9.9 kcal/mol of binding energy with significant interactions with the effective residues of Asp623, Try619, Asp618, Thr556, Ala554, and Arg553 present in the active site of RdRp. Both aglycone and glycone subunits of this molecule formed hydrogen bonds with the target site of RdRp. Narirutin, another flavanone glycoside whose natural source is sweet orange, showed an effective binding energy of –9.7 kcal/mol at the catalytic site of RdRp. Moreover, compounds having similar structures like nirurin and naringin also showed promising binding energies of –9.0 and –8.9 kcal/mol. Myricetin-3-rutinoside, a glycoside flavonol from *Chrysobalanus icaco* L., has a binding energy of –9.5 kcal/mol with numerous hydrogen bonds with catalytic residues of RdRp including Ser682, Asp623, Cys622, Arg555, Ala554, Arg553, Tyr456, and Asp452. Kaempferol and isoginkgetin showed the same binding energy of –9.5 kcal/mol where isoginkgetin interacted with Ala685, Thr687, Thr556, and Arg553 residues of the RdRp catalytic site. Although flavonol glycosides revealed effective inhibition, their bioavailability score was low. Aglycones have been reported to have high bioavailability. Myricetin, an aglycone, has a binding energy of –8.4 kcal/mol and formed hydrogen bonds with RdRp active site residues of Thr680, Arg624, Arg555, and Arg553, making it a promising candidate for drug development against SARS-CoV-2 [81].

Rehman et al., in their *in silico* RdRp inhibition analysis, revealed strong binding with glycyrrhizin at two binding sites, one of which was the catalytic site of the enzyme. The effective residues that interacted to form hydrogen bonds at the catalytic site were Ser814, Cys813, Glu811, Lys798, Asp760, Trp617, Ala554, and Asp452. Further, hesperidin docked with RdRp showed an effective binding energy of –9.53 kcal/mol due to its hydrogen bond interaction with residues lying near the active site, including Ser628, Arg624, Lys621, and Thr556 and π–π interaction with Phe793 [82].

Therefore, flavonoids based on their structural, chemical, and bioactive properties have been extensively exploited for RdRp inhibition to evaluate the most potent candidate that can be effective in treating COVID-19.

5.6 SUMMARY

Flavonoids, a broad class of phytochemicals containing phenolic groups, are categorized into flavone, flavanol, flavonol, flavanone, isoflavone, and anthocyanidin. The diverse chemistry of flavonoids conveys different bioactive properties including antioxidant, anti-inflammatory, antimicrobial, anticancer, and antiviral effects, which are mainly mediated by specific binding of flavonoids to the enzymes responsible for different ailments resulting in their structural inhibition. Structure–function relationship thus provides a basis for the bioactive properties of flavonoids. SARS-CoV-2, a positive-sense RNA virus, has challenged the medical society by causing huge life loss. RdRp is a key player in SARS-CoV-2 replication and transcription and has been considered a potent drug target. Though few synthetic effective drugs targeting or inhibiting RdRp have been investigated, their undesirable side effects increase the adversity of diseased state. Identification and characterization of natural compounds, especially flavonoids, have been done for effectiveness in RdRp inhibition. Numerous *in silico, in vitro*, and *in vivo* evaluations of flavonoids have been conducted to evaluate the mechanism of RdRp inhibition. In this chapter, the peculiarities of flavonoid structures and their bioactive properties have been elaborated to understand their chemistry. The structural insights of RdRp to disclose the effective targetable sites of enzymes have been illustrated. Eventually, the diverse studies revealing the structural inhibition potential of flavonoids against RdRp have been discussed in detail to encourage understanding and further investigation into this concern to develop a natural, efficient, and specific treatment against COVID-19.

REFERENCES

1. Sharma, A., Tiwari, S., Deb, M. K., & Marty, J. L. (2020). Severe acute respiratory syndrome coronavirus-2 (SARS-CoV-2): A global pandemic and treatment strategies. *International Journal of Antimicrobial Agents*, 56(2), 106054. https://doi.org/10.1016/j.ijantimicag.2020.106054
2. Erdogan, T. (2023). In-silico investigation of some recent natural compounds for their potential use against SARS-CoV-2: A DFT, molecular docking and molecular dynamics study. *Journal of Biomolecular Structure & Dynamics*, 41(6), 2448–2465. https://doi.org/10.1080/07391102.2022.2033136
3. Dias, M. C., Pinto, D. C. G. A., & Silva, A. M. S. (2021). Plant flavonoids: Chemical characteristics and biological activity. *Molecules (Basel, Switzerland)*, 26(17), 5377. https://doi.org/10.3390/molecules26175377
4. Ullah, A., Munir, S., Badshah, S. L., Khan, N., Ghani, L., Poulson, B. G., Emwas, A. H., & Jaremko, M. (2020). Important flavonoids and their role as a therapeutic agent. *Molecules (Basel, Switzerland)*, 25(22), 5243. https://doi.org/10.3390/molecules25225243
5. Jiang, Y., Yin, W., & Xu, H. E. (2021). RNA-dependent RNA polymerase: Structure, mechanism, and drug discovery for COVID-19. *Biochemical and Biophysical Research Communications*, 538, 47–53. https://doi.org/10.1016/j.bbrc.2020.08.116
6. Campagnola, G., Govindarajan, V., Pelletier, A., Canard, B., & Peersen, O. B. (2022). The SARS-CoV nsp12 polymerase active site is tuned for large-genome replication. *Journal of Virology*, 96(16), e0067122. https://doi.org/10.1128/jvi.00671-22
7. Yin, W., Mao, C., Luan, X., Shen, D. D., Shen, Q., Su, H., Wang, X., Zhou, F., Zhao, W., Gao, M., Chang, S., Xie, Y. C., Tian, G., Jiang, H. W., Tao, S. C., Shen, J., Jiang, Y., Jiang, H., Xu, Y., Zhang, S., & Xu, H. E. (2020). Structural

basis for inhibition of the RNA-dependent RNA polymerase from SARS-CoV-2 by remdesivir. *Science*, 368(6498), 1499–1504. https://doi.org/10.1126/science.abc1560

8. Eastman, R. T., Roth, J. S., Brimacombe, K. R., Simeonov, A., Shen, M., Patnaik, S., & Hall, M. D. (2020). Remdesivir: A review of its discovery and development leading to emergency use authorization for treatment of COVID-19. *ACS Central Science*, 6(5), 672–683. https://doi.org/10.1021/acscentsci.0c00489
9. Aftab, S. O., Ghouri, M. Z., Masood, M. U., Haider, Z., Khan, Z., Ahmad, A., & Munawar, N. (2020). Analysis of SARS-CoV-2 RNA-dependent RNA polymerase as a potential therapeutic drug target using a computational approach. *Journal of Translational Medicine*, 18(1), 275. https://doi.org/10.1186/s12967-020-02439-0
10. Hasan, M. K., Kamruzzaman, M., Bin Manjur, O. H., Mahmud, A., Hussain, N., Alam Mondal, M. S., Hosen, M. I., Bello, M., & Rahman, A. (2021). Structural analogues of existing anti-viral drugs inhibit SARS-CoV-2 RNA dependent RNA polymerase: A computational hierarchical investigation. *Heliyon*, 7(3), e06435. https://doi.org/10.1016/j.heliyon.2021.e06435
11. Tian, L., Pang, Z., Li, M., Lou, F., An, X., Zhu, S., Song, L., Tong, Y., Fan, H., & Fan, J. (2022). Molnupiravir and its antiviral activity against COVID-19. *Frontiers in Immunology*, 13, 855496. https://doi.org/10.3389/fimmu.2022.855496
12. Xu, X., Chen, Y., Lu, X., Zhang, W., Fang, W., Yuan, L., & Wang, X. (2022). An update on inhibitors targeting RNA-dependent RNA polymerase for COVID-19 treatment: Promises and challenges. *Biochemical Pharmacology*, 205, 115279. https://doi.org/10.1016/j.bcp.2022.115279
13. Salgado-Benvindo, C., Thaler, M., Tas, A., Ogando, N. S., Bredenbeek, P. J., Ninaber, D. K., Wang, Y., Hiemstra, P. S., Snijder, E. J., & van Hemert, M. J. (2020). Suramin inhibits SARS-CoV-2 infection in cell culture by interfering with early steps of the replication cycle. *Antimicrobial Agents and Chemotherapy*, 64(8), e00900. https://doi.org/10.1128/AAC.00900-20
14. Yin, W., Luan, X., Li, Z., Zhou, Z., Wang, Q., Gao, M., Wang, X., Zhou, F., Shi, J., You, E., Liu, M., Wang, Q., Jiang, Y., Jiang, H., Xiao, G., Zhang, L., Yu, X., Zhang, S., & Eric Xu, H. (2021). Structural basis for inhibition of the SARS-CoV-2 RNA polymerase by suramin. *Nature Structural & Molecular Biology*, 28(3), 319–325. https://doi.org/10.1038/s41594-021-00570-0
15. Izcovich, A., Siemieniuk, R. A., Bartoszko, J. J., Ge, L., Zeraatkar, D., Kum, E., Qasim, A., Khamis, A. M., Rochwerg, B., Agoritsas, T., Chu, D. K., McLeod, S. L., Mustafa, R. A., Vandvik, P., & Brignardello-Petersen, R. (2022). Adverse effects of remdesivir, hydroxychloroquine and lopinavir/ritonavir when used for COVID-19: Systematic review and meta-analysis of randomised trials. *BMJ Open*, 12(3), e048502. https://doi.org/10.1136/bmjopen-2020-048502
16. Panche, A. N., Diwan, A. D., & Chandra, S. R. (2016). Flavonoids: An overview. *Journal of Nutritional Science*, 5, e47. https://doi.org/10.1017/jns.2016.41
17. Brunetti, C., Di Ferdinando, M., Fini, A., Pollastri, S., & Tattini, M. (2013). Flavonoids as antioxidants and developmental regulators: Relative significance in plants and humans. *International Journal of Molecular Sciences*, 14(2), 3540–3555. https://doi.org/10.3390/ijms14023540
18. Kumar, S., & Pandey, A. K. (2013). Chemistry and biological activities of flavonoids: An overview. *The Scientific World Journal*, 2013, 162750. https://doi.org/10.1155/2013/162750
19. Zhuang, W. B., Li, Y. H., Shu, X. C., Pu, Y. T., Wang, X. J., Wang, T., & Wang, Z. (2023). The classification, molecular structure and biological biosynthesis of flavonoids, and their roles in biotic and abiotic stresses. *Molecules (Basel, Switzerland)*, 28(8), 3599. https://doi.org/10.3390/molecules28083599
20. Halbwirth H. (2010). The creation and physiological relevance of divergent hydroxylation patterns in the flavonoid pathway. *International Journal of Molecular Sciences*, 11(2), 595–621. https://doi.org/10.3390/ijms11020595
21. Spiegel, M., Andruniów, T., & Sroka, Z. (2020). Flavones' and flavonols' antiradical structure-activity relationship-a quantum chemical study. *Antioxidants (Basel, Switzerland)*, 9(6), 461. https://doi.org/10.3390/antiox9060461
22. Peterson, J., Dwyer, J. T., Beecher, G. R., Bhagwat, S., Gebhardt, S. E., Haytowitz, D. B., & Holden, J. M. (2006b). Flavanones in oranges, tangerines (mandarins), tangors, and tangelos: A compilation and review of the data from the analytical literature. *Journal of Food Composition and Analysis*, 19, S66–S73. https://doi.org/10.1016/j.jfca.2005.12.006
23. De Souza Farias, S. A., Da Costa, K. S., & Martins, J. B. L. (2021). Analysis of conformational, structural, magnetic, and electronic properties related to antioxidant activity: Revisiting flavan, anthocyanidin, flavanone, flavonol, isoflavone, flavone, and flavan-3-ol. *ACS Omega*, 6(13), 8908–8918. https://doi.org/10.1021/acsomega.0c06156
24. Khoo, H. E., Azlan, A., Tang, S. T., & Lim, S. M. (2017). Anthocyanidins and anthocyanins: Colored pigments as food, pharmaceutical ingredients, and the potential health benefits. *Food & Nutrition Research*, 61(1), 1361779. https://doi.org/10.1080/16546628.2017.1361779
25. Rudrapal, M., Khan, J., Dukhyil, A. A. B., Alarousy, R. M. I. I., Attah, E. I., Sharma, T., Khairnar, S. J., & Bendale, A. R. (2021). Chalcone scaffolds, bioprecursors of flavonoids: Chemistry, bioactivities, and pharmacokinetics. *Molecules (Basel, Switzerland)*, 26(23), 7177. https://doi.org/10.3390/molecules26237177
26. Wyse, J., Latif, S., Gurusinghe, S., McCormick, J., Weston, L. A., & Stephen, C. P. (2022). Phytoestrogens: A review of their impacts on reproductive physiology and other effects upon grazing livestock. *Animals: An Open Access Journal from MDPI*, 12(19), 2709. https://doi.org/10.3390/ani12192709
27. Sohn, S. I., Pandian, S., Oh, Y. J., Kang, H. J., Cho, W. S., & Cho, Y. S. (2021). Metabolic engineering of isoflavones: An updated overview. *Frontiers in Plant Science*, 12, 670103. https://doi.org/10.3389/fpls.2021.670103
28. Umer, S. M., Shamim, S., Khan, K. M., & Saleem, R. S. Z. (2023). Perplexing polyphenolics: The isolations, syntheses, reappraisals, and bioactivities of flavonoids, isoflavonoids, and neoflavonoids from 2016 to 2022. *Life*, 13(3), 736. https://doi.org/10.3390/life13030736
29. Sharifi-Rad, M., Anil Kumar, N. V., Zucca, P., Varoni, E. M., Dini, L., Panzarini, E., Rajkovic, J., Tsouh Fokou, P. V., Azzini, E., Peluso, I., Prakash Mishra, A., Nigam, M., El Rayess, Y., Beyrouthy, M. E., Polito, L., Iriti, M., Martins, N., Martorell, M., Docea, A. O., Setzer, W. N., & Sharifi-Rad, J. (2020). Lifestyle, oxidative stress, and

antioxidants: Back and forth in the pathophysiology of chronic diseases. *Frontiers in Physiology*, 11, 694. https://doi.org/10.3389/fphys.2020.00694

30. Caruso, F., Incerpi, S., Pedersen, J. Z., Belli, S., Kaur, S., & Rossi, M. (2022). Aromatic polyphenol π-π interactions with superoxide radicals contribute to radical scavenging and can make polyphenols mimic superoxide dismutase activity. *Current Issues in Molecular Biology*, 44(11), 5209–5220. https://doi.org/10.3390/cimb44110354
31. Tan, B. L., Norhaizan, M. E., Liew, W. P., & Sulaiman Rahman, H. (2018). Antioxidant and oxidative stress: A mutual interplay in age-related diseases. *Frontiers in Pharmacology*, 9, 1162. https://doi.org/10.3389/fphar.2018.01162
32. Kim, T. Y., Leem, E., Lee, J. M., & Kim, S. R. (2020). Control of reactive oxygen species for the prevention of Parkinson's disease: The possible application of flavonoids. *Antioxidants (Basel, Switzerland)*, 9(7), 583. https://doi.org/10.3390/antiox9070583
33. Mucha, P., Skoczyńska, A., Małecka, M., Hikisz, P., & Budzisz, E. (2021). Overview of the antioxidant and anti-inflammatory activities of selected plant compounds and their metal ions complexes. *Molecules (Basel, Switzerland)*, 26(16), 4886. https://doi.org/10.3390/molecules26164886
34. Charlton, N. C., Mastyugin, M., Török, B., & Török, M. (2023). Structural features of small molecule antioxidants and strategic modifications to improve potential bioactivity. *Molecules (Basel, Switzerland)*, 28(3), 1057. https://doi.org/10.3390/molecules28031057
35. Lambert, J. D., & Elias, R. J. (2010). The antioxidant and pro-oxidant activities of green tea polyphenols: A role in cancer prevention. *Archives of Biochemistry and Biophysics*, 501(1), 65–72. https://doi.org/10.1016/j.abb.2010.06.013
36. Speisky, H., Shahidi, F., De Camargo, A. C., & Fuentes, J. (2022). Revisiting the oxidation of flavonoids: Loss, conservation or enhancement of their antioxidant properties. *Antioxidants*, 11(1), 133. https://doi.org/10.3390/antiox11010133
37. Xie, L., Deng, Z., Zhang, J., Dong, H., Wang, W., Xing, B., & Liu, X. (2022). Comparison of flavonoid O-glycoside, C-glycoside and their aglycones on antioxidant capacity and metabolism during in vitro digestion and in vivo. *Foods (Basel, Switzerland)*, 11(6), 882. https://doi.org/10.3390/foods11060882
38. Górniak, I., Bartoszewski, R., & Króliczewski, J. (2018). Comprehensive review of antimicrobial activities of plant flavonoids. *Phytochemistry Reviews*, 18(1), 241–272. https://doi.org/10.1007/s11101-018-9591-z
39. Shamsudin, N. F., Ahmed, Q. U., Mahmood, S., Ali Shah, S. A., Khatib, A., Mukhtar, S., Alsharif, M. A., Parveen, H., & Zakaria, Z. A. (2022). Antibacterial effects of flavonoids and their structure-activity relationship study: A comparative interpretation. *Molecules (Basel, Switzerland)*, 27(4), 1149. https://doi.org/10.3390/molecules27041149
40. Komiazyk, M., Palczewska, M., Sitkiewicz, I., Pikula, S., & Groves, P. (2019). Neutralization of cholera toxin by Rosaceae family plant extracts. *BMC Complementary and Alternative Medicine*, 19(1), 140. https://doi.org/10.1186/s12906-019-2540-6
41. Jomová, K., Hudecova, L., Lauro, P., Simunkova, M., Alwasel, S. H., Alhazza, I. M., & Valko, M. (2019). A switch between antioxidant and prooxidant properties of the phenolic compounds myricetin, morin, 3',4'-dihydroxyflavone, taxifolin and 4-hydroxy-coumarin in the presence of copper(II) ions: A spectroscopic, absorption titration and DNA damage study. *Molecules (Basel, Switzerland)*, 24(23), 4335. https://doi.org/10.3390/molecules24234335
42. Duda-Madej, A., Stecko, J., Sobieraj, J., Szymańska, N., & Kozłowska, J. (2022). Naringenin and its derivatives-health-promoting phytobiotic against resistant bacteria and fungi in humans. *Antibiotics (Basel, Switzerland)*, 11(11), 1628. https://doi.org/10.3390/antibiotics11111628
43. Cushnie, T. P., & Lamb, A. J. (2005). Antimicrobial activity of flavonoids. *International Journal of Antimicrobial Agents*, 26(5), 343–356. https://doi.org/10.1016/j.ijantimicag.2005.09.002
44. Tsukiyama, R., Katsura, H., Tokuriki, N., & Kobayashi, M. (2002). Antibacterial activity of licochalcone A against spore-forming bacteria. *Antimicrobial Agents and Chemotherapy*, 46(5), 1226–1230. https://doi.org/10.1128/AAC.46.5.1226-1230.2002
45. Chen, L., Deng, H., Cui, H., Fang, J., Zuo, Z., Deng, J., Li, Y., Wang, X., & Zhao, L. (2017). Inflammatory responses and inflammation-associated diseases in organs. *Oncotarget*, 9(6), 7204–7218. https://doi.org/10.18632/oncotarget.23208
46. Tungmunnithum, D., Thongboonyou, A., Pholboon, A., & Yangsabai, A. (2018). Flavonoids and other phenolic compounds from medicinal plants for pharmaceutical and medical aspects: An overview. *Medicines (Basel, Switzerland)*, 5(3), 93. https://doi.org/10.3390/medicines5030093
47. Al-Khayri, J. M., Sahana, G. R., Nagella, P., Joseph, B. V., Alessa, F. M., & Al-Mssallem, M. Q. (2022). Flavonoids as potential anti-inflammatory molecules: A review. *Molecules (Basel, Switzerland)*, 27(9), 2901. https://doi.org/10.3390/molecules27092901
48. Bode, A. M., & Dong, Z. (2013). Signal transduction and molecular targets of selected flavonoids. *Antioxidants & Redox Signaling*, 19(2), 163–180. https://doi.org/10.1089/ars.2013.5251
49. Farzaei, M. H., Singh, A. K., Kumar, R., Croley, C. R., Pandey, A. K., Coy-Barrera, E., Kumar Patra, J., Das, G., Kerry, R. G., Annunziata, G., Tenore, G. C., Khan, H., Micucci, M., Budriesi, R., Momtaz, S., Nabavi, S. M., & Bishayee, A. (2019). Targeting Inflammation by flavonoids: Novel therapeutic strategy for metabolic disorders. *International Journal of Molecular Sciences*, 20(19), 4957. https://doi.org/10.3390/ijms20194957
50. Ko, W. C., Shih, C. M., Lai, Y. H., Chen, J. H., & Huang, H. L. (2004). Inhibitory effects of flavonoids on phosphodiesterase isozymes from guinea pig and their structure-activity relationships. *Biochemical Pharmacology*, 68(10), 2087–2094. https://doi.org/10.1016/j.bcp.2004.06.030
51. Karimi, G., Vahabzadeh, M., Lari, P., Rashedinia, M., & Moshiri, M. (2011). "Silymarin", a promising pharmacological agent for treatment of diseases. *Iranian Journal of Basic Medical Sciences*, 14(4), 308–317.
52. Oh, W. J., Endale, M., Park, S. C., Cho, J. Y., & Rhee, M. H. (2012). Dual roles of quercetin in platelets: Phosphoinositide-3-kinase and MAP kinases inhibition, and cAMP-dependent vasodilator-stimulated phosphoprotein stimulation. *Evidence-Based Complementary and Alternative Medicine: eCAM*, 2012, 485262. https://doi.org/10.1155/2012/485262
53. Kopustinskiene, D. M., Jakstas, V., Savickas, A., & Bernatoniene, J. (2020). Flavonoids as anticancer agents. *Nutrients*, 12(2), 457. https://doi.org/10.3390/nu12020457

54. Kubina, R., Iriti, M., & Kabała-Dzik, A. (2021). Anticancer potential of selected flavonols: Fisetin, kaempferol, and quercetin on head and neck cancers. *Nutrients*, 13(3), 845. https://doi.org/10.3390/nu13030845
55. Pang, X., Zhang, X., Jiang, Y., Su, Q., Li, Q., & Li, Z. (2021). Autophagy: Mechanisms and therapeutic potential of flavonoids in cancer. *Biomolecules*, 11(2), 135. https://doi.org/10.3390/biom11020135
56. Coleman, D. T., Bigelow, R., & Cardelli, J. A. (2009). Inhibition of fatty acid synthase by luteolin post-transcriptionally down-regulates c-Met expression independent of proteosomal/lysosomal degradation. *Molecular Cancer Therapeutics*, 8(1), 214–224. https://doi.org/10.1158/1535-7163.MCT-08-0722
57. Asgharian, P., Tazekand, A. P., Hosseini, K., Forouhandeh, H., Ghasemnejad, T., Ranjbar, M., Hasan, M., Kumar, M., Beirami, S. M., Tarhriz, V., Soofiyani, S. R., Kozhamzharova, L., Sharifi-Rad, J., Calina, D., & Cho, W. C. (2022). Potential mechanisms of quercetin in cancer prevention: Focus on cellular and molecular targets. *Cancer Cell International*, 22(1), 257. https://doi.org/10.1186/s12935-022-02677-w
58. Murrill, W. B., Brown, N. M., Zhang, J. X., Manzolillo, P. A., Barnes, S., & Lamartiniere, C. A. (1996). Prepubertal genistein exposure suppresses mammary cancer and enhances gland differentiation in rats. *Carcinogenesis*, 17(7), 1451–1457. https://doi.org/10.1093/carcin/17.7.1451
59. Lee, C. J., Wilson, L., Jordan, M. A., Nguyen, V., Tang, J., & Smiyun, G. (2010). Hesperidin suppressed proliferations of both human breast cancer and androgen-dependent prostate cancer cells. *Phytotherapy Research: PTR*, 24(Suppl 1), S15–S19. https://doi.org/10.1002/ptr.2856
60. Li, B. Q., Fu, T., Dongyan, Y., Mikovits, J. A., Ruscetti, F. W., & Wang, J. M. (2000). Flavonoid baicalin inhibits HIV-1 infection at the level of viral entry. *Biochemical and Biophysical Research Communications*, 276(2), 534–538. https://doi.org/10.1006/bbrc.2000.3485
61. Wang, Y. Q., Li, Q. S., Zheng, X. Q., Lu, J. L., & Liang, Y. R. (2021). Antiviral effects of green tea EGCG and its potential application against COVID-19. *Molecules (Basel, Switzerland)*, 26(13), 3962. https://doi.org/10.3390/molecules26133962
62. Wang, L., Song, J., Liu, A., Xiao, B., Li, S., Wen, Z., Lu, Y., & Du, G. (2020). Research progress of the antiviral bioactivities of natural flavonoids. *Natural Products and Bioprospecting*, 10(5), 271–283. https://doi.org/10.1007/s13659-020-00257-x
63. Zakaryan, H., Arabyan, E., Oo, A., & Zandi, K. (2017). Flavonoids: Promising natural compounds against viral infections. *Archives of Virology*, 162(9), 2539–2551. https://doi.org/10.1007/s00705-017-3417-y
64. Mucsi I. (1984). Combined antiviral effects of flavonoids and 5-ethyl-2'-deoxyuridine on the multiplication of herpesviruses. *Acta Virologica*, 28(5), 395–400.
65. Zandi, K., Teoh, B. T., Sam, S. S., Wong, P. F., Mustafa, M. R., & Abubakar, S. (2011). Antiviral activity of four types of bioflavonoid against dengue virus type-2. *Virology Journal*, 8, 560. https://doi.org/10.1186/1743-422X-8-560
66. Wang, L., Song, J., Liu, A., Xiao, B., Li, S., Wen, Z., Lu, Y., & Du, G. (2020). Research progress of the antiviral bioactivities of natural flavonoids. *Natural Products and Bioprospecting*, 10(5), 271–283. https://doi.org/10.1007/s13659-020-00257-x
67. Gao, Y., Yan, L., Huang, Y., Liu, F., Zhao, Y., Cao, L., Wang, T., Sun, Q., Ming, Z., Zhang, L., Ge, J., Zheng, L., Zhang, Y., Wang, H., Zhu, Y., Zhu, C., Hu, T., Hua, T., Zhang, B., Yang, X., & Rao, Z. (2020). Structure of the RNA-dependent RNA polymerase from COVID-19 virus. *Science*, 368(6492), 779–782. https://doi.org/10.1126/science.abb7498
68. Wang, Q., Wu, J., Wang, H., Gao, Y., Liu, Q., Mu, A., Ji, W., Yan, L., Zhu, Y., Zhu, C., Fang, X., Yang, X., Huang, Y., Gao, H., Liu, F., Ge, J., Sun, Q., Yang, X., Xu, W., Liu, Z., & Rao, Z. (2020). Structural basis for RNA replication by the SARS-CoV-2 polymerase. *Cell*, 182(2), 417–428. https://doi.org/10.1016/j.cell.2020.05.034
69. Munafò, F., Donati, E., Brindani, N., Ottonello, G., Armirotti, A., & De Vivo, M. (2022). Quercetin and luteolin are single-digit micromolar inhibitors of the SARS-CoV-2 RNA-dependent RNA polymerase. *Scientific Reports*, 12(1), 10571. https://doi.org/10.1038/s41598-022-14664-2
70. Rameshkumar, M. R., Indu, P., Arunagirinathan, N., Venkatadri, B., El-Serehy, H. A., & Ahmad, A. (2021). Computational selection of flavonoid compounds as inhibitors against SARS-CoV-2 main protease, RNA-dependent RNA polymerase and spike proteins: A molecular docking study. *Saudi Journal of Biological Sciences*, 28(1), 448–458. https://doi.org/10.1016/j.sjbs.2020.10.028
71. Lung, J., Lin, Y. S., Yang, Y. H., Chou, Y. L., Shu, L. H., Cheng, Y. C., Liu, H. T., & Wu, C. Y. (2020). The potential chemical structure of anti-SARS-CoV-2 RNA-dependent RNA polymerase. *Journal of Medical Virology*, 92(6), 693–697. https://doi.org/10.1002/jmv.25761
72. Leal, C. M., Leitão, S. G., Sausset, R., Mendonça, S. C., Nascimento, P. H. A., de Araujo R Cheohen, C. F., Esteves, M. E. A., Leal da Silva, M., Gondim, T. S., Monteiro, M. E. S., Tucci, A. R., Fintelman-Rodrigues, N., Siqueira, M. M., Miranda, M. D., Costa, F. N., Simas, R. C., & Leitão, G. G. (2021). Flavonoids from Siparuna cristata as potential inhibitors of SARS-CoV-2 replication. *Revista brasileira de farmacognosia : Orgao oficial da Sociedade Brasileira de Farmacognosia*, 31(5), 658–666. https://doi.org/10.1007/s43450-021-00162-5
73. Ogunyemi, O. M., Gyebi, G. A., Elfiky, A. A., Afolabi, S. O., Ogunro, O. B., Adegunloye, A. P., & Ibrahim, I. M. (2020). Alkaloids and flavonoids from African phytochemicals as potential inhibitors of SARS-Cov-2 RNA-dependent RNA polymerase: An in silico perspective. *Antiviral Chemistry & Chemotherapy*, 28, 2040206620984076. https://doi.org/10.1177/2040206620984076
74. da Silva, F. M. A., da Silva, K. P. A., de Oliveira, L. P. M., Costa, E. V., Koolen, H. H., Pinheiro, M. L. B., de Souza, A. Q. L., & de Souza, A. D. L. (2020). Flavonoid glycosides and their putative human metabolites as potential inhibitors of the SARS-CoV-2 main protease (Mpro) and RNA-dependent RNA polymerase (RdRp). *Memorias do Instituto Oswaldo Cruz*, 115, e200207. https://doi.org/10.1590/0074-02760200207
75. Vijayakumar, B. G., Ramesh, D., Joji, A., Jayachandra Prakasan, J., & Kannan, T. (2020). In silico pharmacokinetic and molecular docking studies of natural flavonoids and synthetic indole chalcones against essential proteins of SARS-CoV-2. *European Journal of Pharmacology*, 886, 173448. https://doi.org/10.1016/j.ejphar.2020.173448
76. Metwaly, A. M., Elkaeed, E. B., Alsfouk, B. A., Saleh, A. M., Mostafa, A. E., & Eissa, I. H. (2022). The computational preventive potential of the rare flavonoid, patuletin,

isolated from tagetes patula, against SARS-CoV-2. *Plants (Basel, Switzerland)*, 11(14), 1886. https://doi.org/10.3390/plants11141886
77. Faisal, S., Badshah, S. L., Kubra, B., Sharaf, M., Emwas, A. H., Jaremko, M., & Abdalla, M. (2021). Computational study of SARS-CoV-2 RNA dependent RNA polymerase allosteric site inhibition. *Molecules (Basel, Switzerland)*, 27(1), 223. https://doi.org/10.3390/molecules27010223
78. Raj, V., Lee, J. H., Shim, J. J., & Lee, J. (2022). Antiviral activities of 4H-chromen-4-one scaffold-containing flavonoids against SARS-CoV-2 using computational and in vitro approaches. *Journal of Molecular Liquids*, 353, 118775. https://doi.org/10.1016/j.molliq.2022.118775
79. Zandi, K., Musall, K., Oo, A., Cao, D., Liang, B., Hassandarvish, P., Lan, S., Slack, R. L., Kirby, K. A., Bassit, L., Amblard, F., Kim, B., AbuBakar, S., Sarafianos, S. G., & Schinazi, R. F. (2021). Baicalein and baicalin inhibit SARS-CoV-2 RNA-dependent-RNA polymerase. *Microorganisms*, 9(5), 893. https://doi.org/10.3390/microorganisms9050893
80. Hamdy, R., Mostafa, A., Abo Shama, N. M., Soliman, S. S. M., & Fayed, B. (2022). Comparative evaluation of flavonoids reveals the superiority and promising inhibition activity of silibinin against SARS-CoV-2. *Phytotherapy Research: PTR*, 36(7), 2921–2939. https://doi.org/10.1002/ptr.7486
81. Puttaswamy, H., Gowtham, H. G., Ojha, M. D., Yadav, A., Choudhir, G., Raguraman, V., Kongkham, B., Selvaraju, K., Shareef, S., Gehlot, P., Ahamed, F., & Chauhan, L. (2020). In silico studies evidenced the role of structurally diverse plant secondary metabolites in reducing SARS-CoV-2 pathogenesis. *Scientific Reports*, 10(1), 20584. https://doi.org/10.1038/s41598-020-77602-0
82. Rehman, M. F. U., Akhter, S., Batool, A. I., Selamoglu, Z., Sevindik, M., Eman, R., Mustaqeem, M., Akram, M. S., Kanwal, F., Lu, C., & Aslam, M. (2021). Effectiveness of natural antioxidants against SARS-CoV-2? Insights from the in-silico world. *Antibiotics (Basel, Switzerland)*, 10(8), 1011. https://doi.org/10.3390/antibiotics10081011

ABBREVIATIONS

ETC	Electron Transport Chain
MRSA	Methicillin-Resistant strain of *Staphylococcus aureus*
NiRAN	Nidovirus RdRp-Associated Nucleotidyl transferase
nsp	Non-structural Proteins
RdRp	RNA-dependent RNA polymerase
RNS	Reactive Nitrogen Species
ROS	Reactive Oxygen Species
SARS-CoV-2	Severe Acute Respiratory Syndrome Coronavirus 2

6 Flavonols and Their Derivatives
Pioneer Natural Compounds as Potential Antiviral Drugs against COVID-19

Nermeen Yosri, Heba Alshater, Maria Daglia, Alessandro Di Minno, Shaden A. M. Khalifa, Chao Zhao, Ming Du, Hamud A. Altaleb, Zisheng Luo, Xiaobo Zou, Zhiming Guo, and Hesham R. El-Seedi

6.1 INTRODUCTION

Natural products, including plants, microbes, marines, and bee products, have been utilized in Traditional Chinese, Indian Ayurvedic, and Islamic Medicines since time immemorial as antiviral remedies, as well as representing a crucial origin for a huge number of advanced drugs [1–3]. Flavonoids are natural polyphenolic products bearing an assortment of biological characteristics, and they are a favored framework for levelheaded drug design [4]. Their prominent antiviral, antioxidant, antimicrobial, and anti-inflammatory activities broadly qualified them for examination and implementation via diverse functioning mechanisms. Furthermore, they are able to modify cellular functions, membrane permeability, and viral replication, and they are employed to combat HIV, herpes simplex virus, and influenza virus [5,6]. Moreover, diverse surveys have demonstrated the anti-inflammatory action of polyphenol products in the improvement of chronic diseases [7].

Flavonols have various structural features that cause them to stand out among the other flavonoid subclasses. Generally, flavonols have 3 aromatic rings: A, B and C. On the C ring, there is an unsaturated bond at the C2–C3 position and the existence of a hydroxyl group (OH) in the third position; this provides both hydrogen and electron exchange, in addition to the occurrence of a hydroxyl group (OH) on the ring (B) in a catechol-type plan, enhances the pairing between the aromatic ring B and the other aromatic portion of the compound. There is also a benzoyl moiety on ring A and a carbonyl group on ring C. These features play a fundamental role in the unique activities of this subclass [8,9]. Diverse plant-based foods contain flavonols, mainly present in vegetables including onions, kale, tomatoes, broccoli, lettuce, and olive; fruits including grapefruit, citrus, apples, and berries; and drinks including tea and coffee (Figure 6.1 and Table 6.1) [10,11]. Quercetin and kaempferol are present in therapeutic herbs such as *Desmodium canadense*, *Azadirachta indica*, and *Cannabis sativa*. Glucosides of

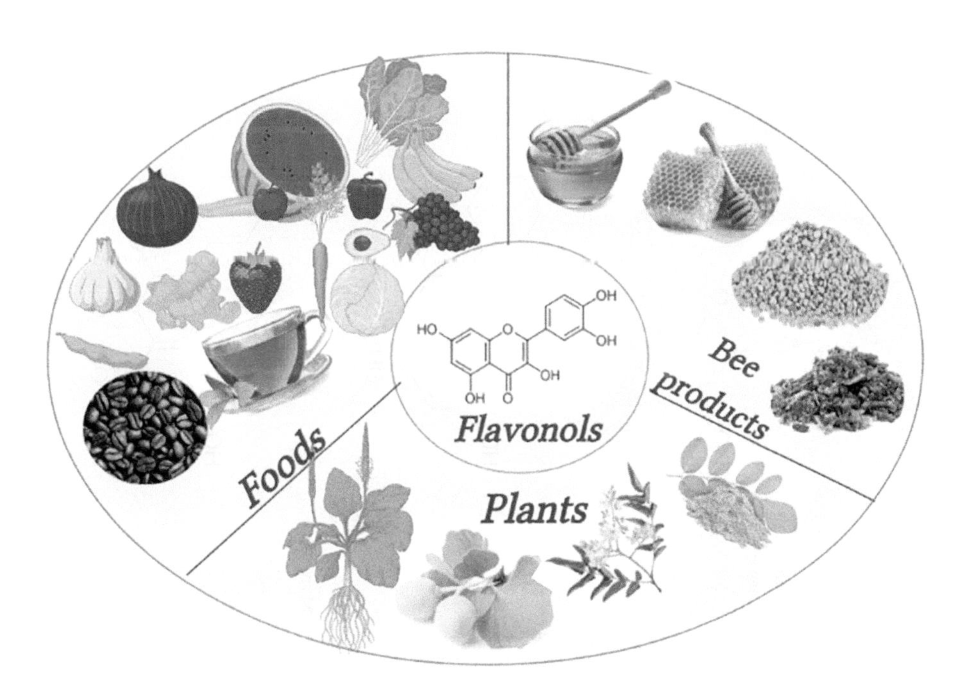

FIGURE 6.1 Sources of flavonols and their derivatives (created with Biorender.com).

DOI: 10.1201/9781003433200-6

TABLE 6.1
Structures of Flavonol, Their Derivatives, and Their Sources

No.	Name	Structure	Source	References
1	Quercetin	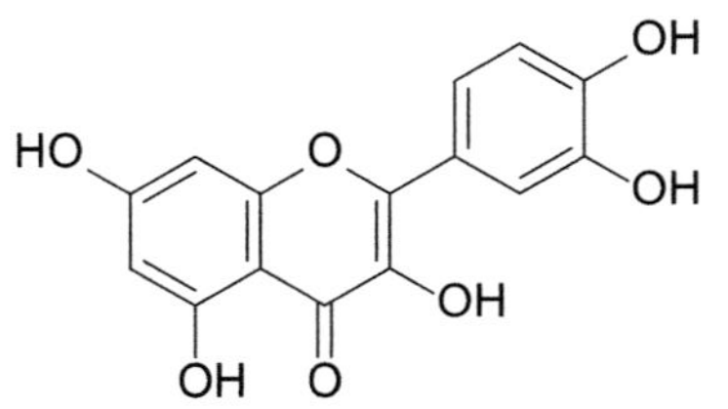	Berries, apples, cauliflower, lovage, capers, cilantro, dill, green pepper, tomatoes, tea, cabbage, nuts, and onions	[101,102]
2	Kaempferol Kaempferide	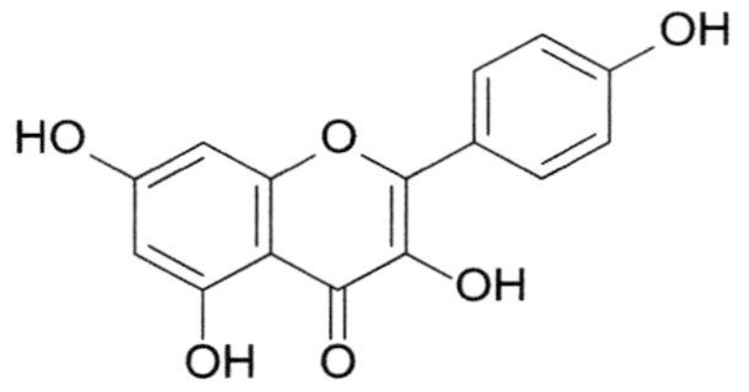 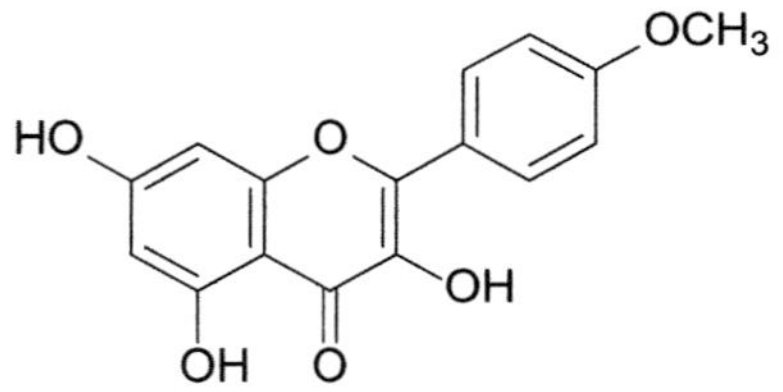	Seeds, leaves, herbs such as strawberries, apples, grapes, broccoli, tea, tomatoes, spinach, kale, beans, leeks, *Equisetum* spp., *Ginkgo biloba* leaves, *Tilia* spp., *Sophora japonica, Moringa oleifera Delphinium, Citrus, Camellia, Brassica, Berberis, Acacia nilotica* (L.), *Allium, Aloe vera* (L.), *Malus Delile* Burm.f., *Hypericum perforatum* L., *Crocus sativus* L., *Euphorbia pekinensis* Rupr., *Ginkgo biloba* L., *Phyllanthus emblica* L., *Rosmarinus officinalis* L., and *Ribes nigrum* L. *Mentha arvensis, Coriandrum sativum*, and *Ocimum sanctum*	[43,103,104]
3	Myricetin	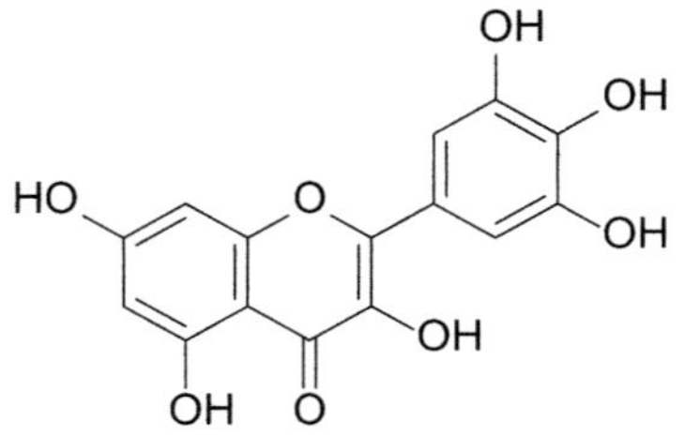	*Myricaceae, Anacardiaceae, Polygonaceae, Pinaceae, Primulaceae, Vitaceae, Leguminosae, Rosaceae, Ericaceae*, and *Fagaceae* and berries, nuts, tea, wine, honey, strawberry, and spinach	[44]
4	Morin	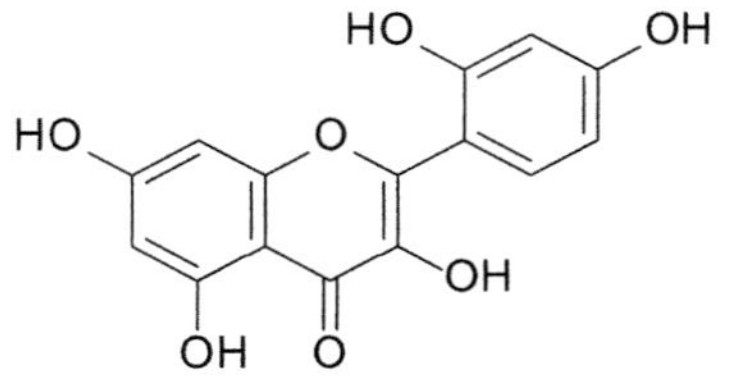	*Moraceae, Myrtaceae, Rosaceae*, and *Fagaceae*; *Maclura pomifera, Maclura tinctoria, Morus alba* L., *Psidium guajava, Castanea sativa, Malus pumila, Artocarpus heterophyllus, Prunus dulcis*, and *Chlorophora tinctoria*; cereal grains, red wine, coffee, tea, and seaweed	[105]

(Continued)

TABLE 6.1 (*Continued*)
Structures of Flavonol, Their Derivatives, and Their Sources

No.	Name	Structure	Source	References
5	Fisetin	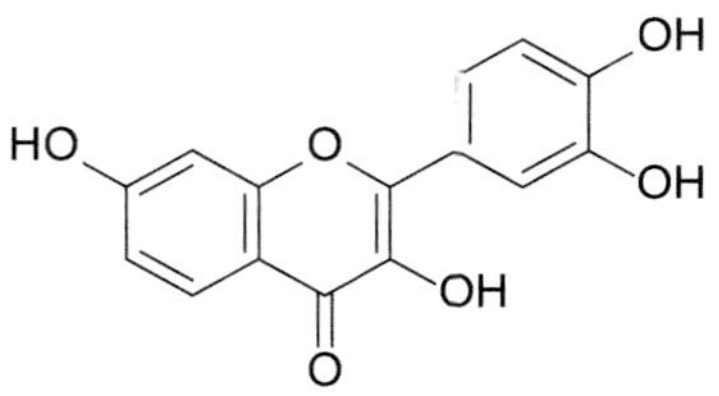 HO, O, OH, OH, OH, O	Persimmon, grapes, strawberry, lotus root, apples, onions, peach, tomatoes, persimmons, kiwi, kale, and cucumbers *Acacia greggii, Acacia berlandieri, Anacardiaceae* parrot tree (Butea fronds) *Gleditsia triacanthos, Quebracho colorado, Rhus cotinus, Rhus verniciflua* Stokes extract, *Cotinus coggygria, Pinophyta* species such as *Callitropsis nootkatensis*, and other trees and shrubs	[106]
6	Galangin	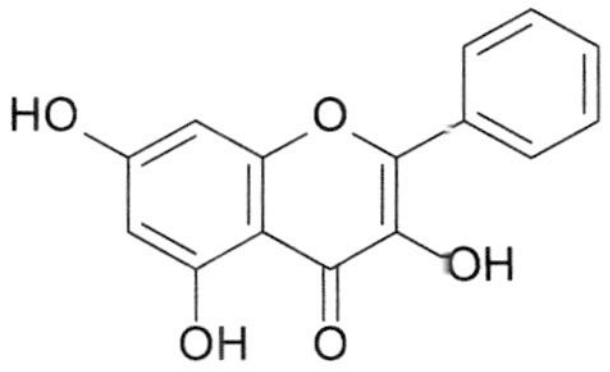 HO, O, OH, OH, O	Bee products and medicinal herbs, i.e., propolis, honey, *Alpinia officinarum* Hance (lesser galangal) shoots of *Helichrysum aureonitens, Alnus pendula* Matsum., *Scutellaria galericulata* L., and *Plantago major* L.	[65,66,104,107]
7	Gossypetin	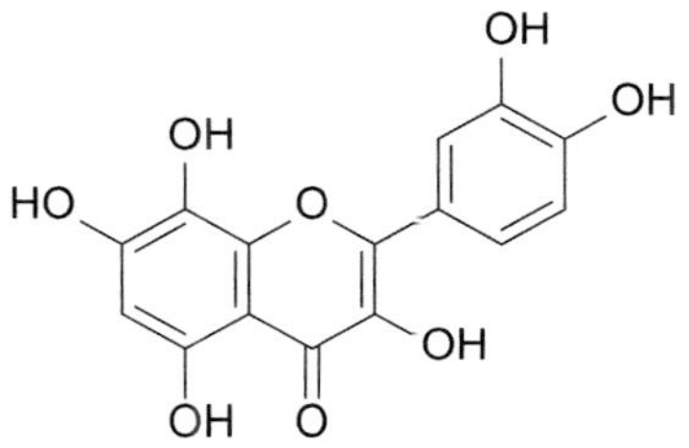 OH, OH, OH, HO, O, OH, OH, O	Flowers and Hibiscus such as *Hibiscus vitifolius* (tropical rose mallow; *Panchavam in Malayalam*), flowers of *Dionysia aretioides, Dionysia bornmuelleri*, and *Dionysia paradoxa. Loiseleuria procumbens, Leiophyllum buxifolium, Bruckenthallia speciliforum*, calyx of *Hibiscus sabdariffa* (Roselle), and *Menziesia lasiophylla*	[108]
8	Rhamnetin	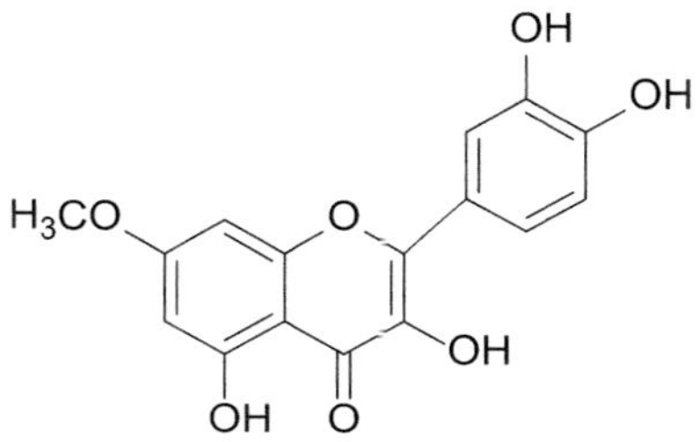 OH, OH, H_3CO, O, OH, OH, O	*Coriandrum sativum, Rhamnus petiolaris, Syzygium aromaticum, Moringa oleifera*, and *Prunus cerasus*	[109]
9	Isorhamnetin	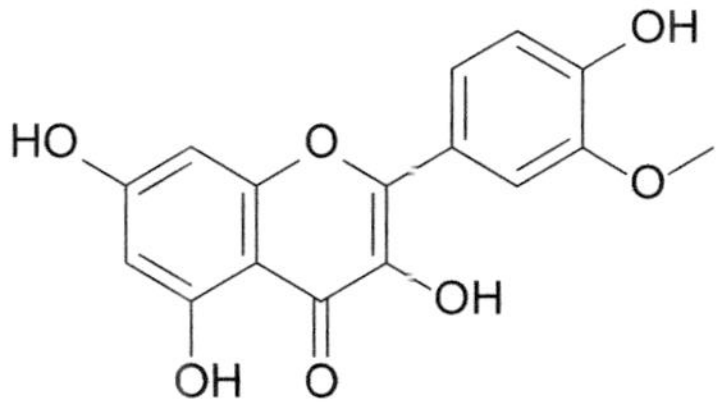 OH, HO, O, O, OH, OH, O	*Ginkgo biloba* L. leaves, *Hippophae rhamnoides* L., *Typha angustifolia* L., and *Calendula officinalis* L. and pears, olive oil, wine, and tomato sauce	[110,111]

(*Continued*)

TABLE 6.1 (*Continued*)
Structures of Flavonol, Their Derivatives, and Their Sources

No.	Name	Structure	Source	References
10	Astragalin	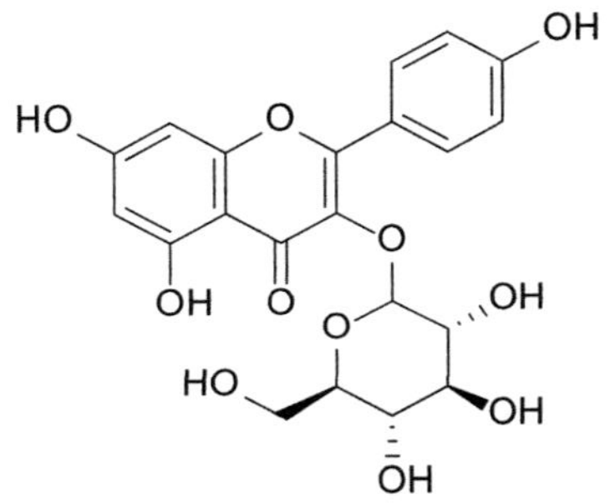	*Cuscuta chinensis* Lam., *Cassia alata* L., *Allium victorialis*, *Aceriphyllum rossii*, *Jasminum subtriplinerve* Blume., *Astragalus hamosus*, *Carthamus lanatus* L., *Corchorus olitorius* L., *Eupatorium cannabinum* L., *Glycyrrhiza uralensis* Fisch., *Hippophae rhamnoides* L., *Moringa oleifera* Lam., *Phaseolus vulgaris* L., *Solidago canadensis* L., and *Urtica cannabina* L.	[112]
11	Hyperoside	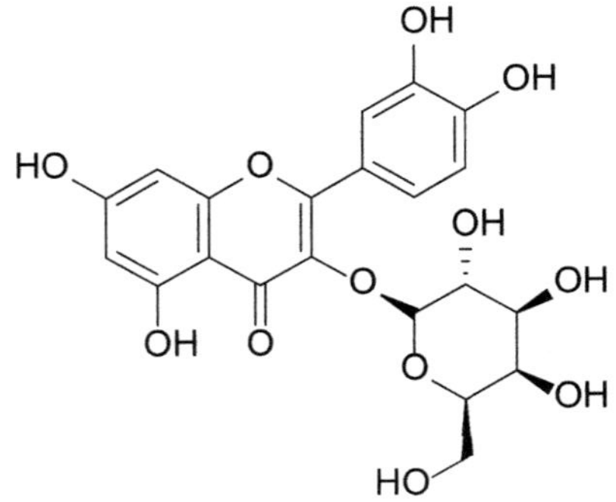	*Abelmoschus manihot*, *Apocynum venetum* L., *Crataegus*, *Semen cuscutae* , St. John's wort, *Hawthorn*, *Balbisia calycina*, and *Alchornea cordifolia*	[84]
12	Isoquercetin	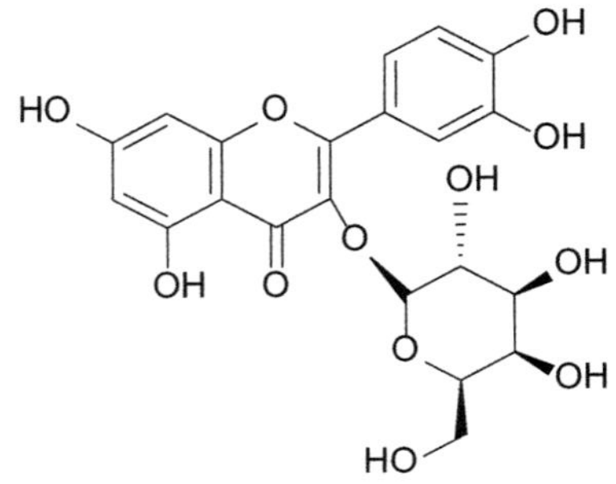	Several plants	[89]

(Continued)

TABLE 6.1 (*Continued*)
Structures of Flavonol, Their Derivatives, and Their Sources

No.	Name	Structure	Source	References
13	Rutin	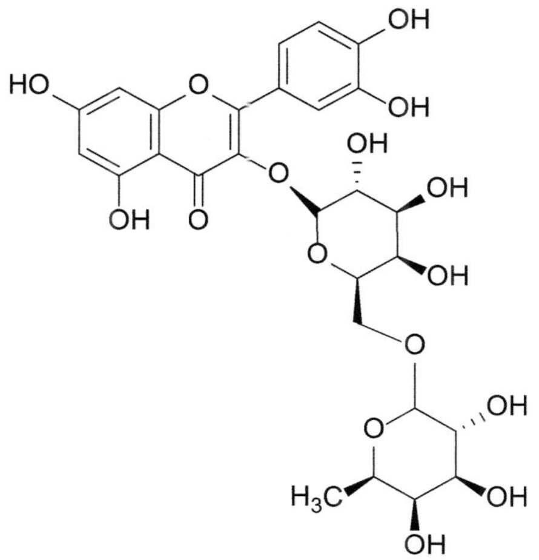	Apricots, asparagus, grapes, buckwheat, apples, cherries, grapefruit, oranges, plums, tea; *Maranta leuconeura*, *Sophora japonica* L., *Eucalyptus* spp., and *Ruta graveolens* L.	[113]
14	Robinin	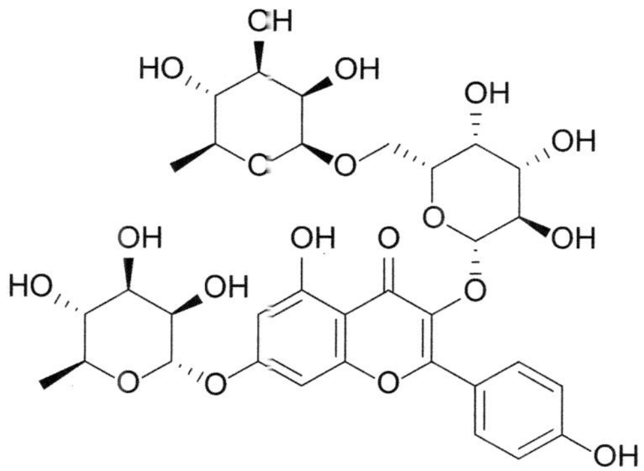	*Astragalus*, *Vigna unguiculata* leaf, *Vinca erecta*, *Astragalus shikokianus*, and *Pueraria hirsuta* L. (*kudzu vine*)	[114,115]

flavonols usually existed in *Acalypha indica*, *Bauhinia monandra*, *Betula pendula*, and *Ginkgo biloba* [12]. They are also present in herbs including Mediterranean oregano, thyme, mint, and dill as well as *Hedeoma patens*, *Lippia palmeri*, and *Lippia graveolens* [13].

The rapid spread and progression of the COVID-19 outbreak in 2019 and 2020 threatened public health and diminished the global economy worldwide, with a severity ranging from mild to critical. Improvement of symptoms in severe patients relies on anti-inflammatory medication, with delays in the removal of the virus increasing the risk of secondary infection, and to date, there are no clinically approved available drugs against this disease. So, it is essential to explore diverse potential antiviral drug strategies for combating COVID-19. Several studies have dedicated their attention to natural plant products, particularly flavonols, due to their anti-inflammatory potential (*in vitro* and *in vivo*), which makes them candidate therapeutic tools in diverse acute and chronic disorders [7,14]. Since flavonols possess antioxidant, antiviral, and anti-inflammatory actions, as well as the ability to bind with SARSCoV-2 spike protein and ACE2 receptors, they are thought to be a good target for enzyme inhibition of SARS-CoV-2 [15].

This work has been conducted as part of our ongoing project focused on the chemical and pharmacological study of plants utilized in traditional medicine [16–20]. The point of this survey is to attract the consideration of scientists to the antiviral and immunomodulatory activity of flavonols and their derivatives. On the whole, the data highlighted in our survey draws attention to the incipient and promising future of the field, which may see a significant increase in the utilization of flavanols.

6.2 FLAVONOLS AS POTENTIAL ANTIVIRAL DRUGS AGAINST COVID-19

In general, flavonols appear to have more effective bioactivities compared to various flavonoid family members. *In vitro*, *in vivo* studies and clinical trials have offered vibrant proof of the bioactivity of these compounds. The movement of antiviral medicine toward flavonols is owing to the expanded struggle of pathogens to antiviral and antimicrobial drugs; therefore, greater efforts are being applied to study the provable properties behind innovative therapeutics to be employed alone or in a mixture with current molecules [8]. A plethora of studies have demonstrated the efficacy of nanoencapsulation of brain-boosting polyphenols, such as caffeine, quercetin, and cocoa flavanols, in treatments responding to the COVID pandemic. Also, the antiviral potential of secondary metabolites including flavonols has been illustrated against a diverse range of viruses such as Coxsackie B virus type 1, respiratory syncytial virus (RSV), poliovirus type 1, HSV-1 and -2, Canine distemper virus, HIV-1, and finally, SARS-CoV. Indeed, these molecules can interfere with pathologic processes and interrupt the entrance and duplication cycle of viruses [15].

Several flavanols, i.e., myricetin, fisetin, rutin, kaempferol, and astragalin, have been shown to suppress cytokine production and synthesis [21]. Notably, flavanols inhibit various enzymes, mainly due to their presence of 2,3-double bonds and carbonyl groups existing at the C-4 position in the ring (C), and galloyl moiety. The existence of the (OH) in the position of C-5′ on the B ring in flavonoid skeletons also decreases the potential for chemical interactions, the formulation of hydrogen bonds, and the electrostatic interactions via the active site of $3CL^{pro}$ (Figure 6.2). Moreover, the existence of a glucose unit in the flavanol skeleton results in a vigorous binding affinity, which enables them to bind with SARSCoV-2 spike protein and ACE2 receptors [22].

Surveys have indicated that flavonol derivatives such as epigallocatechin gallate and gallocatechin gallate verified inhibitory potential toward SARS-CoV $3CL^{pro}$ during proteolytic conditions (FRET-based) test. Molecular docking studies and different assays have demonstrated that flavonols and chalcones are the major collections of flavonoids comprising the maximum number of compounds efficient toward PL and 3CL proteases, and these can inhibit viral replication of MERS-CoV and SARS by blocking the enzymatic activities of 3CL and PL, with SARS-helicase, and prohibit the deubiquitinating and deISGylation process of SARS PL^{pro} [23]. More consideration should thus be given to the effectiveness of flavanols, particularly when applied with other standard antiviral drugs.

6.2.1 Quercetin

Quercetin and its derivatives demonstrate a potent activity toward coronaviruses [24]. In general, quercetin has wide-spectrum antiviral potential, interfering with numerous stages of pathogen virulence, including virus entry, virus reproduction, and protein assembly [14,25]. For example, it was found that quercetin has an inhibitory action against papain-like protease (PL^{pro}) enzymes and 3 chymotrypsin-like protease ($3CL^{pro}$) with IC_{50} values of 8.6 μM and 23.8 μM, respectively, which are essential for SARS-CoV-1 replication [26–28]. Quercetin may play a significant role in the prevention and management of COVID-19 by blocking viral admission into the cell by fluctuating viral proteins, stimulating the nuclear factor erythroid-derived 2-like 2 (NRF2) pathway, and diminishing coagulopathy (inhibiting protein disulfide isomerase) [29]. Additionally, quercetin demonstrates a sufficient binding affinity to the active site of 3C-like protease ($3CL^{pro}$) enzyme and deactivates it with inhibition constant Ki~7 μM [30]. Moreover, current molecular docking research has shown that quercetin is able to inhibit replication of SARS-CoV-2 by targeting the active sites of the main virus protease (3CL; BE= –4.53 kcal/mol) and its specific receptor, angiotensin-converting enzyme 2 (ACE2; BE=–3.78 kcal/mol) [31]. Another *in vitro* study by Munafò et al. indicated that quercetin has inhibition potency against RNA-dependent RNA polymerase (RdRp) of SARS-CoV-2, which is responsible for

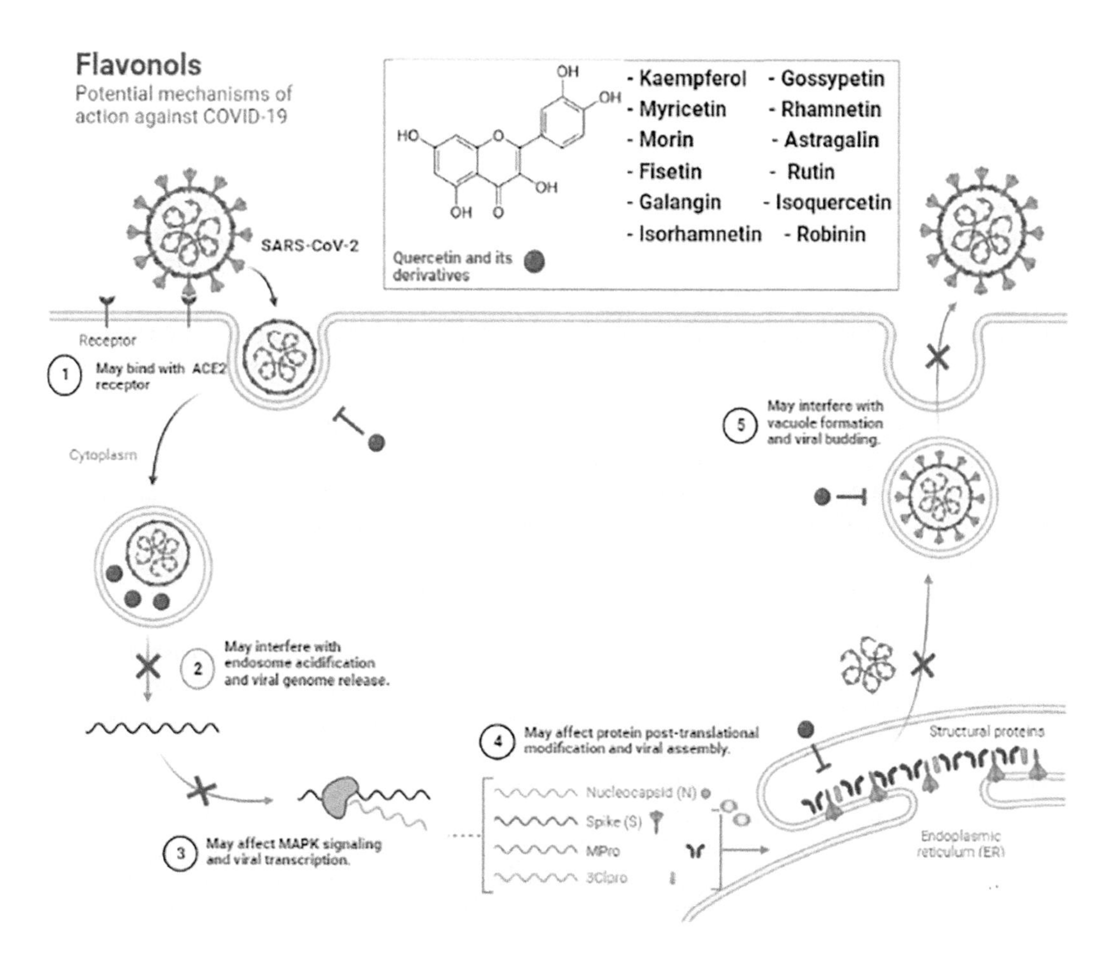

FIGURE 6.2 Scheme illustrating the possible inhibition mechanism of SARS-COV-2 via flavonols and their derivatives (created with Biorender.com).

genome viral replication, with an IC_{50} value of 6.9±1.0 μM [32]. Additionally, it was found that quercetin plays a key role by acting as an ionophore for zinc influx, where the presence of zinc can block RdRp regeneration [33]. Quercetin compounds also inhibit the NLRP3 inflammasome and modulate TH17-related cytokines (IL-17A and IL-21), which together contribute to its anti-inflammatory action [34] (Table 6.1).

6.2.2 Kaempferol

Kaempferol has antiviral efficacy toward flu infections, namely H9N2 and H1N1, since it focuses on the infection's neuraminidase protein and its derivatives. Many studies have demonstrated its activity against pulmonary capillary, pulmonary edema, and myeloperoxidase (MPO), as well as inhibition of many inflammatory cells. Its derivatives exhibit antiviral properties against SARS coronavirus 3a channel protein, which can inhibit virus construction and allow the infested body to set up or reinforce its own immune system. Subsequently, kaempferol and its derivatives can conceivably be chosen as options to battle COVID-19 disease via focusing on coronavirus proteases [35].

The inhibitory efficiency of kaempferol-7-*O*-glucoside toward HIV1/HIV 2, avian influenza H5N1, HSV-1, HCMV, HSV-2, and influenza A viruses was studied utilizing diverse mechanisms [36,37]. Additionally, computational research specified the potential significance of diverse flavonoids such as kaempferol and its derivatives, which have inhibitory effects against the 3a protein of SARS-CoV; *in vitro* and *in vivo* studies also demonstrated that kaempferol glucorhamnoside could inhibit NF-κB and MAP kinase phosphorylation [38,39]. The embarrassment of SARS-CoV 2 reproduction by kaempferol was also investigated using cell-based assays [40]. When cells are infected with SARS-CoV-2, kaempferol can maintain cellular homeostasis while promoting apoptosis by inhibiting cell growth and angiogenesis, as well as preventing inflammasome formation and the death of cells. Consequently, it could serve as a good factor for avoiding the replication of the virus [41].

Moreover, a recent molecular docking study has shown that kaempferol binds to the hACE2–S protein complex via MD simulations on hACE2 and *S* protein binding interfaces, and this was established by molecular mechanic/Poisson–Boltzmann surface area analysis, which showed that kaempferol has a low binding affinity (ΔGbind=–15.07±2.42 kcal/mol) [42]. Kaempferide, a derivative of kaempferol, displayed strong efficacy toward anti-SARS-CoV-2 and its hydrogen bonding with ASN142, LEU141, HIS163, SER144, and GLU166 with a binding energy of –7.83 kcal/mol. It also interacts with active positions on the SARS-CoV-2 spike protein, and this occurs directly or indirectly with the Cys/His dyad of the main protease to impede replication of the virus, which is supported by molecular dynamic simulation studies.

Kaempferide additionally showed a bonding affinity of –9.1 kcal/mol by molecular docking, and this was stabilized through three hydrogen bonds with active site residues of Gln260 (3.00 and 4.66), Tyr333. (6.20), and one hydrophobic bonding between (Pi–Pi stacked) with the active site residue. Thus, these compounds exhibited activity against the protein of SARS-CoV-2 [43]. MD simulation studies revealed that phytochemical activity of kaempferol occurs via binding at the junction of the hACE2–S protein complex with lower binding energy, and *in vitro* and *in vivo* studies indicated that it may be utilized for new anti-COVID-19 agents [37].

6.2.3 Myricetin

Myricetin has anticancer, anti-photoaging, antihypertensive, antiplatelet aggregation, immunomodulatory, anti-allergic, anti-inflammatory, and analgesic activities [44]. Several studies have demonstrated an antioxidant impact on cancer cells, its ability to reduce platelet aggregation by reducing alpha granule secretion and fibrinogen binding, its antioxidant activity, and the regression of neurodegenerative diseases. Also, they have demonstrated the therapeutic effect of myricetin on several ailments, such as thrombosis, atherosclerosis, diabetes, Alzheimer's disease, cerebral ischemia, and pathogenic microbial infections. The aforementioned activities have indicated a role for myricetin in reinforcing immunomodulatory potentials, suppressing cytokine storms, and managing cardiac dysfunction [45].

Myricetin was implemented as a potent entity toward Zika virus, which is causing infants' microcephaly and Guillain-Barre syndrome [36,46]. A molecular docking study showed a lower value for myricetin compared to other flavanols (–8.0), which indicates that it can inhibit SARS-CoV-2 M^{pro} activity, with an IC_{50} value of $3.684 \pm 0.076\,\mu M$ [47]. Myricetin also presented the lowest half-maximal effective concentration (EC_{50}) values of $0.91 \pm 0.05\,\mu M$, which indicates that it is a good candidate to stop SARS-CoV-2 replication, with the molecular docking results indicating that van der Waals and hydrogen bonding are the main intermolecular forces for interaction between amino acid residues in the SARS-CoV-2 nsp14 ExoN active site and myricetin [40]. Despite the high biological activities of myricetin, few clinical trials have been performed [48].

6.2.4 Morin

Morin is known to have bacteriostatic [49], antioxidant, anticancer, and anti-diabetic effects [50], promoting inhibition of acetylcholinesterase [51], management of metabolic syndrome [50], acceleration of bone regeneration (*in vivo*) [52], elevation of Alzheimer disease by enhancement of neural glyoxalase pathway [53], as well as antiepileptic and anti-inflammatory effects [54]. In a study conducted by Hong et al., it was found that morin, both alone and when combined with Oseltamivir, displays potential inhibition toward influenza A/Puerto Rico/8/1934 (A/PR/8; H1N1) and oseltamivir-resistant A/PR/8 influenza viruses by reducing hemagglutination, pro-inflammatory cytokine levels, chemokines (TNF-α and CCL2) levels, and by blocking the entry of viruses [55].

Very recently, Mori et al. found that morin at 100 µg/mL eliminated 53.64% $3CL^{pro}$ of SARS-CoV-2 via establishing H-bonds with viral amino acid residues: Asn142, Gln189, and Thr190 [56]. Another molecular docking study showed that morin interacts with $3CL^{pro}$ and RBD, with binding energies of –9.6 and –8.46, respectively, which resulted in the potential inhibitor agent of SARS-CoV-2 replication [57]. Additionally, Tallei et al. suggested that the morin's docking scores against main protease (M^{pro}) and spike (S) glycoprotein are dependent on the presence of carbonyl group (positions suitable for an electrophilic attack) and hydroxyl group (positions suitable for a nucleophilic attack) [58]. For example, morin can interact with PL^{pro} with a dock score of 6.897 kcal/mol and a binding energy of 33.25 kcal/mol, owing to the establishment of hydrogen bonding with the subsequent amino acid residues: Gly 266 and Asn 267 at (–OH group of rings); Thr 301 with (C=O of the ring), as well as TYR 273, TYR 264, and TYR 268 with (OH groups of aryl substituent) [59]. According to Gupta et al. morin serves as an anti-inflammatory and inhibitory agent against various types of coronaviruses including SARS-CoV-1, SARS-CoV-2, and MERS-CoV, since it is able to suppress proinflammatory cytokines (IL-6, 8, and 10), viral spike glycoprotein, $3CL^{pro}$ and PL^{pro}, in addition to targeting host ACE2, importin-α (IMP-α), and poly (ADP-ribose) polymerase (PARP)-1 [60].

6.2.5 Fisetin

The activity of fisetin on SARs is due to hydroxyl groups on its aromatic rings, particularly the 3,4-dihydroxylation at the B ring, and the substitution of various hydroxyl (OH) groups in fisetin impact its original properties [61]. It can inhibit HIV-1 and CHIKV infections through obstructive viral admission and virus-cell synthesis [37]. It exhibited anti-inflammatory and immunomodulatory efficacy, and it inhibits COX-2 as well as causes a reduction in IL-8, IL-6, CCL5, TNF-*α*, and MCP1 after pre-treatment with IL-1*β*-stimulated human lung epithelial cells. It also has an effect on the inhibitor of protein kinase, suggesting that it could be implemented as a latent immunomodulator in the lung inflammation [38].

Fisetin can connect to the spike protein, making it an appropriate principal molecule against SARS-CoV-2 [62]. The molecular docking and binding affinity calculations for fisetin gave a result of –8.5 kcal/mol, and a simulation study was performed on the hACE2–S protein complex destined with fisetin, with the outcomes indicating that this phytocompound binds to the hACE2–S complex with low binding free energy, so fisetin can restrict with the complex and potentially prohibit entry of the virus [62].

Under *in vivo* pharmacokinetic conditions, fisetin was absorbed speedily after intravenous uptake (10 mg/kg) in Sprague-Dawley rats [61]. Fisetin is currently being inspected preclinically in anti-COVID-19 trials, and the results demonstrate that fisetin can remove senescent cells in β-coronavirus-infected mice [63]. The activity of fisetin against COVID-19 is due to its ability to bind with the hACE2–S protein complex of the virus with a low binding free energy. Thus, it is an innovative anti-COVID-19 agent [37].

6.2.6 Galangin

Galangin can block several pro-inflammatory mediators, including the nucleotide binding and oligomerization domain-like receptor family pyrin domain-containing protein 3 (NLRP3) inflammasome and inducible nitric oxide synthase (iNOS) [64]. Galangin also inhibits numerous viruses, including HSV-1, by eliminating viral replication at $IC_{50}=0.00045\%$ with a selectivity index (SI)=3.3 [65]. Furthermore, galangin isolated from *Helichrysum aureonitens* aerial parts, at doses between 12 – 47 µg/mL, has exhibited potent antiviral action against HSV-1 and coxsackie B virus type 1 (CoxB1) [66].

Notably, some molecular docking studies claim that galangin displayed an inhibitory effect on SARS-CoV-2. For instance, an experimental work established by Guler et al. claims that propolis and its constituent (galangin) eliminate the activity of SARS-CoV-2 by strapping the angiotensin-converting enzyme (ACE)-related carboxypeptidase (ACE-II) and targeting the functional receptor of the coronavirus with binding energy and K_i of –7.35 kcal/mol and 4.10 µM, respectively [67]. Another study claims that galangin inhibits viral replication by targeting the main protease M^{pro} enzyme (Docking score: –6.295 kcal/mol) [68]. Further, compared to the control (hydroxychloroquine), galangin exhibited an excellent record against SARS-CoV-2 main protease with binding scores of –8.066 kcal/mol since this compound has shown a high number of hydrogen bond interactions with PHE140, HIS164, and HIS41 residues [69]. Similarly, galangin derived from Egyptian propolis could interact with an amino acid residue namely GLY482 in the S1 spike protein of COVID-19 with an ICM Score of –59.5 [70].

6.2.7 Gossypetin

Gossypetin derivatives are able to target 3CL^{pro}, a protein-enzyme that shows a vital role in virus replication. Molecular docking results indicate that these compounds interact with an amino acid in the catalytic center at the S1–S5 subsites of COVID-19, as well as binding to the critical conserved Cys145 and His41 of the catalytic center [71]. Gossypetin displays an interaction with VP35 and VP24, which are indispensable for function of the Ebola virus, through docking and simulation techniques utilized to evaluate the binding of gossypetin and hydroxychloroquine on the virus of SARS-CoV-2. Gossypetin was found to bind to the protein of VP24 with a docking value of –8.252 kcal/mol and a binding free energy of –34.633 kcal/mol, and it also formulates 4 bonds of hydrogen with the protein side chain of VP24. Also, it can interact vigorously with VP35, with a docking score of –5.395 and a binding energy of 31.628 kcal/mol. The binding free energy of gossypetin has been calculated in reaction with *S* protein subsites of coronavirus by Molecular Mechanics Poisson–Boltzmann Surface Area [72].

6.2.8 Rhamnetin

Rhamnetin [73] presents antiviral activity against coronaviruses owing to the specific binding sites of 3CL^{pro} and Nsp15 in COVID-19. The docked structure of rhamnetin with 3CL^{pro} and endoribonuclease of SARS-CoV-2 is due to the formation of strong H-bonds with amino acids of the main protease 3CL^{pro} [74]. Also, it binds to RdRp via amino acids. Because these amino acids have significant hydrophilic properties, they can interact with water molecules on the receptor's surface, resulting in the stability of the RdRp's inner structure [75].

6.2.9 Isorhamnetin

According to an *in vitro* investigation by Zhan et al., the isorhamnetin molecule inhibits the duplication of the SARS-CoV-2 spike pseudotyped virus by antagonizing ACE2 and reducing the entrance of the virus into ACE2h cells [76]. Another *in vitro* study assembled by Shahhamzehei et al. showed isorhamnetin's activity against SARS-CoV-1 M^{pro} and SARS-CoV-2 M^{pro} enzymes, with IC_{50} values of 8.42 ± 1.15 µM and 13.13 ± 1.78 µM, respectively. In the same study, isorhamnetin was found to combat SARS-CoV-1, SARS-CoV-2, HCoV-HKU1, MERS-CoV, HCoV-OC43, HCoV-NL63, and HCoV-229E with binding affinity scores ranging from –7.65 to 9.01 kcal/mol [77]. Furthermore, Xu et al. found that some flavonoid molecules, such as quercetin, luteolin, and isorhamnetin, have potential antiviral activity toward COVID-19, by using a molecular docking study documenting the suppression of 3CL^{pro} and ACE2 enzymes with binding affinities ranging from –25.95 to –36.82 kJ/mol [78].

The derivatives of isorhamnetin, particularly isorhamnetin 3,7-*O*-α-L-*di*-rhamnoside, also show the latent to be anti-COVID-19 M^{pro} virus, with this latter showing a binding score of –9.04 kcal/mol with Phe140, Arg188, Thr190, Gln192, and Gly143 amino acids [79]. Moreover, isorhamnetin-3-*O*-β-D-glucose has a binding score of –8.7 kcal/mol, forming H-bonds with the amino acids (Phe140, Asn142, His164, Gly143, Cys145, Glu166, His163, Met165, Gln189, His41, Arg188, Thr26, Thr25, Met49, Ser144, Leu141, Asp187, Ser146, and Leu27) [80].

6.3 FLAVONOL DERIVATIVES AS POTENTIAL ANTIVIRAL DRUGS AGAINST COVID-19

6.3.1 Astragalin

Murugesan et al. reported that astragalin, in addition to other flavonoid compounds, has anti-SARS-CoV-2 via the stopping of M^{pro} enzyme, with scoring values of –5.8 and –7.31 kcal/mol, respectively [81]. Astragalin may inhibit SARS-CoV-2 via elimination of M^{pro}, with a binding energy score of –6.9 to –8.7 kcal/mol [82]. In a similar study, astragalin was shown to be the most effective COVID-19 inhibitor among 198 compounds through interactions with the molecular docking energies –7.6, –8.0, and –8.5, kcal/mol for 6VXX, 6LZG, and 6LU7 proteins of SARS-CoV-2, respectively, based on hydrogen bonding interaction with the active site catalytic residues, HIS-41 or CYS-145, of the main protease SARS-CoV-2 [83].

6.3.2 Hyperoside

Hyperoside displayed pharmacological effectiveness against inflammatory and oxidative activities, as well as protecting cardiomyocytes and liver and against influenza viruses, and recently, in the treatment of COVID-19 [84]. Its antioxidant activity is due to the occurrence of hydroxyl groups (OH) on the A and B rings and glycosides which are linked to the C ring.

Recently, the therapeutic efficacy of hyperoside was studied, resulting in the prohibition of high glucose-induced vascular inflammation *in vitro* and *in vivo* [85]. Hyperoside inhibited N3 and can bind to the substrate binding pockets of new COVID-19 with a docking score of –8.6 kcal/mol, and it forms six H-bonds with Ser144, Leu141, Arg188, Thr190, His163, and Gln192 residues of SARS-CoV-2 M^{pro} [6,86]. The potential inhibition of hyperoside extracted from fresh Neem leaves toward influenza viruses (H9N2, H7N7, H7N3, H7N2, H5N1, H2N3, H2N2, H1N2, and H1N1) was studied [87]. So far, the examination on the pharmacokinetics and pharmacology of hyperoside is not enough, restricting its therapeutic development [88].

6.3.3 Isoquercetin

Isoquercetin possesses antiviral activities [89]. However, isoquercetin activity is lower than that of quercetin *in vitro* and *ex vivo*, but it appears to be the same or greater *in vivo*. Isoquercetin can protect against influenza, Zika, Ebola, and dengue virus infections. The pharmacokinetic profile of isoquercetin after oral administration, including its IC_{50} and CC_{50} values, was investigated to produce selectivity indices. It was found to prohibit (SARS-CoV-2) and diminish the severity and lethality of COVID-19. *In silico* screening of small molecules revealed a high efficacy for isoquercetin in binding to proteins involved in SARS-CoV-2 infestation, with a docking score of –7.7 to –10.3 kcal/mol for SARS-CoV-2 S protein and RdRp, 3CL^{pro}, and PL^{pro} [90].

6.3.4 Rutin

Rutin, like other flavonol compounds, displays anti-COVID behavior as detected in *in silico* studies. For instance, Al-Zahrani reported that rutin inhibits the M^{pro} enzyme of COVID-19 with a docking score value of –9.00 kcal/mol, compared to a positive control (Lopinavir; –8.4 kcal/mol) [91]. It can suppress the viral activity of COVID by building hydrogen bonds with Asn142 (2.1 Å), Cys145 (2.63Å), Thr190 (2.35 Å), and Gly143 (2.3 Å) as well as forming a σ–π stacking interaction with Gln189 [92]. Similarly, rutin isolated from *Berberis asiatica* can eliminate SARS-CoV-2 M^{pro} by intermolecular interactions with His41 and Cys145 (catalytic residues) with free energy and docking scores of –31.12 kcal/mol and –8.4 kcal/mol, respectively [93]. Other studies indicate that rutin compound not only inhibits M^{pro}, but can also eliminate RdRp, PL, and S proteins of SARS-CoV-2 [94].

6.3.5 Robinin

Molecular dynamics studies demonstrate that robinin possesses the highest binding affinity among flavonols (–9 kcal/mol) against HIV-1. It binds with the hydrogen of Asn142, Gly143, Arg188, and Thr190 amino acids and is placed securely in the protein binding pocket. The molecular dynamic study also demonstrated the protein complex stability during the simulation and showed that there are no conformational changes to the proteins attached, which in turn suggests its potential against SASR-CoV-2 M^{pro} [95].

6.4 FLAVONOLS AND THEIR DERIVATIVES IN CLINICAL TRIALS AND DRUGS

Generally, clinical trials are initiated as soon as a drug's efficacy has been investigated in *in vitro*, *in vivo*, and *in silico* research [96]. Notably, The United States Food and Drug Administration (USFDA) declared quercetin (up to a level of 500 mg/day) as a GRAS (generally recognized as safe) ingredient [97]. According to Agrawal et al. and Di Pierro et al., quercetin can inhibit coronavirus entry and replication in various (*in silico*, *in vitro*, and *in vivo*) studies [98,99]. Clinically, quercetin has also been found to inhibit coronavirus replication. For instance, in a study conducted by Di Pierro et al., 152 patients with COVID-19 were recruited and separated into two groups of 76 patients each. The first group was preserved as control (standard care), while the second was administered quercetin supplements (200 mg) at a dose of 2 tablets/day (1 every 12 hours) for 30 days. The results showed that quercetin improved the immune system of patients compared to the standard group, with 15 (19.7%) patients within the standard care group requiring invasive oxygen therapy compared to 1 patient (1.3%) in the quercetin group [99]. Another randomized, double-blinded, placebo-controlled trial conducted by Heinz et al. indicated that digestion of 1,000 mg of quercetin per day for 12 weeks decreased the duration of upper respiratory tract infections [100].

TABLE 6.2
Screening for Flavonol Compounds and Their Derivatives in Clinical Trials Targeting COVID-19

Flavonol Compound	Clinical Trials/ Phase	Dose/Route of Administration/Mechanism of Activity	Reference ClinicalTrials.gov Identifier
Quercetin	Clinical trial/dietary supplement/ phase 3/interventional	200 mg/twice daily for 30 days boosted the natural immunity of the patients and helped in preventing the COVID-19 disease progression	(https://clinicaltrials.gov, NCT04578158) [99]
	Clinical trial/dietary supplement/ not applicable/interventional	Quercetin/1,000 mg/once daily/treatment	(https://clinicaltrials.gov, NCT04377789)
	Clinical trial/dietary supplement/ not applicable/interventional	600 mg/once daily for a week prevents progression of COVID-19 and alleviates symptoms in the initial stage of infection	(https://clinicaltrials.gov, NCT04861298)
	Clinical trial/dietary supplement/ phase 4/interventional	500 mg/once daily for 28 days inhibited viral replication and decreased the severity of the disease	(https://clinicaltrials.gov, NCT04468139)
	Clinical trial/dietary supplement/ not applicable/interventional	500 mg/twice daily for 3 months/prevention of COVID-19 infection	(https://clinicaltrials.gov, NCT05037240)
	Clinical trial/dietary supplement/ phase 1/interventional	One tablet/three times per day	(https://clinicaltrials.gov, NCT04851821)
Isoquercetin	Clinical trial/ drug/ phase 2/ interventional	Twice daily for 28 days alleviate disease progression	(https://clinicaltrials.gov, NCT04536090)
Fisetin	Clinical trial/Dietary supplement/ phase 2/interventional	20 mg/kg per day oral for 2 following days can prevent the progression of the disease and alleviate symptoms of coronavirus due to its anti-inflammatory action	(https://clinicaltrials.gov, NCT04476953)
	Clinical trial/phase 2/interventional	20 mg/kg per day/oral for 2 successive days twice (days 0 and 1; days 8 and 9) to alleviate dysfunction and excessive inflammatory response in elder adults in nursing homes	(https://clinicaltrials.gov, NCT04537299)
	Clinical trial/phase 2/interventional	20 mg/kg per day/oral for 4 days (days 0 and 1; days 8 and 9) alleviated dysfunction and decreased complications in at-risk outpatients	(https://clinicaltrials.gov, NCT04771611)

Furthermore, three different clinical trials demonstrated that oral intake of 20 gm/kg of fisetin daily can improve dysfunction and extreme inflammatory responses in elder adults in nursing homes, decrease complications in at-risk outpatients, prevent the progression of the disease, and subsequently reduce the rate of death (https://clinicaltrials.gov; NCT04476953; NCT04771611; NCT04537299). Finally, Isoquercetin at 500 mg was found to alleviate the progression of the disease in another clinical trial study (https://clinicaltrials.gov, NCT04536090) (Table 6.2)

6.5 CONCLUSION

The current global eruption of SARS-CoV-2, combined with the lack of definite drugs for it, has prompted scientific scholars to inspect some alternate natural treatments. In our present study, we reviewed the efficacy of 14 bioactive flavonol derivatives against SARS-CoV-2. They are found among traditional medicinal plants and other natural products, including bee products, and possess antiviral activities. Our study revealed that fisetin, quercetin, kaempferol, myricetin, astragalin, and rutin have the highest activity among bioactive flavonol derivatives, indicating that they are able to stimulate the inhibition of cytokine expression and synthesis. These results were further established through molecular docking, MD simulation, and dynamic studies of these bioactive flavonol derivatives against SARS-CoV, as well as *in silico* studies and clinical trials. The results indicate that they can inhibit $3CL^{pro}$ and PL^{pro} and interfere with the replication of SARS-CoV-2. Furthermore, quercetin, isoquercetin, and fisetin can bind with the hACE2–S complex with lower binding free energy than other flavonol derivatives. So, due to their probable ability to prevent and manage COVID-19 in patients, more attention should be paid to the use of these flavonol derivatives in clinical research aimed at combating COVID-19.

REFERENCES

[1] N. Gogoi, P. Chowdhury, A. Kumar, G. Aparoop and D. Chetia, Computational guided identification of a citrus flavonoid as potential inhibitor of SARS - CoV - 2 main protease, *Mol. Divers.* 25 (2021), pp. 1745–1759.

[2] H.R. El-Seedi, S.A.M. Khalifa, N. Yosri, A. Khatib, L. Chen, A. Saeed et al., Plants mentioned in the Islamic Scriptures (Holy Qur'ân and Ahadith): Traditional uses and medicinal importance in contemporary times, *J. Ethnopharmacol.* 243 (2019), pp. 112007.

[3] N. Yosri, A.A. Abd El-Wahed, R. Ghonaim, O.M. Khattab, A. Sabry, M.A.A. Ibrahim et al., Anti-viral and immunomodulatory properties of propolis: Chemical diversity, pharmacological properties, preclinical and clinical applications, and in silico potential against SARS-CoV-2, *Foods.* 10 (2021), pp. 1776.

[4] F. Mangiavacchi, P. Botwina, E. Menichetti, L. Bagnoli, O. Rosati, F. Marini et al., Seleno-functionalization of quercetin improves the non-covalent inhibition of M pro and its antiviral activity in cells against SARS-CoV-2, *Int. J. Mol. Sci.* 22 (2021), pp. 7048.

[5] A.O. Fadaka, N. Remaliah, S. Sibuyi, D.R. Martin, A. Klein, A. Madiehe et al., Development of effective therapeutic molecule from natural sources against coronavirus protease, *Int. J. Mol. Sci.* 22 (2021), pp. 9431.

[6] Q. Wei, Q.-Z. Li and R.-L. Wang, Flavonoid components, distribution, and biological activities in Taxus: A review, *Molecules.* 28 (2023), pp. 1713.

[7] S.C. Patients, G. Giovinazzo, C. Gerardi, C. Uberti-foppa and L. Lopalco, Can natural polyphenols help in reducing cytokine, *Molecules.* 25 (2020), pp. 5888.

[8] T. Gervasi, A. Calderaro, D. Barreca, E. Tellone, D. Trombetta, S. Ficarra et al., Biotechnological applications and health-promoting properties of flavonols : An updated view, *Int. J. Mol. Sci.* 23 (2022), pp. 1710.

[9] R.T. Magar and J.K. Sohng, A Review on structure, modifications and structure-activity relation of quercetin and its derivatives, *J. Microbiol. Biotechnol.* 30 (2020), pp. 11–20.

[10] A. Kozłowska and D. Szostak-Węgierek, Targeting cardiovascular diseases by flavonols : An update, *Nutrients.* 14 (2022), pp. 1439.

[11] J. Popiolek-kalisz, The impact of dietary flavonols on central obesity parameters in polish adults, *Nutrients.* 14 (2022), pp. 5051.

[12] E.B. Guglya, Pharmacokinetics of quercetin and other flavonols studied by liquid chromatography and LC-MS (a review), *Pharm. Chem. J.* 48 (2014), pp. 489–498.

[13] R. Lone, R. Shuab and A.N. Kamili, *Plant Phenolics in Sustainable Agriculture*, Vol. 1, Springer, Singapore, 2020.

[14] S.A.M. Khalifa, N. Yosri, M.F. El-Mallah, R. Ghonaim, Z. Guo, S.G. Musharraf et al., Screening for natural and derived bio-active compounds in preclinical and clinical studies: One of the frontlines of fighting the coronaviruses pandemic, *Phytomedicine.* 85 (2020), pp. 153311.

[15] C. Mouffouk, S. Mouffouk, S. Mouffouk, L. Hambaba and H. Haba, Flavonols as potential antiviral drugs targeting SARS-CoV-2 proteases (3CL^{pro} and PL^{pro}), spike protein, RNA-dependent RNA polymerase (RdRp) and angiotensin-converting enzyme II receptor (ACE2), *Eur. J. Pharmacol.* 891 (2020), pp. 173759.

[16] H.R. El-Seedi, S.A.M. Khalifa, A.H. Mohamed, N. Yosri, C. Zhao, N. El-Wakeil et al., Plant extracts and compounds for combating schistosomiasis, *Phytochem. Rev.* 22 (2022), pp. 1691–1806.

[17] N. Yosri, S.A.M. Khalifa, Z. Guo, B. Xu, X. Zou and H.R. El-Seedi, Marine organisms: Pioneer natural sources of polysaccharides/proteins for green synthesis of nanoparticles and their potential applications, *Int. J. Biol. Macromol.* 193 (2021), pp. 1767–1798.

[18] H.R. El-Seedi, N. Yosri, S.A.M. Khalifa, Z. Guo, S.G. Musharraf, J. Xiao et al., Exploring natural products-based cancer therapeutics derived from egyptian flora, *J. Ethnopharmacol.* 269 (2021), pp. 113626.

[19] S.A.M. Khalifa, M.M. Swilam, A.A.A. El-Wahed, M. Du, H.H.R. El-Seedi, G. Kai et al., Beyond the pandemic: COVID-19 pandemic changed the face of life, *Int. J. Environ. Res. Public Health.* 18 (2021), pp. 5645.

[20] M.A.A. Ibrahim, A.H.M. Abdelrahman, T.A. Hussien, E.A.A. Badr, T.A. Mohamed, H.R. El-Seedi et al., In silico drug discovery of major metabolites from spices as SARS-CoV-2 main protease inhibitors, *Comput. Biol. Med.* 126 (2020), pp. 104046.

[21] S. Higa, T. Hirano and M. Kotani, Fisetin, a flavonol, inhibits T H 2-type cytokine production by activated human basophils, *J. Allergy Clin. Immunol.* 111 (2003), pp. 1299–1306.

[22] W. Ngwa, R. Kumar, D. Thompson, W. Lyerly, R. Moore, T. Reid et al., Potential of flavonoid-inspired phytomedicines against COVID-19, *Molecules* 25 (2020), pp. 2707.

[23] J. Solnier and J. Fladerer, Flavonoids : A complementary approach to conventional therapy of COVID-19 ?, *Phytochem. Rev.* 20 (2021), pp. 773–795.

[24] A. Pawar, M. Russo, I. Rani, K. Goswami, G.L. Russo and A. Pal, A critical evaluation of risk to reward ratio of quercetin supplementation for COVID-19 and associated comorbid conditions, *Phyther. Res.* 36 (2022), pp. 2394–2415.

[25] R.M.L. Colunga Biancatelli, M. Berrill, J.D. Catravas and P.E. Marik, Quercetin and vitamin c: An experimental, synergistic therapy for the prevention and treatment of sars-cov-2 related disease (COVID-19), *Front. Immunol.* 11 (2020), pp. 1451.

[26] Y.B. Ryu, H.J. Jeong, J.H. Kim, Y.M. Kim, J.-Y. Park, D. Kim et al., Biflavonoids from *Torreya nucifera* displaying SARS-CoV 3CL^{pro} inhibition, *Bioorg. Med. Chem.* 18 (2010), pp. 7940–7947.

[27] J.-Y. Park, H.J. Yuk, H.W. Ryu, S.H. Lim, K.S. Kim, K.H. Park et al., Evaluation of polyphenols from Broussonetia papyrifera as coronavirus protease inhibitors, *J. Enzyme Inhib. Med. Chem.* 32 (2017), pp. 504–512.

[28] A. Di Petrillo, G. Orrù, A. Fais and M.C. Fantini, Quercetin and its derivates as antiviral potentials: A comprehensive review, *Phyther. Res.* 36 (2022), pp. 266–278.

[29] S.H. Manjunath and R.K. Thimmulappa, Antiviral, immunomodulatory, and anticoagulant effects of quercetin and its derivatives: Potential role in prevention and management of COVID-19, *J. Pharm. Anal.* 12 (2022), pp. 29–34.

[30] O. Abian, D. Ortega-Alarcon, A. Jimenez-Alesanco, L. Ceballos-Laita, S. Vega, H.T. Reyburn et al., Structural stability of SARS-CoV-2 3CL^{pro} and identification of quercetin as an inhibitor by experimental screening, *Int. J. Biol. Macromol.* 164 (2020), pp. 1693–1703.

[31] Y.-Y. Gu, M. Zhang, H. Cen, Y.-F. Wu, Z. Lu, F. Lu et al., Quercetin as a potential treatment for COVID-19-induced acute kidney injury: Based on network pharmacology and molecular docking study, *PLoS One.* 16 (2021), pp. e0245209.

[32] F. Munafò, E. Donati, N. Brindani, G. Ottonello, A. Armirotti and M. De Vivo, Quercetin and luteolin are single-digit micromolar inhibitors of the SARS-CoV-2 RNA-dependent RNA polymerase, *Sci. Rep.* 12 (2022), pp. 10571.

[33] M. Saakre, D. Mathew and V. Ravisankar, Perspectives on plant flavonoid quercetin-based drugs for novel SARS-CoV-2, *Beni-Suef Univ. J. Basic Appl. Sci.* 10 (2021), pp. 21.

[34] A. Saeedi-Boroujeni and M.-R. Mahmoudian-Sani, Anti-inflammatory potential of Quercetin in COVID-19 treatment, *J. Inflamm.* 18 (2021), pp. 3.

[35] T. Efferth and W. Schwarz, Kaempferol derivatives as antiviral drugs against the 3a channel protein of Coronavirus, *Planta Med* (2014), pp. 177–182.

[36] K. Komolafe, T.R. Komolafe and T.H. Fatoki, Coronavirus disease 2019 and herbal therapy : Pertinent issues relating to toxicity and standardization of phytopharmaceuticals, *Rev. Bras. Farmacogn.* 31 (2021), pp. 142–161.
[37] P. Pandey, J.S. Rane, A. Chatterjee, A. Kumar, R. Khan, A. Prakash et al., Targeting SARS-CoV-2 spike protein of COVID-19 with naturally occurring phytochemicals : An in silico study for drug development, *J. Biomol. Struct. Dyn.* 39 (2021), pp. 6306–6316.
[38] M. Mohamed, A. Rania, H. Naglaa, A.M. Hamoda, F. Alkhayat, N. Naser et al., Flavonoids are promising safe therapy against COVID-19, *Phytochem. Rev.* 21 (2022), pp. 291–312.
[39] R. Ahmadian and R. Rahimi, Kaempferol: An encouraging flavonoid for COVID-19, *Blacpma.* 19 (2020), pp. 492–494.
[40] N. Fintelman-rodrigues, X. Wang, C.Q. Sacramento, J.R. Temerozo, A.C. Ferreira, M. Mattos et al., Commercially available flavonols are better SARS-CoV-2 inhibitors than isoflavone and flavones, *Viruses.* 14 (2022), pp. 1458.
[41] A. Firoz and P. Talwar, COVID-19 and retinal degenerative diseases: Promising link "Kaempferol," *Curr. Opin. Pharmacol.* 64 (2022), pp. 102231.
[42] P. Kashyap, M. Thakur, N. Singh, D. Shikha, S. Kumar, P. Baniwal et al., In silico evaluation of natural flavonoids as a potential inhibitor of coronavirus disease, *Molecules.* 27 (2022), pp. 6374.
[43] S. Muthumanickam, A. Kamaladevi, P. Boomi, S. Gowrishankar and S.K. Pandian, Indian ethnomedicinal phytochemicals as promising inhibitors of RNA-binding domain of SARS-CoV-2 nucleocapsid phosphoprotein : An in silico study, *Front. Mol. Biosci.* 8 (2021), pp. 637329.
[44] G. Agraharam, A. Girigoswami and K. Girigoswami, Myricetin : A multifunctional flavonol in biomedicine, *Curr. Pharmacol. Rep.* 8 (2022), pp. 48–61.
[45] X. Song, L. Tan, M. Wang, C. Ren, C. Guo, B. Yang et al., Myricetin : A review of the most recent research, *Biomed. Pharmacother.* 134 (2021), pp. 111017.
[46] B. Patel, S. Sharma, N. Nair, J. Majeed, R.K. Goyal and M. Dhobi, Therapeutic opportunities of edible antiviral plants for COVID - 19, *Mol. Cell. Biochem.* 476 (2021), pp. 2345–2364.
[47] T. Xiao, M. Cui, C. Zheng, M. Wang, R. Sun and D. Gao, Myricetin inhibits SARS-CoV-2 viral replication by targeting M pro and ameliorates pulmonary inflammation, *Front. Pharmacol.* 12 (2021), pp. 669642.
[48] Y. Taheri, H. Ansar, R. Suleria, N. Martins, O. Sytar and A. Beyatli, Myricetin bioactive effects : Moving from preclinical evidence to potential clinical applications, *BMC Complement. Med. Ther.* 20 (2020), pp. 241.
[49] P. Rattanachaikunsopon and P. Phumkhachorn, Bacteriostatic effect of flavonoids isolated from leaves of Psidium guajava on fish pathogens, *Fitoterapia.* 78 (2007), pp. 434–436.
[50] K. Thakur, Y.-Y. Zhu, J.-Y. Feng, J.-G. Zhang, F. Hu, C. Prasad et al., Morin as an imminent functional food ingredient: An update on its enhanced efficacy in the treatment and prevention of metabolic syndromes, *Food Funct.* 11 (2020), pp. 8424–8443.
[51] C. Remya, K. V Dileep, I. Tintu, E.J. Variyar and C. Sadasivan, Design of potent inhibitors of acetylcholinesterase using morin as the starting compound, *Front. Life Sci.* 6 (2012), pp. 107–117.
[52] J. Wan, T. Ma, Y. Jin and S. Qiu, The effects of morin on bone regeneration to accelerate healing in bone defects in mice, *Int. J. Immunopathol. Pharmacol.* 34 (2020), pp. 1–16.
[53] J. Frandsen, S. Choi and P. Narayanasamy, Neural glyoxalase pathway enhancement by morin derivatives in an Alzheimer's disease model, *ACS Chem. Neurosci.* 11 (2020), pp. 356–366.
[54] J.Y. Kwon, M.-T. Jeon, U.J. Jung, D.W. Kim, G.J. Moon and S.R. Kim, Perspective: Therapeutic potential of flavonoids as alternative medicines in epilepsy, *Adv. Nutr.* 10 (2019), pp. 778–790.
[55] E.-H. Hong, J.-H. Song, S.-R. Kim, J. Cho, B. Jeong, H. Yang et al., Morin hydrate inhibits influenza virus entry into host cells and has anti-inflammatory effect in influenza-infected mice, *Immune Netw.* 20 (2020), pp. e32.
[56] M. Mori, D. Quaglio, A. Calcaterra, F. Ghirga, L. Sorrentino, S. Cammarone et al., Natural flavonoid derivatives have pan-coronavirus antiviral activity, *Microorganisms.* 11 (2023), pp. 314.
[57] H. Hadni, A. Fitri, A.T. Benjelloun, M. Benzakour and M. Mcharfi, Evaluation of flavonoids as potential inhibitors of the SARS-CoV-2 main protease and spike RBD: Molecular docking, ADMET evaluation and molecular dynamics simulations, *J. Indian Chem. Soc.* 99 (2022), pp. 100697.
[58] T.E. Tallei, S.G. Tumilaar, N.J. Niode, Fatimawali, B.J. Kepel, R. Idroes et al., Potential of plant bioactive compounds as SARS-CoV-2 main protease (M^{pro}) and spike (S) glycoprotein inhibitors: A molecular docking study, *Scientifica (Cairo).* 2020 (2020), pp. 6307457.
[59] M. Rudrapal, A.R. Issahaku, C. Agoni, A.R. Bendale, A. Nagar, M.E.S. Soliman et al., In silico screening of phytopolyphenolics for the identification of bioactive compounds as novel protease inhibitors effective against SARS-CoV-2, *J. Biomol. Struct. Dyn.* 40 (2022), pp. 10437–10453.
[60] A. Gupta, R. Ahmad, S. Siddiqui, K. Yadav, A. Srivastava, A. Trivedi et al., Flavonol morin targets host ACE2, IMP-α, PARP-1 and viral proteins of SARS-CoV-2, SARS-CoV and MERS-CoV critical for infection and survival: A computational analysis, *J. Biomol. Struct. Dyn.* 40 (2022), pp. 5515–5546.
[61] R. Zhong and M.A. Farag, Recent advances in the biosynthesis, structure - activity relationships, formulations, pharmacology, and clinical trials of fi setin, *eFood.* 3 (2022), pp. e3.
[62] C.S. Sharanya, A. Sabu and M. Haridas, Potent phytochemicals against COVID-19 infection from phyto-materials used as antivirals in complementary medicines : A review, *Futur. J. Pharm. Sci.* 7 (2021), pp. 113.
[63] S. Lee, Y. Yu, J. Trimpert, F. Benthani, M. Mairhofer, P. Richter-pechanska et al., Virus-induced senescence is a driver and therapeutic target in COVID-19, *Nature.* 599 (2021), pp. 283–289.
[64] B.W. Yang, S. Yang, S. Kim, A.R. Baek, B. Sung, Y.-H. Kim et al., Flavonoid-conjugated gadolinium complexes as anti-inflammatory theranostic agents, *Antioxidants.* 11 (2022), pp. 2470.
[65] P. Schnitzler, A. Neuner, S. Nolkemper, C. Zundel, H. Nowack, K.H. Sensch et al., Antiviral activity and mode of action of propolis extracts and selected compounds, *Phyther. Res.* 24 (2010), pp. S20–S28.

[66] J.J.M. Meyer, A.J. Afolayan, M.B. Taylor and D. Erasmus, Antiviral activity of galangin isolated from the aerial parts of Helichrysum aureonitens, *J. Ethnopharmacol.* 56 (1997), pp. 165–169.

[67] H.I. Guler, G. Tatar, O. Yildiz, A.O. Belduz and S. Kolayli, Investigation of potential inhibitor properties of ethanolic propolis extracts against ACE-II receptors for COVID-19 treatment by molecular docking study, *Arch. Microbiol.* 203 (2021), pp. 3557–3564.

[68] H. Hashem, In silico approach of some selected honey constituents as SARS-CoV-2 main protease (COVID-19) inhibitors, *Eurasian J. Med. Oncol.* 4 (2020), pp. 196–200.

[69] A. Bora, L. Pacureanu and L. Crisan, In silico study of some natural flavonoids as potential agents against COVID-19: Preliminary results, *Chem. Proc.* 3 (2021), pp. 25.

[70] H. Refaat, F.M. Mady, H.A. Sarhan, H.S. Rateb and E. Alaaeldin, Optimization and evaluation of propolis liposomes as a promising therapeutic approach for COVID-19, *Int. J. Pharm.* 592 (2021), pp. 120028.

[71] K. Raguette and S. Morin, Gossypetin derivatives are also putative inhibitors of SARS-COV 2 : Results of a computational study, *J. Biomed. Res. Environ. Sci.* 1 (2020), pp. 201–212.

[72] S. Lal, S. Faisal, A. Muhammad, B. Gabriel, A. Hamid and M. Jaremko, Antiviral activities of flavonoids, *Biomed. Pharmacother.* 140 (2021), pp. 111596.

[73] A. Fischer, M. Sellner, M.A. Lill and S. Neranjan, Potential inhibitors for novel coronavirus protease identified by virtual screening of 606 million compounds, *Int. J. Mol. Sci.* 21 (2020), pp. 3626.

[74] R.M.A.Q. Jamhour, A.H. Al-nadaf, F. Wedian and G.M. Al-mazaideh, Phytochemicals as a potential inhibitor of COVID-19 : An in-silico perspective, *Russ. J. Phys. Chem.* 96 (2022), pp. 1589–1597.

[75] M. Firdaus, R. Nurdiani, I.N. Artasasta, S. Mutoharoh and O.N.I. Pratiwi, Potency of three brown seaweeds species as the inhibitor of RNA-dependent RNA polymerase of SARS-CoV-2, *Rev. Chim.* 71 (2020), pp. 80–86.

[76] Y. Zhan, W. Ta, W. Tang, R. Hua, J. Wang, C. Wang et al., Potential antiviral activity of isorhamnetin against SARS-CoV-2 spike pseudotyped virus in vitro, *Drug Dev. Res.* 82 (2021), pp. 1124–1130.

[77] N. Shahhamzehei, S. Abdelfatah and T. Efferth, In silico and in vitro identification of pan-coronaviral main protease inhibitors from a large natural product library, *Pharmaceuticals.* 15 (2022), pp. 308.

[78] J. Xu, L. Gao, H. Liang and S. Chen, In silico screening of potential anti-COVID-19 bioactive natural constituents from food sources by molecular docking, *Nutrition.* 82 (2021), pp. 111049.

[79] A.A. Zaki, A.A. Al-Karmalawy, Y.A. El-Amier and A. Ashour, Molecular docking reveals the potential of Cleome amblyocarpa isolated compounds to inhibit COVID-19 virus main protease, *New J. Chem.* 44 (2020), pp. 16752–16758.

[80] P. Das, R. Majumder, M. Mandal and P. Basak, In-silico approach for identification of effective and stable inhibitors for COVID-19 main protease (M^{pro}) from flavonoid based phytochemical constituents of Calendula officinalis, *J. Biomol. Struct. Dyn.* 39 (2021), pp. 6265–6280.

[81] S. Murugesan, S. Kottekad, I. Crasta, S. Sreevathsan, D. Usharani, M.K. Perumal et al., Targeting COVID-19 (SARS-CoV-2) main protease through active phytocompounds of ayurvedic medicinal plants - Emblica officinalis (Amla), Phyllanthus niruri Linn. (Bhumi Amla) and Tinospora cordifolia (Giloy) - A molecular docking and simulation study, *Comput. Biol. Med.* 136 (2021), pp. 104683.

[82] C. Vicidomini, V. Roviello and G.N. Roviello, In silico investigation on the interaction of chiral phytochemicals from Opuntia ficus-indica with SARS-CoV-2 M^{pro}, *Symmetry (Basel).* 13 (2021), pp. 1041.

[83] I.A. Adejoro, D.D. Babatunde and G.F. Tolufashe, Molecular docking and dynamic simulations of some medicinal plants compounds against SARS-CoV-2: An in silico study, *J. Taibah Univ. Sci.* 14 (2020), pp. 1563–1570.

[84] W. Yuan, J. Wang, X. An, M. Dai, Z. Jiang and L. Zhang, UPLC - MS / MS method for the determination of hyperoside and application to pharmacokinetics study in rat after different administration routes, *Chromatographia.* 84 (2021), pp. 249–256.

[85] J. Zhang, H. Fu, Y. Xu, Y. Niu and X. An, Hyperoside reduces albuminuria in diabetic nephropathy at the early stage through ameliorating renal damage and podocyte injury, *J. Nat. Med.* 70 (2016), pp. 740–748.

[86] P. Halder, U. Pal, P. Paladhi, S. Dutta, P. Paul, S. Pal et al., Evaluation of potency of the selected bioactive molecules from Indian medicinal plants with M Pro of SARS-CoV-2 through in silico analysis, *J. Ayurveda Integr. Med.* 13 (2022), pp. 100449.

[87] P. Isabel, M. Santana, J. Pablo, P. Tivillin, I.A. Chóez-guaranda, A. Délida et al., Potential bioactive compounds of medicinal plants against new Coronavirus (SARS-CoV-2): A review potential bioactive compounds of medicinal plants against new coronavirus (SARS-CoV-2): A review, *Rev. Bionatura.* 6 (2021), pp. 1653–1658.

[88] S. Xu, S. Chen, W. Xia, H. Sui and X. Fu, Hyperoside: A review of its structure, synthesis, pharmacology, pharmacokinetics and toxicity, *Molecules.* 27 (2022), pp. 3009.

[89] L. Wang, J. Song, A. Liu, B. Xiao, S. Li, Z. Wen et al., Research progress of the antiviral bioactivities of natural flavonoids, *Nat. Products Bioprospect.* 10 (2020), pp. 271–283.

[90] M. Mbikay and M. Chrétien, Isoquercetin as an anti-COVID-19 medication : A potential to realize, *Front. Pharmacol.* 13 (2022), pp. 372.

[91] A.A. Al-Zahrani, Rutin as a promising inhibitor of main protease and other protein targets of COVID-19: In silico study, *Nat. Prod. Commun.* 15 (2020), pp. 1.

[92] Z. Xu, L. Yang, X. Zhang, Q. Zhang, Z. Yang, Y. Liu et al., Discovery of potential flavonoid inhibitors against COVID-19 3CL proteinase based on virtual screening strategy, *Front. Mol. Biosci.* 7 (2020), pp. 556481.

[93] T. Joshi, S. Bhat, H. Pundir and S. Chandra, Identification of berbamine, oxyacanthine and rutin from berberis asiatica as anti-SARS-CoV-2 compounds: An in silico study, *J. Mol. Graph. Model.* 109 (2021), pp. 108028.

[94] F. Rahman, S. Tabrez, R. Ali, A.S. Alqahtani, M.Z. Ahmed and A. Rub, Molecular docking analysis of rutin reveals possible inhibition of SARS-CoV-2 vital proteins, *J. Tradit. Complement. Med.* 11 (2021), pp. 173–179.

[95] L. Oktavia and P. Tiwi, Molecular docking, molecular dynamic and drug-likeness studies of natural flavonoids as inhibitors for SARS-CoV-2 main protease (M pro), *J. Res. Pharm.* 25 (2021), pp. 998–1009.

[96] T. Burt, G. Young, W. Lee, H. Kusuhara, O. Langer, M. Rowland et al., Phase 0/microdosing approaches: Time for mainstream application in drug development?, *Nat. Rev. Drug Discov.* 19 (2020), pp. 801–818.
[97] Food and Drug Adminstration, GRAS Notices, 2023. https://www.cfsanappsexternal.fda.gov/scripts/fdcc/index.cfm?set=GRASNotices&id=341&sort=GRN_No&order=DESC&startrow=1&type=basic&search=quercetin. Data accessed 2023-04-10
[98] P.K. Agrawal, C. Agrawal and G. Blunden, Quercetin: antiviral significance and possible COVID-19 integrative considerations, *Nat. Prod. Commun.* 15 (2020), pp. 1–10.
[99] F. Di Pierro, G. Derosa, P. Maffioli, A. Bertuccioli, S. Togni, A. Riva et al., Possible therapeutic effects of adjuvant quercetin supplementation against early-stage COVID-19 infection: A prospective, randomized, controlled, and open-label study, *Int. J. Gen. Med.* 14 (2021), pp. 2359–2366.
[100] S.A. Heinz, D.A. Henson, M.D. Austin, F. Jin and D.C. Nieman, Quercetin supplementation and upper respiratory tract infection: A randomized community clinical trial, *Pharmacol. Res.* 62 (2010), pp. 237–242.
[101] G. D'Andrea, Quercetin: A flavonol with multifaceted therapeutic applications?, *Fitoterapia.* 106 (2015), pp. 256–271.
[102] G.E. Batiha, A.M. Beshbishy, M. Ikram, Z.S. Mulla, M.E.A. El-Hack, A.E. Taha et al., The pharmacological activity, biochemical properties, and pharmacokinetics of the major natural polyphenolic flavonoid: Quercetin, *Foods.* 9 (2020), pp. 374.
[103] J. Xiao, S. Mohammad and M. Daglia, Kaempferol and inflammation: From chemistry to medicine, *Pharmacol. Res.* 99 (2015), pp. 1–10.
[104] A. Khalil and D. Tazeddinova, The upshot of polyphenolic compounds on immunity amid COVID-19 pandemic and other emerging communicable diseases: An appraisal, *Nat. Products Bioprospect.* 10 (2020), pp. 411–429.
[105] S.A. Rajput, X. Wang and H.-C. Yan, Morin hydrate: A comprehensive review on novel natural dietary bioactive compound with versatile biological and pharmacological potential, *Biomed. Pharmacother.* 138 (2021), pp. 111511.
[106] S.C. Gupta, *Anti-Inflammatory Nutraceuticals and Chronic Diseases*, Springer, New York, USA, 2016.
[107] D. Fang, Z. Xiong, J. Xu, J. Yin and R. Luo, Chemopreventive mechanisms of galangin against hepatocellular carcinoma: A review, *Biomed. Pharmacother.* 109 (2019), pp. 2054–2061.
[108] A. Khan, K. Manna, C. Bose, M. Sinha, D.K. Das and S.B. Kesh, Gossypetin, a naturally occurring hexahydroxy flavone, ameliorates gamma radiation-mediated DNA damage, *Int. J. Radiat. Biol.* 89 (2013), pp. 965–975.
[109] H. Lee, M. Krishnan, M. Kim, Y.K. Yoon and Y. Kim, Rhamnetin, A natural flavonoid, ameliorates organ damage in a mouse model of carbapenem-resistant Acinetobacter baumannii -induced sepsis, *Int. J. Mol. Sci.* 23 (2022), pp. 12895.
[110] W.-Q. Li, J. Li, W.-X. Liu, L.-J. Wu, J.-Y. Qin, Z.-W. Lin et al., Isorhamnetin: A novel natural product beneficial for cardiovascular disease, *Curr. Pharm. Des.* 28 (2022), pp. 2569–2582.
[111] J. Popiolek-Kalisz, P. Blaszczak and E. Fornal, Dietary isorhamnetin intake is associated with lower blood pressure in coronary artery disease patients, *Nutrients.* 14 (2022), pp. 4586.
[112] A. Riaz, A. Rasul, G. Hussain, M.K. Zahoor, F. Jabeen, Z. Subhani et al., Astragalin: A bioactive phytochemical with potential therapeutic activities, *Adv. Pharmacol. Pharm. Sci.* 2018 (2018), pp. 9794625.
[113] B. Gullón, T.A. Lú-Chau, M.T. Moreira, J.M. Lema and G. Eibes, Rutin: A review on extraction, identification and purification methods, biological activities and approaches to enhance its bioavailability, *Trends Food Sci. Technol.* 67 (2017), pp. 220–235.
[114] L. Tsiklauri, K. Šv, M. Chrastina, S. Poništ, M. Alania, E. Kemertelidze et al., Bioflavonoid robinin from Astragalus falcatus Lam. mildly improves the effect of metothrexate in rats with adjuvant arthritis, *Nutrients.* 13 (2021), pp. 1268.
[115] L. Tsiklauri, G. An, D.M. Ruszaj, M. Alaniya, E. Kemertelidze and M.E. Morris, Simultaneous determination of the flavonoids robinin and kaempferol in human breast cancer cells by liquid chromatography-tandem mass spectrometry, *J. Pharm. Biomed. Anal.* 55 (2011), pp. 109–113.

7 Anti-SARS-CoV-2 Activity of Flavonols and Their Glycosylated Derivatives

Salar Hafez Ghoran, Fatemeh Taktaz, Pouya Alipour, Mojtaba Ghobadi, Seyedeh Elham Faghih-Shirazi, Alireza Zali, and Seyed Abdulmajid Ayatollahi

ABBREVIATIONS

3CLpro Chymotrypsin-like protease
4CL 4-Coumarate-CoA ligase
ACCase Acetyl-CoA carboxylase
ACE2 Angiotensin-converting enzyme 2
ARDS Acute respiratory distress syndrome
C4H Cinnamic acid 4-hydroxylase
CHI Chalcone isomerase
CHS Chalcone synthase
Cox-B3 Coxsackie B virus type 3
CPE Cytopathic effect
CTSL Cathepsin L
DAMPs Damage-associated molecular patterns
DCs Dendritic cells
DENV-2 Dengue virus
EBV Epstein-Barr virus
ExoN Exoribonuclease
F3H Flavanone-3-hydroxylase
FLS Flavonol synthase
F3′H Flavonoid-3′-hydroxylase
HA Hemagglutinin
HCMV Human Cytomegalovirus
HBV Hepatitis B virus
HCV Hepatitis C virus
HIV Human immunodeficiency virus
HSV-1 Herpes simplex virus 1
IL-6 Interleukin-6
iNOS Inducible nitric oxide synthase
LPS Lipopolysaccharide
MHV Murine coronavirus
NA Neuraminidase
NDV Newcastle disease virus
NF-κB Nuclear factor kappa B
NK Natural killers
PAL Phenylalanine ammonia-lyase
PAMPs Pathogen-associated molecular patterns
PBMCs Peripheral blood mononuclear cells
PEDV Porcine epidemic diarrhea virus
PGE2 Prostaglandin E2
PLpro Papain-like protease
RdRp RNA-dependent RNA polymerase
ROS Reactive oxygen species
rhACE2 Recombinant human angiotensin-converting enzyme 2
SAR Structure–activity relationship
SARS Severe acute respiratory syndrome
SARS-CoV-2 Severe acute respiratory syndrome coronavirus 2
TCM Traditional Chinese medicine
TLR4 Toll-like receptor 4
TMPRSS2 Transmembrane serine protease 2

7.1 SARS-COV-2 AND COVID-19 OUTBREAK

The global emergency known as the COVID-19 pandemic, initiated by the Severe Acute Respiratory Syndrome Coronavirus Type 2 (SARS-CoV-2), has become a widespread crisis impacting millions across the globe. Initially detected in Wuhan, China, in December 2019, the outbreak quickly disseminated across the continents, resulting in substantial illness, loss of life, and socioeconomic upheaval [1]. Belonging to the coronavirus family, SARS-CoV-2 is an enveloped, single-strand RNA virus whose genetic sequence reveals similarities with the previous coronavirus (SARS) initiated in 2003. The virus primarily targets the respiratory system, binding to the human angiotensin-converting enzyme 2 (ACE2) receptors, particularly in lung cells. Regarding viral entry, the spike proteins on the virus surface are important because of their interaction with ACE2, a key determinant of infectivity. On the other hand, several enzymes have significant roles in various stages of SARS-CoV-2 infections. Some of the key enzymes associated with the virus are cysteine proteases (i.e., chymotrypsin- and papain-like proteases; $3CL^{pro}$ and PL^{pro}), transmembrane serine protease 2 (TMPRSS2), RNA-dependent RNA polymerase (RdRp), exoribonuclease (ExoN), and furin. A complete understanding of the functions of these enzymes has helped scientists develop therapeutic interventions either from natural sources or synthetic origins and examine them as suppressor agents against SARS-CoV-2.

Currently, various antivirals are being used or investigated for treating SARS-CoV-2 infection; however, they have yet to be fully approved clinically. For example, remdesivir, chloroquine/hydroxychloroquine, favipiravir, and lopinavir–ritonavir are used for antiviral, antimalarial, and anti-inflammatory purposes. While nutraceuticals, vitamins (i.e., C and D), minerals (i.e., zinc), and probiotics alone do not prevent or cure COVID-19, scientists and

DOI: 10.1201/9781003433200-7

healthcare professionals believe that they can contribute to maintaining a healthy immune system [2].

Plants containing phenolic compounds, particularly flavonoids, have garnered interest for their potential effects against SARS-CoV-2 since there is scientific evidence for these secondary metabolites showing beneficial properties against SARS-CoV-2 infection. To be more specific, flavonoids have demonstrated viral inhibitory activity toward various viruses, such as coronaviruses. They can inhibit host cell-viral entry, replication, and the release of progeny viruses. Moreover, flavonoids may modulate host immune responses, enhancing antiviral defense mechanisms and reducing excessive inflammation, which is a hallmark of severe COVID-19 cases [3,4]. Furthermore, the antioxidant activity of flavonoids can help reduce the oxidative stress induced by viral infections, potentially mitigating the severity of symptoms and reducing tissue damage. Meanwhile, their immunomodulatory effects may contribute to a balanced immune response, preventing the production of pro-inflammatory mediators in acute COVID-19 cases [5]. Therefore, incorporating plant-based foods rich in flavonoids into a balanced diet may provide potential health benefits and support overall well-being during the current COVID-19 pandemic. This chapter aims to highlight naturally occurring flavonols along with their glycosylated derivatives, which act as anti-SARS-CoV-2 components. The experimental, computational, and clinical studies will be covered where available.

7.2 FLAVONOIDS AND THEIR SUBCLASSES

Flavonoids, which are widely distributed in plants, exhibit a range of vibrant hues that contribute to the visual beauty of flowers, fruits, and vegetables. The basic chemical framework consists of two benzene rings, which are connected by a three-carbon-containing chain. This core structure forms the foundation for various captivating subclasses of flavonoids, each distinguished by subtle modifications and substitutions. Flavonols, flavones, flavanones, flavan-3-ols (also known as catechins), anthocyanins, isoflavones, homoisoflavones, chalcones, and aurones are some of the subclasses of flavonoids that possess unique chemical features resulting in distinct biological activities and health benefits [6].

For instance, flavonols (Figure 7.1) are characterized by a 3-OH residue in the C ring and extensively possess antioxidant properties. They are commonly found in onions, apples, and berries [7]. Although these secondary metabolites are not so good at capturing UV-B (290–320 nm), they have a remarkable ability to efficiently regulate reactive oxygen species (ROS) homeostasis within cells. They also regulate the growth of individual plant organs and the development of the plant itself. Surprisingly, flavonols at tiny concentrations, even in terrestrial plants, can totally function as plant-growth regulators [8]. Meanwhile, besides being nutraceuticals, pharmaceuticals, and health promoters, several industrial applications have been reported for flavonols; for instance, these secondary metabolites are significantly applicable in the cosmetic and textile industries [9].

The other common subclasses of flavonoids are as follows: flavones (Figure 7.1), which lack a 3-OH group in the C ring, are abundant in parsley, celery, and *Citrus* fruits [10]. Flavanones (Figure 7.1), which are predominantly found in *Citrus* fruits, exhibit notable antioxidant and anti-inflammatory effects [11]. Meanwhile, flavanones, such as taxifolin (also known as dihydroquercetin), are

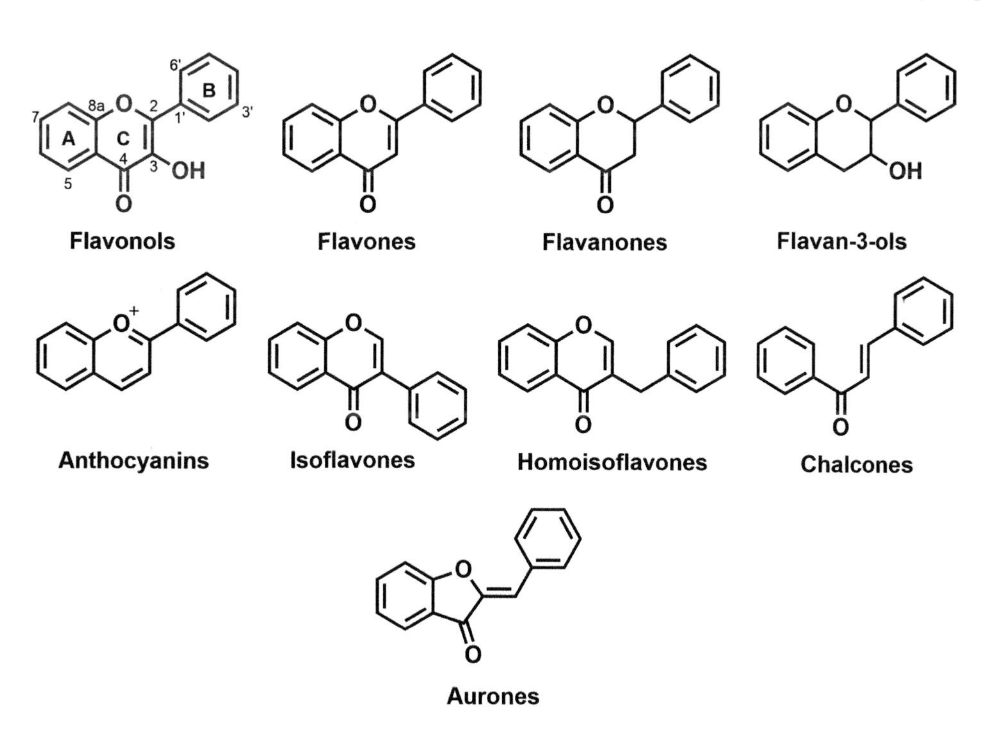

FIGURE 7.1 Common subclass of flavonoids.

recognized for their potential to protect against oxidative stress and promote cardiovascular health [12]. Flavan-3-ols (Figure 7.1), which are abundant in green tea, cocoa, and various fruits, are renowned for their beneficial effects on cardiovascular health and their antioxidant activity [13]. Anthocyanins (Figure 7.1), which are responsible for the vibrant red, purple, and blue pigments in grapes and berries, have gained attention for their potential role in supporting cognitive function and reducing the risk of chronic diseases [14]. Isoflavones (Figure 7.1), mostly found in soybeans and legumes, have been extensively studied because of their estrogenic activities and potential health benefits, particularly in menopausal women [15]. Homoisoflavones (Figure 7.1) are characterized by one extra carbon in the carbon skeleton of isoflavonoids, which can be found in *Scilla* sp., *Polygonatum* sp., and *Ophiopogon japonicus*. These compounds display diverse biological properties, including anti-diabetic and cytotoxic effects against cancer cell lines [16–18]. Chalcones (Figure 7.1), which are found in various plant sources, such as fruits, vegetables, and medicinal herbs, including liquorice, ginger, and turmeric, are identified by a distinctive chemical structure with a central open-chain flavonoid backbone [19]. Lastly, aurones (Figure 7.1), which are significant yellow pigments in plants, are derived from chalcones. These secondary metabolites are relatively uncommon and limited to very few plants, including snapdragons, sunflowers, and coreopsis [20].

7.2.1 Biosynthesis of Flavonols

Remarkably, the biosynthetic pathway of flavonols has remained conserved over millions of years, predominantly serving as a mechanism for plants to respond to diverse stress-inducing factors. This is despite the extensive evolution of flavonoid metabolism, resulting in over 10,000 distinct flavonoid structures [8]. Recent advancements in scientific research have put forward new proposals regarding the multifaceted roles of flavonoids in eukaryotic cells, particularly as regulators of development and/or molecules involved in cellular signaling. These roles have been observed in response to various environmental stimuli. In the realm of plants, these functions are primarily attributed to flavonols. It is noteworthy that liverworts and mosses have the full complement of genes responsible for the biosynthesis of flavonols, encompassing chalcone synthase (CHS), chalcone isomerase (CHI), flavanone-3-hydroxylase (F3H), flavonol synthase (FLS), and flavonoid-3′-hydroxylase (F3′H) [8]. It is important to note that the flavonol pathway starts in two separate directions: (i) phenylalanine is changed to *p*-coumaroyl-CoA by phenylalanine ammonia-lyase (PAL), cinnamic acid 4-hydroxylase (C4H), and 4-coumarate-CoA ligase (4CL); and (ii) acetyl-CoA is changed to malonyl-CoA by acetyl-CoA carboxylase (ACCase), which creates the ring A in flavonoids. Putting these products together by CHS will make naringenin-chalcone, which then produces flavonols through the activity of CHI, F3H, and FLS, respectively (Figure 7.2) [20].

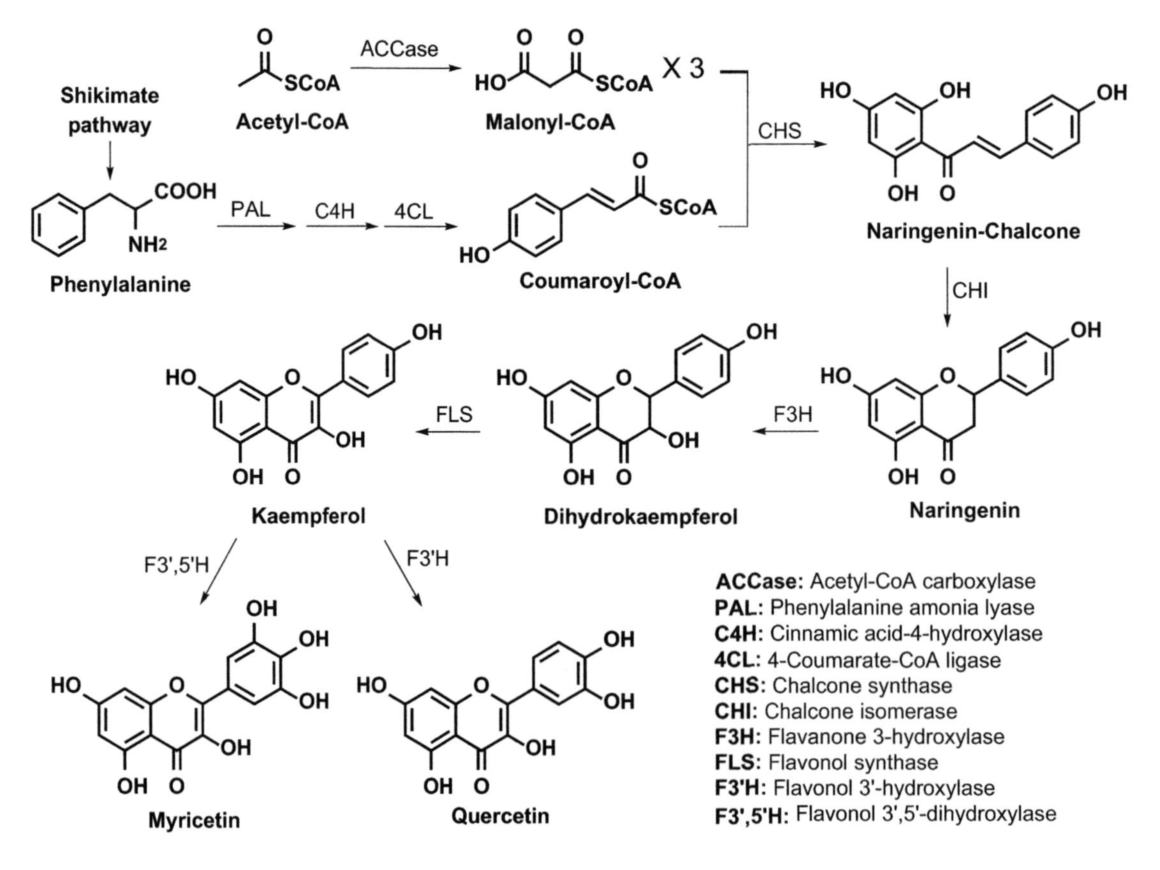

FIGURE 7.2 Biosynthetic pathway of flavonols.

7.3 PLANTS RICH IN FLAVONOLS: ETHNOPHARMACOLOGICAL SIGNIFICANCE

Plants containing high amounts of flavonols and their preparations have garnered attention due to their potential antiviral properties against COVID-19. For example, kaempferol, quercetin, and myricetin are the most well-known flavonols, which can be found in various plant sources, including lettuce, tomatoes, kale, saffron flowers, broccoli, green tea, onions, apples, citrus fruits, berries, grape, and tea. To a lesser extent, fisetin, isorhamnetin, and morin are the other fascinating flavonols that refine human health [21–23]. The plants rich in flavonols and their glycosylated derivatives have a long history of traditional medicine use and may offer potential therapeutic benefits against COVID-19. For instance, traditional Chinese medicine (TCM) and other East Asian healing practices often incorporate plants rich in flavonols for their antiviral and immune-boosting properties. Indigenous communities in South America rely on plant-containing flavonols to treat various ailments and viral infections. In African folk medicine, flavonol-rich plants are often used to boost the immune system, alleviate symptoms, and support overall health. Incorporating these remedies, passed down through generations, into modern research could uncover potential COVID-19 treatments (Table 7.1).

7.4 FLAVONOLS AS IMMUNOMODULATORS AND ANTI-INFLAMMATORY AGENTS

In the presence of different types of damage, be it pathological, cellular, or vascular, arising from physical, chemical, or mechanical trauma, inflammation is an inherent response. It functions as a defensive mechanism to preserve the organism from tissue damage. Typical signs associated with inflammation include redness, pain, reduced functionality, and increased heat in the affected area [69]. Inflammation can be categorized as acute or chronic, depending on the duration and persistence of the response. The induction of the immune response, which is responsible for initiating inflammation, activates various immunity-related cells like neutrophils, natural killers (NK), and B and T cells. Meanwhile, various enzymes, including protein and tyrosine kinases, cyclooxygenases, phosphodiesterase, and phospholipase A, initiate the immune response, activating endothelial cells and subsequent inflammatory response [70]. Flavonols and their sugar-containing derivatives exhibit potent anti-inflammatory effects, which can effectively control the cytokine storm and reduce inflammation in acute COVID-19 patients. They contribute to the overall improvement of patient outcomes. These secondary metabolites can also boost the immune system's ability to fight off viral infections by controlling the production of cytokines and chemokines, which helps the body's defense system [5].

From the cellular and molecular mechanism point of view, Alam et al. summarized that kaempferol (**1**) (Figure 7.3) has a promising anti-inflammatory potency. It inhibits toll-like receptor 4 (TLR4) and DNA from binding to nuclear factor kappa B (NF-κB), suppresses the production of interleukins (i.e., IL-6, IL-1β, and IL-18) and tumor necrosis factor (TNF)-α, and raises the expression of mRNA and proteins in Nrf2-regulated genes. Even in clinical trials, it was evident that using broccoli for 10 days as a kaempferol-rich diet reduced the inflammatory mediators (i.e., IL-6 and TNF-α) in male smokers suffering from lung inflammation [71]. As the most abundant flavonoid in the methanolic fraction of *Semen cuscutae*, kaempferol (**1**) reduced the cytokine and chemokine contents in dendritic cells (DCs) stimulated by lipopolysaccharide (LPS). This reduction was not attributed to the cytotoxic effects of compound **1** on DCs. Moreover, in *in vitro* and *in vivo* experiments, it disrupted the capacity of LPS-stimulated DCs to induce activation of antigen-specific T cells. What the authors found is that compound **1** is not only the immunosuppressive agent on DCs, but it also helps treat autoimmune and chronic inflammatory diseases [72].

Inflammasomes are groups of proteins that get together in the cytosol when they sense pathogen-associated molecular patterns (PAMPs) and damage-associated molecular patterns (DAMPs). They then produce active forms of cytokines like IL-1β and IL-18. Several regulators, including TXNIP, SIRT1, and Nrf2, can affect NLRP3, a specific type of inflammasome. Quercetin (**2**; also known as vitamin P, Figure 7.3), which is of high interest to researchers due to a myriad of biological and pharmacological activities, enhances the innate immune response by diminishing the genetic expression of cytokines, chemokines, and interferons – molecules intricately linked with systematic inflammation [73]. Furthermore, modulating these regulators exerts a suppressive effect on the NLRP3 inflammasome. According to this, quercetin (**2**) may be useful in treating severe inflammation, which is a major cause of death for COVID-19 patients [74].

Acute respiratory distress syndrome (ARDS) in COVID-19 patients is exacerbated by the release of cytokines, which are caused by NF-κB activation, inflammasome, and IL-6 signaling pathways. It is noteworthy to mention that kaempferol (**1**) and quercetin (**2**) (Figure 7.3) have the ability to inhibit these pro-inflammatory signals, thereby potentially mitigating the detrimental effects associated with excessive inflammation in COVID-19 patients. These flavonol compounds can effectively reduce the ROS levels and the expression of various inflammatory indicators, such as inducible nitric oxide synthase (iNOS), TNF-α, IL-1α, IL-1β, IL-6, IL-10, IL-12 p70, and chemokines [75,76]. This make them a worthwhile avenue to control and treat COVID-19.

In another research, myricetin (**3**) (Figure 7.3), derived from the hydromethanolic extract of *Diospyros lotus* leaves, showed a dose-dependent suppression of pro-inflammatory mediators (i.e., COX-2, PGE2, NO, and iNOS) in LPS-stimulated RAW 264.7 macrophage cells. Moreover, its administration in mice led to reduce the production and levels of IL-6, IL-12, NO, and iNOS. In LPS-stimulated

TABLE 7.1
Some Important Medicinal Plants Containing Flavonol Compounds with Anti-inflammatory, Immunomodulatory, and Antiviral Properties

Plant Name	Common Name	Part Used	Main Flavonol	Traditional and Pharmacological Uses	Anti-Inflammatory and Immunomodulatory Effects	Antiviral Activity	References
Lactuca sativa L. (Asteraceae family)	Lettuce	Leaves	Quercetin	Hepatoprotective, prevents cancer, protects neurons, relieves stress, and anti-diabetic activity	Reduction of ROS, NO, iNOS production, and COX-2 expression	Anti-HBV, anti-Cox-B3, and anti-HCMV activity	[24–26]
Solanum lycopersicum L. (Solanaceae family)	Tomatoes	Fruits, seeds, and pulp	Quercetin	Anticancer activity against several types of cancers, treatment of high blood pressure, hepato- and kidney-protective, edema treatment, antioxidant, cathartic effects	Reduction of PGE2 and COX-2 production Immunomodulatory effects and Suppression of mRNA expression of IL-6, IL-1β, and TNF-α	Blocking the interaction of SARS-CoV-2 spike protein and host ACE2 receptor	[27,28]
Different varieties of *Brassica oleracea* (Brassicaceae family)	Broccoli, Kale, Cabbage	Flowering head and young leaves	Quercetin, kaempferol	Gastro-, hepato-, and cardioprotective, anti-obesity, anti-diabetic, hyperglycemic, hyperlipidemic, reducing fever, antimicrobial, anticancer and antioxidant activities	Inhibition of NO release, NF-κB activation, and IκB-α degradation in LPS-stimulated macrophages Downregulation of IL-6, IL-1β, and TNF-α gene expression IL-10 inducer in LPS-stimulated macrophages	Anti-influenza A virus activity	[29–32]
Camellia sinensis (L.) Kuntze (Theaceae family)	Tea	Leaves, flowers	Kaempferol, quercetin, myricetin	Stimulant, diuretic, astringent, cardioprotective, flatulent treatment, controlling the body temperature and blood sugar, refreshing mental and digestion processes	Reduction of inflammation by suppressing immune cell infiltration, cytokines/chemokines release, mucus production, and lung tissue damage Inhibition of IL-6, IL-8, MUC5AC, MUC5B, and neutrophil activity	Anti-HIV-1 and anti-HIV-2 activity Anti-influenza A/H3N2, H9N2, A/H1N1, H2N2, and B type Anti-HBV, anti-HSV-1, and anti-enterovirus 71 activity	[33–35]

(Continued)

TABLE 7.1 (*Continued*)
Some Important Medicinal Plants Containing Flavonol Compounds with Anti-inflammatory, Immunomodulatory, and Antiviral Properties

Plant Name	Common Name	Part Used	Main Flavonol	Traditional and Pharmacological Uses	Anti-Inflammatory and Immunomodulatory Effects	Antiviral Activity	References
Allium cepa L. (Amaryllidaceae family)	Onion	Bulbs	Quercetin	Cardiovascular diseases, ingestion, hemorrhoids and lower gastrointestinal bleeding, fever, flu, sinusitis, diarrhea, jaundice, alopecia, hypotensive, hypoglycemic, hypolipidemic, anthelmintic, antiseptic, stimulant, diuretic, fungal infection	Reduction of NO secretion through downregulation of COX-2 and iNOS mRNA and protein expression Modulation of IL-6, IL-1β, and TNF-α, decrease of lung inflammatory cells, such as monocytes, neutrophils, and eosinophils in asthmatic rats	Anti-influenza type A, anti-influenza virus subtype avian H9N2, anti-poliovirus, anti-hepatitis viruses, anti-SARS-CoV Anti-HSV-1 activity in Vero E6 cells	[36–40]
Malus species (Rosaceae family)	Apples	Fruits, leaves, flowers	Quercetin, rutin	Controlling blood pressure, diuretic, antidiarrheal, soft laxative	Inhibition of COX-2 and 15-lipoxygenase (15-LOX) activity	Anti-influenza activity	[41–43]
Citrus species (Rutaceae family)	Citrus fruits	Fruits, leaves, pulps, flowers	Quercetin, kaempferol, rutin	Antihypertensive, cough and cold, abdominal pain, fever, indigestion, diabetes, anti-stress, cardiac tonic, diuretic, jaundice, flu, rheumatism	Reduction of NO release, induction of IL-6 and TNF-α release, promotion of IL-6, IL-1β, TNF-α, COX-2, and iNOS mRNA expression in SARS-CoV-2 positive patients	Anti-HSV-1 activity in Vero E6 cells Anti-SARS-CoV-2 activity	[4,44–46]
Vitis vinifera L. (Vitaceae family)	Grapes	Fruits, leaves, seeds, latex	Quercetin, kaempferol	Anemia and blood forming, cold and flu, bronchitis, carminatives, wound care, treatment of allergies and stomach diseases, diuretic, antiseptic, and diarrhea	Reduction of IL-1β, IL-6, and IL-8 production in LPS-stimulated macrophages Inhibition of TNF-α-induced IL-8 release in human gastric epithelial cells	Anti-influenza activity Anti-MERS-CoV activity	[47]
Capparis spinosa L. (Capparaceae family)	Caper	Leaves, fruits, buds, roots	Quercetin, isoquercetin, kaempferol, rutin	Treatment of liver, spleen, gastrointestinal, skin, kidney diseases, rheumatism, cough, diuretic, earache, headache, diabetes, analgesic, anti-hemorrhoid, and asthma	Inhibition of NF-κB activation Reduction of TNF-α, IL-1β, LTB4, and superoxide anion in paw edema inflammation	Anti-HSV-2 replication in human PBMCs by upregulating IL-12, TNF-α, and IFN-γ expression	[48,49]

(*Continued*)

TABLE 7.1 (*Continued*)
Some Important Medicinal Plants Containing Flavonol Compounds with Anti-inflammatory, Immunomodulatory, and Antiviral Properties

Plant Name	Common Name	Part Used	Main Flavonol	Traditional and Pharmacological Uses	Anti-Inflammatory and Immunomodulatory Effects	Antiviral Activity	References
Crocus sativus L. (Iridaceae family)	Saffron	Flowers	Kaempferol	Prevents hangovers, diuretic, sedative, treatment of erysipelas, carminative, febrifuge, anti-anxiety, arthritis, asthma, cough and chest pain, tonic and heart stimulant, increases the functionality of liver, kidney, and lungs	Inhibition of IL-6, IL-1β, IL-12, IL-17A, TNF-α, NF-κB, and IFN-γ levels Downregulation of COX-2, iNOS, myeloperoxidase (MPO), phospholipase A2, and prostanoids	Anti-HIV-1 and anti-HSV-1 activity	[50–52]
Petroselinum crispum (Mill.) Fuss (Umbelliferae/ Apiaceae family)	Parsley	Aerial parts, leaves, seeds, roots	Quercetin, kaempferol, myricetin, isorhamnetin	Treatment of hemorrhoid and urethral inflammation, diabetes, increases the brain function and memory, high blood pressure, diuretic, intestinal tonic, demulcent, anti-urolithiasis, cardiac, urinary, and gastrointestinal problems	Reduction of p53 and COX-2 expression in rat liver and heart Refining the antioxidant enzymatic status, increase of GSH levels, and decrease of MDA levels	Anti-HSV-1 activity in Vero E6 cells	[40,53,54]
Ceratonia siliqua L. (Leguminosae/ Fabaceae family)	Carob	Fruit pulp, leaves	Myricetin, rutin	Gastro- and cardioprotective, laxative, antidiarrhea, analgesic, anti-diabetic, hyperlipidemic, hypoglycemic, and hypercholesterolemic activities	Inhibition of NO, IL-6, and TNF-α release in LPS-induced macrophages Deactivation of NF-κB	Anti-HAV activity in Vero E6 cells Anti-NDV (Newcastle disease virus) activity	[55–58]

(Continued)

TABLE 7.1 (*Continued*)
Some Important Medicinal Plants Containing Flavonol Compounds with Anti-inflammatory, Immunomodulatory, and Antiviral Properties

Plant Name	Common Name	Part Used	Main Flavonol	Traditional and Pharmacological Uses	Anti-Inflammatory and Immunomodulatory Effects	Antiviral Activity	References
Ginkgo biloba L. (Ginkgoaceae family)	Gingko	Leaves, seeds, exocarp	Isorhamnetin, myricetin	Treatment of diarrhea, infantile enteritis, asthma, phlegm, wind chill, chronic cough, protection of kidney and spleen	Suppression of NF-κB and MAPK signaling pathways Reduction of inflammatory cells (i.e., neutrophils and IL-8) in mice lung	Anti-NDV activity Anti-PEDV (porcine epidemic diarrhea virus), anti-HIV protease, anti-influenza (H1N1 and H3N2), and anti-HBV activities	[59–61]
Ampelopsis grossedentata (Hand.-Mazz) W.T. Wang (Vitaceae)	Vina tea	Leaves	Myricetin, ampelopsin, quercetin, kaempferol, rutin	Treatment of pyretic fever, cough, pain killer, larynx and pharynx, jaundice, and hepatitis	Immunomodulatory enhancer Inhibition of NO, IL-6, IL-1β, and TNF-α production, and increase of IL-10 levels in LPS-stimulated macrophages Reduction of COX-2, iNOS, and TNF-α protein expression, and suppression of NF-κB and IκB-α phosphorylation in macrophage cells	Anti-Ebola, anti-Marburg, anti-infectious bronchitis, anti-HIV-1, anti-HSV, anti-Bourbon, and anti-African swine fever viruses	[62–64]
Tussilago farfara L. (Asteraceae family)	Coltsfoot	Aerial parts, leaves, flowers, rhizomes	Quercetin, kaempferol, rutin	Treating COVID-19 patients, cough, asthma, old bronchitis, catarrh, lung problems, fistula and hemorrhoid, chronic osteomyelitis, anti-hypertensive activity	Reduction of overproduction of IL-1β and TNF-α). Decrease of intracellular MDA and ROS content in lung homogenate of mice exposed to cigarette smoke Inhibition of NF-κB and NLRP3 inflammasome gene expression in mice	Anti-enterovirus 71 activity in CCFS-1/KMC and RD cells	[65,66]
Acacia species (Leguminosae/Fabaceae family)	Salam	Bark, leaves, fruits, twigs, seeds	Kaempferol, quercetin, myricetin, and their glycosylated derivatives	Cardiac tonic, treatment of diarrhea, toothache, internal pain, analgesic, astringent, dysentery, diuretic, wound healing, gastrointestinal disorders, antiseptic, diabetes, and inflammation	COX-1 and COX-2 suppressor Reduction of IL-6, IL-1β, and TNF-α production in LPS-stimulated macrophages	Anti-hepatitis type C virus (HCV), anti-HCMV, anti-influenza, and anti-HIV-1 activities	[67,68]

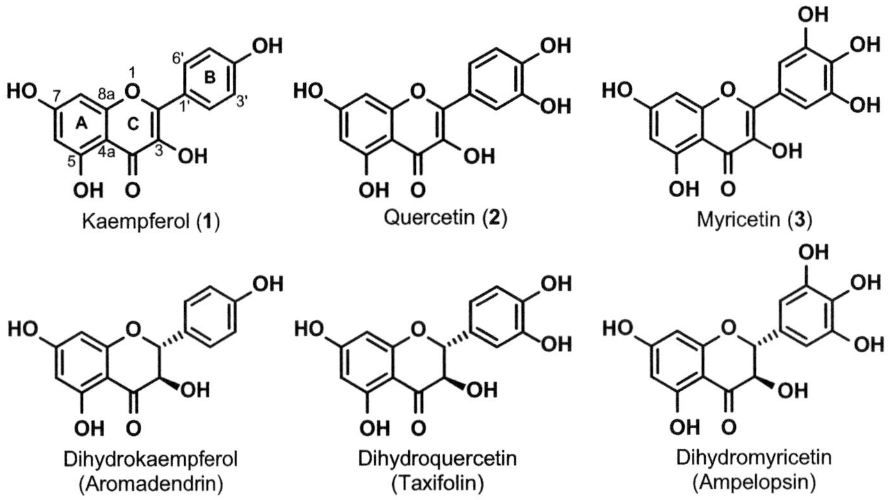

FIGURE 7.3 Chemical structures of some anti-inflammatory flavonols, **1–3**.

RAW 264.7 macrophage cells, myricetin (**3**) impeded NF-κB activation by blocking IκBα degradation, nuclear translocation of the p65 component, and NF-κB DNA binding. This flavonol also attenuated STAT1 phosphorylation and IFN-β production in the same cells. Furthermore, myricetin (**3**) stimulated HO-1 expression through Nrf2 translocation. Therefore, the authors concluded that myricetin (**3**) deserves to be a candidate for anti-inflammatory drugs and for treating inflammatory-related diseases; however, its mechanism of action and clinical studies were recommended [77].

7.5 ANTIVIRAL ACTIVITY OF FLAVONOLS AND THEIR DERIVATIVES

Flavonols have been found to prevent the replication of viruses by disturbing the key steps involved in the life cycle of the virus. They can effectively disturb viral attachment, entry, and replication, reducing the viral load [5]. Khazdair et al. have reviewed that plants containing kaempferol **1** and quercetin **2** (Figure 7.4) can fight viruses. Some of these plants are *Crocus sativus* L. (Iridaceae), *Kaempferia galanga* L. (Zingiberaceae), *Allium cepa* L. (Amaryllidaceae), and *Portulaca oleracea* L. (Portulacaceae), possessing their ability to inhibit protein kinase B and phosphorylation of protein kinase. The plant extracts also exhibit blocking effects on a specific channel (3a channel) found in SARS-CoV-infected cells [75]. The molecular mechanism of the antiviral activity of quercetin **2** and quercetin-rich plants has been reported to inhibit the viral load by blocking virus entry. Likewise, compound **2** can attach to the active regions of viruses and disturb their functionality to utilize the resources within host cells to fulfill their requirements, consequently constraining their ability for replication [78]. Kaempferol **1** also inhibited the Epstein-Barr virus (EBV) in nasopharyngeal carcinoma cells by repressing the most important factor in EBV reactivation, Spl expression, which is targetable. Evaluation of the mechanism of action showed that compound **1** treatment suppresses Spl expression and reduces Spl promoter activity, proposing a viable candidate for anti-EBV therapy and cancer prevention [79]. An *in vitro* study showed that green tea-derived flavonols, for example, quercetin **2** and myricetin **3** (Figure 7.3), inhibit infection by severe fever with thrombocytopenia syndrome virus (SFTSV) [80]. The influenza virus includes eight RNA fragments and surface antigens like neuraminidase (NA) and hemagglutinin (HA). The former enzyme can cleave the sugars bound to mature viruses, and therefore, the virus can be released from the infected cells. In 2009, Jeong and coworkers assessed the *in vitro* anti-influenza properties of flavonols (i.e., kaempferol **1**, herbacetin **4**, rhodiolinin **5**, rhodionin **6**, and rhodiosin **7**; Figure 7.4) derived from *Rhodiola rosea* roots and compared them with those flavonols, which are commercially available (i.e., quercetin **2**, gossypetin **8**, astragalin **9**, linocinarnarin **10**, and nicotiflorin **11**; Figure 7.4). Using a reduction assay in virus-induced cytopathic effect (CPE) in MDCK cell lines, the authors figured out that all flavonols and flavonol glycosides inhibited H9N2 and H1N1 strains. The effective concentration (EC_{50}) ranges for flavonols against H9N2 and H1N1 strains were determined to be 18.5–133.6 μM and 30.2–99.1 μM, respectively. Interestingly, the most active metabolites against both strains were kaempferol **1** (EC_{50} of 18.5 and 30.2 μM, respectively) and herbacetin **4** (EC_{50} values of 23.0 and 35.0 μM, respectively). Furthermore, they tested flavonols and glycosylated flavonols to see if these metabolites could inhibit two types of NA from recombinant influenza virus A (rvH1N1) and *Clostridium perfringens*. The most significant results were recorded for gossypetin **8** (IC_{50} of 2.6 and 0.8 μM, respectively) [81]. Using a plaque reduction assay, the anti-influenza activity-guided isolation of the EtOAc-soluble fraction of *Zanthoxylum piperitum* (L.) DC. yielded flavonol glycosides, such as hyperoside (quercetin-3-*O*-β-D-galactoside **12**), quercitrin (quercetin-3-*O*-α-L-rhamnoside **13**), and afzelin (kaempferol-3-*O*-α-L-rhamnoside **14**) (Figure 7.4).

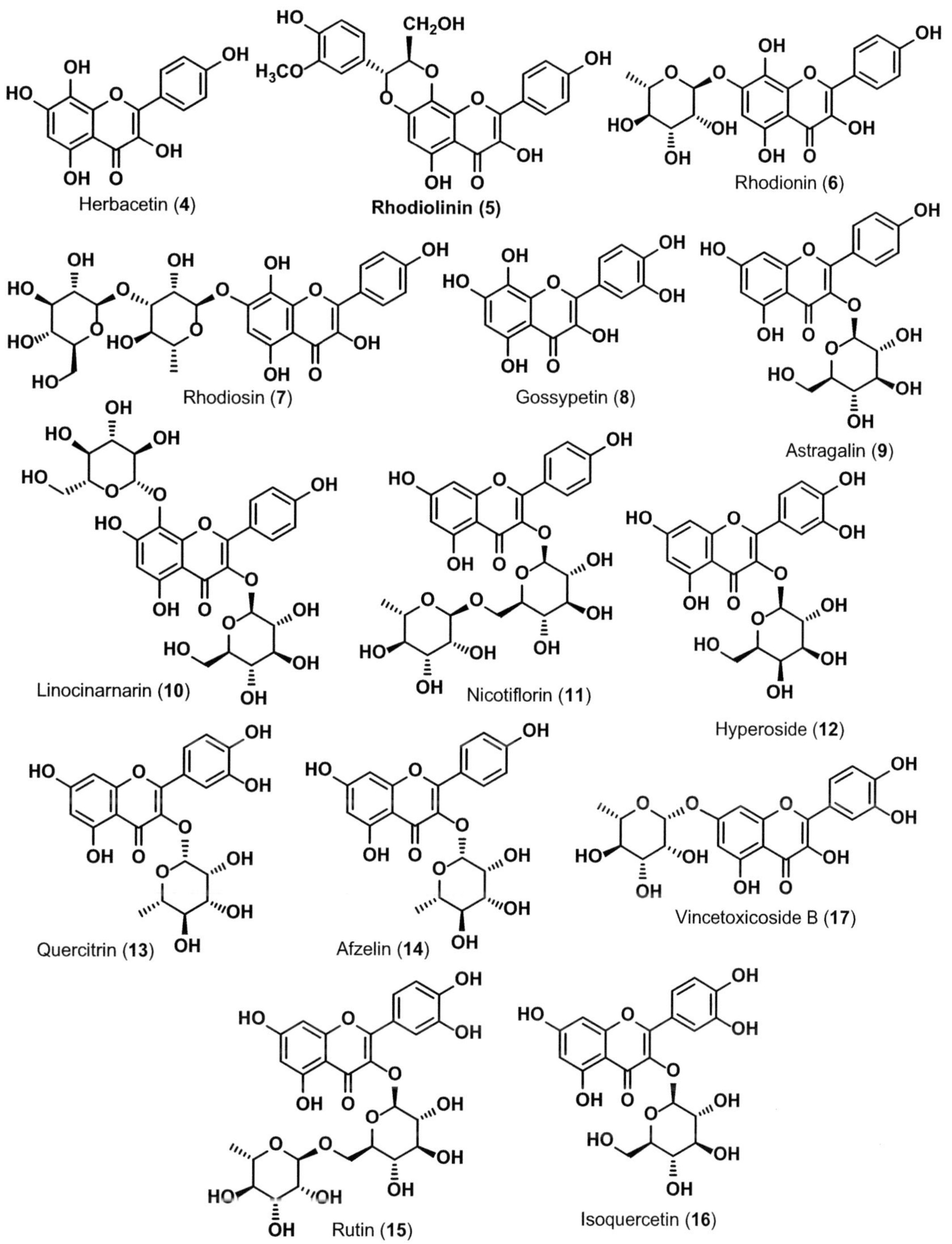

Herbacetin (4) Rhodiolinin (5) Rhodionin (6) Rhodiosin (7) Gossypetin (8) Astragalin (9) Linocinarnarin (10) Nicotiflorin (11) Hyperoside (12) Quercitrin (13) Afzelin (14) Vincetoxicoside B (17) Rutin (15) Isoquercetin (16)

FIGURE 7.4 Chemical structures of some antiviral flavonols, **4–17**.

Moreover, these metabolites inhibited the influenza virus (A/NWS/33 H1N1 strain) and showed anti-NA activity at relatively high concentrations [82]. In another research, the flavonol glycosides and their aglycones, such as kaempferol **1**, quercetin **2**, nicotiflorin (kaempferol-3-*O*-rutoside **11**), rutin (quercetin-3-*O*-rhamnoglucoside or quercetin-3-*O*-rutoside **15**), and isoquercetin (quercetin-3-*O*-glucoside **16**) (Figure 7.4), reported from the polyphenol-rich extracts of *Pistacia vera* L., inhibited the replication of herpes simplex virus type 1 (HSV-1) in Vero cell lines using a plaque reduction assay [83]. The most well-known sugar-containing quercetin is rutin **15**, which effectively acts as an antiviral against HSV, dengue virus type 2 (DENV-2), and human immunodeficiency virus (HIV) [78]. Vincetoxicoside B (quercetin-7-*O*-α-L-rhamnoside **17**) (Figure 7.4), derived from *Houttuynia cordat* aerial parts, inhibited the replication of porcine epidemic diarrhea virus (PEDV), the causal agent responsible for inducing severe entero-pathogenic diarrhea in pigs. In comparison to quercetin **2**, exhibiting an IC_{50} of 1.7 μg/mL, a CC_{50} of 365.2 μg/mL, and a therapeutic index of 214.8, vincetoxicoside B **17** showed better anti-PEDV replication with an IC_{50} of 0.014 µg/mL, a CC_{50} of 100 µg/mL, and a therapeutic index of >7,142. However, both flavonols were more active than ribavirin as the positive control, with an IC_{50} of 4.1 µg/mL, a CC_{50} of 423.3 µg/mL, and a therapeutic index of 103.2 [84].

7.6 VARIOUS DRUGGABLE TARGETS FOR SARS-COV-2

7.6.1 Spike Proteins

The entry process of SARS-CoV-2 is contingent upon the interaction between its spike protein and host receptors located on the cell surface. The spike proteins of SARS-CoV-2 are comprised of 1,273 amino acids and exhibit a structural configuration characterized by two distinct domains. The first domain, known as the *N*-terminal S1 domain, is responsible for the attachment of the virus to the cellular receptor. On the other hand, the second domain, referred to as the *C*-terminal S2 domain, facilitates the fusion between the virus and the membrane of the host cell. Following the interaction between the S1 domain and the host receptor, significant structural alterations occur in the S2 domain [85]. One strategy for drug discovery in SARS-CoV-2 could be looking for fusion inhibitors and anti-adhesive compounds, either natural or synthetic (Figure 7.5).

7.6.2 Extracellular Targets

Until now, some extracellular targets, such as the ACE2 receptor, TMPRSS2, and furin, have been recognized as potential draggable targets for COVID-19 (Figure 7.5).

Host ACE2 serves as the principal binding site for the entry of the SARS-CoV-2 virus into the cells. In the context of biological systems, there is a prevalent occurrence of this phenomenon throughout many bodily tissues, including the respiratory tract, blood vessels, heart, kidneys, and gastrointestinal system. The virus's spike protein on the surface binds to host ACE2, facilitating viral attachment and entry into target cells. Several studies have delineated that the downregulation of the ACE2 receptor is associated with blocking viral attachment and fusion [86].

Host TMPRSS2 is another cell surface protease that is encoded by the TMPRSS2 gene. It is mostly located in the cell membrane of the epithelial cells of the lungs. The action of this molecule enhances the activation of the SARS-CoV-2 spike protein, facilitating the merging of the viral envelope with the membrane of the host cell and expediting the process of viral entry with greater efficiency. Hoffman et al. demonstrated that the entry of SARS-CoV-2, contingent upon the presence of TMPRSS2 and ACE2 receptors, may be effectively inhibited through the utilization of clinically authorized protease inhibitors, such as Camostat [86].

Host Furin is a proprotein convertase enzyme that activates fusion proteins in various viruses. The enzyme cleaves the spike protein of SARS-CoV-2 at the S1 and S2 domains while the virus is attaching and fusing. This is necessary for the spike protein to become activated and for the virus to be infectious. Furin-mediated cleavage enhances the entry process of the virus into host cells [87]. Along with other extracellular proteases like ACE2 and TMPRSS2, furin also makes it possible for SARS-CoV-2 to be proteolytically activated and then spread out through the human airway epithelial cells. These make them salient candidates for drug targets [88].

7.6.3 Intracellular Targets

Identifying intracellular SARS-CoV-2 druggable targets is an interesting aspect of developing therapies to halt viral replication and combat COVID-19. While studies are ongoing, these intracellular targets have shown promise as possible drug targets: $3CL^{pro}$ (or M^{pro}), PL^{pro}, RdRp, exoribonuclease (ExoN), cathepsin L (CTSL), and helicase (nsp13) (Figure 7.5).

SARS-CoV-2 $3CL^{pro}/M^{pro}$ and SARS-CoV-2 PL^{pro} are two proteolytic enzymes belonging to non-structural proteins. These enzymes are highly conserved cysteine proteinases, which effectively process the replicase polyprotein 1ab into functional components that are necessary

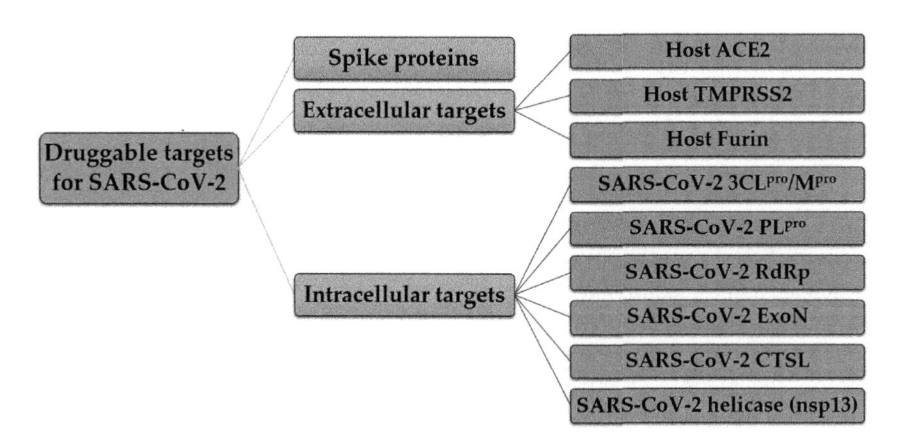

FIGURE 7.5 General targets that could be appropriate toward SARS-CoV-2 activity.

for SARS-CoV-2 to replicate. Meanwhile, PL^{pro} weakens the host's immune response during viral infection by interfering with critical immune pathways (i.e., tool-like receptor 7 and type 1 interferon signaling pathway) within the host cell. This interference helps the virus evade the host's defenses and establish an infection. Inhibiting the activity of these proteases can disrupt viral replication and potentially serve as targets for antiviral therapies [89].

SARS-CoV-2 RdRp sequence, belonging to the non-structural proteins (nsp12), resembles that of SARS-CoV RdRp (96%). As a key enzyme for replication, RdRp is effectively involved in the synthesis of RNA viruses by catalyzing the replication of viral genomes. Therefore, inhibition of RdRp activity can be a potential target for drug discovery. *In vitro* studies showed that synthetic drugs like favipiravir, galidesivir, remdesivir, ribavirin, and penciclovir can efficiently inhibit RdRp activity [90].

SARS-CoV-2 ExoN is responsible for a mechanism for error correction that aids in replicating the extensive RNA genomes of SARS-CoV-2 and MERS-CoV. During the replication of virus RNA, the proofreading and error correction are checked by an nsp14 (i.e., 3′-to-5′ exoribonuclease; ExoN). ExoN removes incorrect nucleotides when viral RdRp synthesizes viral RNA. This helps the virus maintain its genome intact and makes it easier for the virus to prevent the host's immune response. Disturbing such targets will be beneficial for halting viral replication [91].

SARS-CoV-2 CTSL, a ubiquitous cysteine protease, acts as an essential enzyme in various coronaviruses (MERS and SARS) entry into the host cells. Having bonded the virus to the ACE2 receptor, followed by endocytosis, the pH-dependent CTSL enzyme is activated. This activation triggers the fusion of the virus with the membranes of endosomes, resulting in the depletion of viral genetic material in the host cytoplasm. Zhao et al. concluded that CTSL has a significant role in both human and animal model cells infected by SARS-CoV-2. This finding highlights CTSL as a druggable target for antiviral compounds [92].

SARS-CoV-2 helicase (nsp13), one of 16 known non-structural proteins from coronaviruses, is another helpful enzyme during replication and has the most preserved sequence throughout the coronavirus family. It could serve as a viable target for the employment of anti-SARS-CoV-2 therapeutics [93].

7.7 IN VITRO AND IN SILICO ANTI-SARS-COV-2 ACTIVITY OF FLAVONOLS

The reputation of various fruits, vegetables, and medicinal plants containing dietary flavonols along with their methyl- and sugar-containing derivatives against SARS-CoV-2 has been described.

In 2023, Meng et al. experimented to evaluate the anti-entry activity of natural flavonols and their corresponding glycosylated derivatives, including kaempferol **1**, quercetin **2**, myricetin **3** (Figure 7.3), astragalin **9**, hyperoside **12**, rutin **15**, and isoquercetin **16** (Figure 7.4), against the SARS-CoV-2 pseudovirus through targeting spike RDB. Results reported the different behaviors of flavonols to suppress the viral entry into the host HEK-293T-ACE2[h] cells. The IC_{50} values were recorded in the following descending order: **3** (10.27 μM) > **2** (17.00 μM) > **1** (34.65 μM) > **12** (83.12 μM) > **16** (94.39 μM) > **9** (133.05 μM) > **15** (146.87 μM). Based on the SAR studies, the authors realized that glycosylation reduces the antiviral entry of flavonols, so quercetin (**2**) displayed 4.6–8.6 times better inhibition than **12**, **15**, and **16**. Like quercetin, the same results were found for kaempferol in the aglycone form. Since the flavonol aglycones possessed good antiviral activity, compounds **1–3** were selected to evaluate their affinities with the spike RBD. All compounds successfully dose-dependently bound to spike RBD (K_D of 18.51, 11.23, and 9.62 μM, respectively). These data revealed that the more hydroxyl substitutions in the B ring, the greater the binding affinities with spike RBD. According to the CDOCKER energy scores, the significant binding sites for flavonols on the RBD were Pockets 3 and 5. The amino acid residues Gln498, Asn501, and Tyr505 within Pocket 5 were identified as pivotal residues facilitating the viral infection. Moreover, the authors hypothesized that Pocket 3 could potentially serve as an allosteric binding site within the spike RBD [94].

Liu and coworkers *in vitro* investigated the effects of quercetin **2** (Figure 7.3) and its metabolites bearing OH and $-OCH_3$ at 3′ and 4′ positions, including hyperoside **12**, rutin **15**, isoquercetin **16** (Figure 7.4), isorhamnetin **18**, tamarixetin **19**, and quercetin-3-*O*-glucuronide (miquelianin **20**) (Figure 7.6), against the activity of recombinant human angiotensin-converting enzyme 2 (rhACE2), using the fluorogenic substrate Mca-APK(Dnp). It is worth mentioning that the host's gastrointestinal tract and gut microbiota metabolized the sugar-containing quercetin derivatives to free quercetin. Compared to nicotianamine and DX600, the positive controls (with rhACE2 inhibitory activities of 64.5% and 67.7%, respectively), quercetin **2** was the most active flavonol that inhibited rhACE2 activity at a concentration of 5 μM (66.2% rhACE2 inhibition at 2 min and an IC_{50} value of 4.48 μM). Other flavonol glycosides like rutin **15** and isoquercetin **16**, along with tamarixetin **19**, were moderate rhACE2 inhibitors with 41.5%–48.3% inhibition ranges [95].

In Brazilian traditional medicine, *Siparuna* species are employed for managing colds and flu. Using high-speed countercurrent chromatography technique, the dichloromethane extract of *Siparuna cristata* leaves furnished 3,3′,4′-tri-*O*-methylquercetin **21**, retusin (3,7,3′,4′-tetra-*O*-methylquercetin **22**), and kumatakenin (3,7-di-*O*-methylkaempferol **23**) (Figure 7.6). The inhibitory activity of flavonols was assessed against SARS-CoV-2 replication in two cell lines, including Calu-3 (human lung adenocarcinoma) and Vero E6 (African green monkey kidney). The results were in favor of **22** (EC_{50} 0.6 and 0.4 μM, respectively) and **23** (EC_{50} 0.3 and 10.0 μM, respectively), surpassing the selective index of lopinavir/ritonavir and chloroquine, which were used as the positive controls. Molecular docking studies for the most active flavonol (retusin **22**) showed docking scores of –6.3 and –5.5 kcal/mol against SARS-CoV-2 3CL^{pro} and PL^{pro}, respectively. In the case of PL^{pro}, retusin **22** interacted with Tyr268 residue at a distance of 1.9Å. For the 3CL^{pro} hydrogen-bonding interaction, retusin **22** was near the His41 and Cys145 residues of the catalytic site. The authors suggested that applying methylated flavonols might be beneficial in the COVID-19 treatment [96].

Zhu et al. also evaluated the *in vitro* inhibition of five nutraceutical flavonols, such as kaempferol **1**, quercetin **2**, myricetin **3** (Figure 7.3), rutin **15**, and isoquercetin **16** (Figure 7.4), together with their dihydroflavonol derivatives, including dihydrokaempferol (aromadendrin), dihydroquercetin (taxifolin), and dihydromyricetin (ampelopsin) (Figure 7.3) on SARS-CoV-2 M^{pro} activity. Interestingly, all compounds demonstrated effective inhibition with IC_{50} values of 6.83, 6.79, 5.92, 0.125, and 4.03 µM, respectively (for flavonols) and IC_{50} values of 12.04, 12.94, and 20.3 µM, respectively (for dihydroflavonols). On the other hand, only quercetin (**2**), isoquercitrin **16**, and taxifolin *in vitro* inhibited the replication of the HCoV-229E virus, which is the causative agent for the common cold, in Huh-7 cells. Based on the ligand–receptor interaction modeling involving two coronaviruses (i.e., SARS-CoV-2 and HCoV-229E) M^{pro} enzymes, docking simulation results represented effective binding to the multiple subregions, such as S1′, S1, S2, and S4 in the binding pocket. The compounds **2**, **16**, and taxifolin inhibited M^{pro} activity (binding affinities between –8.8 and –7.4 kcal/mol). Furthermore, these flavonols, along with dihydroflavonols, exhibited binding potential and inhibitory effects on HCoV-229E M^{pro} (with binding affinities between –7.8 and –7.1 kcal/mol). It is noteworthy to mention that, in the computational studies, the glycosylation of flavonols resulted in the improvement of the binding affinities of rutin **15** and isoquercitrin **16**, while in the *in vitro* assays, the results were in favor of their respective aglycones [97].

The antiviral properties of morin **24** (Figure 7.6), which is commonly found in medicinal plants, vegetables, and fruits, were compared with hydroxychloroquine, remdesivir, and baricitinib using PASS analysis. The findings indicated that morin **24** complied with Lipinski's rule of five and other drug-likeness filters. It also exhibited no side effects related to tumorigenicity, reproduction, or irritation, demonstrating efficient absorption and permeation characteristics by the gastrointestinal tract (clogP<5). In a three-dimensional (3D) space, principal component analysis positioned morin **24** closely aligned with baricitinib. Molecular docking experiments also showed strong binding of morin **24** to the main proteins (spike protein) and proteases (3CL^{pro} and PL^{pro}) of various coronaviruses (MERS and SARS). Furthermore, morin **24** interacted potently with human ACE2 receptor, importin-α, and poly (ADP-ribose) polymerase (PARP-1). These results represent morin **24** as an efficient viral inhibitor. Meanwhile, it showed strong binding affinities to IL-6, IL-8, and IL-10, covering its anti-inflammatory properties. Considering the above points, morin **24** was suggested as a promising therapeutic for treating COVID-19, SARS-CoV, and MERS; however, an in-depth investigation was recommended [98].

Nguyen et al. tested some commercially available flavonol compounds found in black garlic extract, including

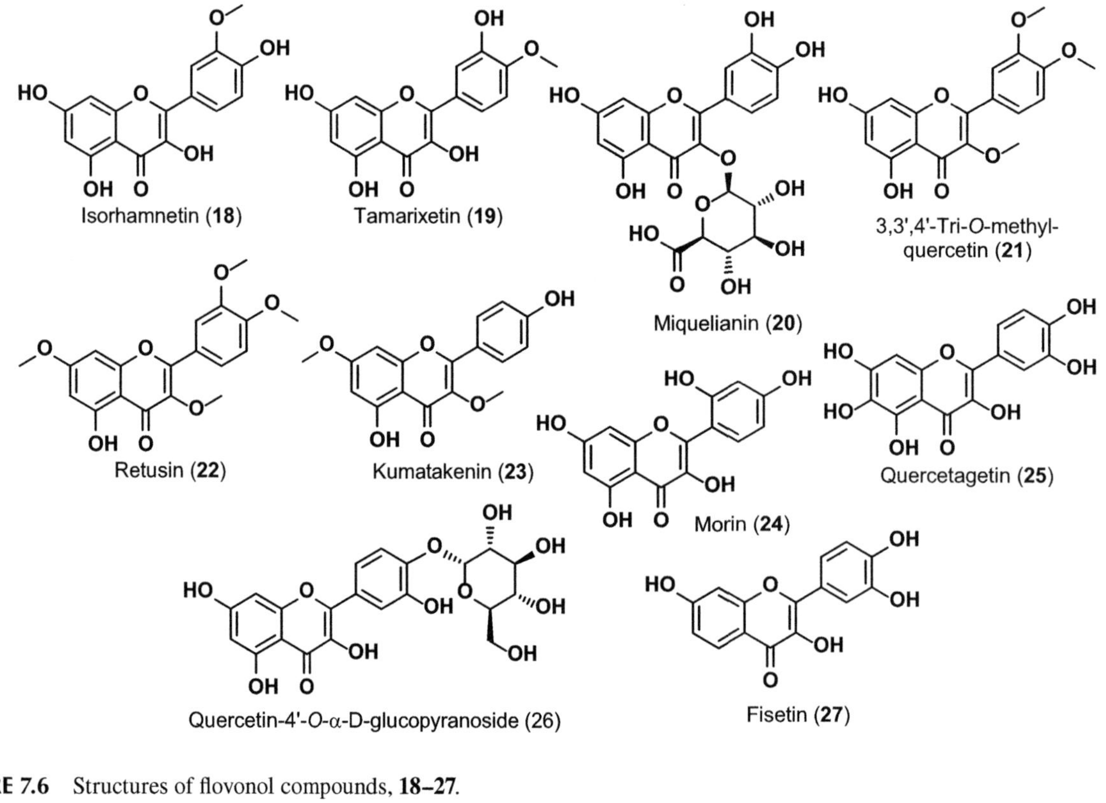

FIGURE 7.6 Structures of flovonol compounds, **18–27**.

kaempferol **1**, quercetin **2**, myricetin **3** (Figure 7.3), astragalin **9**, rutin **15** (Figure 7.4), and quercetagetin **25** together with a synthesized flavonol, quercetin-4′-*O*-α-D-glucoside **26** (Figure 7.6), against M^pro^ obtained through *Escherichia coli* BL21 (DE3) overexpression. Results exhibited that compounds **2**, **3**, **9**, and **25** were able to inhibit M^{pro} at 200 μM (IC_{50} values of 93, 43, 143, and 145 μM, respectively), while compounds **1**, **15**, and **26** showed M^{pro} inhibition less than 50%. According to structure–activity relationship (SAR) studies, the OH residues in the B ring impacted the inhibition of M^{pro}. Kaempferol (**1**, lacking a 3′-OH moiety in the B ring) and quercetin **2** displayed lower inhibitory activity compared to myricetin **3**, which has a pyrogallol moiety in the same ring. Glycosylation of quercetin at C-4′ (compound **26**) and C-3 in the C ring (rutin, **15**) resulted in decreased inhibitory effects. In the case of quercetagetin **25**, bearing a 6-OH in the A ring, the inhibitory effects were also reduced compared to quercetin **2**. Most importantly, ampelopsin (dihydromyricetin) (Figure 7.3) demonstrated lower inhibitory activity than myricetin **3** due to the lack of a C2-C3-double bond in the C ring, a characteristic present in myricetin **3** [99].

In order to decipher the potency of the inhibition of SARS-CoV-2 activity, Cheves et al. compared *in vitro* and *in silico* results of some commercially available flavonoid subclasses, including flavonols (i.e., kaempferol **1**, quercetin **2**, myricetin **3** (Figure 7.3), and fisetin **27** (Figure 7.6)), flavones (i.e., luteolin and apigenin), and an isoflavone (i.e., genistein). Results showed that flavonols proved to be more effective inhibitors rather than other flavonoids. The flavonol compounds also prominently inhibited the replication of SARS-CoV-2 in Calu-3 cells with the following EC_{50}: 3.02 μM (for kaempferol **1**), 2.40 μM (for quercetin **2**), 0.91 μM (for myricetin **3**), and 2.03 μM (for fisetin **27**). The EC_{50} value of the corresponding positive control, remdesivir, was 0.0305 μM. Interestingly, the SARS-CoV-2 inhibitory activity of flavonols is attributed to the influence of the hydroxy groups on the molecular orientation of these secondary metabolites within the enzyme target. Among the flavonols, myricetin **3** and fisetin **27** stand out as promising candidates, possessing anti-inflammatory properties (evidenced by their ability to decrease TNF-α levels in the Calu-3 SARS-CoV-2 infections) and disturbing non-competitively the functionality of SARS-CoV-2 M^{pro}. Compared to the positive control, GC376, low Morrison's inhibitory constant (*Ki*) values were reported for myricetin **3** and fisetin **27** (*Ki* values of 205.0, 11.1, and 41.8 nM, respectively), representing promising candidates to inhibit M^{pro}. To approve the *in silico* results, enzymatic assays exhibited that fisetin **27** can inhibit the ExoN complex (nsp14/nsp10) responsible for RNA cleavage, especially at high concentrations (150 μM). However, the inhibition pattern differed from fisetin's effects observed in the infected Calu-3 cells. This data suggests that ExoN might not be the primary target for fisetin **27**. Overall, the authors concluded that myricetin **3** and fisetin **27** may be starting points for synthetic modifications to enhance effectiveness against SARS-CoV-2 replication. These modifications could extend beyond M^{pro} inhibition and include targeting the 3′-to-5′ exoribonuclease (ExoN) activity, further improving their anti-SARS-CoV-2 properties [100].

As discussed earlier, the $3CL^{pro}$ enzyme is a primary coronavirus protease for replication and is indicated as a druggable target for several antiviral drugs. Su et al. reported that myricetin **3** and its derivatives, including ampelopsin (Figure 7.3), 7-methoxy-myricetin **28**, 7-ethoxy-myricetin **29**, 7-*O*-isoamyl-myricetin **30**, 7-*O*-cyclopentylmethyl-myricetin **31**, myricetin-7-yl 5,5-dimethyl,1,3,2-dioxayl phosphate **32**, myricetin-7-yl diphenyl phosphate **33**, 7-methoxy-dihydromyricetin, and dihydromyricetin-7-yl diphenyl phosphate (Figure 7.7), can inhibit the $3CL^{pro}$ activity (IC_{50} of 0.63, 1.14, 0.30, 0.74, 1.92, 2.45, 6.62, 3.13, 0.26, and 1.84 μM, respectively). In a Vero E6 cell-based antiviral experiment, the compounds also exhibited inhibitory activity against SARS-CoV-2 replication with EC_{50} values of 8.00, 13.56, 11.50, 12.59, 51.01, 31.54, 7.56, 33.45, 3.15, and 9.03 μM, respectively. Comparatively, remdesivir inhibited the replication of SARS-CoV-2 (EC_{50} = 3.68 μM). Besides inhibition of $3CL^{pro}$, myricetin **3** was further assayed on SARS-CoV $3CL^{pro}$ and SARS-CoV-2 PL^{pro}, together with human bovine chymotrypsin, to study its selective behavior. Myricetin **3** similarly inhibited SARS-CoV $3CL^{pro}$ with an IC_{50} of 0.74 μM; however, it was very weak in inhibition of PL^{pro} and chymotrypsin (IC_{50} of 159.10 and 132.30 μM, respectively). These findings confirmed the selective inhibition of myricetin **3** against $3CL^{pro}$ in both virus strains (i.e., SARS-CoV and SARS-CoV-2). Computational studies supported the experimental results so that myricetin **3** and its derivatives were found to be non-peptidomimetic covalent inhibitors by modifying the catalytic cysteine through the pyrogallol group. The authors also concluded that bearing the pyrogallol residue serves as an electrophilic warhead. Furthermore, some myricetin derivatives like 7-methoxy-dihydromyricetin are optimized as potent antiviral agents, showing high potency for oral administration. It possessed a favorable pharmacokinetic (PK) profiling, such as half-time ($T_{1/2}$ value of 1.74 hour), the area under the curve (AUC value of 510 ng h/mL), oral bioavailability of 18.1%, maximal concentration (C_{max} value of 724 ng/mL), and plasma duration (MRT value of 1.89 hour). Overall, the authors concluded regarding the significant insights of the covalent mode of action exhibited by natural products containing pyrogallol. They have also provided a blueprint for the development of non-peptidomimetic covalent $3CL^{pro}$ inhibitors, with a particular focus on the potential of pyrogallol as an alternative warhead in the design of targeted covalent ligands [101].

According to a preliminary $3CL^{pro}$ inhibitory assay, *Ampelopsis grossedentata* extract showed potent inhibition. Further investigation involved the isolation of the phytoconstituents from the flavonoid-rich fraction and assessing them against $3CL^{pro}$. Compared to ebselen (IC_{50} value of 2.62 μM as the positive control), the assayed compounds, such as myricetin **3**, ampelopsin (Figure 7.3), and

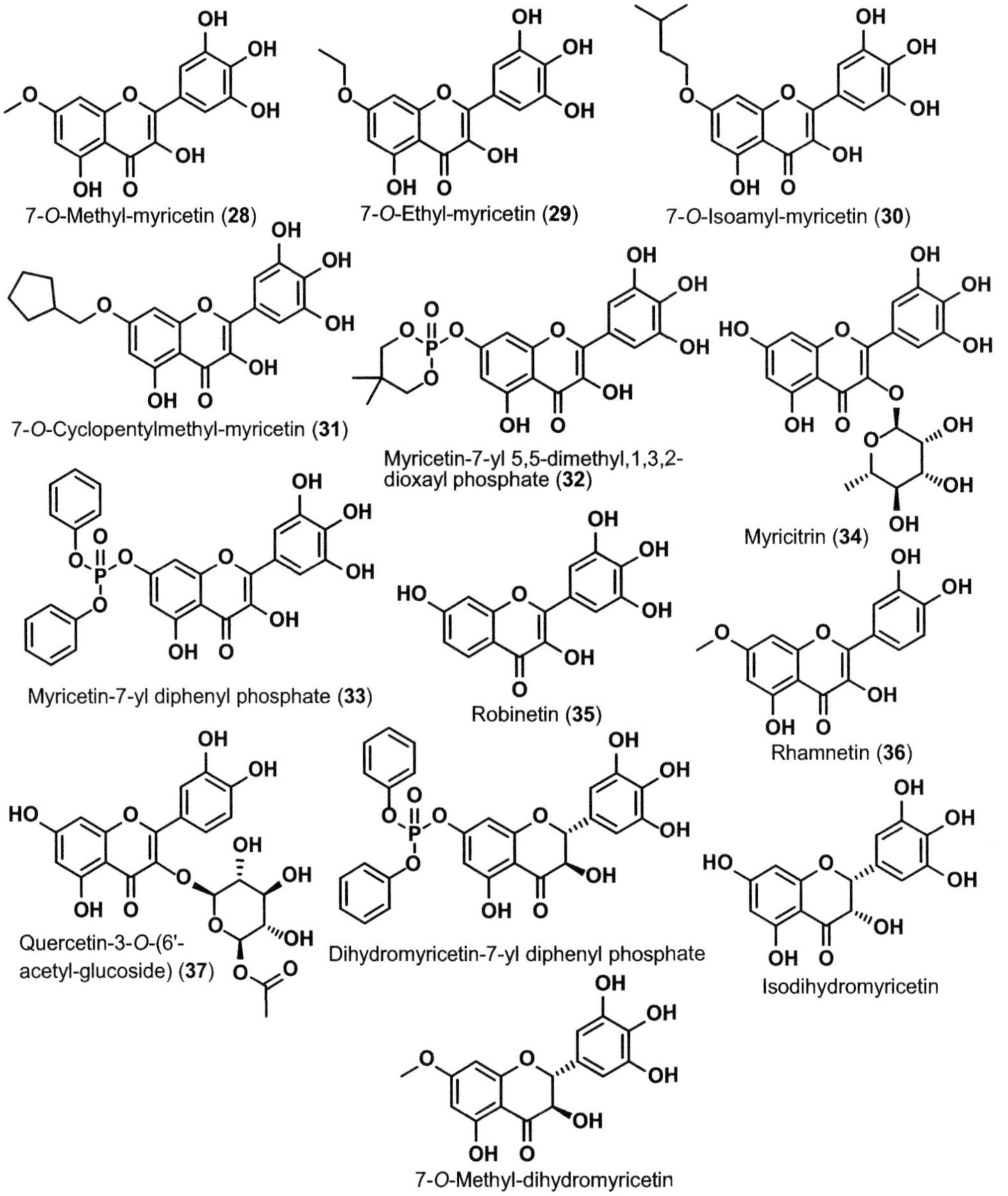

7-*O*-Methyl-myricetin (**28**)
7-*O*-Ethyl-myricetin (**29**)
7-*O*-Isoamyl-myricetin (**30**)
7-*O*-Cyclopentylmethyl-myricetin (**31**)
Myricetin-7-yl 5,5-dimethyl,1,3,2-dioxayl phosphate (**32**)
Myricitrin (**34**)
Myricetin-7-yl diphenyl phosphate (**33**)
Robinetin (**35**)
Rhamnetin (**36**)
Quercetin-3-*O*-(6'-acetyl-glucoside) (**37**)
Dihydromyricetin-7-yl diphenyl phosphate
Isodihydromyricetin
7-*O*-Methyl-dihydromyricetin

FIGURE 7.7 Structures of flovonol compounds, **35–37**.

isodihydromyricetin (Figure 7.7), time-dependently inhibited 3CL[pro] activity with IC_{50}s of 1.21, 4.91, and 3.73 µM, respectively, after 60 minutes of pre-incubation. At the same time, myricitrin **34** (Figure 7.7) and taxifolin (Figure 7.3) had an IC50 range of 10–100 µM and moderately suppressed the enzyme activity. The authors used molecular docking to find that myricetin **3** could connect covalently with 3CL[pro] at specific sites, mainly Cys300 and Cys44. Ampelopsin and isodihydromyricetin, on the other hand, had only covalent interactions with Cys300. Overall, in terms of anti-SARS-CoV-2 3CL[pro] activity, a C2-C3 double bond in myricetin **3** plays a pivotal role in enzyme inhibition, and the combination of these phytochemicals makes *A. grossedentata* a promising plant for the development of antiviral secondary metabolites [102].

Using molecular docking studies, Teli et al. screened 170 types of natural products, including flavonols like kaempferol **1**, myricetin **3** (Figure 7.3), astragalin **9**, quercitrin **13**, rutin **15** (Figure 7.4), morin **24**, quercetagetin **25**, fisetin **27** (Figure 7.6), robinetin **35**, and rhamnetin **36** (Figure 7.7), against M[pro] and spike RBD. In the former computational experiment, rutin **15** was the most active compound with a docking score of –11.19 kcal/mol, followed by quercetagetin **25** (–9.41 kcal/mol) and astragalin **9** (–9.120 kcal/mol). The remaining compounds had binding energies between –8.12 and –6.58 kcal/mol. In the latter molecular docking on spike protein, only rutin **15** and quercitrin **13** showed moderate hydrogen-bonding energy with docking scores of –7.91 and –7.15 kcal/mol, respectively [103].

In another computational research on 46 bioflavonoids, Kumar et al. investigated the glycosylation effects of flavonols, such as astragalin **9** (Figure 7.4), myricitrin **34**, and quercetin-3-*O*-(6″-acetyl-glucoside) **37** (Figure 7.7), compared to their respective aglycones (i.e., kaempferol **1**, quercetin **2**, and myricetin **3**; Figure 7.3) against three main SARS-CoV-2 targets, such as the human TMPRSS2 protein, $3CL^{pro}$, and PL^{pro}. Surprisingly, the molecular docking studies revealed that the sugar moieties increase the binding affinities (translated to decrease the binding energies) of the mentioned targets. For instance, kaempferol **1** and astragalin **9** had docking scores of –7.4 and –7.8 kcal/mol in human TMPRSS2 protein. For myricetin **3** and myricitrin **34**, the binding energies were –7.4 and –8.9 kcal/mol in $3CL^{pro}$. Quercetin-3-*O*-(6″-acetyl-glucoside) **37** showed better binding affinities than quercetin **2** (–6.2 and –7.6 kcal/mol, respectively) in PL^{pro}. In comparison to flavonoids, flavanols, and flavanones that lack a C2-C3 double bond, computational analysis displayed the negative impact of the double bond on binding energies. Overall, the authors proposed that the glycosylated flavonoids not only reduced the binding energies and their toxicity, but also enhanced their solubility and bioavailability [104].

7.8 SYNERGISTIC COMBINATION EFFECTS OF FLAVONOLS AND THE GLYCOSYLATED DERIVATIVES AGAINST SARS-COV-2

Combination therapy against COVID-19 is emerging as a promising approach for effective treatment due to the synergistic behaviors. However, achieving the optimal drug combination and administration as anti-infective agents is sometimes hindered by the low bioavailability and systematic toxicity associated with individual drugs. Plant-containing flavonols and their glycosylated derivatives present therapeutic prospects to improve and manage the SARS-CoV-2 infection and its undesired inflammation consequences.

Salvadora persica L., commonly known as meswak among Muslims and used for oral hygiene, has been noted for its antiviral properties. Owis et al. identified 11 flavonol glycosides, such as astragalin **9** (Figure 7.4), kaempferol-3-*O*-α-L-rhamnosyl(1→6)-*O*-[α-L-rhamnosyl(1→2)]-*O*-β-D-glucoside **38**, kaempferol-3-*O*-α-L-rhamnosyl(1→6)-*O*-[α-L-rhamnosyl(1→2)]-*O*-β-D-galactoside (mauritianin **39**), kaempferol-3-*O*-α-L-rhamnosyl(1→6)-O-β-D-glucoside **40**, kaempferol-3-*O*-α-L-rhamnosyl(1→6)-O-β-D-galactoside **41**, isorhamnetin-3-*O*-α-L-rhamnosyl(1→6)-*O*-[α-L-rhamnosyl(1→2)]-*O*-β-D-glucoside **42**, isorhamnethin-3-*O*-α-L-rhamnosyl(1→6)-*O*-[α-L-rhamnosyl(1→2)]-*O*-β-D-galactoside **43**, isorhamnetin-3-*O*-α-L-rhamnosyl(1→6)-*O*-β-D-glucoside (narcissin **44**), isorhamnetin-3-*O*-α-L-rhamnosyl(1→6)-*O*-β-D-galactoside **45**, isorhamnetin-3-*O*-β-D-glucoside **46**, and isorhamnetin-3-*O*-β-D-galactoside **47** (Figure 7.8), from the aqueous extract of meswak. All glycosylated flavonols showed strong binding energies with both hACE2 and the viral spike protein (docking scores ranging from –9.48 to –5.21 kcal/mol). When the aqueous extract was formulated into liposomes, the glycosylated flavonols inhibited the SARS-CoV-2 activity in A549 cell cultures during *in vitro* tests. Most significantly, the encapsulation of the mixture of flavonol glycosides remarkably enhanced the inhibition of SARS-CoV-2 replication from 38.09% to 85.56%. This improved inhibition approached the effectiveness of remdesivir (91.20% inhibition). Considering the preliminary assessment results, including a total flavonoid assay, molecular docking study, $3CL^{pro}$ inhibition assay (IC_{50} value of 8.59 μg/mL), and safety profile for human use (CC_{50} value of 24.5 μg/mL), the findings highlighted the potential and synergistic capability of meswak-derived flavonol glycosides, an affordable and accessible natural resource, as a promising candidate against SARS-CoV-2. Particularly, their release of phytochemicals into saliva during regular oral use showcases a potential additional benefit [105].

The National Administration of TCM reported that the effectiveness of herbal medicine lies not only in protecting people against COVID-19 but also in improving the health condition of treated patients. Clinical studies mostly utilized herbal extract mixtures. Tannic acid, a phytochemical belonging to the tannin family, is found in various sources like bananas, legumes, sorghum, persimmons, raspberries, red wines, etc., with varied concentrations of tannins estimated to range from 5 to 100 mM. Tannic acid demonstrated inhibitory effects on both SARS-CoV-2 M^{pro} and TMPRSS2 enzymes. However, it was noted that tannic acid, while effective, could form complexes with proteins, digestive enzymes, and starch, potentially reducing nutritional value. Interestingly, to decrease these side effects, Nguyen and coworkers designed an experiment to see the anti-SARS-CoV-2 M^{pro} activity of a combination of tannic acid with myricetin **3** (Figure 7.3), daidzein (an isoflavone), and puerarin (an isoflavone glycoside) on SARS-CoV-2 M^{pro} activity. Tannic acid was kept at a concentration of 5 μM, while myricetin, daidzein, and puerarin were maintained at 20 μM, representing individually 30%, 34%, 27%, and 23% inhibition, respectively. Further inhibitory effects were evaluated, showing that combining tannic acid with these compounds significantly enhanced the inhibition of SARS-CoV-2 M^{pro}. An example of potent inhibitory activity on M^{pro} was observed with a combination of 5 μM tannic acid, 20 μM myricetin, 20 μM daidzein, and 20 μM puerarin, resulting in the highest inhibition rate of 77 ± 1%. These findings revealed the potential synergistic effects of these compounds in combating COVID-19 through SARS-CoV-2 M^{pro} inhibition [99].

In another study, the ethyl acetate fraction of *Houttuynia cordata* Thunb. (Saururaceae) and its three flavonol metabolites (i.e., quercetin **2**, quercitrin **13**, and rutin **15**; Figures 7.3 and 7.4) were assessed against the infection murine coronavirus (MHV) and dengue virus type 2 (DENV-2). The EtOAc fraction effectively inhibited the infection for up to 6 days, while cinanserin hydrochloride

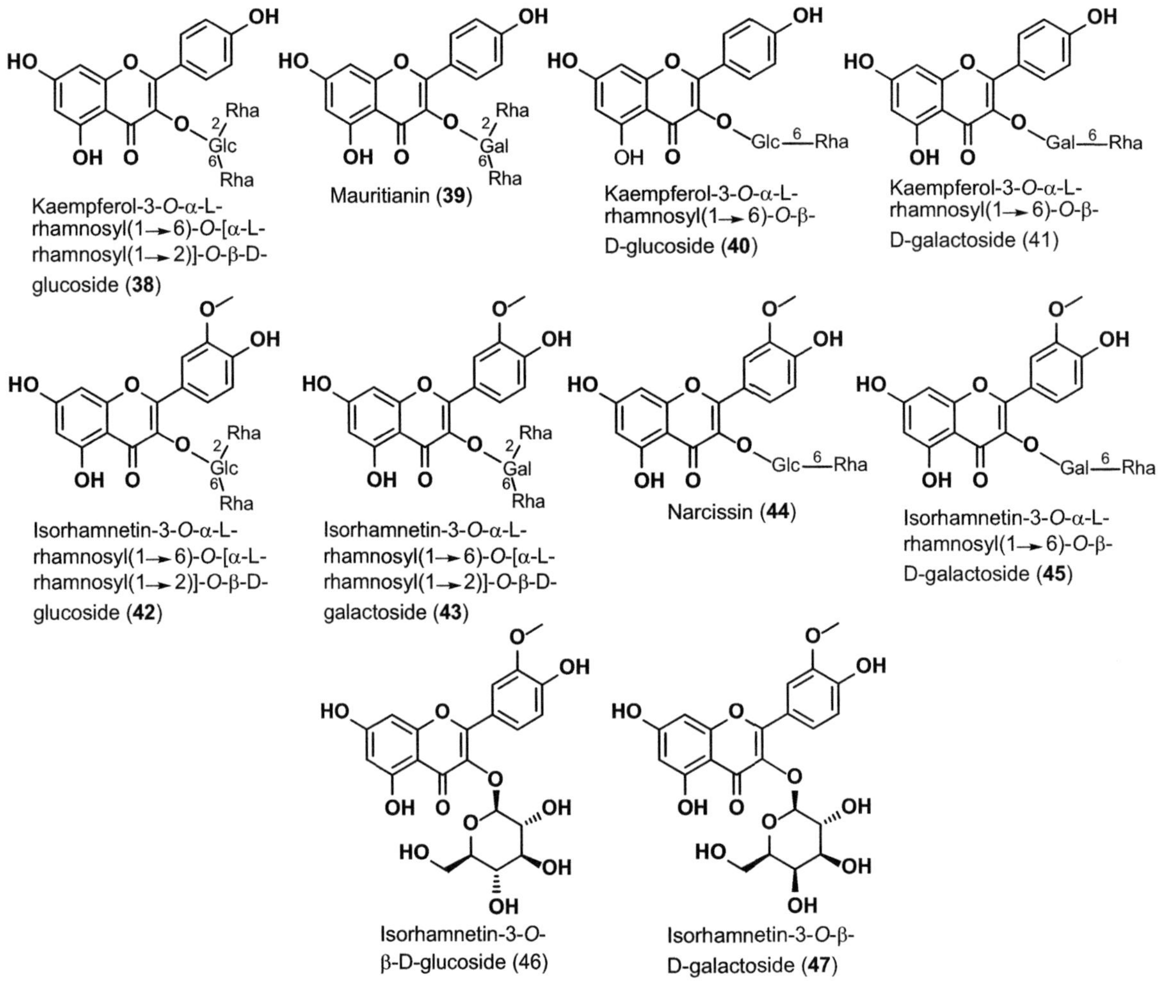

Kaempferol-3-O-α-L-rhamnosyl(1→6)-O-[α-L-rhamnosyl(1→2)]-O-β-D-glucoside (**38**)

Mauritianin (**39**)

Kaempferol-3-O-α-L-rhamnosyl(1→6)-O-β-D-glucoside (**40**)

Kaempferol-3-O-α-L-rhamnosyl(1→6)-O-β-D-galactoside (41)

Isorhamnetin-3-O-α-L-rhamnosyl(1→6)-O-[α-L-rhamnosyl(1→2)]-O-β-D-glucoside (**42**)

Isorhamnetin-3-O-α-L-rhamnosyl(1→6)-O-[α-L-rhamnosyl(1→2)]-O-β-D-galactoside (**43**)

Narcissin (**44**)

Isorhamnetin-3-O-α-L-rhamnosyl(1→6)-O-β-D-galactoside (**45**)

Isorhamnetin-3-O-β-D-glucoside (46)

Isorhamnetin-3-O-β-D-galactoside (**47**)

FIGURE 7.8 Structures of glycosylated flovonols, **38–47**.

(an antiviral compound) inhibited MHV only for 2 days. Administrated prior to the virus adsorption stage, the EtOAc fraction also showed the IC_{50} values of 0.98 and 7.5 μg/mL, respectively, without showing toxicity to the cells. There were no toxicity signs, and all the organs were normal when the mice were fed with 2000 mg/kg of the EtOAc fraction. Among the flavonols, quercetin **2** showed potent antiviral activity against both viruses (IC_{50} value of 125.00 and 176.76 μg/mL, respectively). On the other hand, quercitrin **13** displayed weaker anti-DENV-2 activity, while rutin **15** had no inhibitory activity on either virus. Note that combining quercetin **2** and quercitrin **13** enhanced the inhibition of DENV-2 activity and diminished the cytotoxicity. However, the combined flavonoids were efficiently not as strong as that of the EtOAc fraction, representing the potency of combination compounds of *H. cordata* [106].

7.9 FLAVONOLS AND SARS-COV-2 CLINICAL TRIALS

Flavonols have piqued interest in SARS-CoV-2 clinical trials. Due to their excellent biological properties, such as antioxidant, anti-inflammatory, immunomodulatory, and antiviral activities, researchers are exploring flavonols and their glucosylated derivatives. These compounds could contribute to combating viral infections.

When taken orally, quercetin has a very low bioavailability in the human serum, which makes it very challenging for its antiviral potency. Recently, Quercetin Phytosome®, a special quercetin delivery form, has emerged to enhance human oral absorption. Following the cleavage of quercetin glucuronide in human plasma, the results revealed that there was a considerable 20-fold increase in the overall bioavailability of quercetin. Moreover, specific enzymes like glucuronidase were found to release free quercetin in the body. Considering the anti-inflammatory and thrombin-inhibitory properties, Quercetin Phytosome® holds promise as a valuable drug against COVID-19 clinically [107].

It is also well documented that combining quercetin **2** (Figure 7.3) with vitamins, especially vitamin C as an antioxidant compound, can reduce COVID-19 symptoms. Intracellular zinc levels have been associated with a rise in intracellular pH, effectively hindering SARS-CoV-2 RdRp activity and consequently disturbing viral replication. Quercetin **2**, acting as a zinc ionophore, facilitates the inflow of zinc into cells. Bromelain, derived from pineapples, serves as a dietary supplement and has exhibited the capacity to reduce ACE2 and TMPRSS2 expression, thereby inhibiting viral infection in Vero E6 cells. During the COVID-19 pandemic, the sepsis generated by the SARS-CoV-2 virus led to a meaningful elevation in the levels of pro-inflammatory cytokines. This, in turn, enhances the neutrophil gathering within the pulmonary region, leading to further damage to the capillaries of the alveoli. Vitamin C has the dual purpose of reducing this aggregation and assisting in the maintenance of alveolar fluid levels [108]. Considering the ability of both compounds, some clinical trials have been designed to evaluate the combination therapy to treat SARS-CoV-2-positive patients. One of the clinical studies dealt with the effects of a quadruple therapy containing quercetin **2**, vitamin C, bromelain, and zinc. Results uncovered a promising improvement in clinical findings among COVID-19 patients [109]. (https://classic.clinicaltrials.gov/ct2/show/NCT04468139, assessed on September 22, 2023).

7.10 SUMMARY POINTS

- Flavonols and their sugar-containing derivatives demonstrate potent anti-inflammatory properties that could help people with severe COVID-19 cases deal with cytokine storms and speed up their recovery. Moreover, these secondary metabolites can enhance the immune response against viral infections by regulating the production of cytokines and chemokines, bolstering the body's defense mechanisms.
- Flavonols inhibit viral replication by disturbing key steps in the vial life cycle, including attachment, entry, and replication, thus reducing the viral load.
- One strategy for SARS-CoV-2 drug discovery involves identifying natural or synthetic fusion inhibitors and anti-adhesive compounds.
- So far, several extracellular targets like ACE2, TMPRSS2, and furin have been identified as potential drug targets for COVID-19.
- Promising intracellular targets for potential drugs include $3CL^{pro}$ (or M^{pro}), PL^{pro}, RdRp, ExoN, CTSL, and helicase (nsp13), although ongoing research continues to explore this field.
- Some studies revealed that using methylated flavonols could be beneficial in treating COVID-19.
- Several research studies have declared the covalent mechanism of action in natural products containing pyrogallol. These ideas serve as a foundational guide for designing non-peptidomimetic covalent inhibitors targeting $3CL^{pro}$, highlighting the potential utility of pyrogallol as an alternative warhead for the development of targeted covalent ligands.
- The computational analysis revealed that the presence of a C2-C3 double bond in flavonoids has a negative impact on binding energies when compared to those flavonoids missing this double bond (i.e., flavanones and flavanols).
- Based on preliminary assessment results, including a total flavonoid assay, molecular docking study, $3CL^{pro}$ inhibition assay, and human safety profile, the findings highlighted the potential of meswak-derived flavonol glycosides as a promising and accessible candidate against SARS-CoV-2. Releasing phytochemicals into saliva during regular oral use presents an additional potential benefit.
- The combination of tannic acid, myricetin, daidzein, and puerarin significantly increased SARS-CoV-2 M^{pro} inhibition. This highlights the potential synergistic effects of these compounds in combating COVID-19 through SARS-CoV-2 M^{pro} inhibition.
- Due to the high potency of flavonol compounds against viruses, especially SARS-CoV-2, it is recommended that researchers go through combination therapy to see their synergistic effects.

REFERENCES

1. Lauxmann MA, Santucci NE, Autrán-Gómez AM. The SARS-CoV-2 coronavirus and the COVID-19 outbreak. *International Brazilian Journal of Urology*. 2020;46:6–18.
2. Rabby MII. Current drugs with potential for treatment of COVID-19: A literature review: Drugs for the treatment process of COVID-19. *Journal of Pharmacy & Pharmaceutical Sciences*. 2020;23:58–64.
3. Hafez Ghoran S, El-Shazly M, Sekeroglu N, Kijjoa A. Natural products from medicinal plants with anti-human coronavirus activities. *Molecules*. 2021;26(6):1754.
4. Hafez Ghoran S, Taktaz F, Ayatollahi SA. Plant immunoenhancers: Promising ethnopharmacological candidates for anti-SARS-CoV-2 activity. In: Chen JT, editor. *Ethnopharmacology and Drug Discovery for COVID-19: Anti-SARS-CoV-2 Agents from Herbal Medicines and Natural Products*. Singapore: Springer Nature Singapore; 2023. pp. 39–84.
5. Mouffouk C, Mouffouk S, Mouffouk S, Hambaba L, Haba H. Flavonols as potential antiviral drugs targeting SARS-CoV-2 proteases (3CLpro and PLpro), spike protein, RNA-dependent RNA polymerase (RdRp) and angiotensin-converting enzyme II receptor (ACE2). *European Journal of Pharmacology*. 2021;891:173759.
6. Harborne JB, Marby H, Marby T. The flavonoids. Singapore: Springer; 2013.
7. Kesarkar S, Bhandage A, Deshmukh S, Shevkar K, Abhyankar M. Flavonoids: An overview. *Journal of Pharmacy Research*. 2009;2(6):1148–1154.
8. Pollastri S, Tattini M. Flavonols: Old compounds for old roles. *Annals of Botany*. 2011;108(7):1225–1233.
9. Gervasi T, Calderaro A, Barreca D, Tellone E, Trombetta D, Ficarra S, et al. Biotechnological applications and health-promoting properties of flavonols: An updated view. *International Journal of Molecular Sciences*. 2022;23(3):1710.

10. Sharifi-Rad J, Herrera-Bravo J, Salazar LA, Shaheen S, Abdulmajid Ayatollahi S, Kobarfard F, et al. The therapeutic potential of wogonin observed in preclinical studies. *Evidence-Based Complementary and Alternative Medicine.* 2021;2021:9.
11. Pérez-Cano FJ, Castell M. Flavonoids, inflammation and immune system. *Nutrients.* 2016;8(10):659.
12. Sunil C, Xu B. An insight into the health-promoting effects of taxifolin (dihydroquercetin). *Phytochemistry.* 2019;166:112066.
13. Ferruzzi MG, Bordenave N, Hamaker BR. Does flavor impact function? Potential consequences of polyphenol-protein interactions in delivery and bioactivity of flavan-3-ols from foods. *Physiology & Behavior.* 2012;107(4):591–597.
14. Khoo HE, Azlan A, Tang ST, Lim SM. Anthocyanidins and anthocyanins: Colored pigments as food, pharmaceutical ingredients, and the potential health benefits. *Food & Nutrition Research.* 2017;61(1):1361779.
15. Zaheer K, Humayoun Akhtar M. An updated review of dietary isoflavones: Nutrition, processing, bioavailability and impacts on human health. *Critical Reviews in Food Science and Nutrition.* 2017;57(6):1280–1293.
16. Hafez Ghoran S, Saeidnia S, Babaei E, Kiuchi F, Hussain H. Scillapersicene: A new homoisoflavonoid with cytotoxic activity from the bulbs of *Scilla persica* HAUSSKN. *Natural Product Research.* 2016;30(11):1309–1314.
17. Hafez Ghoran S, Firuzi O, Pirhadi S, Khattab OM, El-Seedi HR, Jassbi AR. Sappanin-type homoisoflavonoids from *Scilla bisotunensis* Speta: Cytotoxicity, molecular docking, and chemotaxonomic significance. *Journal of Molecular Structure.* 2023;1273:134326.
18. Hafez Ghoran S, Saeidnia S, Babaei E, Kiuchi F, Dusek M, Eigner V, et al. Biochemical and biophysical properties of a novel homoisoflavonoid extracted from *Scilla persica* HAUSSKN. *Bioorganic Chemistry.* 2014;57:51–56.
19. Rudrapal M, Khan J, Dukhyil AAB, Alarousy RMII, Attah EI, Sharma T, et al. Chalcone scaffolds, bioprecursors of flavonoids: Chemistry, bioactivities, and pharmacokinetics. *Molecules.* 2021;26(23):7177.
20. Liu W, Feng Y, Yu S, Fan Z, Li X, Li J, et al. The flavonoid biosynthesis network in plants. *International Journal of Molecular Sciences.* 2021;22(23):12824.
21. Panche AN, Diwan AD, Chandra SR. Flavonoids: An overview. *Journal of Nutritional Science.* 2016;5:e47.
22. Sharma A, Sharma P, Tuli HS, Sharma AK. Phytochemical and pharmacological properties of flavonols. *eLS.* 2018;2018:1–12.
23. Hafez Ghoran S, Azadi B, Hussain H. Chemical composition and antimicrobial activities of *Perovskia artemisioides* Boiss. essential oil. *Natural Product Research.* 2016;30(17):1997–2001.
24. Pepe G, Sommella E, Manfra M, De Nisco M, Tenore GC, Scopa A, et al. Evaluation of anti-inflammatory activity and fast UHPLC-DAD-IT-TOF profiling of polyphenolic compounds extracted from green lettuce (*Lactuca sativa* L.; var. *Maravilla de Verano). Food Chemistry.* 2015;167:153–161.
25. Cui X-X, Yang X, Wang H-J, Rong X-Y, Jing S, Xie Y-H, et al. Luteolin-7-O-glucoside present in lettuce extracts inhibits hepatitis B surface antigen production and viral replication by human hepatoma cells *in vitro. Frontiers in Microbiology.* 2017;8:2425.
26. Edziri Há, Smach M, Ammar S, Mahjoub M, Mighri Z, Aouni M, et al. Antioxidant, antibacterial, and antiviral effects of *Lactuca sativa* extracts. *Industrial Crops and Products.* 2011;34(1):1182–1185.
27. Gautam GK. A review on the taxonomy, ethnobotany, chemistry and pharmacology of *Solanum lycopersicum* linn. *International Journal of Chemistry and Pharmaceutical Sciences.* 2013;1(8):521–527.
28. Kumar M, Tomar M, Bhuyan DJ, Punia S, Grasso S, Sa AGA, et al. Tomato (*Solanum lycopersicum* L.) seed: A review on bioactives and biomedical activities. *Biomedicine & Pharmacotherapy.* 2021;142:112018.
29. Owis A. Broccoli; the green beauty: A review. *Journal of Pharmaceutical Sciences and Research.* 2015;7(9):696.
30. Hwang J-H, Lim S-B. Antioxidant and anti-inflammatory activities of broccoli florets in LPS-stimulated RAW 264.7 cells. *Preventive Nutrition and Food Science.* 2014;19(2):89.
31. Cho W-K, Yim N-H, Lee M-M, Han C-H, Ma JY. Broccoli leaves attenuate influenza a virus infection by interfering with hemagglutinin and inhibiting viral attachment. *Frontiers in Pharmacology.* 2022;13:899181.
32. Cicio A, Serio R, Zizzo MG. Anti-inflammatory potential of Brassicaceae-derived phytochemicals: *In vitro* and *in vivo* evidence for a putative role in the prevention and treatment of IBD. *Nutrients.* 2022;15(1):31.
33. Chopade V, Phatak A, Upaganlawar A, Tankar A. Green tea (*Camellia sinensis*): Chemistry, traditional, medicinal uses and its pharmacological activities-a review. *Pharmacognosy Reviews.* 2008;2(3):157.
34. Shin D-U, Eom J-E, Song H-J, Jung SY, Nguyen TV, Lim KM, et al. *Camellia sinensis* L. alleviates pulmonary inflammation induced by porcine pancreas elastase and cigarette smoke extract. *Antioxidants.* 2022;11(9):1683.
35. Saeed M, Naveed M, Arif M, Kakar MU, Manzoor R, Abd El-Hack ME, et al. Green tea (*Camellia sinensis*) and l-theanine: Medicinal values and beneficial applications in humans-A comprehensive review. *Biomedicine & Pharmacotherapy.* 2017;95:1260–1275.
36. Teshika JD, Zakariyyah AM, Zaynab T, Zengin G, Rengasamy KR, Pandian SK, et al. Traditional and modern uses of onion bulb (*Allium cepa* L.): A systematic review. *Critical Reviews in Food Science and Nutrition.* 2019;59(Supp.1):S39–S70.
37. Zhao X-X, Lin F-J, Li H, Li H-B, Wu D-T, Geng F, et al. Recent advances in bioactive compounds, health functions, and safety concerns of onion (*Allium cepa* L.). *Frontiers in Nutrition.* 2021;8:669805.
38. Ademiluyi AO, Oyeniran OH, Oboh G. Tropical food spices: A promising panacea for the novel coronavirus disease (COVID-19). *eFood.* 2020;1(5):347–356.
39. Ahmadi S, Rajabi Z, Vasfi-Marandi M. Evaluation of the antiviral effects of aqueous extracts of red and yellow onions (*Allium Cepa*) against avian influenza virus subtype H9N2. *Iranian Journal of Veterinary Science and Technology.* 2018;10(2):23–27.
40. Romeilah RM, Fayed SA, Mahmoud GI. Chemical compositions, antiviral and antioxidant activities of seven essential oils. *Journal of Applied Sciences Research.* 2010;6(1):50–62.

41. Pandey J, Bastola T, Tripathi J, Tripathi M, Rokaya RK, Dhakal B, et al. Estimation of total quercetin and rutin content in *Malus domestica* of Nepalese origin by HPLC method and determination of their antioxidative activity. *Journal of Food Quality*. 2020;2020:1–13.
42. Hamauzu Y, Yasui H, Inno T, Kume C, Omanyuda M. Phenolic profile, antioxidant property, and anti-influenza viral activity of Chinese quince (*Pseudocydonia sinensis* Schneid.), quince (*Cydonia oblonga* Mill.), and apple (*Malus domestica* Mill.) fruits. *Journal of Agricultural and Food Chemistry*. 2005;53(4):928–934.
43. Abuduaini M, Li J, Ruan JH, Zhao YX, Maitinuer M, Aisa HA. Bioassay-guided preparation of antioxidant, anti-inflammatory active fraction from crabapples (*Malus prunifolia* (Willd.) Borkh.). *Food Chemistry*. 2023;406:135091.
44. Zibaee E, Kamalian S, Tajvar M, Amiri MS, Ramezani M, Moghadam AT, et al. Citrus species: A review of traditional uses, phytochemistry and pharmacology. *Current Pharmaceutical Design*. 2020;26(1):44–97.
45. Mejri H, Aidi Wannes W, Mahjoub FH, Hammami M, Dussault C, Legault J, et al. Potential bio-functional properties of *Citrus aurantium* L. leaf: Chemical composition, antiviral activity on herpes simplex virus type-1, antiproliferative effects on human lung and colon cancer cells and oxidative protection. *International Journal of Environmental Health Research*. 2023;8:1–11.
46. Zalpoor H, Bakhtiyari M, Shapourian H, Rostampour P, Tavakol C, Nabi-Afjadi M. Hesperetin as an anti-SARS-CoV-2 agent can inhibit COVID-19-associated cancer progression by suppressing intracellular signaling pathways. *Inflammopharmacology*. 2022;30(5):1533–1539.
47. Insanu M, Karimah H, Pramastya H, Fidrianny I. Phytochemical compounds and pharmacological activities of *Vitis vinifera* L.: An updated review. *Biointerface Research in Applied Chemistry*. 2021;11(5):13829–13849.
48. Nabavi SF, Maggi F, Daglia M, Habtemariam S, Rastrelli L, Nabavi SM. Pharmacological effects of *Capparis spinosa* L. *Phytotherapy Research*. 2016;30(11):1733–1744.
49. Annaz H, Sane Y, Bitchagno GTM, Ben Bakrim W, Drissi B, Mahdi I, et al. Caper (*Capparis spinosa* L.): An updated review on its phytochemistry, nutritional value, traditional uses, and therapeutic potential. *Frontiers in Pharmacology*. 2022;13:878749.
50. Zeinali M, Zirak MR, Rezaee SA, Karimi G, Hosseinzadeh H. Immunoregulatory and anti-inflammatory properties of *Crocus sativus* (Saffron) and its main active constituents: A review. *Iranian Journal of Basic Medical Sciences*. 2019;22(4):334.
51. Cardone L, Castronuovo D, Perniola M, Cicco N, Candido V. Saffron (*Crocus sativus* L.), the king of spices: An overview. *Scientia Horticulturae*. 2020;272:109560.
52. Soleymani S, Zabihollahi R, Shahbazi S, Bolhassani A. Antiviral effects of saffron and its major ingredients. *Current Drug Delivery*. 2018;15(5):698–704.
53. Farzaei MH, Abbasabadi Z, Ardekani MRS, Rahimi R, Farzaei F. Parsley: A review of ethnopharmacology, phytochemistry and biological activities. *Journal of Traditional Chinese Medicine*. 2013;33(6):815–826.
54. Abdellatief SA, Galal AA, Farouk SM, Abdel-Daim MM. Ameliorative effect of parsley oil on cisplatin-induced hepato-cardiotoxicity: A biochemical, histopathological, and immunohistochemical study. *Biomedicine & Pharmacotherapy*. 2017;86:482–491.
55. El-Zeftawy M, Ghareeb D. Pharmacological bioactivity of *Ceratonia siliqua* pulp extract: *In vitro* screening and molecular docking analysis, implication of Keap-1/Nrf2/NF-κB pathway. *Scientific Reports*. 2023;13(1):12209.
56. Moumou M, Mokhtari I, Milenkovic D, Amrani S, Harnafi H. Carob (*Ceratonia siliqua* L.): A comprehensive review on traditional uses, chemical composition, pharmacological effects and toxicology (2002–2022). *Journal of Biologically Active Products from Nature*. 2023;13(3):179–223.
57. Al-Hadid KJ. Evaluation of antiviral activity of different medicinal plants against Newcastle disease virus. *American Journal of Agricultural and Biological Science*. 2016;11(4):157–163.
58. Aboura I, Nani A, Belarbi M, Murtaza B, Fluckiger A, Dumont A, et al. Protective effects of polyphenol-rich infusions from carob (*Ceratonia siliqua*) leaves and cladodes of *Opuntia ficus-indica* against inflammation associated with diet-induced obesity and DSS-induced colitis in Swiss mice. *Biomedicine & Pharmacotherapy*. 2017;96:1022–1035.
59. Liu Y, Xin H, Zhang Y, Che F, Shen N, Cui Y. Leaves, seeds and exocarp of *Ginkgo biloba* L.(Ginkgoaceae): A comprehensive review of traditional uses, phytochemistry, pharmacology, resource utilization and toxicity. *Journal of Ethnopharmacology*. 2022;298:115645.
60. Tao Z, Jin W, Ao M, Zhai S, Xu H, Yu L. Evaluation of the anti-inflammatory properties of the active constituents in *Ginkgo biloba* for the treatment of pulmonary diseases. *Food & Function*. 2019;10(4):2209–2220.
61. Ibrahim MA, Ramadan HH, Mohammed RN. Evidence that *Ginkgo Biloba* could use in the influenza and coronavirus COVID-19 infections. *Journal of Basic and Clinical Physiology and Pharmacology*. 2021;32(3):131–143.
62. Carneiro RC, Ye L, Baek N, Teixeira GH, O'Keefe SF. Vine tea (*Ampelopsis grossedentata*): A review of chemical composition, functional properties, and potential food applications. *Journal of Functional Foods*. 2021;76:104317.
63. Hou X, Tong Q, Wang W, Shi C, Xiong W, Chen J, et al. Suppression of inflammatory responses by dihydromyricetin, a flavonoid from *Ampelopsis grossedentata*, via inhibiting the activation of NF-κB and MAPK signaling pathways. *Journal of Natural Products*. 2015;78(7):1689–1696.
64. Murakami T, Miyakoshi M, Araho D, Mizutani K, Kambara T, Ikeda T, et al. Hepatoprotective activity of tocha, the stems and leaves of *Ampelopsis grossedentata*, and ampelopsin. *Biofactors*. 2004;21(1-4):175–178.
65. Chen S, Dong L, Quan H, Zhou X, Ma J, Xia W, et al. A review of the ethnobotanical value, phytochemistry, pharmacology, toxicity and quality control of *Tussilago farfara* L. (coltsfoot). *Journal of Ethnopharmacology*. 2021;267:113478.
66. Xu L-T, Wang T, Fang K-L, Zhao Y, Wang X-N, Ren D-M, et al. The ethanol extract of flower buds of *Tussilago farfara* L. attenuates cigarette smoke-induced lung inflammation through regulating NLRP3 inflammasome, Nrf2, and NF-κB. *Journal of Ethnopharmacology*. 2022;283:114694.
67. Subhan N, Burrows GE, Kerr PG, Obied HK. Phytochemistry, ethnomedicine, and pharmacology of *Acacia*. *Studies in Natural Products Chemistry*. 2018;57:247–326.
68. Tekleweyni T, Kidu H, Tesfay W, Chaithanya KK, Hagos Z, Gopalakrishnan V. Inhibitory effect of ethanolic leaf extract of *Acacia etbaica* schweinf on production of pro-inflammatory cytokines in lipopolysaccharide-stimulated RAW 264.7 macrophage cells. *Drug Invent Today*. 2020;13:151–154.

69. Medzhitov R. Origin and physiological roles of inflammation. *Nature*. 2008;454(7203):428–435.
70. Rathee P, Chaudhary H, Rathee S, Rathee D, Kumar V, Kohli K. Mechanism of action of flavonoids as anti-inflammatory agents: A review. *Inflammation & Allergy-Drug Targets (Formerly Current Drug Targets-Inflammation & Allergy)(Discontinued)*. 2009;8(3):229–235.
71. Alam W, Khan H, Shah MA, Cauli O, Saso L. Kaempferol as a dietary anti-inflammatory agent: current therapeutic standing. *Molecules*. 2020;25(18):4073.
72. Lin M-K, Yu Y-L, Chen K-C, Chang W-T, Lee M-S, Yang M-J, et al. Kaempferol from *Semen cuscutae* attenuates the immune function of dendritic cells. *Immunobiology*. 2011;216(10):1103–1109.
73. Li Y, Yao J, Han C, Yang J, Chaudhry MT, Wang S, et al. Quercetin, inflammation and immunity. *Nutrients*. 2016;8(3):167.
74. Saeedi-Boroujeni A, Mahmoudian-Sani M-R. Anti-inflammatory potential of quercetin in COVID-19 treatment. *Journal of Inflammation*. 2021;18:1–9.
75. Khazdair MR, Anaeigoudari A, Agbor GA. Anti-viral and anti-inflammatory effects of kaempferol and quercetin and COVID-2019: A scoping review. *Asian Pacific Journal of Tropical Biomedicine*. 2021;11(8):327.
76. Manjunath SH, Thimmulappa RK. Antiviral, immunomodulatory, and anticoagulant effects of quercetin and its derivatives: Potential role in prevention and management of COVID-19. *Journal of Pharmaceutical Analysis*. 2022;12(1):29–34.
77. Cho BO, Yin HH, Park SH, Byun EB, Ha HY, Jang SI. Anti-inflammatory activity of myricetin from *Diospyros lotus* through suppression of NF-κB and STAT1 activation and Nrf2-mediated HO-1 induction in lipopolysaccharide-stimulated RAW264.7 macrophages. *Bioscience, Biotechnology, and Biochemistry*. 2016;80(8):1520–1530.
78. Septembre-Malaterre A, Boumendjel A, Seteyen A-LS, Boina C, Gasque P, Guiraud P, et al. Focus on the high therapeutic potentials of quercetin and its derivatives. *Phytomedicine Plus*. 2022;2(1):100220.
79. Wu C-C, Lee T-Y, Cheng Y-J, Cho D-Y, Chen J-Y. The dietary flavonol kaempferol inhibits epstein-barr virus reactivation in nasopharyngeal carcinoma cells. *Molecules*. 2022;27(23):8158.
80. Ogawa M, Shimojima M, Saijo M, Fukasawa M. Several catechins and flavonols from green tea inhibit severe fever with thrombocytopenia syndrome virus infection *in vitro*. *Journal of Infection and Chemotherapy*. 2021;27(1):32–39.
81. Jeong HJ, Ryu YB, Park S-J, Kim JH, Kwon H-J, Kim JH, et al. Neuraminidase inhibitory activities of flavonols isolated from *Rhodiola rosea* roots and their *in vitro* anti-influenza viral activities. *Bioorganic & Medicinal Chemistry*. 2009;17(19):6816–6823.
82. Ha S-Y, Youn H, Song C-S, Kang SC, Bae JJ, Kim HT, et al. Antiviral effect of flavonol glycosides isolated from the leaf of *Zanthoxylum piperitum* on influenza virus. *Journal of Microbiology*. 2014;52:340–344.
83. Musarra-Pizzo M, Pennisi R, Ben-Amor I, Smeriglio A, Mandalari G, Sciortino MT. In vitro anti-HSV-1 activity of polyphenol-rich extracts and pure polyphenol compounds derived from pistachios kernels (*Pistacia vera* L.). *Plants*. 2020;9(2):267.
84. Choi H-J, Kim J-H, Lee C-H, Ahn Y-J, Song J-H, Baek S-H, et al. Antiviral activity of quercetin 7-rhamnoside against porcine epidemic diarrhea virus. *Antiviral Research*. 2009;81(1):77–81.
85. Wrapp D, Wang N, Corbett KS, Goldsmith JA, Hsieh C-L, Abiona O, et al. Cryo-EM structure of the 2019-nCoV spike in the prefusion conformation. *Science*. 2020;367(6483):1260–1263.
86. Hoffmann M, Kleine-Weber H, Schroeder S, Krüger N, Herrler T, Erichsen S, et al. SARS-CoV-2 cell entry depends on ACE2 and TMPRSS2 and is blocked by a clinically proven protease inhibitor. *Cell*. 2020;181(2):271–280.
87. Garten W. Characterization of proprotein convertases and their involvement in virus propagation. *Activation of Viruses by Host Proteases*. 2018;16:205–248.
88. Bestle D, Heindl MR, Limburg H, Pilgram O, Moulton H, Stein DA, et al. TMPRSS2 and Furin are both essential for proteolytic activation of SARS-CoV-2 in human airway cells. *Life Science Alliance*. 2020;3(9):e202000786.
89. Faheem S, Kumar BK, Sekhar KVGC, Kunjiappan S, Jamalis J, Balaña-Fouce R, et al. Druggable targets of SARS-CoV-2 and treatment opportunities for COVID-19. *Bioorganic Chemistry*. 2020;104:104269.
90. Wang M, Cao R, Zhang L, Yang X, Liu J, Xu M, et al. Remdesivir and chloroquine effectively inhibit the recently emerged novel coronavirus (2019-nCoV) *in vitro*. *Cell Research*. 2020;30(3):269–271.
91. Moeller NH, Passow KT, Harki DA, Aihara H. SARS-CoV-2 nsp14 exoribonuclease removes the natural antiviral 3′-Deoxy-3′, 4′-didehydro-cytidine nucleotide from RNA. *Viruses*. 2022;14(8):1790.
92. Zhao M-M, Yang W-L, Yang F-Y, Zhang L, Huang W-J, Hou W, et al. Cathepsin L plays a key role in SARS-CoV-2 infection in humans and humanized mice and is a promising target for new drug development. *Signal Transduction and Targeted Therapy*. 2021;6(1):134.
93. White MA, Lin W, Cheng X. Discovery of COVID-19 inhibitors targeting the SARS-CoV-2 Nsp13 helicase. *The Journal of Physical Chemistry Letters*. 2020;11(21):9144–9151.
94. Meng J-R, Liu J, Fu L, Shu T, Yang L, Zhang X, et al. Anti-entry activity of natural flavonoids against SARS-CoV-2 by targeting spike RBD. *Viruses*. 2023;15(1):160.
95. Liu X, Raghuvanshi R, Ceylan FD, Bolling BW. Quercetin and its metabolites inhibit recombinant human angiotensin-converting enzyme 2 (ACE2) activity. *Journal of Agricultural and Food Chemistry*. 2020;68(47):13982–13989.
96. Leal CM, Leitão SG, Sausset R, Mendonça SC, Nascimento PH, de Araujo R. Cheohen CF, et al. Flavonoids from *Siparuna cristata* as potential inhibitors of SARS-CoV-2 replication. *Revista Brasileira de Farmacognosia*. 2021;31:1–9.
97. Zhu Y, Scholle F, Kisthardt SC, Xie D-Y. Flavonols and dihydroflavonols inhibit the main protease activity of SARS-CoV-2 and the replication of human coronavirus 229E. *Virology*. 2022;571:21–33.
98. Gupta A, Ahmad R, Siddiqui S, Yadav K, Srivastava A, Trivedi A, et al. Flavonol morin targets host ACE2, IMP-α, PARP-1 and viral proteins of SARS-CoV-2, SARS-CoV and MERS-CoV critical for infection and survival: A computational analysis. *Journal of Biomolecular Structure and Dynamics*. 2022;40(12):5515–5546.

99. Nguyen TTH, Jung J-H, Kim M-K, Lim S, Choi J-M, Chung B, et al. The inhibitory effects of plant derivate polyphenols on the main protease of SARS coronavirus 2 and their structure-activity relationship. *Molecules*. 2021;26(7):1924.
100. Chaves OA, Fintelman-Rodrigues N, Wang X, Sacramento CQ, Temerozo JR, Ferreira AC, et al. Commercially available flavonols are better SARS-CoV-2 inhibitors than isoflavone and flavones. *Viruses*. 2022;14(7):1458.
101. Su H, Yao S, Zhao W, Zhang Y, Liu J, Shao Q, et al. Identification of pyrogallol as a warhead in design of covalent inhibitors for the SARS-CoV-2 3CL protease. *Nature Communications*. 2021;12(1):3623.
102. Xiong Y, Zhu G-H, Zhang Y-N, Hu Q, Wang H-N, Yu H-N, et al. Flavonoids in *Ampelopsis grossedentata* as covalent inhibitors of SARS-CoV-2 3CLpro: Inhibition potentials, covalent binding sites and inhibitory mechanisms. *International Journal of Biological Macromolecules*. 2021;187:976–987.
103. Teli DM, Shah MB, Chhabria MT. *In silico* screening of natural compounds as potential inhibitors of SARS-CoV-2 main protease and spike RBD: Targets for COVID-19. *Frontiers in Molecular Biosciences*. 2021;7:599079.
104. Kumar S, Paul P, Yadav P, Kaul R, Maitra S, Jha SK, et al. A multi-targeted approach to identify potential flavonoids against three targets in the SARS-CoV-2 life cycle. *Computers in Biology and Medicine*. 2022;142:105231.
105. Owis AI, El-Hawary MS, El Amir D, Refaat H, Alaaeldin E, Aly OM, et al. Flavonoids of *Salvadora persica* L. (meswak) and its liposomal formulation as a potential inhibitor of SARS-CoV-2. *RSC Advances*. 2021;11(22):13537–13544.
106. Chiow KH, Phoon MC, Putti T, Tan BKH, Chow VT. Evaluation of antiviral activities of *Houttuynia cordata* Thunb. extract, quercetin, quercetrin and cinanserin on murine coronavirus and dengue virus infection. *Asian Pacific Journal of Tropical Medicine*. 2016;9(1):1–7.
107. Di Pierro F, Khan A, Bertuccioli A, Maffioli P, Derosa G, Khan S, et al. Quercetin Phytosome(r) as a potential candidate for managing COVID-19. *Minerva Gastroenterology*. 2021;67(2):190–195.
108. Gasmi A, Mujawdiya PK, Lysiuk R, Shanaida M, Peana M, Gasmi Benahmed A, et al. Quercetin in the prevention and treatment of coronavirus infections: A focus on SARS-CoV-2. *Pharmaceuticals*. 2022;15(9):1049.
109. Ahmed AK, Albalawi YS, Shora HA, Abdelseed HK, Al-Kattan AN. Effects of quadruple therapy: Zinc, quercetin, bromelain and vitamin C on the clinical outcomes of patients infected with COVID-19. *Research International Journal of Endocrinology and Diabetes*. 2020;1(1):018–021.

8 Isoflavonoids As Bioactive Molecules against SARS-CoV-2

S Jayashree, K Sonia, G Mayakkannan,
S Shruthi, S Aruna Sharmili, and Sekar Vijayakumar

8.1 INTRODUCTION

COVID-19 has afflicted about 190 million people worldwide and generated global problems. Due to fast viral genome mutation, host immunity must be improved, while viral proteins are targeted to minimize infection severity. The single-stranded virus causing COVID-19 belongs to the Coronaviridae family (Bahadur et al., 2020). Pathological studies show that inflammatory cytokines/signaling pathways cause pulmonary edema and respiratory problems in COVID-19 patients (Merad and Martin, 2020). Complex pathophysiological mechanisms, lack of adequate therapy, and the associated negative side effects of traditional medications necessitate alternate treatment for viral infections. Phytochemicals from plants could be advantageous for the treatment/management of viral diseases as they possess potential effects in targeting several dysregulated mediators (Majnooni et al., 2020). Numerous plants have demonstrated anti-inflammatory, antiviral, and antioxidant properties, and they are used during this pandemic as well. The Ministry of Ayush (India) also performed clinical trials with medicinal plants (Kotecha, 2021). The phytochemicals can target proinflammatory and oxidative mediators, including tumor necrosis factors, interleukins, proteinases, cyclooxygenases, and ROS (reactive oxygen species) (Majnooni et al., 2020).

In recent years, human nutrition research has focused more and more on the importance of phenols and flavonoids as protective dietary components. Polyphenols and flavonoids comprise carbon skeletons and aromatic rings combined with pyran ring (Dias et al., 2021). Over 6,000 flavonoids have been identified and are widely distributed in plants and fungi. Within this family, researchers have focused on its subclass, isoflavonoids, primarily found in legumes, particularly soybeans and soy foods.

Isoflavonoids are secondary metabolites with a chroman skeleton (Sajid et al., 2021). From 300 plant sources, over 2,400 isoflavonoids have been documented. Isoflavonoids are referred to as phytoestrogens due to their structural similarity with 17β-estradiol (sex hormone) (Sirotkin and Harrath, 2014). Consequently, isoflavonoids exhibit both estrogen agonist and estrogen antagonist activity. Phytoestrogens are regarded as non-steroidal, natural substances with estrogenic and antiestrogenic properties capable of binding to estrogen receptors in mammals (Barros and Gustafsson, 2011; Vitale et al., 2012). Leguminous plants synthesize the majority of isoflavonoids via the phenylpropanoid pathway (Du et al., 2010; Sharma and Ramawat, 2013). The significant dietary source of isoflavonoids is soybeans which produce the maximum amount of isoflavones of all the leguminous crops. Other sources include kidney beans, small white beans, red beans, and mung beans. The biological activity of isoflavonoids originates from their aglycones, which are present in soybeans in glycosylated form. When soy foods are consumed, β glucosidase from gut microflora converts the soy isoflavonoids to their aglycones (Tsuchihashi et al., 2008). The main isoflavone phytoestrogens are daidzein (7,4¢-dihydroxy isoflavone), genistein (4¢,5,7- trihydroxy isoflavone), glycitein (7,4¢-dihydroxy-6-methoxy isoflavone), biochanin A (5,7-dihydroxy-4¢-methoxy-isoflavone), and formononetin (7-hydroxy-4¢-methoxy isoflavone) (Křížová et al., 2019) (Figure 8.1). These isoflavonoids have antioxidant, antimicrobial, and anti-inflammatory properties (Garcia-Lafuente et al., 2009; Crozier et al., 2009).

8.2 SOURCES OF ISOFLAVONES

Isoflavones predominantly occur in the Leguminosae family (Dixon and Sumner, 2003) specifically the soybean (*Glycine max*), which contains the isoflavones daidzein, genistein, and glycitein, and red clover (*Trifolium pratense*), which contains the isoflavones biochanin A and formononetin. Soy-derived products such as soybean, soy flakes, soy flour, soy drinks, and fermented soy products (miso and temp) are consumed by humans and are rich in isoflavones (Table 8.1) (Zaheer and Humanyoun, 2015). Soybean, red clover, white clover (*Medicago lupulina*), and lucerne (*Medicago sativa* L.) are the significant isoflavones in the diet of farm animals (Křížová et al., 2019). These plants are utilized as food supplements or medicinal teas in phytotherapy. Isoflavone content in soybean ranges from 1.2 to 4.2 mg/g dry weight, with daidzein, genistein, and its conjugates being the primary isoflavones (Kurzer and Xu, 1997). The phytoestrogen content in red clover is 10–25 mg/g dry weight, while in white clover, it is 0.5–0.6 mg/g (Hannu et al., 1995). The isoflavone content of lucerne is only 0.05–0.3 mg/g dry weight (Butkutė et al., 2018). Red clover is used as a food supplement, despite

DOI: 10.1201/9781003433200-8

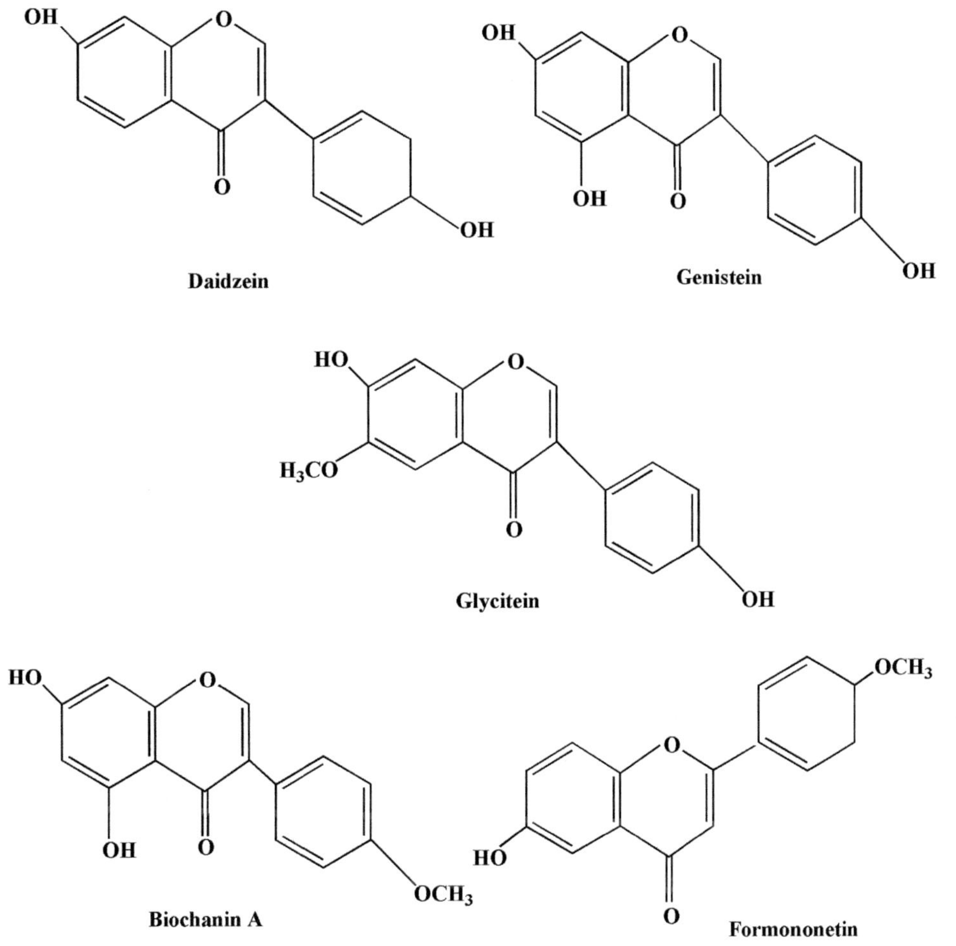

FIGURE 8.1 Chemical structures of isoflavones.

TABLE 8.1
Isoflavonoids Contents in Main Sources

Sources	Contents (Approx.) mg/100 g	References
Soybean	26–381	Cassidy et al. (2000); Sakai and Kogiso (2008)
Roasted soybean	246	Pérez-Jiménez et al. (2010)
Soy flour	83–466	Cassidy et al. (2000); Pérez-Jiménez et al. (2010)
Soy Tempe	148	Pérez-Jiménez et al. (2010)
Tofu	8–67	Cassidy et al. (2000); Sakai and Kogiso (2008)
Tempeh	86.5	Sakai and Kogiso (2008)
Miso	25–89	Cassidy et al. (2000)

the fact that it contains a greater amount of isoflavones than other plants and is not an edible plant. In preparation for dietary supplements for the reduction of menopausal syndrome in women, isoflavonoids from red clover are used as an ingredient (Andres et al., 2015). In addition to soybeans and legumes, isoflavones are found in kudzu root, peanuts, chickpeas, berries, wine, and cereals nuts (Reynaud et al., 2005). Furthermore, the consumption of dairy products and cow's milk was found to contribute to the total intake of isoflavones in the western (mainly American) population (Frankenfeld, 2011).

8.3 EFFECT OF ISOFLAVONOIDS ON IMMUNE SYSTEM

The immunoregulatory effects of isoflavones are well-known (Wei et al., 2012). Most immune cells possess receptors for estrogen, which regulate the immune system (Kovats, 2015). Isoflavones have a structure similar to estrogen and are weak estrogen receptor agonists that regulate the immune system. Isoflavones are demonstrated with estrogenic activity primarily through the estrogen receptor ER-β, while estradiol signals through ER-α. The intestinal epithelial cells present in the gastrointestinal tract can absorb dietary isoflavone aglycones (Masilamani et al., 2012). Isoflavone-rich fractions of soy extract were able to inhibit the production of IL-8 (an inflammatory cytokine) in Caco-2 cells induced by TNF-α (Tumor Necrosis Factor-alpha) (Satsu et al., 2009). IL-8 (Interleukin-8) is needed for the chemotaxis of immune cells such as eosinophils, neutrophils, and T cells (Baggiolini and Lewis, 1992). Daidzein and genistein induce T-cell response (Tyagi et al., 2011; Ghaemi et al., 2012; Huang et al., 2019). T cells recognize not only the spike proteins but also other SARS-CoV-2 proteins (Grifoni et al., 2020). Isoflavone therapy significantly reduced CRP (C-reactive proteins) levels (Chan et al., 2008; Li and Zhang, 2017) and IL-6 levels (Li and Zhang, 2017) in ischemic stroke patients. Treatment of thioacetamide-induced liver injury in rats by genistein increased the serum albumin levels and decreased elevated serum levels of AST (Aspartate Aminotransferase), ALT (Alanine Aminotransferase), and total and direct bilirubin (Saleh et al., 2014). A decrease in blood glucose, CRP, percentage of glycosylated hemoglobin (HbA1c), and an increase in antioxidant reserve in the hearts of streptozotocin-induced rats were observed on treatment of *G. max* genistein (300 mg/kg/day) (Gupta et al., 2015). An increase in the percentage of CD4+ and CD28+T cells was recorded in adult BALB/c mice treated with Daidzein (DAZ). It also regulated B lymphopoiesis (Tyagi et al., 2011). Administration of DAZ in B6C3F1 mice increased the T-cell population, whereas the activities of cytotoxic T cell and natural killer (NK) cells were not altered. Antibody production was modulated in NOD (Non-obese diabetic) mice treated with DAZ. In NOD female rats, an increase in CD8+ CD25+ was observed on administration of DAZ (Huang et al., 2019). Equol-treated BALB/c mice showed significantly high levels of OVA-specific IgE and IL-13 (Sakai et al., 2010). *In vitro* studies of isoflavones showed significant suppression of the activation-induced expression of DC maturation markers, viz., CD83, CD80, CD86, and MHC I (Major Histocompatibility class I) molecule. NK cell degranulation and the percentage death of DCs (Dendritic cells) increased significantly (Wei et al., 2012).

8.4 ANTIVIRAL ACTIVITY OF ISOFLAVONOIDS

Isoflavones are shown with antiviral activity against a wide variety of viruses in both *in vitro* and *in vivo* experimental conditions. Isoflavones, puerarin from black garlic, exhibited good protease activity against SARS-CoV-2. It was found that daidzein and genistein had moderate activity against SARS-CoV-2 (Solnier and Fladerer, 2021). A study of a few isoflavonoids against $3CL^{pro}$ by Su et al. (2021) indicated that the substitution of methoxy group accounted for higher inhibition by formononetin compared to daidzein and genistein. *In vitro* analysis using the cell-free method indicated that isoflavone puerarin showed higher inhibitory activity than daidzein and genistein against SARS-CoV-2 $3CL^{pro}$. The higher activity by puerarin is attributed to the A-ring C8-glycosylation. The activity of these three isoflavonoids was greater by twofold than their flavone analogs (Nguyen et al., 2021).

Genistein is an isoflavonoid with a 15C skeleton belonging to the class of aglycones (Spagnuolo et al., 2015) and has been shown to have antiviral properties against various viruses that affect humans and animals such as adenovirus, herpes simplex virus, rotavirus, HIV, and SARS-CoV (Andres et al., 2009). There was a decrease in CRP levels, an increase in T cells, and an increase in the levels of IFN-γ (Interferon gamma) on administration of genistein in an *in vivo* study conducted by Chan et al. (2008) on an HPV tumor model. There was an increase in serum albumin levels and APO A-1 (Apolipoprotein A -1) concentration in healthy young people on administration of genistein (Sanders et al., 2002; Saleh et al., 2014). Low serum levels were associated with an increase in the risk of death in SARS-CoV-2 patients (Liu et al., 2020). Albumin downregulates ACE2 receptors and improves the ratio of partial pressure arterial oxygen in patients (Uhlig et al., 2014). *In silico* studies on daidzein and genistein have proven as potent inhibitors of M^{pro} and RdRp of COVID-19 when compared to antiviral drugs (Pendyala and Patras, 2020). In a study conducted by Nguyen et al. (2021), a mixture of tannic acid in combination with puerarin, daidzein, and/or myricetin enhanced the inhibitory effects on M^{pro}. When compared with daidzein, puerarin exhibited better inhibitory effects on M^{pro} as it contained the 8-C-glucoside of daidzein (Nguyen et al., 2021). At pharmacological dosage, genistein inhibits protein tyrosine kinase (PTK) (Jafari et al., 2022). The tyrosine kinase signaling cascade can be a crucial target for pharmacologic intervention to prevent airway inflammation (Duan et al., 2003). PTK has a key role in cytokine production and many of the cytokines (IL-10, IL-6, TNF-alpha) use it in their signaling pathway (Hsu et al., 2001; Dahle et al., 2004; Page et al., 2009). Genistein inhibits tyrosine kinase, reducing monocyte chemoattractant protein (MCP-1) excretion by inhibiting TNF α-induced NF-κB activation (Coma et al., 2021; Kim et al., 2021). Thus, the anti-inflammatory effect of genistein due to the inhibition of PTK may be a mechanism for preventing SARS-CoV-2-induced lung injury (Jafari et al., 2022). Protein kinase B (Akt) has been identified as a potential pharmacological target for advanced-stage SARS-CoV-2 infection and inhibition of it will suppress pathological fibroproliferation, cytokine storm, platelet activation, injury, and scarring of lung tissues in ARDS, and inflammation linked to COVID-19 (Wang et al., 2021; Basile et al., 2022). It has been suggested that Akt inhibitors such as genistein could be used to treat ARDS in COVID patients in advanced stages (Li and Sarkar, 2002; Somnath, 2020). P13K/Akt/mTOR pathway is involved in clot formation (Abu-Eid and Ward, 2021; Meng et al., 2021) and inhibition of this pathway by genistein could prevent thrombosis in severe COVID patients (Sahin et al., 2012; Tan et al., 2014; Lee et al., 2016; Malloy et al., 2018).

8.5 MECHANISM OF ANTIVIRAL ACTIVITY

There is a reduction in PTK activity in the host cell, as isoflavones affect certain transcription factors and cytokine secretion. Reorganization of the cytoskeleton takes place whereby the virus-induced actin changes and recruitment of dynamin II to the membrane-bound virus particles are blocked (Dangoria et al., 1996; Li et al., 2000; Pelkmans et al., 2002; Kubo et al., 2003). This inhibition decreases the entry of viruses (SV40, HHV-8, MoMLV, and adenovirus) into the host cell. At later stages of virus infection, viral replication decreased as there was a decrease in phosphorylation of HSV-1 polypeptides and BHV-1 glycoprotein E (Yura et al., 1993; Akula et al., 2002). Genistein inhibits PTK whereby there is a reduction in the excretion of MCP-1 (Monocyte Chemoattractant Protein-1) through the inhibition of TNF-α-induced NF-κB activation (Coma et al., 2021; Kim et al., 2021) which could be a potent mechanism for preventing the entry of SARS-CoV-2 (Jafari et al., 2022). Genistein blocks the replication of ALV-J (Avian Leucosis virus subgroup J) by inhibiting virus transcription (Qian et al., 2014) and activates the adaptive immune pathway in PRRSV (Porcine reproductive and respiratory syndrome virus; Smith et al., 2019). In the case of the African Swine fever virus, genistein disrupted viral DNA replication and blocked transcription of late viral genes and synthesis of late viral protein thereby reducing viral progeny (Arabyan et al., 2018). Rotavirus and Herpes simplex virus replication was inhibited by genistein (Huang et al., 2015; Argenta et al., 2015), whereas Dengue virus type-2 replication was inhibited by daidzein (Zandi et al., 2011). In the case of CMV (cytomegalovirus) and arenavirus-infected host cells, genistein inhibited protein phosphorylation and reduced ATF-2 (Fe transcription 2), cyclic AMP response element-binding protein (CREB), and NF-κB transcription factors (Evers et al., 2005; Vela et al., 2008). Genistein inhibited DNA replication in human CMV post-infection and blocked early and late viral gene expressions and protein synthesis. Viral entry was not prevented by genistein. It was found that in cells treated with genistein, transcription factor NF-κB induction and viral cell cycle perturbation were absent (Evers et al., 2005).

Protein kinase B (Akt) has been identified as a potential pharmacological target for advanced-stage SARS-CoV-2 infection (Basile et al., 2021). Inhibition of Akt has been reported to reduce ACE2 expression which is pivotal for the entry of the virus into the respiratory system. Genistein is an Akt inhibitor used in the treatment of ARDS (Acute Respiratory Distress Syndrome) in advanced stages of COVID-19 patients (Somanath, 2020). Inhibition of the PI3K/Akt/mTOR pathway suppresses the replication of MERS-CoV (Middle East Respiratory Syndrome Coronavirus) and influenza virus (Murray et al., 2012; Kindrachuk et al., 2015). Genistein inhibits the PI3K/Akt/mTOR pathway by preventing the entry of SARS-CoV-2 and decreasing the viral load (Sahin et al., 2012; Tan et al., 2014; Lee et al., 2016; Malloy et al., 2018). As COVID-19 progresses, there is an uncontrolled immune response, and targeting the PI3K/Akt pathway by genistein suppresses regulatory T cells and cytokine storm and inhibits neutrophil recruitment. It also exerts its impact on glucose transporter-4 (GLUT-4) (Langfort et al., 2003). There is rapid translocation of GLUT-4 to the cell surface on Akt activation which increases cellular glucose transport activity. Inhibition of glucose uptake by GLUT-4 inhibits glucose transport and viral replication (Yu et al., 2011). SARS-CoV-2's activation of the PI3K/AKT signaling pathway is believed to induce glucose uptake through GLUT-4, leading to increased glycolysis and viral replication in host cells (Malgotra and Sharma, 2021). Genistein is an inhibitor of GLUT-4 (Lewicki et al., 2018). Viral replication can be prevented by the inhibition of GLUT-4 by genistein and the inhibition of the PI3K/AKT signaling pathway (Jafari et al., 2022). Thus, in general, it is seen that isoflavones affect virus binding, entry, replication, protein translation, and formation of envelop glycoprotein complexes in virus. In the case of the host cell, isoflavones affect the cell signaling process (Wang et al., 2020).

8.6 ANTIOXIDANT PROPERTIES OF ISOFLAVONOIDS

Genistein and daidzein are known metal chelators and radical scavengers, with the former being superior due to its three hydroxyl groups (Han et al., 2009). Genistein enhances the production of superoxide dismutase (SOD), a crucial enzyme that effectively eliminates free radicals (Kuriyama et al., 2013). Oxidative stress-induced tissue injury has been reported in COVID patients as the cells were not able to neutralize the excessive reactive oxygen species (ROS) formed (Laforge et al., 2020; Ghasemnejad-Berenji et al., 2021). This excess production of ROS can flare up inflammation and induce tissue damage (Zorov et al., 2014). ROS is a potential target for developing new therapeutic agents to manage COVID-19-induced injuries (Beltran-Garcia et al., 2020). Genistein protects cells from the over-production of ROS by scavenging free radicals, inhibiting NF-κB activation, a crucial role in cytokine storm and inflammation (Li et al., 2013; Nazari-Khanamiri and Ghasemnejad-Berenji, 2021). The 24-hour treatment with genistein significantly enhances antioxidant enzyme activities by enhancing the expression of phosphatase and tensin homolog (PTEN) and AMPK (AMP-activated protein kinase) (Park et al., 2010). Genestein ability to activate cell signaling pathways, which can inhibit ROS generation, makes it a promising candidate for COVID-19 treatment (Jafari et al., 2022). Isoflavones in healthy mice reduced 8-hydroxydeoxyguanosine content, a marker of somatic cell oxidation and cancer progression, and promoted SOD expression to prevent oxidative damage (Ma et al., 2014).

8.7 ANTI-INFLAMMATORY ACTIVITY OF ISOFLAVONOIDS

It was found that genistein exerted an anti-inflammatory response affecting the lymphocytes, monocytes, and granulocytes (Verdrengh et al., 2003). In mice models injected with endotoxin lipopolysaccharide (LPS), it was found that isoflavone-containing diet prevented inflammation-associated induction of metallothionein in the intestine, while the liver was prevented from induction of manganese superoxide dismutase (Mn-SOD). It was also seen that there was modulation in the action of IL-6 (proinflammatory cytokine interleukin-6) which suppresses the intestinal response to inflammation (Paradkar et al., 2004). Pre-treatment of Wistar rats with genistein has shown a reduction in NO (Nitric Oxide) and PGE2 (Prostaglandin E2), as well as a suppression in the production of D-galactosamine-induced proinflammatory cytokines, TNF-α, IL-1β (Ganai et al., 2015). In the Guinea pig model of asthma, it was found that genistein inhibited ovalbumin-induced bronchoconstriction and reduced ovalbumin-induced pulmonary eosinophilia and eosinophil peroxidase activity (Duan et al., 2003). The methanolic fraction containing soy isoflavone showed anti-inflammatory activity in Swiss mice with Croton oil-induced ear edema (inflammation) (Carrara et al., 2008). Genistein pre-treatment reduces the LPS-stimulated COX-2 protein levels in human chondrocytes cultures. COX-2 suppression leads to reduced production of proinflammatory molecules (Hooshmand et al., 2007). Endothelial cell inflammatory injury is blocked by genistein as seen from the production of ROS and prevention of vascular endothelial cell death (Han et al., 2015). In murine lung epithelial cells, daidzein prevented the expression and activity of TNF-α-induced Cxcl2 (pro-inflammatory cytokine). TNF-α-induced protein polyadenosine diphosphate ribosylation was inhibited significantly (Li et al., 2014). A study with isoflavone powder and standard genistein reported the effective inhibition of LPS-induced inflammation, reduction in leukocyte count, and lower production of IL-1β, IL-6, NO, and PGE2 (Kao et al., 2007). Isoflavone puerarin exerted an anti-inflammatory effect by inhibiting COX-2 expression in astrocytes and microglia in rats with middle cerebral artery occlusion (Lim et al., 2013). The reduction in the inflammatory markers (IL-8, CRP) and the increase in plasma nitric oxide levels were seen in postmenopausal women with metabolic syndrome on consumption of soy nut diet (340 mg isoflavone/100 g soy nut) (Azadbakht et al., 2007). In another study, CRP was reduced in end-stage renal failure patients when administered isoflavone-rich soy foods (Fanti et al., 2006).

8.8 OTHER BIOACTIVITY OF ISOFLAVONES

Dietary isoflavones have been shown in numerous epidemiological and clinical research studies to be protective against the development of specific menopause symptoms and a number of chronic diseases, such as cardiovascular disease, osteoporosis, cognitive impairment, and hormone-dependent malignancies (Franco et al., 2016; Zhang et al., 2016). Hence, isoflavones can be employed as an alternative treatment for several pathologies.

8.9 BONE HEALTH

Osteoporosis is characterized by reduced bone mass and impaired bone tissue formation. It primarily affects women and is associated with aging and a deficiency of hormones. Loss of bone density during menopause might result in osteoporosis. In this regard, soy flavones have been touted to be beneficial because they may contribute to the maintenance of good bone mass in women who are at this stage of their life. Clinical studies on genistein showed a positive effect on osteoporotic bones. Genistein promotes osteoblastic factors (e.g., bone alkaline phosphatase) and reduces osteoclastic factors (e.g., collagen C-telopeptide). Furthermore, genistein specifically counteracts parathormones' catabolic effects on osteoblasts (Onoe et al., 1997). In the systematic review conducted by Perna et al. (2016), nine studies targeting menopausal women and focusing on bone health were analyzed. To determine the impact of soy flavone extracts on bone mineral density in postmenopausal women, Abdi et al. (2016) performed another comprehensive analysis of 23 randomized controlled trials. Despite the fact that not all research observed the same effect, it was concluded that isoflavones had little effect on bone health and bone mineral density during menopause. They additionally indicated that the duration of the treatment, the variety of isoflavone, its dosage, and the diet could all contribute to differences in results between studies. They suggested that genistein, either alone or in combination with daidzein, enhanced bone density and bone turnover in women after menopause with regard to the effects of specific isoflavones. Isoflavones increase bone density in the lumbar spine, femoral neck, and distal radius in menopausal women, according to a meta-analysis by Sansai et al. (2020) that included 6,427 postmenopausal women and 63 controlled studies. These positive effects were associated with 54 mg/day of genistein and 600 mg/day and synthetic isoflavones. Even though the exact mechanisms by which isoflavones act are not fully understood, it seems that they not only reduce bone resorption but also increase bone formation. The increased bone formation can be attributed to the stimulation of osteoblastic activity, which occurs primarily by (i) activating estrogen receptors because they bind to nuclear estrogen receptors and exhibit estrogenic activity, because of their similarity with 17β-estradiol, and (ii) promoting the production of insulin-like growth factor-I (Arjmandi and Smith, 2002; Chadha et al., 2017).

8.10 ESTROGENIC AND ANTIESTROGENIC ACTIVITIES

There are numerous mechanisms by which isoflavones exert both estrogenic and antiestrogenic effects. Although they have the ability to attach to nuclear ERs because of a structure that is similar to that of 17β-estradiol, their affinity for these receptors is not very strong. Soy flavones have the ability to preferentially bind to and transactivate estrogen receptor-β (ERβ), rather than estrogen receptor-α mimicking estrogen's actions in some tissues while antagonistically blocking estrogen's effects in others (Wang, 2002). Only genistein exhibits a greater affinity for the ERβ, where it binds preferentially. Its relative affinity (0.87) is nearer to that of the reference hormone, 17β-estradiol. Daidzein has an affinity of 0.005 for these receptors, but its metabolite, equol, has an affinity that is 5.7 times higher, boosting its estrogenic potential. Depending on the amount of endogenous estrogen present, isoflavones induce agonist or antagonist actions. For people with high levels of estrogen, such as premenopausal women, particularly during the follicular phase of the menstrual cycle, the isoflavones bind to the estrogen receptors. They block the action of endogenous estrogens on their receptors. The estrogenic action of isoflavones becomes apparent and exhibits an additive agonist effect (Hwang et al., 2006) in cases of low endogenous estrogen concentration (women in menopause, women who have had ovarian surgery, or males). This is the reason for the use of isoflavones as a long-term supplemental or alternative hormone therapy (Pilšáková et al., 2010). The activity of some of the enzymes involved in the metabolism of the sex steroid hormones is also influenced by isoflavones. They do this by inhibiting the activity of 5-reductase, an enzyme that converts testosterone into 5-dihydrotestosterone, and aromatase, an enzyme that converts testosterone into estradiol, in low concentrations, while increasing the activity of aromatase in high concentrations. Thus, isoflavones have an affinity for sex hormone-binding globulin (SHBG), and they promote its expression (Pilšáková et al., 2010).

8.11 MENOPAUSAL SYMPTOMS

Menopause is characterized by a decrease in estrogen levels and is often accompanied by an array of symptoms. The most noticeable and troublesome of these are vasomotor symptoms, which include hot flushes and night sweats. These symptoms have an adverse influence on women's quality of life. It has been demonstrated that hormone replacement therapy reduces vasomotor symptoms, although isoflavones have gained popularity as an alternative treatment to hormone replacement therapy to relieve menopausal symptoms. This is mainly a result of their potential adverse consequences, which include increased coronary heart disease, stroke, and cancer (Chen et al., 2019; Daily et al., 2019). Thus, a high dose of isoflavones may be prescribed to menopausal women. Even though numerous research suggests that genistein supplements may reduce menopausal symptoms (Lethaby et al., 2013), the results are uncertain. In general, using isoflavone-containing dietary supplements leads to a modest (10%–20%) reduction in the frequency of hot flashes.

8.12 CARDIOVASCULAR DISEASES

Heart disease incidence was found to be lower in the population that consumes large amounts of soy products. Although there are numerous risk factors for cardiovascular illnesses, high cholesterol levels are the main reasons. Low-density lipoproteins (LDL) are oxidized by free radicals after penetrating the blood vessel walls. LDL then builds up and blocks the blood arteries, resulting in thrombosis. LDL cholesterol was observed to be reduced by soy protein (Zhan and Ho, 2005). According to Zhan and Ho (2005) and Sacks et al. (2006), isoflavone-rich soy protein is preferable to isoflavone-depleted soy protein or isoflavone-extracted soybeans in terms of reducing LDL cholesterol. Following a meta-analysis of the available data, Liu et al. (2012) observed that ingesting 65–153 mg of soy isoflavones daily for 1–12 months along with soy protein lowered blood pressure in the hypertensive population. In the population with normal blood pressure, this effect was not shown evident. They block the action of endogenous estrogens on their receptors. The estrogenic action of isoflavones becomes apparent and exhibits an additive agonist effect in cases of low endogenous estrogen concentration (women in menopause, women who have had ovarian surgery, or males).

8.13 CANCER

Cancer diseases are the biggest challenge of modern medicine. The incidence and mortality rates of cancer continue to rise globally despite the notable advancements made in recent years. One of the most significant epidemiological issues affecting women today is breast cancer. The primary isoflavone in soy, genistein, has a two-phase action on the ER present in breast cancer cells. It enhances the growth of positive-ER breast cancer cells at low concentrations, but it inhibits the growth of breast cancer cells at higher concentrations. The inhibition of a tyrosine kinase system, which might prevent excessive cell proliferation or abnormal angiogenesis by inhibiting signaling pathways related to tyrosine kinase receptors, is another mechanism that may be involved in the prevention of cancer. In addition, genistein prevents the activity of DNA topoisomerase (Varinska et al., 2015; Ziaei and Halaby, 2017). Studies conducted by Shu (2009) proved that in women who fought off breast cancer, the consumption of foods rich in soy isoflavones lowered both the risk of death and the possibility of cancer recurrence. Prostate cancer is the second most commonly diagnosed cancer and the sixth most common cause of cancer death in men worldwide. Studies on animals and in cell culture suggest that soybean isoflavones may be able to inhibit the progression of prostate cancer (Lund et al., 2011). Despite the fact that consuming a soy isoflavone supplement for up to a year didn't significantly lower the prostate-specific antigen (PSA) level in men's blood serum that had not yet confirmed prostate cancer, the soy isoflavone supplement tends to reduce the increase in blood serum PSA concentration, a condition associated with the development of prostate tumor growth in men with prostate cancer (Adams et al., 2004; Fischer at al., 2004). According to Yan and Spitznagel's (2009) meta-analysis of eight trials, consuming isoflavones was associated with a lower probability of developing prostate cancer. In the case of uterine cancer, it is generally accepted that it is largely influenced by a long-term imbalance in the levels of the hormones estrogen and progesterone contributing significantly to cancer formation. As a result, high isoflavone dosage with antiestrogenic action could serve as a preventative measure for endometrial cancer (Horn-Ross et al., 2003). According to Murray et al. (2003), daily isoflavone supplementation of 120 mg over the course of 6 months did not cease postmenopausal women's endometrial hyperplasia, which was caused by exogenous estradiol injection.

8.14 ISOFLAVONES IN TYPE II DIABETES

According to epidemiological research, increased dietary soy isoflavone intake is associated with a lower incidence of diabetes and increased tissue insulin sensitivity (Jayagopal et al., 2002). Based on *in vitro* experiments, Ademiluyi and Oboh (2013) observed that the phenolic compounds found in soy have the capacity to inhibit the activity of amylase and glucosidase, two enzymes crucial for the hydrolysis of carbohydrates in the gastrointestinal tract. As a result, they can slow the absorption of glucose in the small intestine and prevent postprandial hyperglycemia. However, they observed that the phenolic compounds from the soy extract can regulate glucose absorption by means of mechanisms other than inhibiting the activity of those enzymes. A prospective population-based study of 64,227 Chinese women in their middle years showed a correlation between higher soy intake and a reduced likelihood of type 2 diabetes. Additionally, the researchers did not discover any interactions between the menopause status, total consumption of soy protein or soy, and risk of type II DM (Villegas et al., 2008).

8.15 THYROID FUNCTION

In studies carried out on cell cultures and animals, it has been demonstrated that soybean isoflavones suppress thyroid peroxidase activity, the enzyme required for the generation of thyroid hormone (Doerge and Sheehan, 2002). However, a high isoflavone intake does not seem to enhance the risk of hypothyroidism as long as an adequate iodine intake is maintained. Since the 1960s, when iodine was added to soy-based formulas, there have been no more cases of infants developing hypothyroidism via soy formula-fed infants (Messina and Redmond, 2006). Studies on pre- and postmenopausal women who ingested sufficient amounts of iodine also reported similar findings. High isoflavone consumption has no detrimental effects on these women's levels of thyroid hormones in the blood (Dillingham et al., 2007).

8.16 CONCLUSION

Leguminous plants produce ecophysiologically active isoflavonoids from the phenylpropanoid pathway. In humans, digestive tract bacteria and the enzymes hydrolyze them into aglycones. Isoflavonoids offer estrogenic activity-related properties including antioxidant, anti-inflammatory, anticancer, anti-diabetic, and neuroprotective activities, which have been studied for decades. *In vitro* and *in vivo* studies have shown that a wide range of viruses are inhibited by isoflavonoids and related flavonoids. Since isoflavonoids are documented with a broad spectrum of benefits, they may enhance the immune response against viral infections and alleviate infection symptoms. However, more research in isoflavone areas to comprehend this complex issue is needed. Additional research on pharmacological studies, with a particular emphasis on bioavailability, could lead to the discovery of supplementary therapy for COVID-19 or other new viruses.

REFERENCES

Abdi, F.; Alimoradi, Z.; Haqi, P.; Mahdizad, F. Effects of phytoestrogens on bone mineral density during the menopause transition: A systematic review of randomized, controlled trials. *Climacteric*. 2016, 19, 535–545. https://doi.org/10.1080/13697137.2016.1238451.

Abu-Eid, R.; Ward, F.J. Targeting the PI3K/Akt/mTOR pathway: A therapeutic strategy in COVID-19 patients. *Immunol. Lett*. 2021, 240, 1–8. https://doi.org/10.1016/j.imlet.2021.09.005.

Adams, K.F.; Chen, C.; Newton, K.M.; Potter, J.D.; Lampe, J.W. Soy isoflavones do not modulate prostate-specific antigen concentrations in older men in a randomized controlled trial. *Cancer Epidemiol. Biomark. Prev*. 2004, 13, 644–648. https://doi.org/10.1158/1055-9965.644.13.4.

Ademiluyi, A.O.; Oboh, G. Soybean phenolic-rich extracts inhibit key-enzymes linked to type 2 diabetes (α-amylase and α-glucosidase) and hypertension (angiotensin i converting enzyme) in vitro. *Exp. Toxicol. Pathol*. 2013, 65, 305–309. https://doi.org/10.1016/j.etp.2011.09.005.

Akula, S.M.; Hurley, D.J.; Wixon, R.L.; Wang, C.; Chase, C.C. Effect of genistein on replication of bovine herpesvirus type 1. *Am. J. Vet. Res*. 2002, 63, 1124–1128. https://doi.org/10.2460/ajvr.2002.63.1124.

Andres, A.; Donovan, S.M.; Kuhlenschmidt, M.S. Soy isoflavones and virus infections. *J. Nutr. Biochem*. 2009, 20, 563–569. https://doi.org/10.1016/j.jnutbio.2009.04.004.

Andres, S.; Hansen, U.; Niemann, B.; Palavinskas, R.; Lampen, A. Determination of the isoflavone composition and estrogenic activity of commercial dietary supplements based on soy or red clover. *Food. Funct*. 2015, 6, 2017–2025. https://doi.org/10.1039/c5fo00308c.

Arabyan, E.; Hakobyan, A.; Kotsinyan, A.; Karalyan, Z.; Arakelov, V.; Arakelov, G.; Nazaryan, K.; Simonyan, A.; Aroutiounian, R.; Ferreira, F.; Zakaryan, H. Genistein inhibits African swine fever virus replication in vitro by disrupting viral DNA synthesis. *Antiviral. Res*. 2018, 156, 128–137. https://doi.org/10.1016/j.antiviral.2018.06.014.

Argenta, D.F.; Silva, I.T.; Bassani, V.L.; Koester, L.S.; Teixeira, H.F.; Simoes, C.M. Antiherpes evaluation of soybean isoflavonoids. *Adv. Virol*. 2015, 160, 2335–2342. https://doi.org/10.1007/s00705-015-2514-z.

Arjmandi, B. H.; Smith, B. J. Soy isoflavones' osteoprotective role in postmenopausal women: Mechanism of action. *J. Nutr. Biochem.* 2002, 13, 130–137. https://doi.org/10.1016/s0955-2863(02)00172-9.

Azadbakht, L.; Kimiagar, M.; Mehrabi, Y.; Esmaillzadeh, A.; Hu, F.B.; Willett, W.C. Soy consumption, markers of inflammation, and endothelial function: A cross-over study in postmenopausal women with the metabolic syndrome. *Diabetes Care.* 2007, 30, 967–973. https://doi.org/10.2337/dc06-2126.

Baggiolini, M.; Clark-Lewis, I. Interleukin-8, a chemotactic and inflammatory cytokine. *FEBS Lett.* 1992, 307, 97–101. https://doi.org/ 10.1016/0014-5793(92)80909-z.

Bahadur, S.; Long, W.; Shuaib, M. Human coronaviruses with emphasis on the COVID-19 outbreak. *Virus Disease.* 2020, 31, 80–84. https://doi.org/10.1007/s13337-020-00594-y.

Barros, R.P.A.; Gustafsson, J.-Å. Estrogen receptors and the metabolic network. *Cell Metab.* 2011, 14, 289–299. https://doi.org/10.1016/j.cmet.2011.08.005.

Basile, M.S.; Cavalli, E.; McCubrey, J.; Hernández-Bello, J.; MuñozValle, J.F.; Fagone, P.; Nicoletti, F. The PI3K/Akt/mTOR pathway: A potential pharmacological target in COVID-19. *Drug Discov. Today*, 2021, 27, 848–856. https://doi.org/10.1016/j.drudis.2021.11.002.

Beltrán-García, J.; Osca-Verdegal, R.; Pallardó, F.V.; Ferreres, J.; Rodríguez, M.; Mulet, S.; Sanchis-Gomar, F.; Carbonell, N.; García-Giménez, J.L. Oxidative stress and inflammation in COVID-19-associated sepsis: The potential role of anti-oxidant therapy in avoiding disease progression. *Antioxid.* 2020, 9, 936. https://doi.org/ 10.3390/antiox9100936.

Butkutė, B.; Padarauskas, A.; Cesevičienė, J.; Taujenis, L.; Norkevičienė, E. Phytochemical composition of temperate perennial legumes. *Crop. Pasture. Sci.* 2018, 69, 1020. https://doi.org/10.1071/cp18206.

Carrara, V.S.; Melo, J.O.; Bersani-Amado, C.A.; Nakamura, C.V.; Mandarino, J.M.G.; Cortez, L.E.R.; Cortez, D.A.G. Anti-inflammatory activity of the soybean methanolic fraction containing isoflavones. *Planta Med.* 2008, 74, PH32. https://doi.org/10.1055/s-0028-1084877.

Cassidy, A.; Hanley, B.; Lamuela-Raventos, R.M. Isoflavones, lignans and stilbenes - origins, metabolism and potential importance to human health. *J. Sci. Food Agric.* 2000, 80, 1044–1062. https://doi.org/10.1002/(sici)1097-0010(20000515)80:7%3C1044::aid-jsfa586%3E3.0.co;2-n.

Chadha, R.; Bhalla, Y.; Jain, A.; Chadha, K.; Karan, M. Dietary soy isoflavone: A mechanistic insight. *Nat. Prod. Commun.* 2017, 12, 1934578X1701200. https://doi.org/10.1177/1934578x1701200439.

Chan, Y.H.; Lau, K.K.; Yiu, K.H.; Li, S.W.; Chan, H.T.; Fong, D.Y; Tam, S.; Lau, C.P; Tse, H.F. Reduction of C-reactive protein with isoflavone supplement reverses endothelial dysfunction in patients with ischaemic stroke. *Eur. Heart J.* 2008, 29, 2800–2807. https://doi.org/10.1093/eurheartj/ehn409.

Chen, L.-R.; Ko, N.-Y.; Chen, K.H. Isoflavone supplements for menopausal women: A systematic review. *Nutrients.* 2019, 11, 2649. https://doi.org/10.3390/nu11112649.

Čoma, M.; Lachová, V.; Mitrengová, P.; Gál, P. Molecular changes underlying Genistein treatment of wound healing: A review. *Curr. Issues Mol. Biol.* 2021, 43, 127–141. https://doi.org/10.3390/cimb43010011.

Crozier, A.; Jaganath, I.B.; Clifford, M.N. Dietary phenolics: Chemistry, bioavailability and effects on health. *Nat. Prod. Rep.* 2009, 26, 1001. https://doi.org/10.1039/b802662a.

Dahle, M.K.; Øverland, G.; Myhre, A.E.; Stuestøl, J.F.; Hartung, T.; Krohn, C.D.; Mathiesen, Ø.; Wang, J.E.; Aasen, A.O. The phosphatidylinositol 3-kinase/protein kinase B signaling pathway is activated by lipoteichoic acid and plays a role in Kupffer cell production of interleukin-6 (IL-6) and IL-10. *Infect. Immun.* 2004, 72, 5704–5711. https://doi.org/10.1128/iai.72.10.5704-5711.2004.

Daily, J. .; Ko, B.-S.; Ryuk, J.; Liu, M.; Zhang, W.; Park, S. Equol decreases hot flashes in postmenopausal women: A systematic review and meta-analysis of randomized clinical trials. *J. Med. Food.* 2019, 22, 127–139. https://doi.org/10.1089/jmf.2018.4265.

Dangoria, N.S.; Breau, W.C.; Anderson, H.A.; Cishek, D.M.; Norkin, L.C. Extracellular simian virus 40 induces an ERK/MAP kinase-independent signalling pathway that activates primary response genes and promotes virus entry. *J. Gen. Virol.* 1996, 77, 2173–2182. https://doi.org/10.1099/0022-1317-77-9-2173.

Dias, M.C.; Pinto, D.C.G.A.; Silva, A.M.S. Plant flavonoids: Chemical characteristics and biological activity. *Molecules.* 2021, 26, 5377. https://doi.org/10.3390/molecules26175377.

Dillingham, B.L.; McVeigh, B.L.; Lampe, J.W.; Duncan, A.M. Soy protein isolates of varied isoflavone content do not influence serum thyroid Hormones in healthy young men. *Thyroid.* 2007, 17, 131–137. https://doi.org/10.1089/thy.2006.0206.

Dixon, R.A.; Sumner, L.W. Legume natural products: Understanding and manipulating complex pathways for human and animal health. *Plant Physiol.* 2003, 131, 878–885. https://doi.org/10.1104/pp.102.017319.

Doerge, D.R.; Sheehan, D.M. Goitrogenic and estrogenic activity of soy isoflavones. *Environ. Health Perspect.* 2002, 110, 349–353. https://doi.org/10.1289/ehp.02110s3349.

Du, H.; Huang, Y.; Tang, Y. Genetic and metabolic engineering of isoflavonoid biosynthesis. *Appl. Microbiol. Biotechnol.* 2010, 86, 1293–1312. https://doi.org/10.1007/s00253-010-2512-8.

Duan, W.; Kuo, I.C.; Selvarajan, S.; Chua, K.Y.; Bay, B.H.; Wong, W.F. Antiinflammatory effects of genistein, a tyrosine kinase inhibitor, on a guinea pig model of asthma. *Am. J. Respir. Crit. Care Med.* 2003, 167, 185–192. https://doi.org/10.1164/rccm.200205-420OC.

Evers, D.L.; Chao, C. F.; Wang, X.; Zhang, Z.; Huong, S.M.; Huang, E.S. Human cytomegalovirus-inhibitory flavonoids: Studies on antiviral activity and mechanism of action. *Antiviral. Res.* 2005, 68, 124–134. https://doi.org/10.1016/j.antiviral.2005.08.002.

Fanti, P.; Asmis, R.; Stephenson, T.J.; Sawaya, B.P.; Franke, A.A. Positive effect of dietary soy in ESRD patients with systemic inflammation-correlation between blood levels of the soy isoflavones and the acute-phase reactants. *Nephrol. Dial. Transplant.* 2006, 21, 2239–2246. https://doi.org/10.1093/ndt/gfl169.

Fischer, L.; Mahoney, C.; Jeffcoat, A.R.; Koch, M.A.; Thomas, B.F.; Valentine, J.L.; Stinchcombe, T.; Boan, J.; Crowell, J.A.; Zeisel, S.H. Clinical characteristics and pharmacokinetics of purified soy isoflavones: Multiple-dose administration to men with prostate neoplasia. *Nutr. Cancer.* 2004, 48, 160–170. https://doi.org/10.1207/s15327914nc4802_5.

Franco, O.H.; Chowdhury, R.; Troup, J.; Voortman, T.; Kunutsor, S.; Kavousi, M.; Oliver-Williams, C.; Muka, T. Use of plant-based therapies and menopausal symptoms. *J. Am. Med. Assoc.* 2016, 315, 2554. https://doi.org/10.1001/jama.2016.8012.

Frankenfeld, C.L. Dairy consumption is a significant correlate of urinary equol concentration in a representative sample of US adults. *Am. J. Clin. Nutr.* 2011, 93, 1109–1116. https://doi.org/10.3945/ajcn.111.011825.

Ganai, A.A.; Khan, A.A.; Malik, Z.A.; Farooqi, H. Genistein modulates the expression of NF-κB and MAPK (p-38 and ERK1/2), thereby attenuating d-Galactosamine induced fulminant hepatic failure in Wistar rats. *Toxicol. Appl. Pharmacol.* 2015, 283, 139–146. https://doi.org/ 10.1016/j.taap.2015.01.012.

García-Lafuente, A.; Guillamón, E.; Villares, A.; Rostagno, M.A.; Martínez, J.A. Flavonoids as anti-inflammatory agents: Implications in cancer and cardiovascular disease. *Inflamm. Res.* 2009, 58, 537–552. https://doi.org/10.1007/s00011-009-0037-3.

Ghasemnejad-Berenji, M.; Pashapour, S.; Sadeghpour, S. Pentoxifylline: A drug with antiviral and anti-inflammatory effects to be considered in the treatment of coronavirus disease 2019. *Med. Princ. Pract.* 2021, 30, 98–100. https://doi.org/10.1159/000512234.

Grifoni, A.; Weiskopf, D.; Ramirez, S.I.; Mateus, J.; Dan, J.M.; Moderbacher, C.R.; Rawlings, S.A.; Sutherland, A.; Premkumar, L.; Jadi, R.S.; Marrama, D. Targets of T cell responses to SARS-CoV-2 coronavirus in humans with COVID-19 disease and unexposed individuals. *Cell.* 2020, 181, 1489–1501. https://doi.org/10.1016/j.cell.2020.05.015.

Gupta, S.K.; Dongare, S.; Mathur, R.; Mohanty, I.R.; Srivastava, S.; Mathur, S.; Nag, T.C. Genistein ameliorates cardiac inflammation and oxidative stress in streptozotocin-induced diabetic cardiomyopathy in rats. *Mol. Cell Biochem.* 2015, 408, 63–72. https://doi.org/10.1007/s11010-015-2483-2.

Han, R.M.; Tian, Y.X.; Liu, Y.; Chen, C.H.; Ai, X.C.; Zhang, J.P.; Skibsted, L.H. Comparison of flavonoids and isoflavonoids as antioxidants. *J. Agric. Food Chem.* 2009, 57, 3780–3785. https://doi.org/10.1021/jf803850p.

Han, S.; Wu, H.; Li, W.; Gao, P. Protective effects of genistein in homocysteine-induced endothelial cell inflammatory injury. *Mol. Cell. Biochem.* 2015, 403, 43–49. https://doi.org/10.1007/s11010-015-2335-0.

Hannu, S.; Wähälä, K.; Nykänen-Kurki, P.; Kallela, K.; Saastamoinen, I. Phytoestrogen content and estrogenic effect of legume fodder. *Exp. Biol. Med.* 1995, 208, 13–17. https://doi.org/10.3181/00379727-208-43825.

Hooshmand, S.; Soung, D.Y.; Lucas, E.A.; Madihally, S.V.; Levenson, C.W.; Arjmandi, B.H. Genistein reduces the production of proinflammatory molecules in human chondrocytes. *J. Nutr. Biochem.* 2007, 18, 609–614. https://doi.org/10.1016/j.jnutbio.2006.11.006.

Horn-Ross, P.L.; John, E.M.; Canchola, A.J.; Stewart, S.L.; Lee, M.M. Phytoestrogen intake and endometrial cancer risk. *Obstet. Gynecol. Surv.* 2003, 58, 726–728. https://doi.org/10.1097/01.ogx.0000093676.70200.51.

Hsu, H.Y.; Chiu, S.L.; Wen, M.H.; Chen, K.Y.; Hua, K.F. Ligands of macrophage scavenger receptor induce cytokine expression via differential modulation of protein kinase signaling pathways. *J. Biol. Chem.* 2001, 276, 28719–28730. https://doi.org/10.1074/jbc.M011117200.

Huang, G.; Xu, J.; Guo, T.L. Isoflavone daidzein regulates immune responses in the B6C3F1 and non-obese diabetic (NOD) mice. *Int. Immunopharmacol.* 2019, 71, 277–284. https://doi.org/10.1016/j.intimp.2019.03.046.

Huang, H.; Liao, D.; Liang, L.; Song, L.; Zhao, W. Genistein inhibits rotavirus replication and upregulates AQP4 expression in rotavirus-infected Caco-2 cells. *Adv. Virol.* 2015, 160, 1421–1433. https://doi.org/10.1007/s00705-015-2404-4.

Hwang, C.S.; Kwak, H.S.; Lim, H.J.; Lee, S.H.; Kang, Y.S.; Choe, T.B.; Hur, H.G.; Han, K.O. Isoflavone metabolites and their *in vitro* dual functions: They can act as an estrogenic agonist or antagonist depending on the estrogen concentration. *J. Steroid Biochem. Mol. Biol.* 2006, 101, 246–253. https://doi.org/10.1016/j.jsbmb.2006.06.020.

Jafari, A.; Esmaeilzadeh, Z.; Khezri, M.R.; Ghasemnejad-Berenji, H.; Pashapour, S.; Sadeghpour, S.; Ghasemnejad-Berenji, M. An overview of possible pivotal mechanisms of Genistein as a potential phytochemical against SARS-CoV-2 infection: A hypothesis. *J. Food Biochem.* 2022, 46, 14345. https://doi.org/10.1111/jfbc.14345.

Jayagopal, V.; Albertazzi, P.; Kilpatrick, E.S.; Howarth, E.M.; Jennings, P.E.; Hepburn, D.A.; Atkin, S.L. Beneficial effects of soy phytoestrogen intake in postmenopausal women with type 2 diabetes. *Diabetes Care.* 2002, 25, 1709–1714. https://doi.org/10.2337/diacare.25.10.1709.

Kao, T.H.; Wu, W.M.; Hung, C.F.; Wu, W.B.; Chen, B.H. Anti-inflammatory effects of isoflavone powder produced from soybean cake. *J. Agric. Food Chem.* 2007, 55, 11068–11079. https://doi.org/10.1021/jf071851u.

Kim, I.S. Current perspectives on the beneficial effects of soybean isoflavones and their metabolites for humans. *Antioxid.* 2021, 10, 1064. https://doi.org/10.3390/antiox10071064.

Kindrachuk, J.; Ork, B.; Mazur, S.; Holbrook, M.R.; Frieman, M.B.; Traynor, D.; Johnson, R.F.; Dyall, J.; Kuhn, J.H.; Olinger, G.G.; Hensley, L.E. Antiviral potential of ERK/MAPK and PI3K/ AKT/mTOR signaling modulation for Middle East respiratory syndrome coronavirus infection as identified by temporal kinome analysis. *Antimicrob. Agents Chemother.* 2015, 59, 1088–1099. https://doi.org/10.1128/AAC.03659-14.

Kotecha, R. The journey with COVID-19: Initiatives by ministry of AYUSH. *J. Ayurveda Integr. Med.* 2021, 12, 1–3. https://doi.org/10.1016/j.jaim.2021.03.009.

Kovats, S. Estrogen receptors regulate innate immune cells and signaling pathways. *Cell Immunol.* 2015, 294, 63–69. https://doi.org/10.1016/j.cellimm.2015.01.018.

Křížová, L.; Dadáková, K.; Kašparovská, J.; Kašparovský, T. Isoflavones. *Molecules.* 2019, 24, 1076. https://doi.org/10.3390/molecules24061076.

Kubo, Y.; Ishimoto, A.; Amanuma, H. Genistein, a protein tyrosine kinase inhibitor, suppresses the fusogenicity of Moloney murine leukemia virus envelope protein in XC cells. *Arch. Virol.* 2003, 148, 1899–1914. https://doi.org/10.1007/s00705-003-0164-z.

Kuriyama, I.; Takahashi, Y.; Yoshida, H.; Mizushina, Y. Inhibitory effect of isoflavones from processed soybeans on human dna topoisomerase II activity. *J. Plant Biochem. Physiol.* 2013, 1, 106–112. https://doi.org/10.3389/fbioe.2021.673270.

Kurzer, M.S.; Xu, X. Dietary phytoestrogens. *Annu. Rev. Nutr.* 1997, 17, 353–381. https://doi.org/10.1146/annurev.nutr.17.1.353.

Laforge, M.; Elbim, C.; Frère, C.; Hémadi, M.; Massaad, C.; Nuss, P.; Benoliel, J.J; Becker, C. Tissue damage from neutrophil-induced oxidative stress in COVID-19. *Nat. Rev. Immunol.* 2020, 20, 515–516. https://doi.org/10.1038/s41577-020-0407-1.

Langfort, J.; Viese, M.; Ploug, T.; Dela, F. Time course of GLUT4 and AMPK protein expression in human skeletal muscle during one month of physical training. *Scand. J. Med. Sci. Sports*. 2003, 13, 169–174. https://doi.org/10.1034/j.1600-0838.2003.20120.x.

Lee, K.Y.; Kim, J.-R.; Choi, H.C. Genistein-induced LKB1-AMPK activation inhibits senescence of VSMC through autophagy induction. *Vasc. Pharmacol.* 2016, 81, 75–82. https://doi.org/10.1016/j.vph.2016.02.007.

Lethaby, A.; Marjoribanks, J.; Kronenberg, F.; Roberts, H.; Eden, J.; Brown, J. Phytoestrogens for menopausal vasomotor symptoms. *Cochrane Database Syst. Rev*. 2013, 2013, CD001395. https://doi.org/10.1002/14651858.cd001395.pub4.

Lewicki, S.; Lewicka, A.; Kalicki, B.; Sobolewska-Ruta, A.; Debski, B.; Zdanowski, R.; Syryło, T.; Kloc, M.; Kubiak, J.Z. Effects of genistein on insulin pathway-related genes in mouse differentiated myoblast C2C12 cell line: Evidence for two independent modes of action. *Folia Histochem. Cytobiol.* 2018, 56, 123–132. https://doi.org/10.5603/FHC.a2018.0014.

Li, Y.; Sarkar, F.H. 2002. Inhibition of nuclear factor κB activation in PC3 cells by genistein is mediated via Akt signaling pathway. *Clin. Cancer Res*. 2002, 8, 2369–2377.

Li, Y.; Zhang, H. Soybean isoflavones ameliorate ischemic cardiomyopathy by activating Nrf2-mediated antioxidant responses. *Food Funct.* 2017, 8, 2935–2944. https://doi.org/10.1039/c7fo00342k.

Li, E.; Stupack, D.G.; Brown, S.L.; Klemke, R.; Schlaepfer, D.D.; Nemerow, G.R. Association of p130CAS with phosphatidylinositol-3-OH kinase mediates adenovirus cell entry. *J. Biol. Chem.* 2000, 275, 14729–14735. https://doi.org/10.1074/jbc.275.19.14729.

Li, J.; Gang, D.; Yu, X.; Hu, Y.; Yue, Y.; Cheng, W.; Pan, X.; Zhang, P. Genistein: The potential for efficacy in rheumatoid arthritis. *Clin. Rheumatol.* 2013, 32, 535–540. https://doi.org/10.1007/s10067-012-2148-4.

Li, H.Y.; Pan, L.; Ke, Y.S.; Batnasan, E.; Jin, X.Q.; Liu, Z.Y.; Ba, X.Q. Daidzein suppresses pro-inflammatory chemokine Cxcl2 transcription in TNF-α-stimulated murine lung epithelial cells via depressing PARP-1 activity. *Acta Pharmacol. Sin.* 2014, 35, 496–503. https://doi.org/10.1038/aps.2013.191.

Lim, D.W.; Lee, C.; Kim, I.H; Kim, Y.T. Anti-inflammatory effects of total isoflavones from Pueraria lobata on cerebral ischemia in rats. *Molecules.* 2013, 18, 10404–10412. https://doi.org/10.3390/molecules180910404.

Liu, X.X.; Li, S.H.; Chen, J.Z.; Sun, K.; Wang, X.J.; Wang, X.G.; Hui, R.T. Effect of soy isoflavones on blood pressure: A meta-analysis of randomized controlled trials. *Nutr. Metab. Cardiovasc. Dis.* 2012, 22, 463–470. https://doi.org/10.1016/j.numecd.2010.09.006.

Liu, Y.; Yang, Y.; Zhang, C.; Huang, F.; Wang, F.; Yuan, J.; Wang, Z.; Li, J.; Li, J.; Feng, C.; Zhang, Z. Clinical and biochemical indexes from 2019-nCoV infected patients linked to viral loads and lung injury. *Sci. China Life Sci.* 2020, 63, 364–374. https://doi.org/10.1007/s11427-020-1643-8.

Lund, T.D.; Blake, C.; Bu, L.; Hamaker, A.; Lephart, E.D. Equol an isoflavonoid: Potential for improved prostate health, in vitro and in vivo evidence. *Reprod. Biol. Endocrinol.* 2011, 9, 4. https://doi.org/10.1186/1477-7827-9-4.

Ma, D.; Zhang, Y.; Yang, T.; Xue, Y.; Wang, P. Isoflavone intake inhibits the development of 7, 12-dimethylbenz (a) anthracene (DMBA)-induced mammary tumors in normal and ovariectomized rats. *J. Clin. Biochem. Nutr.* 2014, 54, 31–38. https://doi.org/10.3164/jcbn.13-33.

Majnooni, M.B.; Fakhri, S.; Shokoohinia, Y.; Kiyani, N.; Stage, K.; Mohammadi, P.; Gravandi, M.M.; Farzaei, M.H.; Echeverría, J. Phytochemicals: Potential therapeutic interventions against coronavirus-associated lung injury. *Front. Pharmacol.* 2020, 11, 588467. https://doi.org/10.3389/fphar.2020.588467.

Malgotra, V.; Sharma, V. 2-Deoxy-d-glucose inhibits replication of novel coronavirus (SARS-CoV-2) with adverse effects on host cell metabolism. *Clin. Trials*. 2021, 7, 10. https://doi.org/10.20944/preprints202106.0333.v1.

Malloy, K.M.; Wang, J.; Clark, L.H.; Fang, Z.; Sun, W.; Yin, Y.; Kong, W.; Zhou, C.; Bae-Jump, V.L. Novasoy and genistein inhibit endometrial cancer cell proliferation through disruption of the AKT/mTOR and MAPK signaling pathways. *Am. J. Transl. Res*. 2018, 10, 784–795.

Masilamani, M.; Wei, J.; Sampson, H.A. Regulation of the immune response by soybean isoflavones. *Immunol. Res.* 2012, 54, 95–110. https://doi.org/10.1007/s12026-012-8331-5.

Meng, Y.; Yin, Q.; Ma, Q.; Qin, H.; Zhang, J.; Zhang, B.; Pang, H.; Tian, H. FXII regulates the formation of deep vein thrombosis via the PI3K/AKT signaling pathway in mice. *Int. J. Mol. Med.* 2021, 47, 1–13. https://doi.org/10.3892/ijmm.2021.4920.

Merad, M.; Martin, J.C. Pathological inflammation in patients with COVID-19: A key role for monocytes and macrophages. *Nat. Rev. Immunol.* 2020, 20, 6, 355–362. https://doi.org/10.1038/s41577-020-0331-4.

Messina, M.; Redmond, G. Effects of soy protein and soybean isoflavones on thyroid function in healthy adults and hypothyroid patients: A review of the relevant literature. *Thyroid.* 2006, 16, 249–258. https://doi.org/10.1089/thy.2006.16.249.

Murray, M.J.; Meyer, W.R.; Lessey, B.A.; Oi, R.H.; DeWire, R. E.; Fritz, M.A. Soy protein isolate with isoflavones does not prevent estradiol-induced endometrial hyperplasia in postmenopausal women: A pilot trial. *Menopause*. 2003, 10, 456–464. https://doi.org/10.1097/01.gme.0000063567.84134.d1.

Murray, J.L.; McDonald, N.J.; Sheng, J.; Shaw, M.W.; Hodge, T.W.; Rubin, D.H.; O'Brien, W.A.; Smee, D.F. Inhibition of influenza A virus replication by antagonism of a PI3K-AKT-mTOR pathway member identified by gene-trap insertional mutagenesis. *Antivir. Chem. Chemother.* 2012, 22, 205–215. https://doi.org/10.3851/IMP2080.

Nazari-Khanamiri, F.; Ghasemnejad-Berenji, M. Cellular and molecular mechanisms of genistein in prevention and treatment of diseases: An overview. *J. Food Biochem.* 2021, 45, e13972. https://doi.org/10.1111/jfbc.13972.

Nguyen, T.T.H.; Jung, J.H.; Kim, M.K.; Lim, S.; Choi, J.M.; Chung, B.; Kim, D.W.; Kim, D. The inhibitory effects of plant derivate polyphenols on the main protease of SARS coronavirus 2 and their structure-activity relationship. *Molecules*. 2021, 26, 1924. https://doi.org/10.3390/molecules26071924.

Onoe, Y.; Miyaura, C.; Ohta, H.; Nozawa, S.; Suda, T. Expression of estrogen receptor β in rat bone. *Endocrinol.* 1997, 138, 4509–4512. https://doi.org/10.1210/endo.138.10.5575.

Page, T.H.; Smolinska, M.; Gillespie, J.; Urbaniak, A.M.; Foxwell, B.M. Tyrosine kinases and inflammatory signaling. *Curr. Mol. Med.* 2009, 9, 69–85. https://doi.org/10.2174/156652409787314507.

Paradkar, P.N.; Blum, P.S.; Berhow, M.A.; Baumann, H.; Kuo, S.M. Dietary isoflavones suppress endotoxin-induced inflammatory reaction in liver and intestine. *Cancer Lett.* 2004, 215, 21–28. https://doi.org/10.1016/j.canlet.2004.05.019.

Park, C.E.; Yun, H.; Lee, E.-B.; Min, B.I.; Bae, H.; Choe, W.; Kang, I.; Kim, S.S.; Ha, J. The antioxidant effects of genistein are associated with AMP-activated protein kinase activation and PTEN induction in prostate cancer cells. *J. Med. Food.* 2010, 13, 815–820. https://doi.org/10.1089/jmf.2009.1359.

Pelkmans, L.; Puntener, D.; Helenius, A. Local actin polymerization and dynamin recruitment in SV40-induced internalization of caveolae. *Science.* 2002, 296, 535–539. https://doi.org/10.1126/science.1069784.

Pendyala, B.; Patras, A. In silico screening of food bioactive compounds to predict potential inhibitors of COVID-19 main protease (Mpro) and RNA-dependent RNA polymerase (RdRp). *Chem. Rxiv.* 2020, 2, 1–11. https://doi.org/10.26434/chemrxiv.12051927.v2.

Pérez-Jiménez, J.; Neveu, V.; Vos, F.; Scalbert, A. Systematic analysis of the content of 502 polyphenols in 452 foods and beverages: An application of the phenol-explorer database. *J. Agric. Food Chem.* 2010, 58, 4959–4969. https://doi.org/10.1021/jf100128b.

Perna, S.; Peroni, G.; Miccono, A.; Riva, A.; Morazzoni, P.; Allegrini, P.; Preda, S.; Baldiraghi, V.; Guido, D.; Rondanelli, M. Multidimensional effects of soy isoflavone by food or supplements in menopause women: A systematic review and bibliometric analysis. *Nat. Prod. Commun.* 2016, 11, 1934578X1601101. https://doi.org/10.1177/1934578x1601101127.

Pilšáková, L.; Riečanský, I.; Jagla, F. The physiological actions of isoflavone phytoestrogens. *Physiol. Res.* 2010, 651–664. https://doi.org/10.33549/physiolres.931902.

Qian, K.; Gao, A. J.; Zhu, M.Y.; Shao, H.X.; Jin, W.J.; Ye, J.Q.; Qin, A.J. Genistein inhibits the replication of avian leucosis virus subgroup J in DF-1 cells. *Virus Res.* 2014, 192, 114–120. https://doi.org/10.1016/j.virusres.2014.08.016.

Reynaud, J.; Guilet, D.; Terreux, R.; Lussignol, M.; Walchshofer, N. Isoflavonoids in non-leguminous families: An update. *Nat. Prod. Rep.* 2005, 22, 504. https://doi.org/10.1039/b416248j.

Sacks, F.M.; Lichtenstein, A.; Van Horn, L.; Harris, W.; Kris-Etherton, P.; Winston, M. Soy protein, isoflavones, and cardiovascular health. *Arterioscler. Thromb. Vasc. Biol.* 2006, 26, 1689–1692. https://doi.org/10.1161/01.atv.0000227471.00284.ef.

Sahin, K.; Tuzcu, M.; Basak, N.; Caglayan, B.; Kilic, U.; Sahin, F.; Kucuk, O. Sensitization of cervical cancer cells to cisplatin by genistein: The role of NF B and Akt/mTOR signaling pathways. *J. Oncol.* 2012, 2012, 200–205. https://doi.org/10.1155/2012/461562.

Sajid, M.; Stone, S.R.; Kaur, P. Recent advances in heterologous synthesis paving way for future green-modular bioindustries: A review with special reference to isoflavonoids. *Front. Bioeng. Biotechnol.* 2021, 9, 673270. https://doi.org/10.3389/fbioe.2021.673270.

Sakai, T.; Kogiso, M. Soy isoflavones and immunity. *J. Med. Invest.* 2008, 55, 167–173. https://doi.org/10.2152/jmi.55.167.

Sakai, T.; Furoku, S.; Nakamoto, M.; Shuto, E.; Hosaka, T.; Nishioka, Y.; Sone, S. The soy isoflavone equol enhances antigen-specific IgE production in ovalbumin-immunized BALB/c mice. *J. Nutr. Sci. Vitaminol.* 2010, 56, 72–76. https://doi.org/10.3177/jnsv.56.72.

Saleh, D.O.; Abdel Jaleel, G.A.R.; El-Awdan, S.A.; Oraby, F.; Badawi, M. Thioacetamide-induced liver injury: Protective role of genistein. *Can. J. Physiol. Pharmacol.* 2014, 92, 965–973. https://doi.org/10.1139/cjpp-2014-0192.

Sanders, T.A.; Dean, T.S.; Grainger, D.; Miller, G.J.; Wiseman, H. Moderate intakes of intact soy protein rich in isoflavones compared with ethanol-extracted soy protein increase HDL but do not influence transforming growth factor β1 concentrations and hemostatic risk factors for coronary heart disease in healthy subjects. *Am. J. Clin.* Nutr. 2002, 76, 373–377. https://doi.org/10.1093/ajcn/76.2.373.

Sansai, K.; Na Takuathung, M.; Khatsri, R.; Teekachunhatean, S.; Hanprasertpong, N.; Koonrungsesomboon, N. Effects of isoflavone interventions on bone mineral density in postmenopausal women: A systematic review and meta-analysis of randomized controlled trials. *Osteoporos. Int.* 2020, 31, 1853–1864. https://doi.org/10.1007/s00198-020-05476-z.

Satsu, H.; Hyun, J.S.; Shin, H.S; Shimizu, M. Suppressive effect of an isoflavone fraction on tumor necrosis factor-α-induced interleukin-8 production in human intestinal epithelial Caco-2 cells. *J. Nutr. Sci. Vitaminol.* 2009, 55, 442–446. https://doi.org/10.3177/jnsv.55.442.

Sharma, V.; Ramawat, K.G. Isoflavonoids. *Nat. Prod.* 2013, 1849–1865. https://doi.org/10.1007/978-3-642-22144-6_61.

Shu, X.O. Soy food intake and breast cancer survival. *J. Am. Med. Assoc.* 2009, 302, 2437. https://doi.org/10.1001/jama.2009.1783.

Sirotkin, A.V.; Harrath, A.H. Phytoestrogens and their effects. *Eur. J. Pharmacol.* 2014, 741, 230–236. https://doi.org/10.1016/j.ejphar.2014.07.057.

Smith, B.N.; Morris, A.; Oelschlager, M.L.; Connor, J.; Dilger, R.N. Effects of dietary soy isoflavones and soy protein source on response of weanling pigs to porcine reproductive and respiratory syndrome viral infection. *J. Anim. Sci.* 2019, 97, 2989–3006. https://doi.org/10.1093/jas/skz135.

Solnier, J.; Fladerer, J.P. Flavonoids: A complementary approach to conventional therapy of COVID-19?. *Phytochem. Rev.* 2021, 20, 773–795. https://doi.org/10.1007/s11101-020-09720-6.

Somanath, P.R, Is targeting Akt a viable option to treat advanced-stage COVID-19 patients?. *Am. J. Physiol. Lung Cell. Mol. Physiol.* 2020, 319, L45–L47. https://doi.org/10.1152/ajplung.00124.2020.

Spagnuolo, C.; Russo, G.L.; Orhan, I.E.; Habtemariam, S.; Daglia, M.; Sureda, A.; Nabavi, S.F.; Devi, K.P.; Loizzo, M.R.; Tundis, R.; Nabavi, S.M. Genistein and cancer: current status, challenges, and future directions. *Adv. Nutr.* 2015, 15, 408–419. https://doi.org/10.3945/an.114.008052.

Su, H.; Yao, S.; Zhao, W.; Zhang, Y.; Liu, J.; Shao, Q.; Wang, Q.; Li, M.; Xie, H.; Shang, W.; Ke, C. Identification of pyrogallol as a warhead in design of covalent inhibitors for the SARS-CoV-2 3CL protease. *Nat. Commun.* 2021, 12, 3623. https://doi.org/10.1038/s41467-021-23751-3.

Tan, H.K.; Moad, A.I.H.; Tan, M.L. The mTOR signaling pathway in cancer and the potential mTOR inhibitory activities of natural phytochemicals. *Asian Pac. J. Cancer Prev.* 2014, 15, 6463–6475. https://doi.org/10.7314/apjcp.2014.15.16.6463.

Tsuchihashi, R.; Sakamoto, S.; Kodera, M.; Nohara, T.; Kinjo, J. Microbial metabolism of soy isoflavones by human intestinal bacterial strains. *J. Nat. Med.* 2008, 62, 456–460. https://doi.org/10.1007/s11418-008-0271-y.

Tyagi, A.M.; Srivastava, K.; Sharan, K.; Yadav, D.; Maurya, R.; Singh, D. Daidzein prevents the increase in CD4+ CD28null T cells and B lymphopoesis in ovariectomized mice: A key mechanism for anti-osteoclastogenic effect. *PLoS One*. 2011, 6, 21216. https://doi.org/10.1371/journal.pone.0021216.

Uhlig, C.; Silva, P.L.; Deckert, S.; Schmitt, J.; de Abreu, M.G. Albumin versus crystalloid solutions in patients with the acute respiratory distress syndrome: A systematic review and meta-analysis. *Crit. Care*. 2014, 18, 1–8. https://doi.org/10.1186/cc13187.

Varinska, L.; Gal, P.; Mojzisova, G.; Mirossay, L.; Mojzis, J. Soy and breast cancer: Focus on angiogenesis. *Int. J. Mol. Sci.* 2015, 16, 11728–11749. https://doi.org/10.3390/ijms160511728.

Vela, E.M.; Bowick, G.C.; Herzog, N.K.; Aronson, J.F. Genistein treatment of cells inhibits arenavirus infection. *Antiviral. Res.* 2008, 77, 153–156. https://doi.org/10.1016/j.antiviral.2007.09.005.

Verdrengh, M.; Jonsson, I.M.; Holmdahl, R.; Tarkowski, A. Genistein as an anti-inflammatory agent. *Inflamm. Res.* 2003, 52, 341–346. https://doi.org/10.1007/s00011-003-1182-8.

Villegas, R.; Gao, Y.-T.; Yang, G.; Li, H.-L.; Elasy, T.A.; Zheng, W.; Shu, X.O. Legume and soy food intake and the incidence of type 2 diabetes in the shanghai women's health study. *Am. J. Clin. Nutr.* 2008, 87, 162–167. https://doi.org/10.1093/ajcn/87.1.162.

Vitale, D.C.; Piazza, C.; Melilli, B.; Drago, F.; Salomone, S. Isoflavones: Estrogenic activity, biological effect and bioavailability. *Eur. J. Drug Metab. Pharmacokinet.* 2012, 38, 15–25. https://doi.org/10.1007/s13318-012-0112-y.

Wang, L.-Q. Mammalian phytoestrogens: Enterodiol and enterolactone. *J Chromatogr. B. Analyt. Technol. Biomed. Life Sci.* 2002, 777, 289–309. https://doi.org/10.1016/s1570-0232(02)00281-7.

Wang, L.; Song, J.; Liu, A.; Xiao, B.; Li, S.; Wen, Z.; Lu, Y.; Du, G. Research progress of the antiviral bioactivities of natural flavonoids. *Nat. Prod. Bioprospect.* 2020, 10, 271–283. https://doi.org/10.1007/s13659-020-00257-x.

Wang, Y.; Lin, S.; Jiang, P.; Song, Y.; Zhao, Y.; Zheng, Y. Focal adhesion kinase inhibitor inhibits the oxidative damage induced by central venous catheter via abolishing focal adhesion kinase-protein kinase B pathway activation. *BioMed. Res. Int.* 2021, 2021, 1–11. https://doi.org/10.1155/2021/6685493.

Wei, J.; Bhatt, S.; Chang, L.M.; Sampson, H.A; Masilamani, M. Isoflavones, genistein and daidzein, regulate mucosal immune response by suppressing dendritic cell function. *PLoS One*. 2012, 7, e47979. https://doi.org/10.1371/journal.pone.0047979.

Yan, L.; Spitznagel, E.L. Soy consumption and prostate cancer risk in men: A revisit of a meta-analysis. *Am. J. Clin. Nutr.* 2009, 89, 1155–1163. https://doi.org/10.3945/ajcn.2008.27029.

Yu, Y.; Maguire, T.G.; Alwine, J.C. Human cytomegalovirus activates glucose transporter 4 expression to increase glucose uptake during infection. *J. Virol.* 2011, 85, 1573–1580. https://doi.org/10.1128/JVI.01967-10.

Yura, Y.; Yoshida, H.; Sato, M. Inhibition of herpes simplex virus replication by genistein, an inhibitor of protein-tyrosine kinase. *Arch. Virol.* 1993, 132, 451–461. https://doi.org/10.1007/BF01309554.

Zaheer, K.; Humayoun Akhtar, M. An updated review of dietary isoflavones: Nutrition, processing, bioavailability and impacts on human health. *Crit. Rev. Food Sci. Nutr.* 2015, 57, 1280–1293. https://doi.org/10.1080/10408398.2014.989958.

Zandi, K.; Teoh, B.T.; Sam, S.S.; Wong, P.F.; Mustafa, M.R.; Abubakar, S. Antiviral activity of four types of bioflavonoid against dengue virus type-2. *Virol. J.* 2011, 8, 560. https://doi.org/10.1186/1743-422X-8-560.

Zhan, S.; Ho, S. . Meta-analysis of the effects of soy protein containing isoflavones on the lipid profile. *Am. J. Clin. Nutr.* 2005, 81, 397–408. https://doi.org/10.1093/ajcn.81.2.397.

Zhang, Q.; Feng, H.; Qluwakemi, B.; Wang, J.; Yao, S.; Cheng, G.; Xu, H.; Qiu, H.; Zhu, L.; Yuan, M. Phytoestrogens and risk of prostate cancer: An updated meta-analysis of epidemiologic studies. *Int. J. Food Sci. Nutr.* 2016, 68, 28–42. https://doi.org/10.1080/09637486.2016.1216525.

Ziaei, S.; Halaby, R. Dietary isoflavones and breast cancer risk. *Medicines.* 2017, 4, 18. https://doi.org/10.3390/medicines4020018.

Zorov, D.B.; Juhaszova, M.; Sollott, S.J. Mitochondrial reactive oxygen species (ROS) and ROS-induced ROS release. *Physiol. Rev.* 2014, 94, 909–950. https://doi.org/10.1152/physrev.00026.2013.

9 Anti-SARS-CoV-2 Activity of Flavones

Molecular Mechanisms and Its Role in COVID-19 Management

Arun Bahadur Gurung and Atanu Bhattacharjee

9.1 INTRODUCTION

Coronavirus disease 2019 (COVID-19) has been one of the serious global outbreaks brought on by the Severe Acute Respiratory Syndrome Coronavirus 2 (SARS-CoV-2) [1]. Previously, three zoonotic outbreaks of coronaviruses within the betacoronavirus family were recorded including the SARS-CoV and the Middle East Respiratory Syndrome Coronavirus (MERS-CoV) [2]. SARS-CoV-2 infections can lead to serious health issues, especially in individuals who are immunocompromised or have underlying medical conditions. SARS-CoV-2 comprises four structural proteins: the membrane glycoprotein (M), spike protein (S), envelope protein (E), and nucleocapsid protein (N) [3] (Figure 9.1). Additionally, there are non-structural proteins (nsps), such as RNA-dependent RNA polymerase (RdRp), papain-like protease (PL^{pro}), and the viral main protease (M^{pro}) [4]. The S protein binds to the host cell surface receptor angiotensin-converting enzyme 2 (ACE2) and facilitates the entry of SARS-CoV-2 into human cells [5]. Figure 9.2 illustrates the molecular mechanisms involved in the pathogenesis of SARS-CoV-2. Several vaccines and drugs have been approved or are undergoing clinical trials [6] such as mRNA-based vaccines, viral vector vaccines, inactivated virus vaccines, and antigen-based vaccines. On the treatment options, there are only a handful of drug candidates for COVID-19 including the FDA-approved remdesivir [7].

Natural products, such as flavonoids, which are phenolic phytochemicals, have demonstrated antioxidant, anti-inflammatory, and antiviral properties [8]. Flavonoids have several benefits such as lack of systemic toxicity, can work synergistically with conventional drugs, and possess "pleiotropic" characteristics (modulating multiple cellular targets) [9]. These characteristics make flavonoids attractive options for disrupting the coronavirus life cycle. Flavonoids are largely characterized by their structural class, degree of hydroxylation, etc. [10]. They are synthesized by plants in reaction to a variety of stressors and are essential for defending plant cells from insects and pathogens [11]. Flavonoids have several pharmacological characteristics like antimicrobial, antioxidant, anti-inflammatory, and antiviral effects [12]. Within different groups of flavonoids, flavones have diverse roles. Flavones also have shown anti-SARS-CoV-2 activities in a significant amount

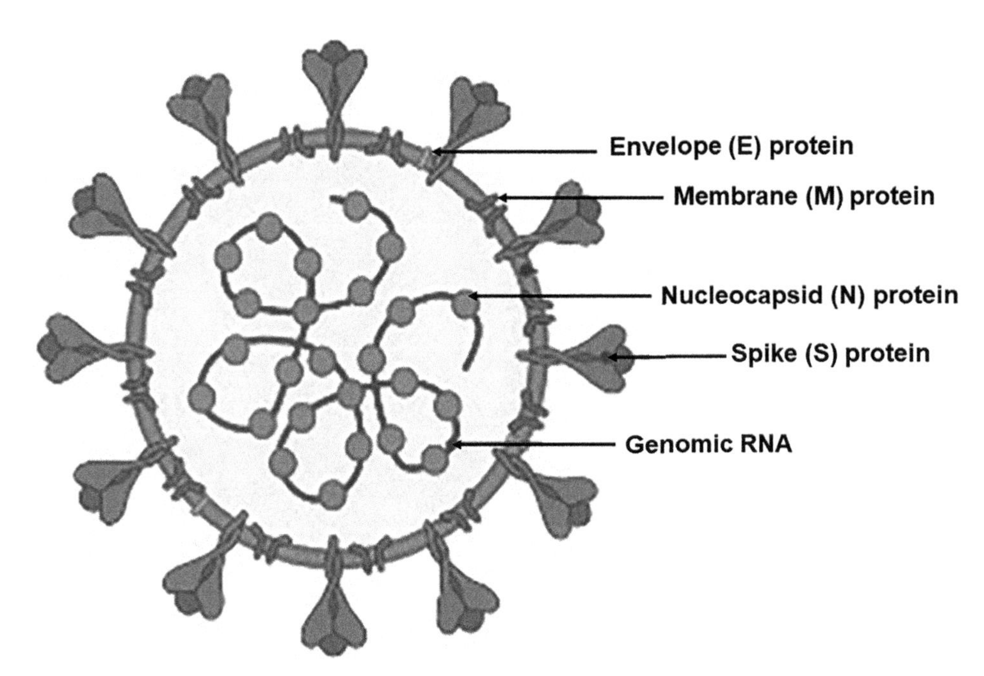

FIGURE 9.1 Structural proteins of SARS-CoV-2 (created with https://biorender.com/).

DOI: 10.1201/9781003433200-9

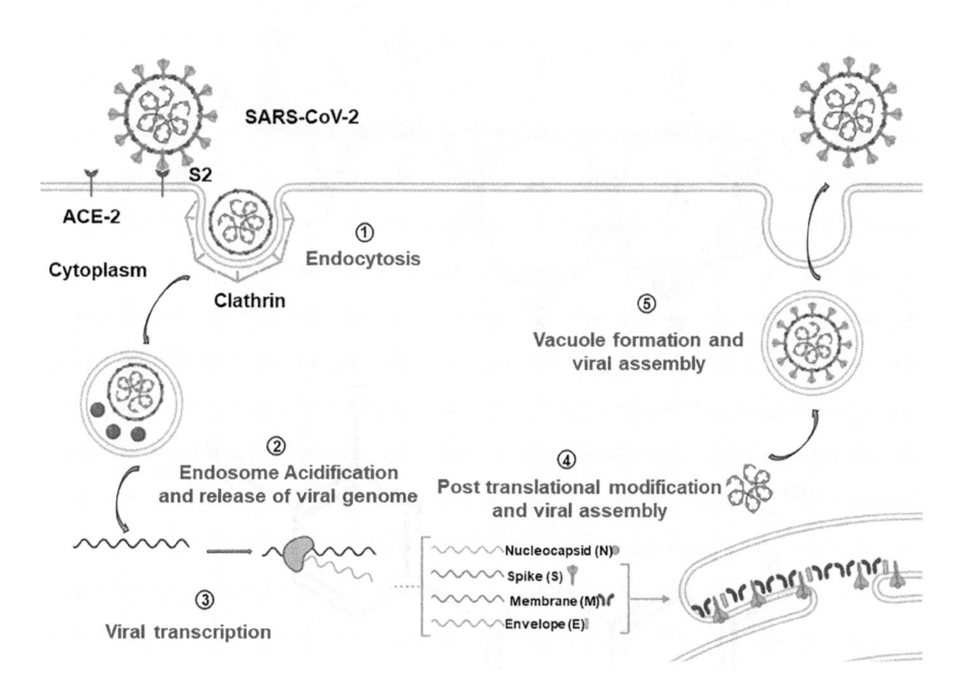

FIGURE 9.2 Molecular mechanisms of SARS-CoV-2 pathogenesis (created with https://biorender.com/).

of research, which raises the possibility of using them to treat COVID-19. We provide an overview of flavones, a prominent subclass of flavonoids, in this chapter. We examine their biological functions, structures, and the molecular mechanisms underlying their anti-SARS-CoV-2 activities. We also discuss how they might be used, singly or in combination, to treat COVID-19.

9.2 AN OVERVIEW OF FLAVONOIDS-BASIC STRUCTURE, BIOSYNTHESIS, AND FUNCTIONS

Back in 1930, a novel substance was isolated from oranges, initially termed vitamin P. Subsequently, it was established that this substance was a flavonoid (specifically rutin) [13]. With more than 9,000 identified structures, flavonoids are a class of naturally occurring compounds distinguished by their varied phenolic structures, frequently found in higher plants and make up a significant group of plant secondary metabolites [14]. From a chemical perspective, flavonoids are constructed around a 15-carbon framework, featuring two benzene rings (referred to as A and B in Figure 9.3a) connected by a heterocyclic pyrane ring (C). Flavonoids are classified into various subgroups, such as flavanes, flavanols, flavanones, flavanonols, flavones, and flavonols. Each of these subgroups shares a basic flavan (2-phenylchroman) structure [15] (Figure 9.3a). The pattern of substitutions on the C ring and the degree of oxidation distinguish these different classes of flavonoids from one another. Furthermore, different compounds in each class have different substitution patterns on the A and B rings [13]. Naturally, flavonoids can be found in a variety of forms, including methylated derivatives, glycosides, and aglycones. Flavonoids frequently undergo hydroxylation at positions on the three rings and show the presence of methyl ethers and acetyl esters in the hydroxyl group [11]. When glycosides are formed, they usually attach to a carbohydrate (L-rhamnose, D-glucose, galactose, or arabinose, etc.) through glycosidic bonds at position 3 or 7 [16]. Flavonoids biosynthetically originate from L-phenylalanine through a sequence of enzymatic reactions. In particular, the phenylalanine ammonia-lyase enzyme helps transform L-phenylalanine into phenyl-propenoyl-S-CoA [17]. The presence of reactive ketones and hydroxyl groups (within *o*-hydroxychalcones) enables their cyclization, giving rise to a variety of flavonoids [18]. A wide spectrum of biological functions are exhibited by flavonoids, including their roles as antioxidants, anticancer, antiviral, and anti-inflammatory activities [19]. Among different categories of flavonoids, flavones are potent antioxidants and have diverse biological activities.

9.3 FLAVONES—BASIC STRUCTURE AND FUNCTIONS

Flavones are distinct from other flavonoids due to a double bond between C2 and C3 in their structure, the absence of substitution at the C3 position, and oxidation at the C4 position [20] (Figure 9.3b). These molecules have several functions in plants. They serve as primary pigments in white and cream-coloured flowers and as copigments with

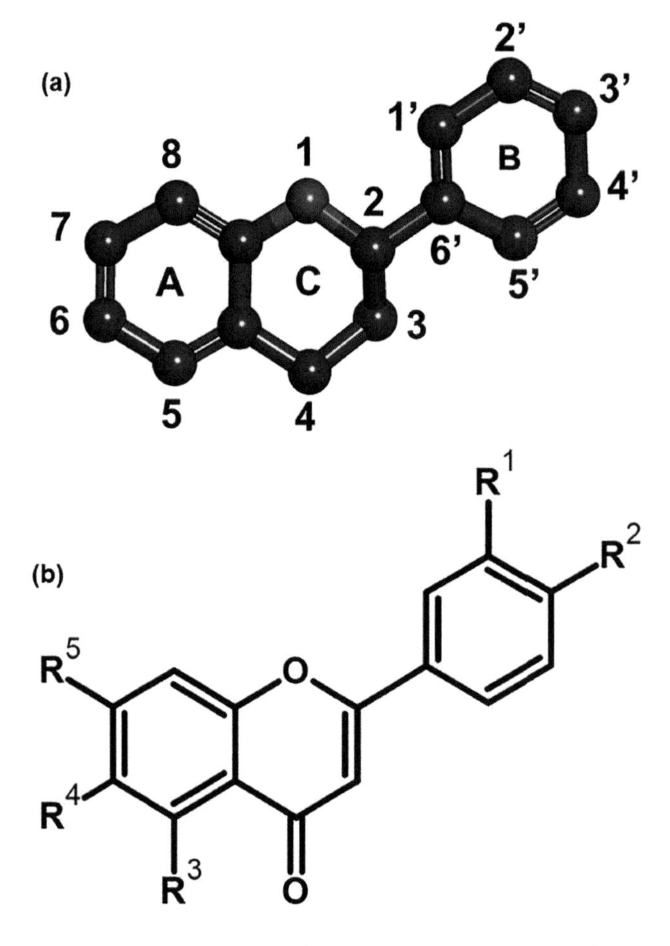

Flavone	R1	R2	R3	R4	R5
Amentoflavone	3,8-Biapigenin	-	-	-	-
Apigenin	H	OH	OH	H	OH
Luteolin	OH	OH	OH	H	OH
Pectolinarin	H	OCH_3	OH	OCH_3	6-O-(6-deoxy-α L mannopyranosyl) β-D glucopyranosyl)oxy
Rhoifolin	H	OH	OH	H	O-rhamnoglucoside
Scutellarein	H	OH	OH	OH	OH

FIGURE 9.3 Basic structure of (a) flavonoid and (b) flavones with examples.

anthocyanins in blue flowers [21]. Flavones can function as UVB protectants in plants [21]. Additionally, they act as natural pesticides against insects [22] and fungal diseases [23]. Furthermore, flavones play a role as signalling molecules in plants, encouraging the colonization of roots by N_2-fixing bacteria [24] and mycorrhizal fungi [25]. In plants, flavones are usually found as 7-O-glycosides [20]. C-glycosides of flavones are primarily identified as 6-C and 8-C-glucosides. Although flavones are synthesized via the phenylpropanoid pathway in plants, they can also be chemically synthesized in the laboratory through different chemical methods.

9.4 CHEMICAL METHODS FOR SYNTHESIS OF FLAVONES

9.4.1 Von Kostanecki Method

This method was developed by Stanislaus von Kostanecki around 1898–1899, which generates 2′-hydroxychalcone (or 2′-acetoxychalcone) as intermediates and the resulting chalcone is transformed into flavone by bromination and dehydrobromination reactions in an alkaline alcoholic solution [26] (Figure 9.4a).

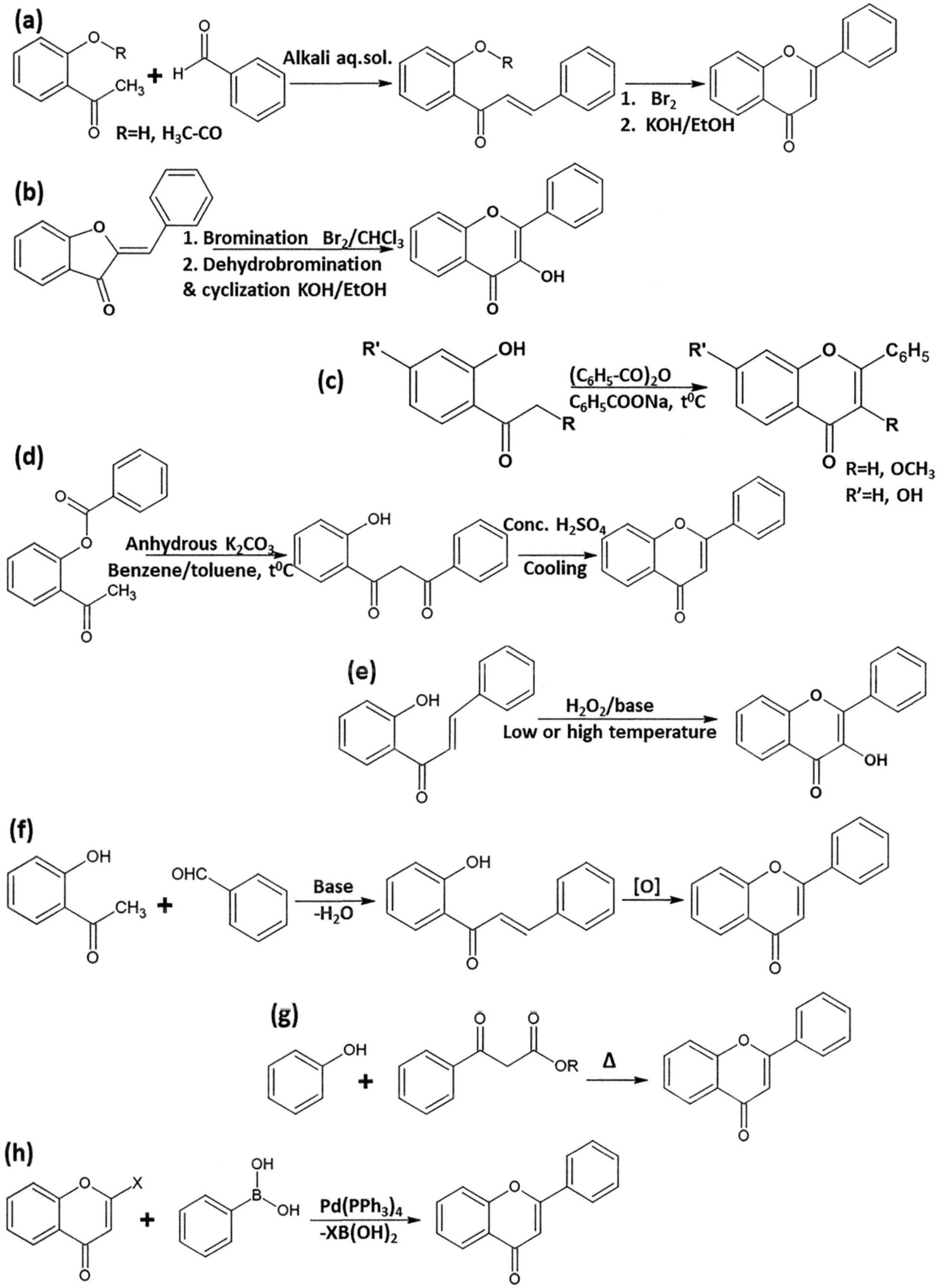

FIGURE 9.4 Chemical methods for synthesis of flavones: (a) Von Kostanecki method, (b) Von Auwers–Müller method, (c) Allan–Robinson method, (d) Baker–Venkataraman method, (e) Algar–Flynn–Oyamada method, (f) Claisen–Schmidt method, (g) Mentzer method, and (h) Suzuki–Miyaura method.

9.4.2 Von Auwers–Müller Method

In 1908, Karl von Auwers, working in collaboration with Müller, introduced a method for transforming aurones into 3-hydroxyflavones. After bromination and dehydrohalogenation, the intermediate compound undergoes ring rearrangement to generate 3-hydroxyflavones in the presence of potassium hydroxide in an ethanol solution [27,28] (Figure 9.4b).

9.4.3 Allan–Robinson Method

This method, which was initially developed in 1924, is a process for transforming 2-hydroxyacetophenones into flavones. This transformation is achieved by subjecting them to treatment with aromatic carboxylic acid anhydrides and the corresponding sodium salts of these acids, typically under heating conditions [29] (Figure 9.4c).

9.4.4 Baker–Venkataraman Method

This method, which was developed by Baker and Venkataraman in 1933, serves the purpose of extracting flavones from *o*-acyloxyacetophenones. The *o*-hydroxydibenzoylmethane derivatives generated as intermediate compounds are converted into flavones through cold-concentrated H_2SO_4 treatment [27,30] (Figure 9.4d).

9.4.5 Algar–Flynn–Oyamada Method

This method combines the efforts of Algar, Flynn, and Oyamada (1934–1935). It serves as a means to produce 3-hydroxyflavones from *o*-hydroxychalcones by employing H_2O_2 in an aqueous NaOH solution while maintaining a cool temperature [31]. Algar and Flynn adopted a similar technique using a hot alcoholic solution of potassium hydroxide, and both achieved favourable yields [32], as illustrated in Figure 9.4e.

9.4.6 Claisen–Schmidt Method

This method was developed in 1962 and comprises two steps: (i) a Claisen–Schmidt condensation occurring in a basic medium and (ii) the oxidative cyclization using a diverse range of conditions and catalysts [27] (Figure 9.4f).

9.4.7 Mentzer Method

This approach entails the production of flavones through a chemical reaction between phenolic compounds, specifically phenol, resorcinol, or phloroglucinol, and β-ketoesters as depicted in Figure 9.4g. Mentzer and coworkers successfully synthesized flavones by subjecting a mixture of resorcinol and ethyl 2-benzylacetoacetate to a 48-hour heating process at 250°C [33].

9.4.8 Suzuki–Miyaura Method

This method involves introducing palladium into sp2-hybridized C-X bonds and employing diverse organoboron precursors in gentle reaction conditions [34] (Figure 9.4h). While this method traditionally relies on 2-bromo- or 2-iodochromones, Kraus and Gupta observed that these precursors result in the formation of aurones rather than flavones. They successfully generated flavones (68%–74% yield) by using 2-chlorochromone instead [35].

9.5 BIOLOGICAL ACTIVITIES OF FLAVONES

9.5.1 Anticancer Activity

Flavones exhibit anti-tumour effects by different mechanisms such as interfering with tubulin polymerization or activating specific enzymes, such as caspases, which trigger apoptosis in tumour cells [17]. Apigenin and nobiletin, which are natural flavones, can control significant inflammatory signalling pathways [36]. Genkwanin, another natural flavone, possesses antioxidant properties and has displayed encouraging anticancer effects against several cancer cell lines, including human MCF-7 breast cancer, HepG-2 human hepatocellular carcinoma, and HCT-116 by modulating the phosphoinositide 3-kinase (PI3K) and phospho-protein kinase B (AKT) signalling pathways [37]. Flavopiridol, a semisynthetic flavone, is clinically used as an anticancer agent for treating acute myeloid leukaemia and inhibits cyclin-dependent kinases (CDKs) [38]. Recently, a set of newly synthesized C-dimethylated flavones was examined for their anticancer potential. Four of the studied flavones exhibited anticancer properties against a human adenocarcinoma cell line (A549). The *in silico* docking study uncovered that these compounds exhibited enhanced molecular interactions with the epidermal growth factor receptor (EGFR) [39].

9.5.2 Antibacterial and Antifungal Activities

The fundamental framework of natural flavones has served as a source of inspiration for scientists seeking to design novel antimicrobial substances that offer enhanced bioavailability as well as improved antibacterial and antifungal characteristics [17]. Mannich base derivatives of flavones were used to develop novel antibacterial flavone analogues against *Staphylococcus aureus*, *Listeria monocytogenes*, *Escherichia coli*, and *Salmonella gallinarum* [40]. Novel hydroxyflavone derivatives, featuring the addition of a dimethylamino group at the fourth position of the B ring, showed notably potent antifungal properties when tested against specific fungal strains [41]. Dimethylated and dimethoxylated flavones, containing fluoro and dimethylamino substitutions, exhibited noteworthy antibacterial activity against replicating *Mycobacterium tuberculosis*, particularly the H37Rv strain [39].

9.5.3 ANTIVIRAL ACTIVITY

Fujimoto and coworkers demonstrated that 6-F-tricin, a flavone derivative, showed significantly higher anti-cytomegalovirus (CMV) activity compared to ganciclovir and had no cytotoxicity in the embryonic lung cells [42]. Flavones have also shown promising results as antiviral agents against tropical diseases like Chikungunya fever [43]. The antiviral activities of flavones against SARS-CoV-2 can be attributed to the inhibition of molecular targets such as $3CL^{pro}$, helicase, S protein, non-specific inhibition, and immunomodulation, which are discussed further with suitable examples.

9.5.3.1 Flavones as SARS-CoV-2 $3CL^{pro}$ Inhibitors

Amentoflavone, a biflavone extracted from *Torreya nucifera*, showed higher inhibitory activity against SARS-CoV $3CL^{pro}$ (IC_{50}=8.3 μM) (Table 9.1) than its precursor compound apigenin (IC_{50}=280.8 μM) and other flavonoids like luteolin (IC_{50}=20 μM) and quercetin (IC_{50}=23.8 μM) [44]. *In silico* docking experiment reveals that apigenin's binding energy to SARS-CoV-2 $3CL^{pro}$ is –7.2 kcal/mol [45]. When luteolin and quercetin were compared to apigenin, compounds with a hydroxyl group substitution at position C-3 clearly showed more inhibitory activity, whereas amentoflavone derivatives with methylation at the 7-, 4'-, and 4'''-OH positions demonstrated reduced activity [44]. The *in silico* molecular docking experiment further confirms that amentoflavone shows a binding energy of –12.43 kcal/mol with SARS-CoV-2 $3CL^{pro}$ [46]. Pectolinarin and rhoifolin (belonging to the flavone group) showed inhibition against SARS-CoV $3CL^{pro}$ enzyme in biological assays and docking experiments (IC_{50} values of 37.78 μM and 27.45 μM, respectively) (Table 9.1). The high affinity of rhoifolin and pectolinarin for SARS-CoV $3CL^{pro}$ was attributed to their carbohydrate groups [47]. Tallei and coworkers used molecular docking experiments to show that pectolinarin and rhoifolin have an equivalent binding energy of –8.2 kcal/mol with SARS-CoV-2 $3CL^{pro}$ [48].

9.5.3.2 Flavones as an Inhibitor of SARS-CoV-2 Helicase

Scutellarein was one of the potent inhibitors of SARS-CoV nsp13 helicase identified amongst the screened natural compounds [49,50]. Scutellarein, a flavone, effectively inhibited the ATPase activity of helicase, with IC_{50} value of 0.86 μM (Table 9.1). This inhibition primarily affected the ATPase activity, but did not interfere with the dsRNA-unwinding activity of the SARS-CoV helicase [50]. Samdani and coworkers using molecular docking demonstrated the effective binding of scutellarein to SARS-CoV-2 nsp13 helicase (binding energy = –9.9 kcal/mol) [51]. Scutellarein is naturally abundant in *Scutellaria baicalensis*, an important traditional Chinese plant used for the treatment of various inflammatory and respiratory diseases [52].

9.5.3.3 Flavones as an Inhibitor of SARS-CoV-2 S Protein

In addition to its ability to inhibit the SARS-CoV $3CL^{pro}$ (IC_{50} value of 20 μM) (Table 9.1), luteolin (a flavone) was also observed to effectively hinder the SARS-CoV entry into host cells, which was determined using a pseudotyped virus-infection assay (EC_{50}=10.6 μM) [53]. Through the use of a competitive binding assay conducted in a laboratory setting, Zhu and coworkers discovered that luteolin effectively hindered the interaction between the S protein RBD and ACE2 (IC_{50} value of 0.61 mM). This finding was further validated through *in vivo* experiments involving the neutralization of SARS-CoV-2 pseudovirus infection [54].

9.5.3.4 Flavones in Non-Specific Inhibition of SARS-CoV-2

Baicalin (a flavone glycoside) exhibited antiviral activity against clinical isolates of SARS-CoV in a plaque reduction assay (at concentrations of 12.5–25 μg/mL after 48 hours) [55]. Baicalin functions as an immune modulator that stimulates the production of IFN-γ in T cells [56]. Some HIV

TABLE 9.1
Anti-Coronavirus Activity of Flavones

Flavone	IC_{50} (μM)			Reference
	$3CL^{pro}$	nsp13	S Protein	
Amentoflavone	8.3			[44,47]
Apigenin	280.8			[44,47]
Luteolin	20.0		10.6*	[44,47,53]
Pectolinarin	37.78			[47]
Rhoifolin	27.45			[47]
Scutellarin		0.86		[49,50]

The asterisk indicates EC_{50} value.

protease inhibitors like lopinavir, ritonavir, or nelfinavir have shown the ability to disrupt SARS-CoV replication [57]. It is worth exploring the potential of flavones such as baicalein known for their anti-HIV properties [52,58] in the context of combating SARS-CoV-2. Consequently, it might be regarded as a promising candidate for drug development targeting SARS-CoV-2.

9.5.4 Immunomodulatory Activities of Flavones

A thorough analysis of the anti-inflammatory and immunomodulatory properties of chrysin, a naturally occurring flavone, indicates that it functions via a variety of pathways. Chrysin inhibits NF-κB, a critical regulator of genes that code for pro-inflammatory cytokines, and also functions as an agonist for peroxisome proliferator-activated receptor gamma PPAR-γ [59]. Pre-treatment with chrysin effectively alleviated inflammation by reducing the release of tumour necrosis factor-alpha (TNF-α), interleukin-1β (IL-1β), interleukin-8 (IL-8), and MPO expression in lung tissue of mice exposed to cigarette smoke [60]. Similarly, pre-treating inflamed human macrophages with apigenin led to significant inhibition of interleukin-6 (IL-6) secretion and inhibition of inflammatory chemokines and adhesion molecules [61]. In a mouse model of lipopolysaccharide (LPS)-induced ALI, luteolin exhibited a protective effect by inhibiting the MAPK and NF-κB pathways [62]. Caflanone exhibits anti-inflammatory activity by inhibiting 5-lipoxygenase and prostaglandin E synthase 1 [63].

9.6 CONCLUSION AND FUTURE PERSPECTIVES

Although flavones have shown encouraging results against SARS- and MERS-CoV in controlled laboratory settings, it is still unclear how well these compounds will work *in vivo* and in clinical settings to treat SARS-CoV-2 infections [15]. A significant challenge arises from the propensity of polyphenolic compounds to aggregate, which affects the reliability of *in vitro* test results. Because phenolic compounds bind to proteins non-specifically, these aggregates may cause denaturation and sequestering of enzymes, which could result in false-positive results [64,65]. However, problems with complexation and aggregate formation can be lessened in proteolytic assays by using the solubilizing agent Triton X-100 [47,64]. It is crucial to understand that *in vitro* molecular and cell-based assays cannot reveal information about these compounds' bioavailability, which is crucial for their therapeutic application [15]. Resolving bioavailability constraints is essential for the therapeutic development of flavones, which can be improved by co-crystallization, nanotechnology, and glycosylation [66]. Combining flavones with other approved antiviral drugs can be a novel strategy to prevent SARS-CoV-2 infections [67]. For the treatment of COVID-19, it is imperative to develop a robust, quick, and affordable drug screening system. Furthermore, the potential therapeutic application of natural products like flavones must take into account the growing problem of drug resistance in antiviral treatments. To ensure that flavones remain relevant as possible antiviral agents, it is imperative to develop efficient strategies to address this problem.

REFERENCES

[1] Ge, X.-Y.; Li, J.-L.; Yang, X.-L.; Chmura, A. A.; Zhu, G.; Epstein, J. H.; Mazet, J. K.; Hu, B.; Zhang, W.; Peng, C. J. N. Isolation and characterization of a bat SARS-like coronavirus that uses the ACE2 receptor. *Nature*, **2013**, *503*(7477), 535–538

[2] Boopathi, S.; Poma, A. B.; Kolandaivel, P. Novel 2019 coronavirus structure, mechanism of action, antiviral drug promises and rule out against its treatment. *J Biomol Struct Dyn*, **2021**, *39*(9), 3409–3418.

[3] Mahmoud, I. S.; Jarrar, Y. B.; Alshaer, W.; Ismail, S. J. B. SARS-CoV-2 entry in host cells-multiple targets for treatment and prevention. *Biochimie*, **2020**, *175*, 93–98.

[4] Dai, W.; Zhang, B.; Jiang, X.-M.; Su, H.; Li, J.; Zhao, Y.; Xie, X.; Jin, Z.; Peng, J.; Liu, F. J. S. Structure-based design of antiviral drug candidates targeting the SARS-CoV-2 main protease. *Science*, **2020**, *368*(6497), 1331–1335.

[5] Kirchdoerfer, R. N.; Cottrell, C. A.; Wang, N.; Pallesen, J.; Yassine, H. M.; Turner, H. L.; Corbett, K. S.; Graham, B. S.; McLellan, J. S.; Ward, A. B. J. N. Pre-fusion structure of a human coronavirus spike protein. *Nature*, **2016**, *531*(7592), 118–121.

[6] Wang, Y.; Zhang, D.; Du, G.; Du, R.; Zhao, J.; Jin, Y.; Fu, S.; Gao, L.; Cheng, Z.; Lu, Q. J. T. L. Remdesivir in adults with severe COVID-19: A randomised, double-blind, placebo-controlled, multicentre trial. *Randomized Controlled Trial*, **2020**, *395*(10236), 1569–1578.

[7] Beigel, J. H.; Tomashek, K. M.; Dodd, L. E.; Mehta, A. K.; Zingman, B. S.; Kalil, A. C.; Hohmann, E.; Chu, H. Y.; Luetkemeyer, A.; Kline, S. J. N. Remdesivir for the treatment of Covid-19. *N Engl J Med*, **2020**, *383*(19), 1813–1826.

[8] Zhang, X.; Huang, H.; Zhao, X.; Lv, Q.; Sun, C.; Li, X.; Chen, K. J. J. Effects of flavonoids-rich Chinese bayberry (Myrica rubra Sieb. et Zucc.) pulp extracts on glucose consumption in human HepG2 cells. *Food Funct*, **2015**, *14*, 144–153.

[9] Russo, M.; Moccia, S.; Spagnuolo, C.; Tedesco, I.; Russo, G. L. J. C. Roles of flavonoids against coronavirus infection. *Chem Biol Interact*, **2020**, *328*, 109211.

[10] Kaul, R.; Paul, P.; Kumar, S.; Büsselberg, D.; Dwivedi, V. D.; Chaari, A. J. I. Promising antiviral activities of natural flavonoids against SARS-CoV-2 targets: Systematic review. *Int J Mol Sci*, **2021**, *22*(20), 11069.

[11] Kumar, S.; Pandey, A. K. J. T. Chemistry and biological activities of flavonoids: An overview. *Sci World J*, **2013**, *2013*, 162750.

[12] Muchtaridi, M.; Fauzi, M.; Khairul Ikram, N. K.; Mohd Gazzali, A.; Wahab, H. A. J. M. Natural flavonoids as potential angiotensin-converting enzyme 2 inhibitors for anti-SARS-CoV-2. *Molecules*, **2020**, *25*(17), 3980.

[13] Middleton Jr, E. J. F. Effect of plant flavonoids on immune and inflammatory cell function. *Adv Exp Med Biol*, **1998**, *439*, 175–182.

[14] Wang, T.-Y.; Li, Q.; Bi, K.-S. J. A. Bioactive flavonoids in medicinal plants: Structure, activity and biological fate. *Asian J Pharm Sci*, **2018**, *13*(1), 12–23.

[15] Solnier, J.; Fladerer, J.-P. J. P. R. Flavonoids: A complementary approach to conventional therapy of COVID-19?. *Phytochem Rev*, **2021**, *20*(4), 773–795.

[16] Yao, L. H.; Jiang, Y.-M.; Shi, J.; Tomas-Barberan, F.; Datta, N.; Singanusong, R.; Chen, S. J. P. Flavonoids in food and their health benefits. *Plant Foods Hum Nutr*, **2004**, *59*, 113–122.

[17] Leonte, D.; Ungureanu, D.; Zaharia, V. J. M. Flavones and related compounds: Synthesis and biological activity. *Molecules*, **2023**, *28*(18), 6528.

[18] Berim, A.; Gang, D. R. J. P. R. Methoxylated flavones: Occurrence, importance, biosynthesis. *Phytochem Rev*, **2016**, *15*, 363–390.

[19] Panche, A. N.; Diwan, A. D.; Chandra, S. R. J. J. Flavonoids: An overview. *J Nutr Sci*, **2016**, *5*, e47.

[20] Martens, S.; Mithöfer, A. J. P. Flavones and flavone synthases. *Plants (Basel)*, **2005**, *66*(20), 2399–2407.

[21] Harborne, J. B.; Williams, C. A. J. P. Advances in flavonoid research since 1992. *Phytochemistry*, **2000**, *55*(6), 481–504.

[22] Harborne, J. B.; Grayer, R. J. Flavonoids and insects. In *The Flavonoids Advances in Research Since 1986*, London: Routledge, **2017**; pp. 589–618.

[23] Lattanzio, V.; Di Venere, D.; Lima, G.; Salerno, M. J. I. J. F. S. Antifungal activity of phenolics against fungi commonly encoijntered dljring storage. *Postharvest Biol Technol*, **1994**, *23*, 325–334.

[24] Rolfe, B. G. J. B. Flavones and isoflavones as inducing substances of legume nodulation. *Biofactors*, **1988**, *1*(1), 3–10.

[25] Siqueira, J.; Safir, G.; Nair, M. J. N. P. Stimulation of vesicular-arbuscular mycorrhiza formation and growth of white clover by flavonoid compounds. *New Phytol*, **1991**, *118*(1), 87–93.

[26] Zemplén, G.; Bognár, R. J. B. Umwandlung des Hesperetins in Diosmetin, des Hesperidins in Diosmin und des Isosakuranetins in Acacetin. *Berichte Dtsch Chem*, **1943**, *76*(5), 452–457.

[27] Kshatriya, R.; Jejurkar, V. P.; Saha, S. J. T. In memory of Prof. Venkataraman: Recent advances in the synthetic methodologies of flavones. *Tetrahedron*, **2018**, *74*(8), 811–833.

[28] Auwers, K. V.; Müller, K. J. B. Umwandlung von Benzalcumaranonen in Flavonole. *EurJIC*, **1908**, *41*(3), 4233–4241.

[29] Allan, J.; Robinson, R. J. J. CCXC.-An accessible derivative of chromonol. *J Chem Soc Trans*, **1924**, *125*, 2192–2195.

[30] Baker, W. J. J. 322 Molecular rearrangement of some o-acyloxyacetophenones and the mechanism of the production of 3-acylchromones. *J Chem Soc*, **1933**, *1933*, 1381–1389.

[31] Oyamada, T. J. B. A new general method for the synthesis of the derivatives of flavonol. *BCSJ*, **1935**, *10*(5), 182–186.

[32] Algar, J.; Flynn, J. P. A new method for the synthesis of flavonols. In *Proceedings of the Royal Irish Academy. Section B: Biological, Geological, and Chemical Science*, Ireland: Royal Irish Academy, **1934**; Vol. 42, pp. 1–8.

[33] Mentzer, C.; Molho, D.; Vercier, P. Sur un nouveau mode de condensation desters beta-cetoniques et de phenols en chromones. *Compt Rend*, **1951**, *232*(16), 1488–1490.

[34] Selepe, M. A.; Van Heerden, F. R. J. M. Application of the Suzuki-Miyaura reaction in the synthesis of flavonoids. *Molecules*, **2013**, *18*(4), 4739–4765.

[35] Kraus, G. A.; Gupta, V. J. O. l. Divergent approach to flavones and aurones via dihaloacrylic acids. Unexpected dependence on the halogen atom. *Org Lett*, **2010**, *12*(22), 5278–5280.

[36] Kariagina, A.; Doseff, A. I. J. I. J. Anti-inflammatory mechanisms of dietary flavones: Tapping into nature to control chronic inflammation in obesity and cancer. *Int J Mol Sci*, **2022**, *23*(24), 15753.

[37] El Menyiy, N.; Aboulaghras, S.; Bakrim, S.; Moubachir, R.; Taha, D.; Khalid, A.; Abdalla, A. N.; Algarni, A. S.; Hermansyah, A.; Ming, L. C. J. B.; et al. Genkwanin: An emerging natural compound with multifaceted pharmacological effects. *Biomed Pharmacother*, **2023**, *165*, 115159.

[38] Sedlacek, H.; Czech, J.; Naik, R.; Kaur, G.; Worland, P.; Losiewicz, M.; Parker, B.; Carlson, B.; Smith, A.; Senderowicz, A. J. I. Flavopiridol (L86 8275; NSC 649890), a new kinase inhibitor for tumor therapy. *Int J Oncol*, **1996**, *9*(6), 1143–1168.

[39] Bollikolla, H. B.; Anandam, R.; Chinnam, S.; Varala, R.; Khandapu, B. M. K.; Kapavarapu, R.; Syed, K. S.; Dubasi, N.; Syed, M. A. J. C.; Biodiversity. C-dimethylated flavones as possible potential anti-tubercular and anticancer agents. *Chem Biodivers*, **2023**, *20*(4), e202201201.

[40] Lv, X.-H.; Liu, H.; Ren, Z.-L.; Wang, W.; Tang, F.; Cao, H.-Q. J. M. D. Design, synthesis and biological evaluation of novel flavone Mannich base derivatives as potential antibacterial agents. *Mol Divers*, **2019**, *23*, 299–306.

[41] Khdera, H. A.; Saad, S. Y.; Moustapha, A.; Kandil, F. J. H. Synthesis of new flavonoid derivatives based on 3-hydroxy-4′-dimethylamino flavone and study the activity of some of them as antifungal. *Heliyon*, **2022**, *8*(12), e12062.

[42] Fujimoto, K. J.; Nema, D.; Ninomiya, M.; Koketsu, M.; Sadanari, H.; Takemoto, M.; Daikoku, T.; Murayama, T. J. A. R. An in silico-designed flavone derivative, 6-fluoro-4′-hydroxy-3′, 5′-dimetoxyflavone, has a greater anti-human cytomegalovirus effect than ganciclovir in infected cells. *Antiviral Res*, **2018**, *154*, 10–16.

[43] Badavath, V. N.; Jadav, S. S.; Pastorino, B.; de Lamballerie, X.; Sinha, B. N.; Jayaprakash, V. J. L. Synthesis and antiviral activity of 2-aryl-4H-chromen-4-one derivatives against chikungunya virus. *Lett Drug Des Discov*, **2016**, *13*, 1–6.

[44] Ryu, Y. B.; Jeong, H. J.; Kim, J. H.; Kim, Y. M.; Park, J.-Y.; Kim, D.; Naguyen, T. T. H.; Park, S.-J.; Chang, J. S.; Park, K. H. J. B.; et al. Biflavonoids from Torreya nucifera displaying SARS-CoV 3CLpro inhibition. *Bioorg Med Chem*, **2010**, *18*(22), 7940–7947.

[45] Farhat, A.; Ben Hlima, H.; Khemakhem, B.; Ben Halima, Y.; Michaud, P.; Abdelkafi, S.; Fendri, I. J. B. Apigenin analogues as SARS-CoV-2 main protease inhibitors: In-silico screening approach. *Bioengineered*, **2022**, *13*(2), 3350–3361.

[46] Hadni, H.; Fitri, A.; Benjelloun, A. T.; Benzakour, M.; Mcharfi, M. J. J. Evaluation of flavonoids as potential inhibitors of the SARS-CoV-2 main protease and spike RBD: Molecular docking, ADMET evaluation and molecular dynamics simulations. *J Indian Chem Soc*, **2022**, *99*(10), 100697.

[47] Jo, S.; Kim, S.; Shin, D. H.; Kim, M.-S. J. J. Inhibition of SARS-CoV 3CL protease by flavonoids. *J Enzyme Inhib Med Chem*, **2020**, *35*(1), 145–151.

[48] Tallei, T. E.; Tumilaar, S. G.; Niode, N. J.; Kepel, B. J.; Idroes, R.; Effendi, Y.; Sakib, S. A.; Emran, T. B. J. S. Potential of plant bioactive compounds as SARS-CoV-2 main protease (M pro) and spike (S) glycoprotein inhibitors: A molecular docking study. *Scientifica (Cairo)*, **2020**, *2020*, 6307457.

[49] Keum, Y.-S.; Jeong, Y.-J. J. B. Development of chemical inhibitors of the SARS coronavirus: viral helicase as a potential target. *Biochem Pharmacol*, **2012**, *84*(10), 1351–1358.

[50] Yu, M.-S.; Lee, J.; Lee, J. M.; Kim, Y.; Chin, Y.-W.; Jee, J.-G.; Keum, Y.-S.; Jeong, Y.-J. J. B. Identification of myricetin and scutellarein as novel chemical inhibitors of the SARS coronavirus helicase. *Bioorg Med Chem Lett*, **2012**, *22*(12), 4049–4054.

[51] Samdani, M. N.; Morshed, N.; Reza, R.; Asaduzzaman, M.; Islam, A. B. M. M. K. J. M. D. Targeting SARS-CoV-2 non-structural protein 13 via helicase-inhibitor-repurposing and non-structural protein 16 through pharmacophore-based screening. *Mol Divers*, **2023**, *27*(3), 1067–1085.

[52] Zhao, Q.; Chen, X.-Y.; Martin, C. J. S. B. Scutellaria baicalensis, the golden herb from the garden of Chinese medicinal plants. *Sci Bull (Beijing)*, **2016**, *61*(18), 1391–1398.

[53] Yi, L.; Li, Z.; Yuan, K.; Qu, X.; Chen, J.; Wang, G.; Zhang, H.; Luo, H.; Zhu, L.; Jiang, P. J. J. Small molecules blocking the entry of severe acute respiratory syndrome coronavirus into host cells. *J Virol*, **2004**, *78*(20), 11334–11339.

[54] Zhu, J.; Yan, H.; Shi, M.; Zhang, M.; Lu, J.; Wang, J.; Chen, L.; Wang, Y.; Li, L.; Miao, L. J. P. R. Luteolin inhibits spike protein of severe acute respiratory syndrome coronavirus-2 (SARS-CoV-2) binding to angiotensin-converting enzyme 2. *Phytother Res*, **2023**, *37*(8), 3508–3521.

[55] Chen, F.; Chan, K.; Jiang, Y.; Kao, R.; Lu, H.; Fan, K.; Cheng, V.; Tsui, W.; Hung, I.; Lee, T. J. J. In vitro susceptibility of 10 clinical isolates of SARS coronavirus to selected antiviral compounds. *J Clin Virol*, **2004**, *31*(1), 69–75.

[56] Chu, M.; Xu, L.; Zhang, M.-b.; Chu, Z.-Y.; Wang, Y.-D. J. B. Role of Baicalin in anti-influenza virus A as a potent inducer of IFN-gamma. *Biomed Res Int*, **2015**, *2015*, 263630.

[57] Zhang, X. W.; Yap, Y. L. J. B. Old drugs as lead compounds for a new disease? Binding analysis of SARS coronavirus main proteinase with HIV, psychotic and parasite drugs. *Bioorg Med Chem*, **2004**, *12*(10), 2517–2521.

[58] Hu, J. Z.; Bai, L.; Chen, D.-G.; Xu, Q.-T.; Southerland, W. M. J. I. S. C. L. S. Computational investigation of the anti-HIV activity of Chinese medicinal formula three-huang powder. *Interdiscip Sci*, **2010**, *2*, 151–156.

[59] Zeinali, M.; Rezaee, S. A.; Hosseinzadeh, H. J. B. An overview on immunoregulatory and anti-inflammatory properties of chrysin and flavonoids substances. *Pharmacother*, **2017**, *92*, 998–1009.

[60] Shen, Y.; Tian, P.; Li, D.; Wu, Y.; Wan, C.; Yang, T.; Chen, L.; Wang, T.; Wen, F. J. I. Chrysin suppresses cigarette smoke-induced airway inflammation in mice. *Int J Clin Exp Med*, **2015**, *8*(2), 2001.

[61] Zhang, X.; Wang, G.; Gurley, E. C.; Zhou, H. J. P. O. Flavonoid apigenin inhibits lipopolysaccharide-induced inflammatory response through multiple mechanisms in macrophages. *PLoS One*, **2014**, *9*(9), e107072.

[62] Kuo, M.-Y.; Liao, M.-F.; Chen, F.-L.; Li, Y.-C.; Yang, M.-L.; Lin, R.-H.; Kuan, Y.-H. J. F.; Toxicology, C. Luteolin attenuates the pulmonary inflammatory response involves abilities of antioxidation and inhibition of MAPK and NFκB pathways in mice with endotoxin-induced acute lung injury. *Food Chem Toxicol*, **2011**, *49*(10), 2660–2666.

[63] Erridge, S.; Mangal, N.; Salazar, O.; Pacchetti, B.; Sodergren, M. H. J. F. Cannflavins-from plant to patient: A scoping review. *Fitoterapia*, **2020**, *146*, 104712.

[64] Pohjala, L.; Tammela, P. J. M. Aggregating behavior of phenolic compounds-a source of false bioassay results? *Molecules*, **2012**, *17*(9), 10774–10790.

[65] Coan, K. E.; Maltby, D. A.; Burlingame, A. L.; Shoichet, B. K. J. J. Promiscuous aggregate-based inhibitors promote enzyme unfolding. *J Med Chem*, **2009**, *52*(7), 2067–2075.

[66] Zhao, J.; Yang, J.; Xie, Y. J. I. Improvement strategies for the oral bioavailability of poorly water-soluble flavonoids: An overview. *Int J Pharm*, **2019**, *570*, 118642.

[67] Shohan, M.; Nashibi, R.; Mahmoudian-Sani, M.-R.; Abolnezhadian, F.; Ghafourian, M.; Alavi, S. M.; Sharhani, A.; Khodadadi, A. J. E. The therapeutic efficacy of quercetin in combination with antiviral drugs in hospitalized COVID-19 patients: A randomized controlled trial. *Eur J Pharmacol*, **2022**, *914*, 174615.

10 Anti-SARS-CoV-2 Activity of Chalcones and Their Synthetic Derivatives

Salar Hafez Ghoran, Fatemeh Taktaz, Pouya Alipour, Farah Khameis Farag Teia, Pascal D. Douanla, and Amna Hamad Abdallah Atia

ABBREVIATIONS

ACE-2	Angiotensin-converting enzyme 2
3CLpro	Chymotrypsin-like protease
CHR	Chalcone reductase
CHS	Chalcone synthase
COVID-19	Coronavirus disease
COX	Cyclooxygenase
DENV	Dengue virus
DUBs	Deubiquitinating enzyme
EGCG	Epigallocatechin-3-gallate
EtOAc	Ethyl acetate
FQs	Fluoroquinolones
GPx	Glutathione peroxidase
HBV	Hepatitis B virus
HCMV	Human cytomegalovirus
HCV	Hepatitis C virus
HCQ	Hydroxychloroquine
HIV	Human immunodeficiency virus
HLT	Human Lung Tissue
HSV	Herpes simplex virus
HTS	High-Throughput Screening
IFN-γ	Interferon γ
IL-1β	Interleukin 1β
IL-6	Interleukin 6
IN	Integrase
iNOS	Inducible NO synthase
IRF-3	Interferon regulatory factor-3
LOX	Lipoxygenase
LPS	Lipopolysaccharide
Malonyl CoA	Malonyl-coenzyme A
MERS-CoV	Middle East Respiratory Syndrome Coronavirus
Mpro	SARS-CoV-2 main protease
NF-κB	Nuclear factor kappa B
NO	Nitric oxide
Nrf2	Nuclear factor erythroid-derived 2-related factor
PAC	Proanthocyanidin
PGE2	Prostaglandin E2
PGG	1,2,3,4,6-*O*-pentagalloylglucose
Pro	Protease
RVFV	Rift Valley fever virus
RT	Reverse transcriptase
RdRp	RNA-dependent RNA polymerase
RBD	Receptor-binding domain
ROS	Reactive oxygen species
SAR	Structure–activity relationship
SARS-CoV	Severe Acute Respiratory Syndrome Coronavirus
SARS-CoV-2	Severe Acute Respiratory Syndrome Coronavirus Type 2
SOD	Superoxide dismutase
TF3	Theaflavin-3,3′-digallate
TNF-α	Tumor necrosis factor-alpha
TxB2	Thromboxane B2
UPS	Ubiquitin–proteasome system
VEEV	Venezuelan equines encephalitis virus
WHO	World Health Organization

10.1 PHENOLIC COMPOUNDS: POTENTIAL ANTIVIRALS AGAINST SARS-COV-2 AND COVID-19

Coronavirus disease, officially COVID-19, is an extremely contagious respiratory disease resulting from the Severe Acute Respiratory Syndrome Coronavirus type 2 (SARS-CoV-2). This virus emerged in Wuhan, China, in December 2019, and swiftly became a worldwide pandemic. COVID-19 has left a profound mark on the global stage, causing widespread ailments, fatalities, and economic upheaval. On March 11, 2020, the World Health Organization (WHO) formally categorized COVID-19 as a pandemic, and it remains a substantial and pressing public health issue [1].

Nutraceuticals are defined as foods or food components that provide health benefits beyond their nutritional content. The nutraceuticals' effects against COVID-19 have been the subject of much research and debate. Several studies have evaluated the potential of various nutraceuticals for either preventing or treating COVID-19. Shakoor and his coworkers suggested that specific nutraceuticals containing minerals (i.e., Zinc and Selenium) and vitamins (i.e., C, D, and E), as well as omega-3 and omega-6 fatty acids, might have the potential to prevent and treat COVID-19. These nutrients were found to have antiviral and immunomodulatory effects, which could help to reduce the severity and duration of the illness [2]. In another study, the combination of nutraceuticals, including vitamins C and D, Zinc, and melatonin, showed a synergistic effect against COVID-19.

DOI: 10.1201/9781003433200-10

The study suggested that these nutrients could help to enhance immune function and reduce inflammation, which could improve outcomes for COVID-19 patients [3].

Moving forward, polyphenols are naturally occurring compounds found in plants and have been studied for their health benefits for centuries [4,5]. Recently, research has shown that phenolic compounds have antiviral properties and can be used against various viral infections, including COVID-19. Moreover, these compounds not only bind to the viral envelope or viral proteins, preventing the virus from attaching to host cells and entering them, but also can interfere with viral replication, thereby reducing the viral load in the body [6]. Mehany et al. (2021), in their review, suggested that polyphenols could be antiviral compounds since they regulate the body's immunity. Polyphenolic-based drugs, along with triterpenoids, anthraquinones, tannins, and flavonoids, could potentially inhibit SARS-CoV-2-related enzymes, which are necessary for viral replication and further infection. Therefore, these secondary metabolites can be utilized in order to design anti-SARS-CoV-2 therapeutics [7,8]. Studies have shown that phenolic compounds can inhibit SARS-CoV-2 replication. For instance, Mhatre et al. (2021) reported that the phenolic compounds found in green tea (epigallocatechin-3-gallate, EGCG) and theaflavins in black teak (theaflavin-3,3′-digallate; TF3) showed inhibitory activity against different types of SARS-CoV-2 proteins and enzymes, including angiotensin-converting enzyme 2 (ACE2) receptor, spike receptor-binding domain (RBD), chymotrypsin-like protease ($3CL^{pro}$), and RNA-dependent RNA polymerase (RdRp), which are essential for the virus replication process [9]. Another *in vitro* study in 2021 showed that resveratrol, a phenolic compound found in grapes and red wine, potentially reduces the viral load in COVID-19 patients with low cytotoxicity and high selectivity index (SI) and inhibits HCoV-229E and SARS-CoV-2 replication [10]. Jin et al. (2022) examined the inhibitory effects of 1,2,3,4,6-*O*-pentagalloylglucose (PGG) together with proanthocyanidin (PAC) on SARS-CoV-2 infection and explored their mechanisms. The authors found that both PGG and PAC dose-dependently suppressed the infection of SARS-CoV-2 in Vero cells, thereby making PGG more effective than PAC. Meanwhile, PGG and PAC showed inhibitory activity against SARS-CoV-2 main protease (M^{pro}) and RdRp [11].

In addition, phenolic compounds are able to reduce oxidative stress, alleviate inflammation, and modulate the immunity of the human body, all of which are necessary for COVID-19 treatment. As reported, COVID-19 causes inflammation in the body, particularly in the lungs, and in severe cases, it can bring about a cytokine storm, resulting in severe lung damage and other complications. These natural metabolites can reduce inflammation by inhibiting the production of certain pro-inflammatory cytokines like interleukins (i.e., IL-1β and IL-6) and tumor necrosis factor-alpha (TNF-α) [12]. A review by Giovinazzo et al. (2020) represented the effects of polyphenols against inflammation and SARS-CoV-2 infection. Despite inhibiting the production of pro-inflammatory cytokines, the authors found that polyphenols can enhance immunity against viral infection by increasing the interferon's production. *Interferons* are proteins that provide the body's immunity when a viral infection happens [13]. The general function of polyphenols as antiviral and anti-inflammatory agents has been mentioned in Figure 10.1.

Taking together, phenolic compounds are promising antiviral agents against COVID-19, and incorporating food rich in phenolic compounds into one's diet provides a natural and effective way to fight against the virus since they are able to reduce inflammation and inhibit viral replication. Therefore, using the mixtures containing polyphenols is considered safe, with minimal side effects; however, further randomized controlled trials are necessary to confirm their efficiency in preventing COVID-19. In this chapter, the authors have tried to share the current knowledge on chalcones and their synthetic derivatives as an important subclass of flavonoids against SARS-CoV-2 activity. Furthermore, the mechanism of action and their efficiency will be discussed, where available.

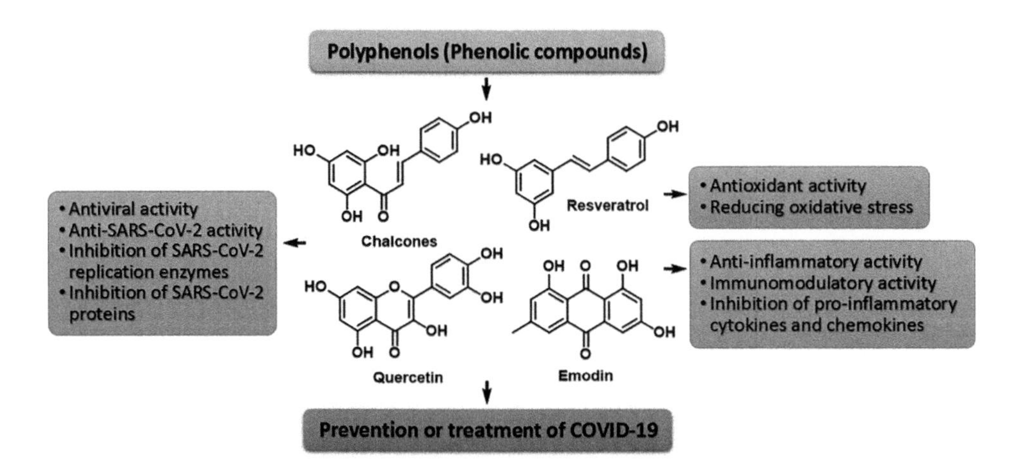

FIGURE 10.1 The significant function of polyphenols as antiviral and anti-inflammatory agents.

10.2 CHALCONES AND THEIR RESOURCES

Chalcones (benzalacetophenones or 1,3-diaryl-2-propen-1-ones; Figure 10.2) are a class of plant-derived polyphenolic metabolites structurally related to the flavonoids, of which an α,β-unsaturated chain connects two phenyl rings. As chalcones are readily converted to the corresponding flavanones by the action of chalcone isomerase, they are called flavonoid precursors [14].

The most significant number of naturally occurring chalones have been derived from the Asteraceae, Leguminosae/Fabaceae, and Moraceae families. The therapeutic potential of chalcone-based compounds, such as antimicrobial, antifungal, antioxidant, anti-inflammatory, cytotoxic, anti-tumor, and chemopreventive properties, has called for researchers to embark on the synthesis of these compounds, which are an important class of natural products. Hence, several synthetic methods have been developed to synthesize chalcones and even modify the structural core. These include chalcones substituted with hydroxy and/or methoxy groups, as well as methylated, prenylated, geranylated, and other monomeric and dimeric derivatives. Karanjapin, derived from *Pongamia pinnata* Pierre (Leguminosae/Fabaceae) root bark, paratocarpin B, obtained from *Glycyrrhiza glabra* L. (Leguminosae/Fabaceae) roots, and 2-geranyl-2′,4′,3,4-tetrahydroxy-dihydrochalcone, isolated from *Artocarpus altilis* (Moraceae) leaves, are some of the examples of bioactive chalcones that have encouraged scientists to synthesize chalcone-based compounds (Figure 10.2) [14].

10.2.1 Biosynthesis of Chalcones

Chalcones are the obligatory intermediates in the biosynthesis process of flavonoids. A ring formation of chalcones involves the utilization of three malonyl-coenzyme A (malonyl CoA) molecules, which are derived from glucose metabolism. The B ring also originates from glucose metabolism, but via the shikimate pathway, which converts phenylalanine to *p*-coumarate and then to *p*-coumaroyl-CoA (Figure 10.3). Subsequently, the A and B rings conjoin through a single enzymatic step, leading to the formation of a chalcone molecule, namely, 4,2′,4′,6′-tetrahydroxychalcone. Chalcone synthase facilitates the reaction [15]. In the biosynthesis process, there are two primary categories of chalcones characterized by the presence or absence of a hydroxyl group at the C-6′ position. While only the gene encoding chalcone synthase (CHS, EC 2.3.1.74) is expressed, the production of 6′-hydroxychalcone (naringenin-chalcone **I**) occurs. However, 6′-deoxychalcones are produced by the simultaneous activation of the gene encoding chalcone reductase (CHR, EC 2.3.1.170) (Figure 10.3). It is believed that CHR functions by reducing the enzyme-bound polyketide intermediate, ultimately resulting in the production of 4,2′,4′-trihydroxychalcone (isoliquiritigenin **II**) [16].

10.3 ANTIVIRAL ACTIVITY OF CHALCONES

Plants biosynthesize chalcones on a large scale, and these natural products have a notable function in protecting the human body from harmful microorganisms. Despite various biological and pharmacological properties of natural and synthetic chalcones, these open-chain flavonoids have also displayed promising antiviral activities against an array of viral species, such as MERS-CoV, SARS-CoV, dengue virus (DEN), human cytomegalovirus (HCMV), human immunodeficiency virus (HIV), hepatitis B and C viruses (HBV and HCV), herpes simplex virus (HSV), human rhinovirus, influenza virus, Venezuelan equines encephalitis virus (VEEV), and Rift Valley fever (RVF) virus. Most importantly, chalcones effectively target virus enzymes, which are involved in the viral replication cycle, including reverse transcriptase (RT), integrase (IN), protease (Pro), neuraminidase, aminotransferases, superoxide dismutase (SOD), and glutathione peroxidase (GPx), besides blocking important viral receptors [17]. Some chalcone-type natural products possessing antiviral activity have been addressed in Figure 10.4.

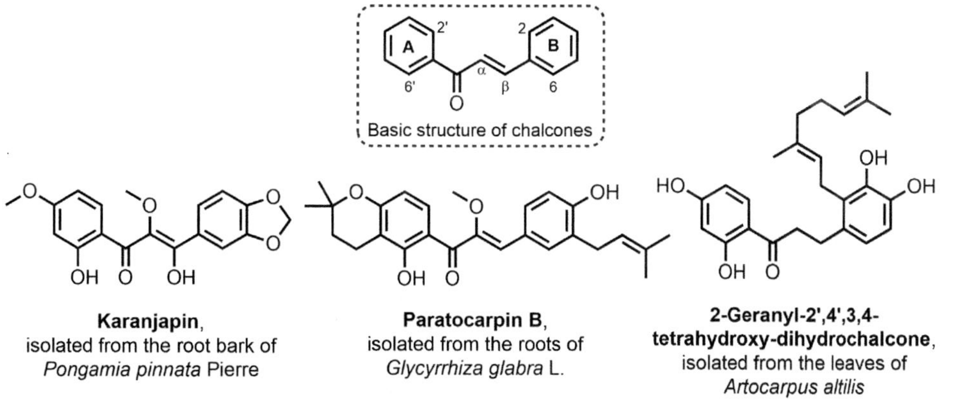

Basic structure of chalcones

Karanjapin, isolated from the root bark of *Pongamia pinnata* Pierre

Paratocarpin B, isolated from the roots of *Glycyrrhiza glabra* L.

2-Geranyl-2′,4′,3,4-tetrahydroxy-dihydrochalcone, isolated from the leaves of *Artocarpus altilis*

FIGURE 10.2 Basic structure and some plant-derived chalcones.

Glycolysis cycle
Acetyl CoA
Erythrose-4-phosphate
+
Phosphoenolpyruvic acid
Shikimic acid
Phenylalanine
1) PAL
2) C4H, [O]/NAD(P)H
Malonyl CoA
x 3
CHS (I)
CHS/CHR (II)
4CL
(I)
(II)
p-Coumaryl CoA
p-Coumaric acid
PAL: Phenylalanine ammonia-lyase
C4H: Cinnamic acid 4-hydroxylase
4CL: 4-Coumarate coenzyme A

FIGURE 10.3 Biosynthesis of chalcones from acetyl CoA and phenylalanine units.

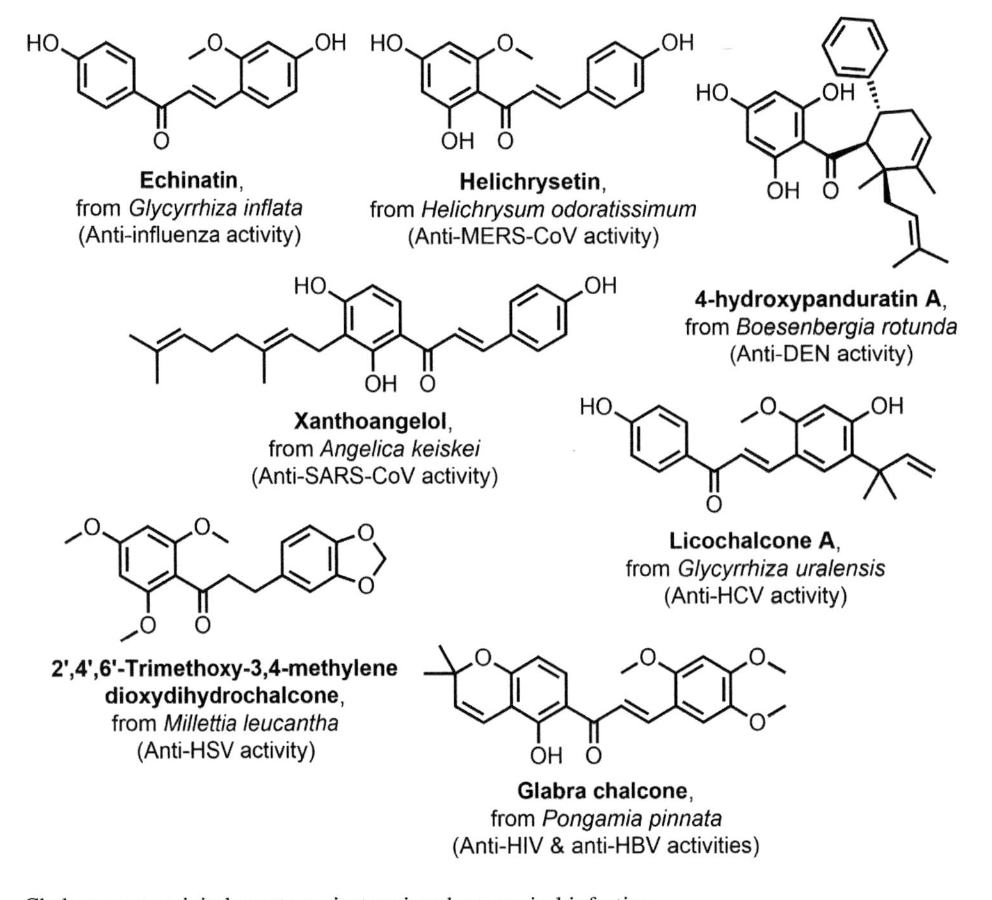

FIGURE 10.4 Chalcones as antiviral agents against various human viral infection.

10.4 ETHNOPHARMACOLOGY OF THE PLANT CONTAINING CHALCONES AGAINST SARS-COV

Chalcones are abundant specialized metabolites in nature and can be found in different parts of plant species, including fruits, leaves, flowers, stem barks, roots, vegetables, tea, and spices. Some plants containing chalcone compounds with high ethnopharmacological impacts include *Angelica keiskei* (Apiaceae), *Butea monosperma* (Lam.) Taub (Fabaceae), *Desmodium gangeticum* (Fabaceae), *Glycyrrhiza inflata* (Fabaceae), *Helichrysum rugulosum* (Asteraceae), *Humulus lupulus* (Cannabaceae), *Mallotus philippinensis* (Euphorbiaceae), *Neoraputia magnifica* (Rutaceae), *Piper hispidum* Sw. (Piperaceae), *Pongamia pinnata* (L.) Pierre (Fabaceae), *Stevia lucida* (Asteraceae), *Tarenna attenuata* (Rubiaceae), and *Uvaria siamensis* (Annonaceae), which are used traditionally and even have come into the laboratories to be explored for SARS-CoV/anti-SARS-CoV-2 inhibitors [18]. Among these potential plants, only *A. keiskei* (also known as ashitaba, a native perennial herb to the southeast Pacific coast of Japan) has been extensively investigated for SARS-CoV and SARS-CoV-2 species since plant contains a high content of natural products like chalcones, coumarins, and flavanones. In addition to wide usage in folk medicine and health-promoting foods, the plant is a potential candidate for following the anti-SARS-CoV-2 strategy because previous studies on SARS-CoV showed that the isolated chalcones have potent anti-inflammatory, immunomodulatory, and anti-SARS-CoV $3CL^{pro}$ (the most crucial enzyme in the viral replication process) activity [19].

10.5 ANTI-INFLAMMATORY AND IMMUNOMODULATORY EFFECTS OF CHALCONES

Inflammation is a biological response observed in higher organisms, whereby the immune system recognizes and responds to damaged cells, irritants, pathogens, and other deleterious stimuli by initiating a complex cascade of cellular and molecular events that promote the healing process. Chronic inflammation denotes a persistent inflammatory state that lasts for an extended period and is often associated with tissue damage, fibrosis, and altered tissue architecture [20]. More specifically, hypercytokinemia is a condition where pro-inflammatory mediators are excessively produced without control. This phenomenon is linked to an overactive immune response that damages vascular endothelial cells and alveolar epithelial cells. At the same time, it is recognized as a specific marker for the severity of COVID-19, particularly in older and co-morbid individuals, where it is often referred to as a "cytokine storm" [21]. On the other hand, due to the dual effects of the COVID-19 virus on the immune system, it initiates a response from the innate immune system, leading to cytokines and chemokines secretion, and also activates the adaptive immune system, which releases the inflammatory cytokines by T and B cells [22].

Nowadays, the potential therapeutic uses of chalcones and their synthetic analogs are increasing because they can effectively combat viral infection and related inflammations. In this concern, several in-depth evaluations have been reported in order to manifest the potency of chalcones against inflammation, modulation of cytokines, and SARS-CoV-2/COVID-19 since they efficiently activate lymphocytes, increase natural killer cell count, and enhance the macrophages' functions. Therefore, these naturally occurring compounds and even some synthetic derivatives possess anti-inflammatory effects by inhibiting lipoxygenase (LOX) and cyclooxygenase (COX) activities and regulating the pro-inflammatory cytokines, including nitric oxide (NO), prostaglandin E_2 (PGE_2), and thromboxane B_2 (TxB_2) in tissues. Moreover, they hinder nuclear factor (NF)-κB activation, interleukins (i.e., IL-6 and IL-1β) release, tumor necrosis factor (TNF)-α production, and reduce the levels of reactive oxygen species (ROS) [23]. Meanwhile, chalcones, for example, isoliquiritigenin, found in *Glycyrrhiza uralensis*, modulate the immune system by inhibiting the pyrin domain-containing 3 inflammasome (NLRP3), on the other hand, by nuclear factor erythroid-derived 2-related factor (Nrf2) activation and the bromodomain-containing protein 4 (BRD4) pathways [24].

As the SARS-CoV-2 virus continues to mutate, finding a successful cure has become more challenging. Studies have revealed that the virus acquires a mutation in the S protein; it increases its affinity for the ACE2 receptor, resulting in the enhancement of cytokines (i.e., TNF-α and NF-κB) levels [25]. In the following, there are some examples of plant-derived chalcones, which have the ability to reduce inflammation and act as immunomodulators.

Ahmad et al. evaluated the effects of cardamonin (**1**) (Figure 10.5), derived from *Alpinia rafflesiana* Wall. Ex Baker (Zingiberaceae) fruits, on various inflammatory markers in macrophage RAW 264.7 cells and whole blood. Results showed the inhibitory activity of compound **1** in NO and PGE2 production in induced cells and whole blood as well as TxB2 production through COX-1 and COX-2 pathways, with more selectivity for COX-2 inhibition. Also, compound **1** suppressed intracellular ROS and TNF-α production in a dose-dependent manner. The authors suggested that **1** acts on pivotal pro-inflammatory factors in the same way as other closely related synthetic hydroxychalcones, thus supporting the efficiency of the methoxy groups at the C-4′ and C-6′ positions of A phenyl ring [26].

In another study, the phytochemical investigation of the roots and stem barks of *Toussaintia orientalis* Verdc. furnished 2′-hydroxy-3′,4′,6′-trimethoxychalcone (**2**) (Figure 10.5). Compound **2** exhibited 50% inhibition of COX-1 at a concentration of 30 μg/mL (IC_{50} of 95.5 μM), which was better than indomethacin as the positive control. Moreover, **2** displayed significant inhibition (88%) of COX-2 at the same concentration [27].

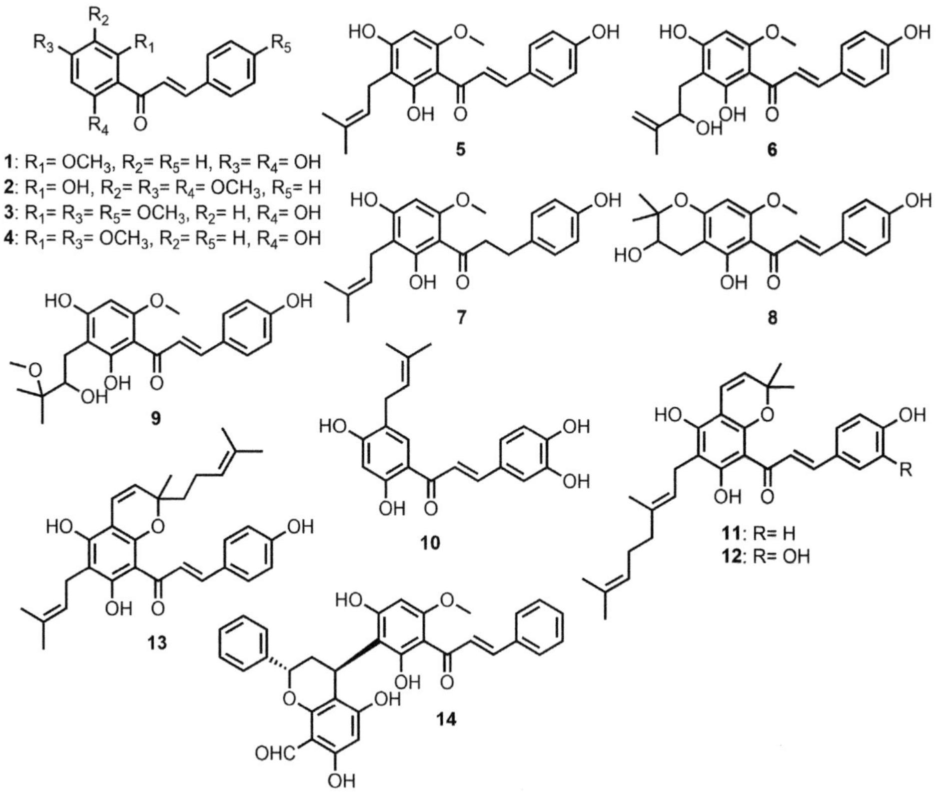

1: R_1= OCH_3, R_2= R_5= H, R_3= R_4= OH
2: R_1= OH, R_2= R_3= R_4= OCH_3, R_5= H
3: R_1= R_3= R_5= OCH_3, R_2= H, R_4= OH
4: R_1= R_3= OCH_3, R_2= R_5= H, R_4= OH

11: R= H
12: R= OH

FIGURE 10.5 Structures of some plant-derived chalcones **1–14** possessing anti-inflammatory and immunomodulatory activities.

Flavokawains A (**3**) and B (**4**) (Figure 10.5), found in the root extracts of the kava plant (*Piper methysticum*; Piperaceae), showed the immunomodulatory effects on Balb/c mice by increasing the secretion of IL-2 and TNF-α and enhancing the population of T cells without significantly altering biochemical parameters in serum [28].

Hop plant (*Humulus lupulus* L.; Cannabinaceae) contains prenylated chalcones, including xanthohumol (**5**), xanthohumol D (**6**), dihydroxanthohumol (**7**), xanthohumol B (**8**), and xanthohumol derivative (**9**) (Figure 10.5). These compounds effectively suppressed NO production stimulated by lipopolysaccharide (LPS) and interferon (IFN)-γ. These chalcones did not show cytotoxicity at doses less than 10 μM (with cell viability more significant than 95%). The inhibitory activity of these compounds ranged from 5.6 to 9.4 μM. Considering the preliminary structure–activity relationship (SAR), the authors suggested that the prenyl chain may not be a requisite element for their inhibitory effects on NO production. Nonetheless, it was evident that chalcones without the double bond between the α- and β-positions demonstrated a lower inhibitory potential compared to other chalcones, affirming the necessity of this double bond in chalcones for inhibition of NO production [29].

Further, a prenylated chalcone, broussochalcone A (**10**) (Figure 10.5), demonstrated a suppressive effect on NO secretion in LPS-induced macrophage cells, exhibiting a dose-dependent behavior (IC_{50} = 11.3 μM). Cheng et al. findings revealed that **10** effectively inhibited NO production by suppressing the expression of iNOS (rather than directly inhibiting the enzymatic activity of iNOS), IκBα phosphorylation, IκBα degradation, and NF-κB activation [30].

In addition, mallotophilippens C (**11**), D (**12**), and E (**13**) (Figure 10.5), derived from *Mallotus philippinensis* (Lam.) Muell. Arg. (Euphorbiaceae) fruits, were found to suppress NO production and iNOS expression in murine macrophage RAW 264.7 cell lines, which was LPS- and IFN-γ-induced. It is noteworthy to mention that these chalcone derivatives inhibited IL-6, IL-1β, and COX-2 gene expression. Daikonya et al. proposed that the primary mode of inhibition could be the deactivation of NF-κB, leading to anti-inflammatory and immunomodulatory properties [31].

Desmosflavan A (**14**) (Figure 10.5), a flavan-chalcone hybrid isolated from the leaves of *Desmos cochinchinensis* Lour. (Annonaceae), showed antioxidant, anti-tumor, and LOX (IC_{50} of 4.4 μM; positive control, nordihydroguaiaretic acid with IC_{50} of 2.9 μM) inhibitory activities [32].

10.6 INTERACTION OF CHALCONES WITH SARS-COV-2 ENZYMES AND PROTEINS

Considering the necessity of identifying anti-infective drugs against SARS-CoV-2, researchers are paying attention to repurposing existing drugs, finding bioactive compounds from natural resources, designing new therapeutics, and improving their potency. To reach this purpose, several SARS-CoV-2 druggable targets, including spike proteins, extracellular proteases (i.e., ACE-2, transmembrane protease serine 2; TMPRSS2, and Furin), intracellular proteases (i.e., cathepsin L), RNA-dependent RNA polymerase (RdRp), 3-chymotrypsin-like protease/main protease ($3CL^{pro}/M^{pro}$), and papain-like protease (PL^{pro}), have been reported and confirmed by computational analyses [33]. Therefore, chalcone-based compounds have become an attractive option for treating COVID-19 because of their biological properties and various antiviral attitudes. Accordingly, we will address some natural chalcones that interact with the viable enzymes and proteins in SARS-CoV and SARS-CoV-2. At last, the potential synthetic anti-SARS-CoV/anti-SARS-CoV-2 compounds containing chalcone scaffolds will be discussed.

10.6.1 Interaction of Chalcones with SARS-CoV-2 Spike Protein (S) and ACE2 Receptor

The SARS-CoV-2 virion's surface features a homotrimeric glycoprotein called the spike (S) protein. This protein serves many functions in viral infection, such as attaching to host receptors, determining cell tropism, and influencing pathogenesis [34]. Furthermore, the spike protein represents the central focus of the host's immune system. These protein varieties influence virulence and are highly immunogenic agents capable of triggering the humoral immune response. Therefore, they are valuable epitopes for developing vaccines and antibody-based therapeutic interventions [35]. To a greater extent, the S protein recognizes viruses during their attachment and host entry. The RBD located on the S1 subunit of the S protein forms a dynamic, high-affinity connection with the ACE2 receptor, inducing a range of structural transformations and conformational modifications in the S protein, for example, alteration in a proteolytic site [36]. One potential approach to developing effective drugs against SARS-CoV-2 is using molecules capable of disturbing and destabilizing the interaction of the viral spike protein and the host ACE2 receptor. Such interference would impede the merging of the viral and host cell membranes, as well as the conveyance of the viral nucleocapsid into the host cell. Thus, natural products may serve as promising therapeutic candidates to block not only the initial binding but also the virus endocytosis into host cells [37].

According to Campos et al., the ethanol extract derived from the bark of *Ampelozizyphus amazonicus* (Rhamnaceae) with high polyphenol content demonstrated the ability to inhibit the *in vitro* interaction of SARS-CoV-2 spike protein with host ACE2 receptor in Calu-3 cells. The presence of saponins along with glycosylated chalcone, 3′,5′-di-*C*-glucosyl phloretin (**15**) (Figure 10.6), was verified through LC-DAD-APCI-MS/MS analysis of the extract [38].

Using computational-based techniques, such as molecular docking, molecular dynamics, and evaluation of physicochemical properties, butein (**16**) (Figure 10.6), a simple chalcone found in *Toxicodendron verniciflua* (Anacardiaceae), showed potential behavior as a binder of ACE2 receptor with a docking score value of –7.38 kcal/mol. Concerning the ACE2 interaction with the spike protein of SARS-CoV-2, compound **16** functioned as a ligand for the ACE2 receptor. Therefore, it interfered with the recognition process and caused structural alterations in the native form of the ACE2 receptor, inhibiting viral entry. Moreover, compound **16** exhibited a favorable ADMET (Adsorption, Distribution, Metabolism, Excretion, and Toxicity) profile, making it a hit natural product to develop novel therapeutics for SARS-CoV-2 infection [39].

Forty-six chalcone derivatives were computationally studied with dual functional activity against spike protein and boosting the body's immunity. Among them, mallotophilippen D (**12**) (Figure 10.5) interfered with the ACE2 and spike protein interaction with a low binding free energy score of –39.2 kcal/mol. Compound **12** effectively regulated the proteins, which are involved in regulating the immune system and alleviating inflammatory mediators, but exhibited value engagement in inhibiting SARS-CoV-2 activity [40].

10.6.2 3-Chymotrypcin-Like Cysteine Protease ($3CL^{pro}$) and Papain-Like Protease (PL^{pro})

The SARS-CoV-2 genome encodes a non-structural protein called chemotrypsin-like protease ($3CL^{pro}$) or main protease (M^{pro}). The protease's 3D structure exhibits uniformity among various coronaviruses and is functional solely in a dimeric state, with individual units or monomers of SARS-CoV-2 M^{pro} lacking independent enzymatic activity. The $3CL^{pro}$ is potentially involved in the proteolytic maturation of viral polyproteins (pp1a and pp1ab), eventually constituting the RNA replicase-transcriptase complex. This complex is vital for viral transcriptive and replicative processes. Hence, antiviral drugs, either with a natural or synthetic origin, which are able to target the SARS-CoV-2 M^{pro}, have efficient potency to target viral infections by preventing the post-transitional processing of SARS-CoV-2 polyproteins, thereby decreasing the likelihood of drugs resistance caused by mutations. Likewise, the M^{pro} enzyme exhibits a substantial sequence similarity (>96%) with the corresponding protease of SARS-CoV and MERS. Therefore, this enzyme could be a significant drug target [41].

Another druggable target in the COVID-19 virus is a papain-like protease (PL^{pro}), which is responsible for viral replication. This enzyme facilitates the cleavage of N-terminal viral polyproteins, producing multiple Nsps such as Nsp1, Nsp2, and Nsp3. Most importantly, PL^{pro} significantly contributes to suppressing the host's innate immunity. Indeed, SARS-CoV viruses employ the enzyme as an antagonist to impede the activation of the interferon regulatory factor-3 (IRF-3) pathway, reducing interferon production and consequently decreasing its antiviral properties. In this concern, the PL^{pro} inhibition could typically prompt the significant immune response mediated by

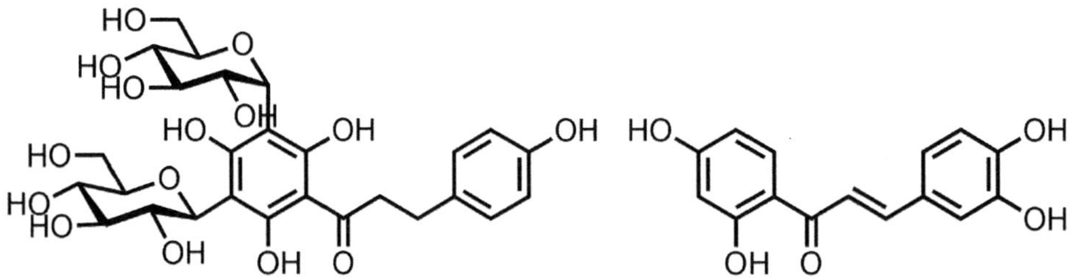

FIGURE 10.6 Structures of plant-derived chalcones **15** and **16**, inhibiting the interaction of SARS-CoV-2 spike protein and host ACE2 receptor.

interferons, disturb the replication process, and trigger the antiviral behavior of host cells [42].

Chalcones have the potential to act as promoter natural products that can trigger the inhibition of various enzymes that play a role in coronavirus infection. As a result, they represent a viable source of drugs and vaccine candidates for combating COVID-19 pandemic. Based on a study conducted by Park et al., prenylated and geranylated chalcones, including isobavachalcone (**17**), 4-hydroxyderricin (**18**), xanthoangelol (**19**), xanthoangelols B (**20**), D (**21**), E (**22**), F (**23**), G (**24**), and xanthoheistal A (**25**) (Figure 10.7), derived from the ethyl acetate (EtOAc)-soluble portion of *Angelica keiskei* leaves (Apiaceae), were assayed *in vitro* against coronavirus 3CL^{pro} and PL^{pro} enzymes. Although all assayed chalcones inhibited the SARS-CoV enzymes, **22** containing peroxyl group showed better inhibitory activity against both enzymes (IC_{50}=11.4 and 1.2 μM, in a competitive and non-competitive manner, respectively). Noteworthy to mention that the C-4′ methoxy group in **17** and **24** decreased the PL^{pro} inhibitory potential [43].

In another study, the SARS-CoV PL^{pro} exhibited significant susceptibility to the ethanol extract derived from *Psoralea corylifolia* (Fabaceae) seeds, as evidenced by its low IC_{50} value of 15 μg/mL. This prompted further isolation through bioactivity-directed purification of the plant extract, which ultimately afforded isobavachalcone (**17**) and 4′-*O*-methyl isobavachalcone (**26**) (Figure 10.7) as the active chalcones responsible for its potent bioactivity. Compounds **17** and **26** dose-dependently inhibited the SARS-CoV PL^{pro} (IC_{50}=7.3 and 10.1 μM, respectively). Kinetic studies showed that **17** and **26** were mixed PL^{pro} inhibitors determined by Lineweaver–Burk and Dixon plots [44].

Park et al. evaluated the effectiveness of polyphenols, including chalcones, isolated from *Broussonetia papyrifera* (L.) Vent. (Moraceae), in inhibiting the activity of cysteine proteases, including 3CL^{pro} and M^{pro}, which are associated with coronavirus replication. Broussochalcone A (**10**) (Figure 10.5), broussochalcone B (**27**), and 4-hydroxyisolonchocarpin (**28**) (Figure 10.7) were found to exhibit higher levels of inhibition against SARS-CoV PL^{pro} (ranging from 9.2 to 35.4 μM) compared to SARS-CoV 3CL^{pro} (ranging from 57.8 to 202.7 μM). It seems that the cyclization of the isoprenyl unit hurts the inhibitory activity of compound **28** when compared with compound **27**. Meanwhile, these chalcones also inhibited MERS PL^{pro} and MERS 3CL^{pro} [45].

Previous studies have shown that the polyphenols found in *B. papyrifera* effectively inhibited the catalytic activity of SARS-CoV and MERS M^{pro}. However, it is currently unknown whether these polyphenols also have inhibitory effects on SARS-CoV-2 M^{pro}. For this purpose, Ghosh et al. computationally investigated the *Broussonetia* polyphenols that demonstrated appropriate drug-like properties, as well as two repurposed drugs (i.e., lopinavir and darunavir) by docking stimulation with SARS-CoV-2 M^{pro} to evaluate their binding properties. In addition to five polyphenols, broussochalcone A (**10**) (Figure 10.5) interacted with two SARS-CoV-2 M^{pro} catalytic residues (i.e., His41 and Cys145) and showed strong binding affinity (–8.1 kcal/mol), serving as a promising anti-SARS-CoV-2 drug [46].

Khanal et al. computationally investigated 46 chalcone derivatives reported as immune enhancers and SARS-CoV-2 enzyme inhibitors, including PL^{pro} and 3CL^{pro}, and spike protein. Results showed that among these chalcones, only 6 chalcone analogs, including xanthohumol (**5**) (Figure 10.5), abyssinone VI (**29**; isolated from the stem of *Erythrina abyssinica*; Fabaceae), isobutrin (**30**; found in *Butea monosperma*; Fabaceae), obochalcolactone (**31**; obtained from *Cryptocarya obovata*; Lauraceae), 4-hydroxycordoin (**32**; found in *Lonchocarpus neuroscapha* Benth.; Fabaceae), and 3′-(3-methyl-2-butyl)-4′-*O*-β-D-glucopyranosyl-4,2′-dihydroxychalcone (**33**) (Figure 10.7), potentially targeted the COVID-19 enzymes and proteins involved in viral replication and viral entry. However, compounds **32** and **33** demonstrated the strongest binding affinity toward PL^{pro} and 3CL^{pro}, respectively. During molecular stimulation, these compounds exhibited stable interactions and achieved low binding free energy scores of –26.09 and –16.28 kcal/mol, respectively. Compounds **29–33** were found to engage with proteins that directly regulate the immune system and modulate the cytokine storm. Additionally, they activated multiple pathways that combat viral and bacterial infections while addressing endocrine dysregulation, a condition that can weaken the immune system. Taken together, the authors suggested that chalcone derivatives may offer a promising therapeutic approach for treating COVID-19 [40].

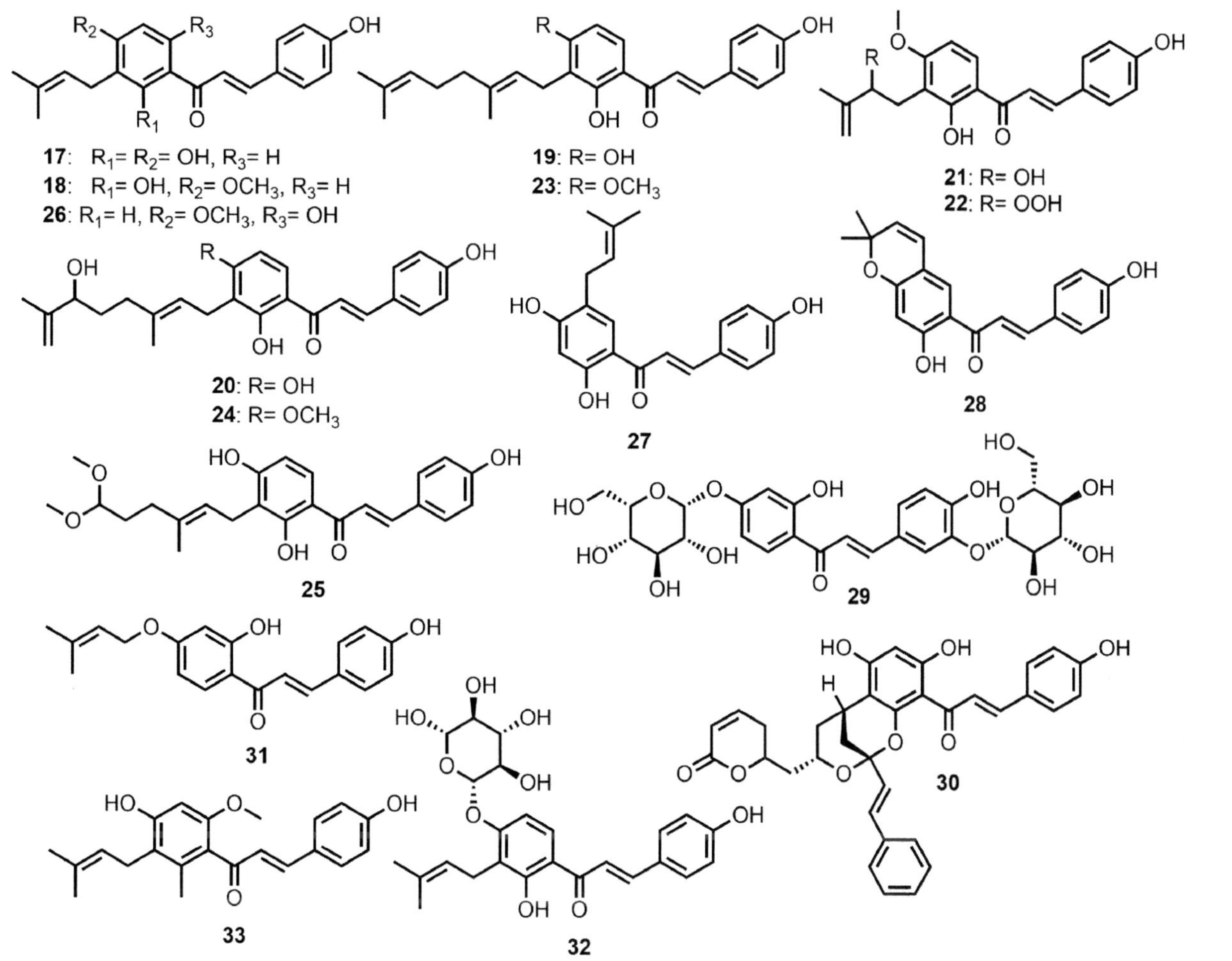

FIGURE 10.7 Structures of plant-derived chalcones **17–33**, inhibiting the coronavirus 3CL^{pro} and PL^{pro}.

10.7 SYNTHETIC CHALCONES AGAINST SARS-COV-2

During the COVID-19 pandemic, several natural products have been identified to inhibit the primary proteins and enzymes of SARS-CoV and MERS-CoV. Recent evidence suggests that the chalcone scaffold holds great promise to develop anti-SARS-CoV-2 drugs, mainly by modifying the substitutions to inhibit irreversibly the COVID-19 main enzymes (i.e., PL^{pro} and 3CL^{pro}) and disturb the interaction of coronavirus spike protein and host ACE2 receptor. Most significantly, chalcones have gained considerable recognition as Michael acceptors in various chemical reactions because of their interaction with thiol (–SH) residues in protein targets, particularly cysteine residues. The α,β-unsaturated carbonyl group can form a covalent bond. As discussed, the reactivity of chalcones assumes greater significance in SARS-CoV-2 pathogenesis. However, due to insufficient research on the chalcone framework in drug discovery and developments against SARS-CoV-2, a new window was opened in front of researchers to consider the chalcone scaffolding in developing protease inhibitors [47].

Following a virtual screening approach on the most druggable SARS-CoV-2 targets, like the RdRp enzyme, Duran et al. synthesized a series of fluorine-containing chalcones. Of all assayed compounds, (*E*)-1-(2,5-diflurophenyl)-3-(2,4,6-trimethoxyphenyl) prop-2-en-1-one (**34**) (Figure 10.8) showed the most significant efficiency toward SARS-CoV-2 activity at 1.6 μg/mL concentration [48].

In another research, the synthesis of 2′-hydroxychalcone analogs was designed using a microwave-assisted method and then computationally compared with reference SARS-CoV-2 M^{pro} inhibitors (i.e., remdesivir and hydroxychloroquine; HCQ). *In silico* studies showed that 1-(2′-hydroxyphenyl)-3-(4-*N*,*N*-dimethylaminophenyl)-prop-2-en-1-one (**35**) (Figure 10.8) had a better affinity to 7BQY, the active site of SARS-CoV-2 M^{pro}, which was comparable to remdesivir and had more potential than HCQ. Meanwhile, ADMET prediction classified compound **35** as non-cytotoxic and safe [49].

Based on the antiviral data of fluoroquinolones (FQs), for example, ciprofloxacin, Alaaeldin et al. decided to improve the SARS-CoV-2 M^{pro} inhibitory effects of ciprofloxacin by reaction with a simple chalcone molecule, leading to the synthesis of 7-(4-(*N*-substituted-carbamoyl-methyl)piperazin-1-yl)-chalcone (**36**) (Figure 10.8). The study involved infecting Vero cells with SARS-CoV-2 and treating them

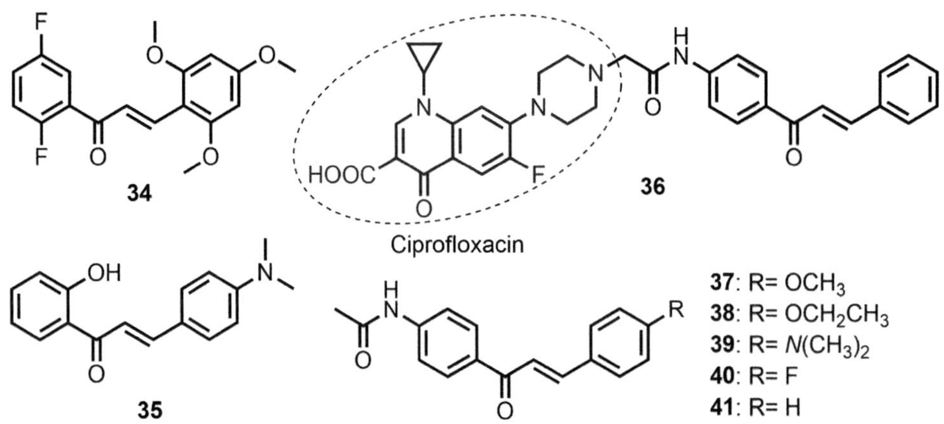

FIGURE 10.8 Structures of synthetic chalcones, **34–41**, against coronaviruses.

with various doses of ciprofloxacin and compound **36** for 48 hours. Results at a concentration of 160 nM revealed that compound **36** inhibited the replication of SARS-CoV-2 in Vero cells, almost 12.7-fold improved when compared with ciprofloxacin alone (EC_{50} values of 3.93 and 50.07 nM, respectively). The positive control, remdesivir, had an EC_{50} of 1.55 nM. In the case of $3CL^{pro}$ inhibition, the following inhibition order was dose-dependently obtained: compound **36**>remdesivir>ciprofloxacin (EC_{50} of 0.6, 1.01, and 5.13 μM, respectively). Molecular docking affirmed the inhibitory activity of compound **36** against SARS-CoV-2 (energy score of −8.9 kcal/mol), and the appropriate ADMET properties were suggested. Eventually, compound **36** considerably decreased the SARS-CoV-2 plaque formation by 86.8% [50].

Almedia-Neto et al. computationally evaluated the effects of synthetic 4′-acetamidechalcones in interaction with the SARS-CoV-2 spike protein and the human ACE2 receptor. Synthetic chalcones, *N*-[4-(2*E*)-3-(4-methoxyphenyl)-1-(phenyl)prop-2-en-1-one]-acetamide (PAAPM **37**) and *N*-[4-(2*E*)-3-(4-ethfoxyphenyl)-1-(phenyl)prop-2-en-1-one]-acetamide (PAAPE **38**) (Figure 10.8), strongly interacted with the spike protein, while *N*-[4-(2*E*)-3-(4-dimethylaminophenyl)-1-(phenyl)prop-2-en-1-one]-acetamide (PAAPA **39**) (Figure 10.8) demonstrated better ACE2 affinity (binding score=−7.0 kcal/mol), forming strong hydrogen interactions. In the case of protease inhibitors, 4′-acetamidechalcones were evaluated against SARS-CoV-2 M^{pro} and Nsp16–Nsp10 heterodimer methyltransferase. For the former target, *N*-[4-(2*E*)-3-(4-fluorophenyl)-1-(phenyl)prop-2-en-1-one]-acetamide (PAAPF **40**) (Figure 10.8) strongly bonded with the same domain that FJC (a natural inhibitor) interacted through hydrogen binding, while for the latter target, *N*-[4-(2*E*)-3-(phenyl)-1-(phenyl)prop-2-en-1-one]-acetamide (PAAB **41**) (Figure 10.8) had two substantial hydrogen bonds. These findings indicated that 4′-acetamidechalcones hinder the post-translational enzymes and the virus–host cell interaction (spike protein-ACE2 binding), potentially creating a steric hindrance and making feasible candidates for *in vitro* assays [51].

10.7.1 Synthetic Chalcone-Amides against SARS-CoV and SARS-CoV-2

Despite the naturally occurring chalcone backbone, a very close scaffold known as chalcone-amide (Figure 10.9)

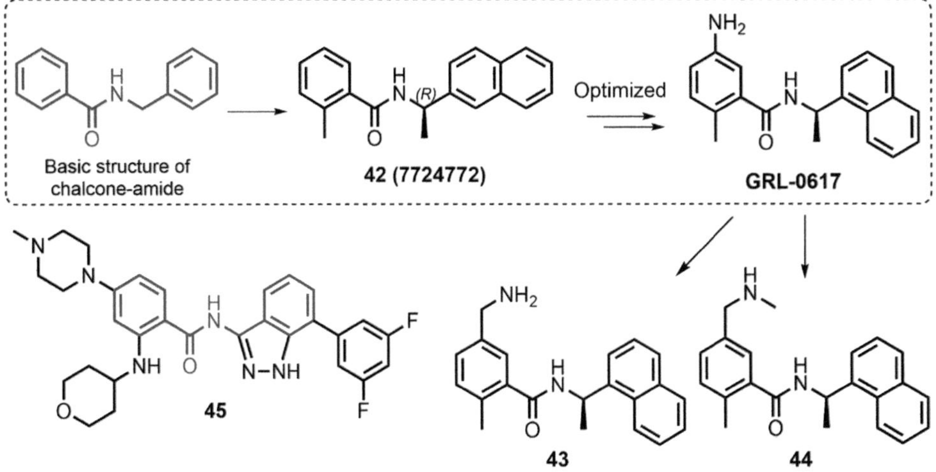

FIGURE 10.9 Basic structure of chalcone-amides **42–45** and its related hit molecules against SARS-CoV and SARS-CoV-2.

was proposed to have the ability to suppress SARS-CoV and SARS-CoV-2 enzymes [52]. The initial finding of chalcone-amide-based compounds that can inhibit viral PL^{pro} dates back to 2008 when Ratia et al. screened 50,080 structurally diverse compounds using an extensive fluorescence-based high-throughput screening (HTS) approach. Interestingly, these endeavors uncovered a racemic chalcone-amide, 2-methyl-*N*-[1-(2-naphthalenyl)ethyl]-benzamide (**42** 7,724,772) (Figure 10.9) as a non-covalent lead molecule, which selectively inhibited the SARS-CoV PL^{pro} with an IC_{50} of 20.1 μM. Furthermore, the authors synthesized both enantiomers (*R*) and (*S*) of **42** to investigate whether SARS-CoV PL^{pro} exhibits any stereospecificity. Results established *R*-**42** as a lead compound (IC_{50} value of 8.7 μM) compared with *S*-**42**, showing low PL^{pro} inhibitory activity (> 14%). The potency of *R*-**42** was improved by synthetic optimization (GRL-0617; IC_{50} value of 0.6 μM). This compound displayed inhibitory activity against viral replication of SARS-CoV in Vero E6 cells with an EC_{50} of 15 μM without cytotoxicity [53]. Since then, the backbone of this compound has been raised as the building block for drug discovery and development.

Later, Gosh et al. refined the inhibitory effects of *R*-**42** by structure-based modification, resulting in the synthesis of compounds **43** and **44** (Figure 10.9). Compared with *R*-**42**, the better anti-SARS-CoV PL^{pro} (IC_{50} of 0.46 μM) and antiviral (EC_{50} of 6 μM in Vero E6 cells) activities were recorded for **43**. On the other hand, compound **44** containing methyl amine residue moderately inhibited SARS-CoV PL^{pro} enzyme (IC_{50} of 1.3 μM), while its antiviral property was better than *R*-**42** (EC_{50} of 5.2 μM in Vero E6 cells) [54].

In the case of chalcone-amide-based compounds, Peralta-Garcia et al. employed the drug repurposing approach by use of an anticancer drug, entrectinib (**45**) (Figure 10.9), to discover drug candidates that disturb viral–host interactions. A combination of virtual screening and experimental assays confirmed **45** as a prospective antiviral drug by remarkably reducing cellular infection in the human lung tissue (HLT) model without stimulating apoptosis (EC_{50} value of <1 μM) [55].

10.7.2 Synthetic *Pseudo*-Chalcone-Amides against SARS-CoV and SARS-CoV-2

The primary HTS approach also was able to propose another hit molecule with a pseudo-chalcone-amide skeleton, *N*-[(4-methoxyphenyl)methyl]-1-(1-naphthalenylmethyl)-2-piperidine carboxamide (**46**; 6,577,871) (Figure 10.10), as an anti-SARS-CoV PL^{pro} agent with an IC_{50} of 59.2 μM. It is worth mentioning that through lead optimization and SAR studies, another lead derivative of compound **46** was synthesized (GRL-0667; IC_{50} of 0.32 μM) and consequently exhibited potent antiviral activity with an EC_{50} of 9.1 μM in Vero E6 cells [56]. Since then, the backbone of this compound has been raised as the building block for drug discovery and development.

In 2014, Baez-Santos et al. were able to synthesize a series of non-covalent anti-SARS-CoV PL^{pro} from the initial benzodioxolane lead (GRL-0667) backbone using a structure-guided design. Among the generated derivatives, **47** and **48** (Figure 10.10), containing fluorine, inhibited SARS-CoV replication with improved IC_{50} values (0.15 and 0.49 μM, respectively), representing the effect of *meta*- and *para*-positioned monofluoro moieties at the phenyl ring. These compounds also inhibited viral replication in Vero E6 cells with EC_{50} of 5.4 and 11.6 μM, respectively [57].

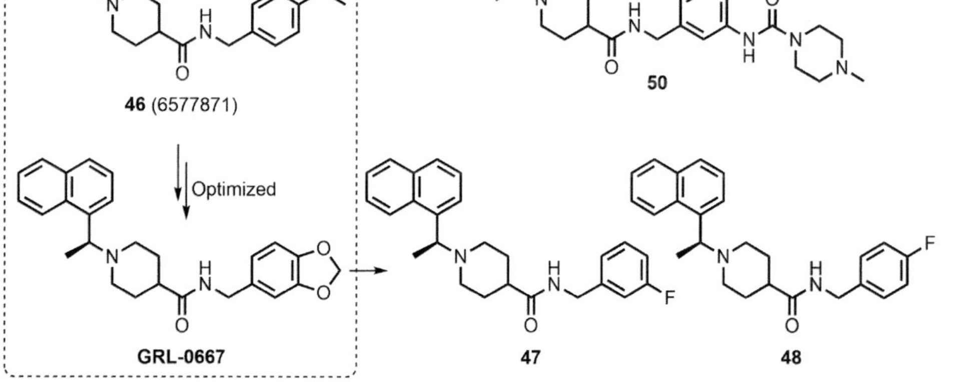

FIGURE 10.10 Basic structure of *pseudo* chalcone-amides **46–50** and its related hit molecules against SARS-CoV and SARS-CoV-2.

The replication cycle of coronaviruses is heavily reliant on the ubiquitin-proteasome system (UPS), especially its deubiquitinating enzymes (DUBs). The SARS-CoV-2 PL^{pro} serves as a viral protease and deubiquitinating enzyme. A study conducted by Grobe et al. revealed that certain DUB-inhibitors (DIs), such as HBX41108 and PR-619, can effectively suppress replication of SARS-CoV-2 by restricting viral replication in infected cells (i.e., Vero B4 and human Calu-3 lung cells) with a low multiplicity of infection (MOI) of 0.02 [58]. To identify and develop the potential anti-PL^{pro} compounds using the HTS approach, Shan et al. synthesized *pseudo*-chalcone-amides and assayed them on both sensitive probe 7 and UB-Rho. Compared with GRL-0617, the anti-PL^{pro} activity of these compounds effectively improved (IC_{50} ranging from 2.64 to 4.32 µM for probe 7 and IC_{50} ranging from 1.75 to 2.00 µM for UB-Rho). The authors tried to refine the activity by modifying the substitutions. Eventually, they reached the most potent ones, including *N*-(3-(aminomethyl)-5-fluorophenyl)-acetamide (**49**) and *N*-(3-(aminomethyl)-5-fluorophenyl)-4-methylpiperazine-1-carboxamide (**50**) (Figure 10.10) with $IC_{50\text{-}PLpro}$ of 1.76 and 0.44 µM (for probe 7) and $IC_{50\text{-}PLpro}$ of 1.13 and 0.26 µM (for UB-Rho), respectively. In conclusion, the fluorine group at the *meta* position of the phenyl ring was vital for bioactivity. Meanwhile, compound **50** can offer a promising lead for drug discovery efforts and development, targeting SARS-CoV-2 PL^{pro}, due to its optimized pharmacophore [59].

Last but not least, in the course of SARS-CoV-2 drug discovery and development, the potential chalconic scaffolds, including azachalcones, indole-chalcones, and chalcone-amide-like derivatives, have been proposed in Figure 10.11 because these scaffolds possess substantial antiviral activities against various human viral infections [52,60,61].

10.8 SUMMARY POINTS AND PERSPECTIVE

- Chalcones are naturally occurring antiviral compounds against various human viruses, including MERS-CoV, SARS-CoV, DENV, HCMV, HIV, HBV, HCV, HSV, RVFV, VEEV, human rhinovirus, and influenza virus.
- Polyphenols, including chalcones as antioxidants, modulate the body's immunity and reduce inflammation during SARS-CoV-2 infection.
- Chalcones can reduce the release of pro-inflammatory markers (i.e., IL-1β, IL-6, and TNF-α) and activate NF-κB, resulting in anti-inflammatory properties.
- Chalcones are considered a promising class of bioactive secondary metabolites for blocking the enzymatic activity of SARS-CoV-2 proteases and preventing COVID-19.
- Experimental and computational studies revealed that chalcones inhibit the viable SARS-CoV-2 enzymes (i.e., $3CL^{pro}$, PL^{pro}, and RdRp) and disturb the interaction between the ACE2 receptor and spike protein.
- Plant-derived chalcones, for example, xanthoangelol E from *Angelica keiskei* (Apiaceae), effectively inhibit SARS-CoV proteases.
- Using the HTS method, the initial backbones of synthetic chalcones, showing excellent anti-SARS-CoV activity, were chalcone-amides and pseudo-chalcone-amides.
- Fluorine-containing synthetic chalcones are better SARS-CoV-2 inhibitors.
- Synthetic chalcone-like compounds, including indole-chalcones and azachalcones, were proposed to receive more attention to improve their activity against SARS-CoV-2.

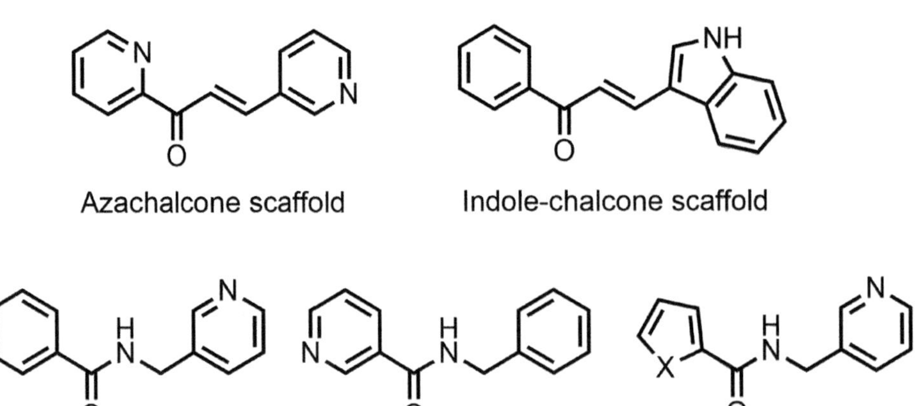

FIGURE 10.11 Basic structure of chalcone-like scaffolds against SARS-CoV/SARS-CoV-2.

REFERENCES

1. World Health Organization WHO Director-General's opening remarks at the media briefing on COVID-19-11 March 2020. https://www.who.int/director-general/speeches/detail/who-director-general-s-opening-remarks-at-the-media-briefing-on-covid-19---11-march-2020
2. Shakoor, H.; Feehan, J.; Al Dhaheri, A. S.; Ali, H. I.; Platat, C.; Ismail, L. C.; Apostolopoulos, V.; Stojanovska, L., Immune-boosting role of vitamins D, C, E, zinc, selenium and omega-3 fatty acids: Could they help against COVID-19? *Maturitas* **2021**, 143, 1–9.
3. Corrao, S.; Mallaci Bocchio, R.; Lo Monaco, M.; Natoli, G.; Cavezzi, A.; Troiani, E.; Argano, C., Does evidence exist to blunt inflammatory response by nutraceutical supplementation during covid-19 pandemic? An overview of systematic reviews of vitamin D, vitamin C, melatonin, and zinc. *Nutrients* **2021**, 13(4), 1261.
4. Sharifi-Rad, J.; Herrera-Bravo, J.; Salazar, L. A.; Shaheen, S.; Abdulmajid Ayatollahi, S.; Kobarfard, F.; Imran, M.; Imran, A.; Custódio, L.; Dolores López, M., The therapeutic potential of wogonin observed in preclinical studies. *Evidence-Based Complementary and Alternative Medicine* **2021**, 2021, 9935451.
5. Hafez Ghoran, S.; Saeidnia, S.; Babaei, E.; Kiuchi, F.; Dusek, M.; Eigner, V.; Khalaji, A. D.; Soltani, A.; Ebrahimi, P.; Mighani, H., Biochemical and biophysical properties of a novel homoisoflavonoid extracted from *Scilla persica* HAUSSKN. *Bioorganic Chemistry* **2014**, 57, 51–56.
6. Hafez Ghoran, S.; El-Shazly, M.; Sekeroglu, N.; Kijjoa, A., Natural products from medicinal plants with anti-human coronavirus activities. *Molecules* **2021**, 26(6), 1754.
7. Mehany, T.; Khalifa, I.; Barakat, H.; Althwab, S. A.; Alharbi, Y. M.; El-Sohaimy, S., Polyphenols as promising biologically active substances for preventing SARS-CoV-2: A review with research evidence and underlying mechanisms. *Food Bioscience* **2021**, 40, 100891.
8. Hafez Ghoran, S.; Taktaz, F.; Ayatollahi, S. A., Plant immunoenhancers: promising ethnopharmacological candidates for anti-SARS-CoV-2 activity. In *Ethnopharmacology and Drug Discovery for COVID-19: Anti-SARS-CoV-2 Agents from Herbal Medicines and Natural Products*, Chen, J.- T., Ed. Springer Nature Singapore: Singapore, 2023; pp. 39–84.
9. Mhatre, S.; Srivastava, T.; Naik, S.; Patravale, V., Antiviral activity of green tea and black tea polyphenols in prophylaxis and treatment of COVID-19: A review. *Phytomedicine* **2021**, 85, 153286.
10. Pasquereau, S.; Nehme, Z.; Haidar Ahmad, S.; Daouad, F.; Van Assche, J.; Wallet, C.; Schwartz, C.; Rohr, O.; Morot-Bizot, S.; Herbein, G., Resveratrol inhibits HCoV-229E and SARS-CoV-2 coronavirus replication *in vitro*. *Viruses* **2021**, 13(2), 354.
11. Jin, Y.-H.; Lee, J.; Jeon, S.; Kim, S.; Min, J. S.; Kwon, S., Natural polyphenols, 1,2,3,4,6-*O*-pentagalloyglucose and proanthocyanidins, as broad-spectrum anticoronaviral inhibitors targeting Mpro and RdRp of SARS-CoV-2. *Biomedicines* **2022**, 10(5), 1170.
12. Dejani, N. N.; Elshabrawy, H. A.; Bezerra Filho, C. D . S. M.; de Sousa, D. P., Anticoronavirus and immunomodulatory phenolic compounds: Opportunities and pharmacotherapeutic perspectives. *Biomolecules* **2021**, 11(8), 1254.
13. Giovinazzo, G.; Gerardi, C.; Uberti-Foppa, C.; Lopalco, L., Can natural polyphenols help in reducing cytokine storm in COVID-19 patients? *Molecules* **2020**, 25(24), 5888.
14. Rozmer, Z.; Perjési, P., Naturally occurring chalcones and their biological activities. *Phytochemistry Reviews* **2016**, 15, 87–120.
15. Heller, W.; Forkmann, G., Biosynthesis of flavonoids. In *The Flavonoids*, Routledge: London, 2017; pp. 499–535.
16. Welle, R.; Grisebach, H., Isolation of a novel NADPH-dependent reductase which coacts with chalcone synthase in the biosynthesis of 6′-deoxychalcone. *FEBS Letters* **1988**, 236(1), 221–225.
17. Elkhalifa, D.; Al-Hashimi, I.; Al Moustafa, A.-E.; Khalil, A., A comprehensive review on the antiviral activities of chalcones. *Journal of Drug Targeting* **2021**, 29(4), 403–419.
18. Rajendran, G.; Bhanu, D.; Aruchamy, B.; Ramani, P.; Pandurangan, N.; Bobba, K. N.; Oh, E. J.; Chung, H. Y.; Gangadaran, P.; Ahn, B.-C., Chalcone: A promising bioactive scaffold in medicinal chemistry. *Pharmaceuticals* **2022**, 15(10), 1250.
19. Yang, F.; Jiang, X.-l.; Tariq, A.; Sadia, S.; Ahmed, Z.; Sardans, J.; Aleem, M.; Ullah, R.; Bussmann, R. W., Potential medicinal plants involved in inhibiting 3CLpro activity: A practical alternate approach to combating COVID-19. *Journal of Integrative Medicine* **2022**, 20(6), 488–496.
20. Ahmed, A. U., An overview of inflammation: Mechanism and consequences. *Frontiers in Biology* 2011, 6(4), 274.
21. Pelaia, C.; Tinello, C.; Vatrella, A.; De Sarro, G.; Pelaia, G., Lung under attack by COVID-19-induced cytokine storm: Pathogenic mechanisms and therapeutic implications. *Therapeutic Advances in Respiratory Disease* **2020**, 14, 1753466620933508.
22. Crisci, C. D.; Ardusso, L. R.; Mossuz, A.; Müller, L., A precision medicine approach to SARS-CoV-2 pandemic management. *Current Treatment Options in Allergy* **2020**, 7, 422–440.
23. Alzaabi, M. M.; Hamdy, R.; Ashmawy, N. S.; Hamoda, A. M.; Alkhayat, F.; Khademi, N. N.; Al Joud, S. M. A.; El-Keblawy, A. A.; Soliman, S. S., Flavonoids are promising safe therapy against COVID-19. *Phytochemistry Reviews* **2021**, 21, 1–22.
24. Zeng, J.; Chen, Y.; Ding, R.; Feng, L.; Fu, Z.; Yang, S.; Deng, X.; Xie, Z.; Zheng, S., Isoliquiritigenin alleviates early brain injury after experimental intracerebral hemorrhage via suppressing ROS-and/or NF-κB-mediated NLRP3 inflammasome activation by promoting Nrf2 antioxidant pathway. *Journal of Neuroinflammation* **2017**, 14, 1–19.
25. Padhi, A. K.; Tripathi, T., Can SARS-CoV-2 accumulate mutations in the S-protein to increase pathogenicity? *ACS Pharmacology & Translational Science* **2020**, 3(5), 1023–1026.
26. Ahmad, S.; Israf, D. A.; Lajis, N. H.; Shaari, K.; Mohamed, H.; Wahab, A. A.; Ariffin, K. T.; Hoo, W. Y.; Aziz, N. A.; Kadir, A. A., Cardamonin, inhibits pro-inflammatory mediators in activated RAW 264.7 cells and whole blood. *European Journal of Pharmacology* **2006**, 538(1–3), 188–194.
27. Nyandoro, S.; Nkunya, M.; Josepha, C.; Odalo, J.; Sattler, I., New glucopyranosylglyceryl-N-octenyl adipate and bioactivity of retro and condensed chalcones from *Toussaintia orientalis*. *Tanzania Journal of Science* **2012**, 38(3), 108–126.
28. Abu, N.; Mohamed, N. E.; Tangarajoo, N.; Yeap, S. K.; Akhtar, M. N.; Abdullah, M. P.; Omar, A. R.; Alitheen, N. B., *In vitro* toxicity and *in vivo* immunomodulatory

effects of flavokawain A and flavokawain B in Balb/C mice. *Natural Product Communications* **2015**, 10(7), 1934578X1501000716.
29. Zhao, F.; Watanabe, Y.; Nozawa, H.; Daikonnya, A.; Kondo, K.; Kitanaka, S., Prenylflavonoids and phloroglucinol derivatives from hops (*Humulus lupulus*). *Journal of Natural Products* **2005**, 68(1), 43–49.
30. Cheng, Z.-J.; Lin, C.-N.; Hwang, T.-L.; Teng, C.-M., Broussochalcone A, a potent antioxidant and effective suppressor of inducible nitric oxide synthase in lipopolysaccharide-activated macrophages. *Biochemical Pharmacology* **2001**, 61(8), 939–946.
31. Daikonya, A.; Katsuki, S.; Kitanaka, S., Antiallergic agents from natural sources 9. Inhibition of nitric oxide production by novel chalcone derivatives from *Mallotus philippinensis* (Euphorbiaceae). *Chemical and Pharmaceutical Bulletin* **2004**, 52(11), 1326–1329.
32. Bajgai, S. P.; Prachyawarakorn, V.; Mahidol, C.; Ruchirawat, S.; Kittakoop, P., Hybrid flavan-chalcones, aromatase and lipoxygenase inhibitors, from *Desmos cochinchinensis*. *Phytochemistry* **2011**, 72(16), 2062–2067.
33. Faheem, S., Kumar, B. K.; Sekhar, K. V. G. C.; Kunjiappan, S.; Jamalis, J.; Balaña-Fouce, R.; Tekwani, B. L.; Sankaranarayanan, M., Druggable targets of SARS-CoV-2 and treatment opportunities for COVID-19. *Bioorganic Chemistry* **2020**, 104, 104269.
34. Sigrist, C. J.; Bridge, A.; Le Mercier, P., A potential role for integrins in host cell entry by SARS-CoV-2. *Antiviral Research* **2020**, 177, 104759.
35. Lv, Z.; Deng, Y.-Q.; Ye, Q.; Cao, L.; Sun, C.-Y.; Fan, C.; Huang, W.; Sun, S.; Sun, Y.; Zhu, L., Structural basis for neutralization of SARS-CoV-2 and SARS-CoV by a potent therapeutic antibody. *Science* **2020**, 369(6510), 1505–1509.
36. Letko, M.; Marzi, A.; Munster, V., Functional assessment of cell entry and receptor usage for SARS-CoV-2 and other lineage B betacoronaviruses. *Nature Microbiology* **2020**, 5(4), 562–569.
37. Bongini, P.; Trezza, A.; Bianchini, M.; Spiga, O.; Niccolai, N., A possible strategy to fight COVID-19: Interfering with spike glycoprotein trimerization. *Biochemical and Biophysical Research Communications* **2020**, 528(1), 35–38.
38. Campos, M. F.; Mendonça, S. C.; Peñaloza, E. M. C.; de Oliveira, B. A.; Rosa, A. S.; Leitão, G. G.; Tucci, A. R.; Ferreira, V. N. S.; Oliveira, T. K. F.; Miranda, M. D., Anti-SARS-CoV-2 activity of *Ampelozizyphus amazonicus* (Saracura-Mirá): Focus on the modulation of the spike-ACE2 interaction by chemically characterized bark extracts by LC-DAD-APCI-MS/MS. *Molecules* **2023**, 28 (7), 3159.
39. Kapoor, N.; Ghorai, S. M.; Khuswaha, P. K.; Bandichhor, R.; Brogi, S., Butein as a potential binder of human ACE2 receptor for interfering with SARS-CoV-2 entry: A computer-aided analysis. *Journal of Molecular Modeling* **2022**, 28(9), 270.
40. Khanal, P.; Patil, V. S.; Bhandare, V. V.; Dwivedi, P. S.; Shastry, C.; Patil, B.; Gurav, S. S.; Harish, D. R.; Roy, S., Computational investigation of benzalacetophenone derivatives against SARS-CoV-2 as potential multi-target bioactive compounds. *Computers in Biology and Medicine* **2022**, 146, 105668.
41. Goyal, B.; Goyal, D., Targeting the dimerization of the main protease of coronaviruses: A potential broad-spectrum therapeutic strategy. *ACS Combinatorial Science* **2020**, 22(6), 297–305.
42. Chen, X.; Yang, X.; Zheng, Y.; Yang, Y.; Xing, Y.; Chen, Z., SARS coronavirus papain-like protease inhibits the type I interferon signaling pathway through interaction with the STING-TRAF3-TBK1 complex. *Protein & Cell* **2014**, 5(5), 369–381.
43. Park, J.-Y.; Ko, J.-A.; Kim, D. W.; Kim, Y. M.; Kwon, H.-J.; Jeong, H. J.; Kim, C. Y.; Park, K. H.; Lee, W. S.; Ryu, Y. B., Chalcones isolated from *Angelica keiskei* inhibit cysteine proteases of SARS-CoV. *Journal of Enzyme Inhibition and Medicinal Chemistry* **2016**, 31(1), 23–30.
44. Kim, D. W.; Seo, K. H.; Curtis-Long, M. J.; Oh, K. Y.; Oh, J.-W.; Cho, J. K.; Lee, K. H.; Park, K. H., Phenolic phytochemical displaying SARS-CoV papain-like protease inhibition from the seeds of *Psoralea corylifolia*. *Journal of Enzyme Inhibition and Medicinal Chemistry* **2014**, 29(1), 59–63.
45. Park, J.-Y.; Yuk, H. J.; Ryu, H. W.; Lim, S. H.; Kim, K. S.; Park, K. H.; Ryu, Y. B.; Lee, W. S., Evaluation of polyphenols from *Broussonetia papyrifera* as coronavirus protease inhibitors. *Journal of Enzyme Inhibition and Medicinal Chemistry* **2017**, 32(1), 504–512.
46. Ghosh, R.; Chakraborty, A.; Biswas, A.; Chowdhuri, S., Identification of polyphenols from *Broussonetia papyrifera* as SARS CoV-2 main protease inhibitors using in silico docking and molecular dynamics simulation approaches. *Journal of Biomolecular Structure and Dynamics* **2021**, 39(17), 6747–6760.
47. Valipour, M., Recruitment of chalcone's potential in drug discovery of anti-SARS-CoV-2 agents. *Phytotherapy Research* **2022**, 36(12), 4477–4490.
48. Duran, N.; Polat, M. F.; Aktas, D. A.; Alagoz, M. A.; Ay, E.; Cimen, F.; Tek, E.; Anil, B.; Burmaoglu, S.; Algul, O., New chalcone derivatives as effective against SARS-CoV-2 agent. *International Journal of Clinical Practice* **2021**, 75(12), e14846.
49. Uddin, M. N.; Ahmed, S. S.; Uzzaman, M.; Knock, M. N. H.; Shumi, W.; Sanaullah, A. F. M.; Bhuyain, M. M. H., Characterization, molecular modeling and pharmacology of some 2′-hydroxychalcone derivatives as SARS-CoV-2 inhibitor. *Results in Chemistry* **2022**, 4, 100329.
50. Alaaeldin, R.; Mustafa, M.; Abuo-Rahma, G. E. D. A.; Fathy, M., *In vitro* inhibition and molecular docking of a new ciprofloxacin-chalcone against SARS-CoV-2 main protease. *Fundamental & Clinical Pharmacology* **2022**, 36(1), 160–170.
51. Almeida-Neto, F. W. Q.; Matos, M. G. C.; Marinho, E. M.; Marinho, M. M.; Sampaio, T. L.; Bandeira, P. N.; Fernandes, C. F. C.; Teixeira, A. M. R.; Marinho, E. S.; de Lima-Neto, P., *In silico* study of the potential interactions of 4′-acetamidechalcones with protein targets in SARS-CoV-2. *Biochemical and Biophysical Research Communications* **2021**, 537, 71–77.
52. Valipour, M., Chalcone-amide, a privileged backbone for the design and development of selective SARS-CoV/SARS-CoV-2 papain-like protease inhibitors. *European Journal of Medicinal Chemistry* **2022**, 240, 114572.

53. Ratia, K.; Pegan, S.; Takayama, J.; Sleeman, K.; Coughlin, M.; Baliji, S.; Chaudhuri, R.; Fu, W.; Prabhakar, B. S.; Johnson, M. E., A noncovalent class of papain-like protease/deubiquitinase inhibitors blocks SARS virus replication. *Proceedings of the National Academy of Sciences* **2008**, 105(42), 16119–16124.
54. Ghosh, A. K.; Takayama, J.; Aubin, Y.; Ratia, K.; Chaudhuri, R.; Baez, Y.; Sleeman, K.; Coughlin, M.; Nichols, D. B.; Mulhearn, D. C., Structure-based design, synthesis, and biological evaluation of a series of novel and reversible inhibitors for the severe acute respiratory syndrome–coronavirus papain-like protease. *Journal of Medicinal Chemistry* **2009**, 52(16), 5228–5240.
55. Peralta-Garcia, A.; Torrens-Fontanals, M.; Stepniewski, T. M.; Grau-Expósito, J.; Perea, D.; Ayinampudi, V.; Waldhoer, M.; Zimmermann, M.; Buzón, M. J.; Genescà, M., Entrectinib-A SARS-CoV-2 inhibitor in human lung tissue (HLT) cells. *International Journal of Molecular Sciences* **2021**, 22(24), 13592.
56. Ghosh, A. K.; Takayama, J.; Rao, K. V.; Ratia, K.; Chaudhuri, R.; Mulhearn, D. C.; Lee, H.; Nichols, D. B.; Baliji, S.; Baker, S. C., Severe acute respiratory syndrome coronavirus papain-like novel protease inhibitors: Design, synthesis, protein– ligand X-ray structure and biological evaluation. *Journal of Medicinal Chemistry* **2010**, 53(13), 4968–4979.
57. Báez-Santos, Y. M.; Barraza, S. J.; Wilson, M. W.; Agius, M. P.; Mielech, A. M.; Davis, N. M.; Baker, S. C.; Larsen, S. D.; Mesecar, A. D., X-ray structural and biological evaluation of a series of potent and highly selective inhibitors of human coronavirus papain-like proteases. *Journal of Medicinal Chemistry* **2014**, 57(6), 2393–2412.
58. Große, M.; Setz, C.; Rauch, P.; Auth, J.; Morokutti-Kurz, M.; Temchura, V.; Schubert, U., Inhibitors of deubiquitinating enzymes interfere with the SARS-CoV-2 papain-like protease and block virus replication *in vitro*. *Viruses* **2022**, 14(7), 1404.
59. Shan, H.; Liu, J.; Shen, J.; Dai, J.; Xu, G.; Lu, K.; Han, C.; Wang, Y.; Xu, X.; Tong, Y., Development of potent and selective inhibitors targeting the papain-like protease of SARS-CoV-2. *Cell Chemical Biology* **2021**, 28(6), 855–865.
60. Vijayakumar, B. G.; Ramesh, D.; Joji, A.; Kannan, T., *In silico* pharmacokinetic and molecular docking studies of natural flavonoids and synthetic indole chalcones against essential proteins of SARS-CoV-2. *European Journal of Pharmacology* **2020**, 886, 173448.
61. Tumskiy, R. S.; Tumskaia, A. V., Multistep rational molecular design and combined docking for discovery of novel classes of inhibitors of SARS-CoV-2 main protease 3CLpro. *Chemical Physics Letters* **2021**, 780, 138894.

11 Plant-Derived Polyphenols in Modulating the Immune Response against COVID-19

Amaresh Mishra, Km Shivangi, Manju Yadav, Dolly Sharma, Jyoti Upadhyay, Yamini Pathak, and Vishwas Tripathi

11.1 INTRODUCTION

An unprecedented global health crisis has been caused by the coronavirus disease (SARS-CoV-2; also known as COVID-19) [1]. As the globe faces the ongoing difficulties posed by this highly transmissible virus, it is crucial to investigate viable therapeutic interventions urgently. Traditional and herbal medicines have long been recognized for their potential in combating various diseases [2,3]. In this book chapter, we aim to bridge the gap between the fields of cancer biology and COVID-19 research by exploring the potential of dietary polyphenols in combating inflammation and modulating the immune response to COVID-19. Inflammation and immune dysregulation play a crucial role in the pathogenesis of COVID-19 [4]. The virus can cause an overactive immune response, leading to a cytokine storm with excessive release of pro-inflammatory cytokines [5]. This overactive immune response leads to severe COVID-19 complications, such as ARDS and multi-organ failure [6]. Therefore, interventions that can modulate the immune response and mitigate inflammation are of great interest.

Polyphenols, naturally occurring chemicals in plant-based diets, have gained attention for their potential health benefits [7,8]. These bioactive chemicals have tremendous therapeutic potential because of their anti-inflammatory and immunomodulatory effects [8]. The ability of dietary polyphenols to inhibit inflammatory pathways, lower oxidative stress, and control immune cell functions has been shown in previous research [7,8]. Regarding COVID-19, exploring the potential benefits of dietary polyphenols takes on even greater significance. Numerous pathogens, including influenza and herpes viruses, have been the focus of research into polyphenols' possible antiviral effects. Their ability to inhibit viral replication, modulate immune responses, and reduce inflammatory cytokine production suggests a potential role in combating viral infections, including SARS-CoV-2.

In this chapter, we will introduce the immunomodulatory properties of dietary polyphenols and their potential application in combating COVID-19. We will review the link between inflammation, immune response, and COVID-19 pathogenesis, highlighting the critical role of immune dysregulation in disease severity. Subsequently, we will explore the diverse chemical structures of dietary polyphenols and discuss their mechanisms of action in modulating the immune response and alleviating inflammation. Furthermore, we will evaluate the evidence on the potential benefits of dietary polyphenols in terms of COVID-19. We aim to learn more about polyphenols' potential against SARS-CoV-2 and their capacity to reduce the immunological dysregulation associated with severe COVID-19 by reviewing research examining their effects on viral infections. In addition to their direct impact on viral infections and immune responses, we will also integrate findings from the field of cancer biology. Specifically, we will discuss the relevance of targeting inflammation and immune dysregulation in breast cancer stem cells with dietary polyphenols regarding COVID-19. This integration will highlight potential overlaps between cancer stem cell biology and viral infections and underscore the relevance of our research on COVID-19 and cancer patients in broader terms.

This book chapter aims to offer valuable perspectives on the effectiveness of dietary polyphenols in fighting inflammation and regulating the immune response in the context of COVID-19. By bridging the fields of cancer biology and COVID-19 research, we aim to explore the multifaceted benefits of dietary polyphenols, not only as adjunctive therapies for COVID-19 but also as potential candidates for future drug development and therapeutic strategies.

11.2 THE LINK BETWEEN INFLAMMATION, IMMUNE RESPONSE, AND COVID-19

Inflammation and immune dysregulation play a pivotal role in the development of COVID-19. Understanding how the immune system interacts with the SARS-CoV-2 virus is crucial to appreciating the course of the disease and developing appropriate therapy options. The angiotensin-converting enzyme 2 (ACE2) receptor is the primary entrance point for SARS-CoV-2 after infection, and the respiratory epithelial cells are the primary target [10]. Nevertheless, the virus's impact extends beyond the respiratory system, affecting various organs and systems in the body. Both innate and adaptive immune responses play a crucial role

DOI: 10.1201/9781003433200-11

in COVID-19 pathogenesis, highlighting the importance of the immune system in recognizing and responding to the virus [9]. Pro-inflammatory cytokines and chemokines are released in response to an infection's first activation of the innate immune response (Figure 11.1). Immune signalling molecules, such as interleukin-6 (IL-6), tumor necrosis factor-alpha (TNF-α), and interleukin-1 beta (IL-1β), attract immune cells to infection sites and enhance inflammation [11]. This initial immune response is crucial in controlling viral replication and thwarting the spread of the virus.

Furthermore, immune dysregulation in COVID-19 involves the impaired function of immune cells, particularly T cells. According to research findings, critically ill COVID-19 patients manifest lymphopenia, a decrease in lymphocyte count [12,13]. This decline in T cells, which play a vital role in orchestrating the adaptive immune response, can impede the elimination of the virus and contribute to the advancement of the disease. Poorer clinical outcomes have been linked to lower T cell numbers and worse T cell activity. The dysregulated immune response of COVID-19 is

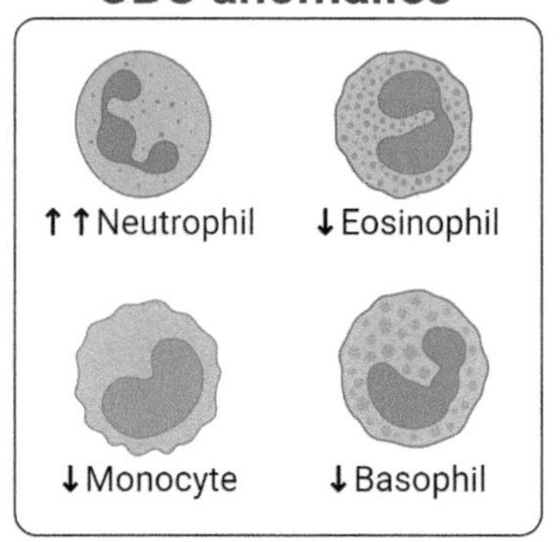

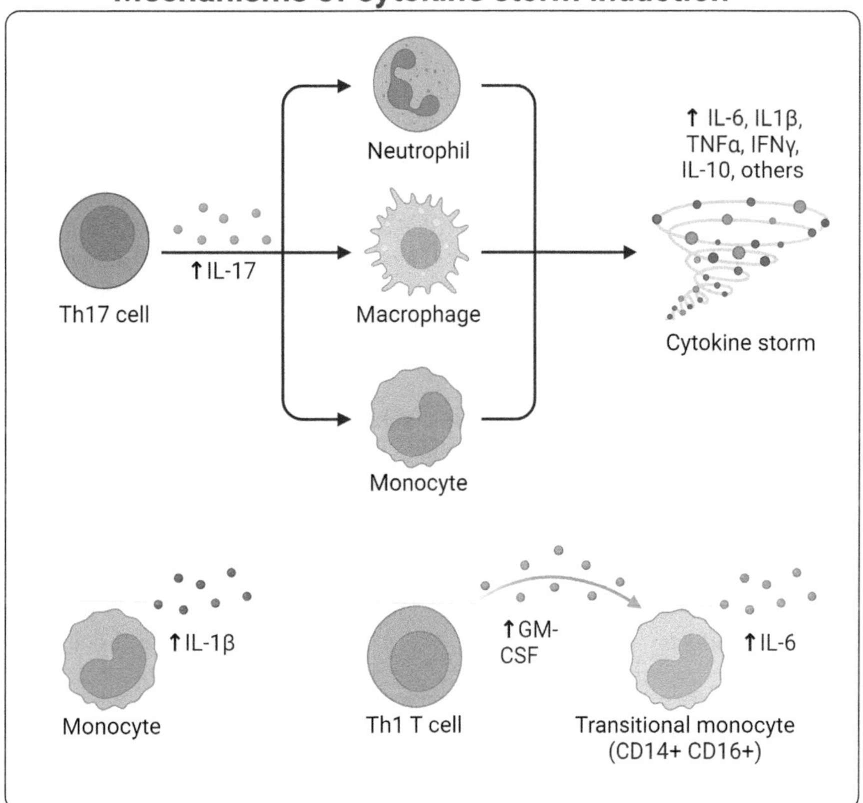

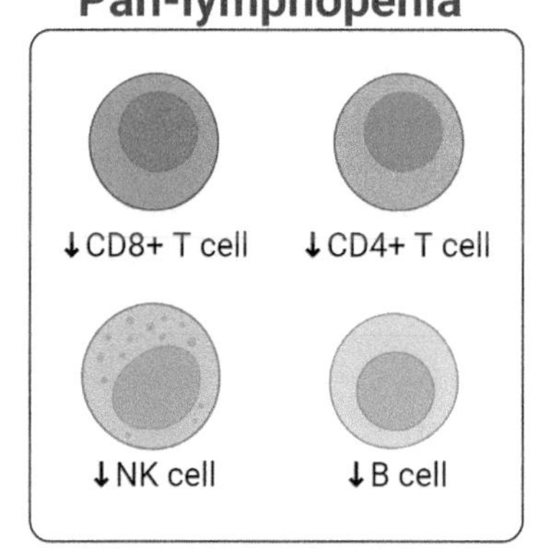

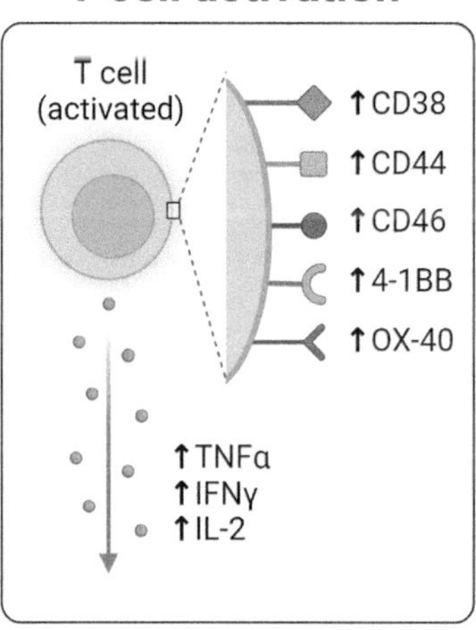

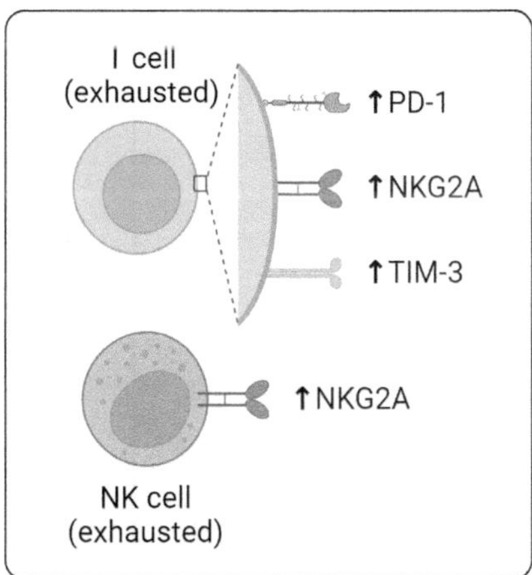

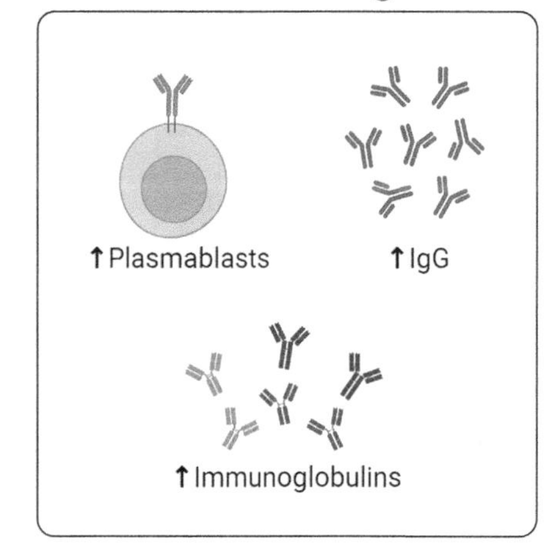

FIGURE 11.1 Overview of the immune mechanisms affected by COVID-19 infection (created with BioRender.com).

defined by an unfavourable ratio among pro-inflammatory and anti-inflammatory mediators [14]. Anti-inflammatory cytokines, such as interleukin-10 (IL-10) and transforming growth factor-beta (TGF-β), are produced at lower levels than pro-inflammatory ones. This imbalance further perpetuates the inflammatory cascade, disrupts the standard resolution of inflammation, and contributes to tissue damage. The interplay between inflammation and coagulation is another critical aspect of COVID-19 pathogenesis [15]. COVID-19 patients often exhibit a hypercoagulable state, leading to an increased risk of thrombosis and subsequent complications. Inflammatory mediators released during the immune response can activate coagulation pathways, forming blood clots [16]. The endothelial dysfunction caused by the virus further contributes to the prothrombotic state observed in severe cases.

In severe cases of COVID-19, the cytokine storm can contribute to devastating consequences as part of the immune response [17]. Pro-inflammatory cytokines are immunological signalling molecules that help coordinate the immune response against infections, and their excessive and unregulated production is the phenomenon that the term refers to (Figure 11.2). While cytokines are essential for combating diseases, an exaggerated and dysregulated cytokine response can harm the body.

Overproduction of pro-inflammatory cytokines, including IL-6, TNF-α, and IL-1β, among others, is a hallmark of the cytokine storm, which is associated with COVID-19. [11]. When immune cells detect an outbreak of Severe Acute Respiratory Syndrome Coronavirus 2 (SARS-CoV-2) virus, these cytokines are produced, specifically by macrophages and T cells. The cytokine storm in COVID-19 has many detrimental effects:

Inflammation and tissue damage: The excessive release of pro-inflammatory cytokines leads to widespread inflammation throughout the body [11]. This inflammation can damage tissues and organs, including the lungs, heart, liver, and kidneys [18]. The resultant tissue damage contributes to the disease's severity and can potentially cause organ dysfunction.

1. **Acute respiratory distress syndrome (ARDS)**: The cytokine storm is a critical factor in developing ARDS, a potentially fatal condition linked to severe COVID-19 [19]. In ARDS, excessive lung inflammation leads to increased vascular permeability, fluid accumulation, and impaired oxygen exchange. This results in severe respiratory failure and the need for mechanical ventilation.
2. **Hypercoagulability and thrombosis**: There is an elevated risk of blood clot development during the cytokine storm seen in COVID-19 [20]. Pro-inflammatory cytokines may stimulate the coagulation system and facilitate blood clot formation, resulting in complications such as deep vein thrombosis, pulmonary embolism, and stroke [11,15]. The presence of blood clots further exacerbates organ damage and contributes to worse clinical outcomes.
3. **Immune cell dysfunction**: The cytokine storm can disrupt the normal functioning of immune cells, particularly T cells [21]. T cells are crucial in coordinating the adaptive immune response and clearing the body of viruses. However, in severe COVID-19 cases, there is evidence of reduced T cell counts and impaired T cell function [22,23].

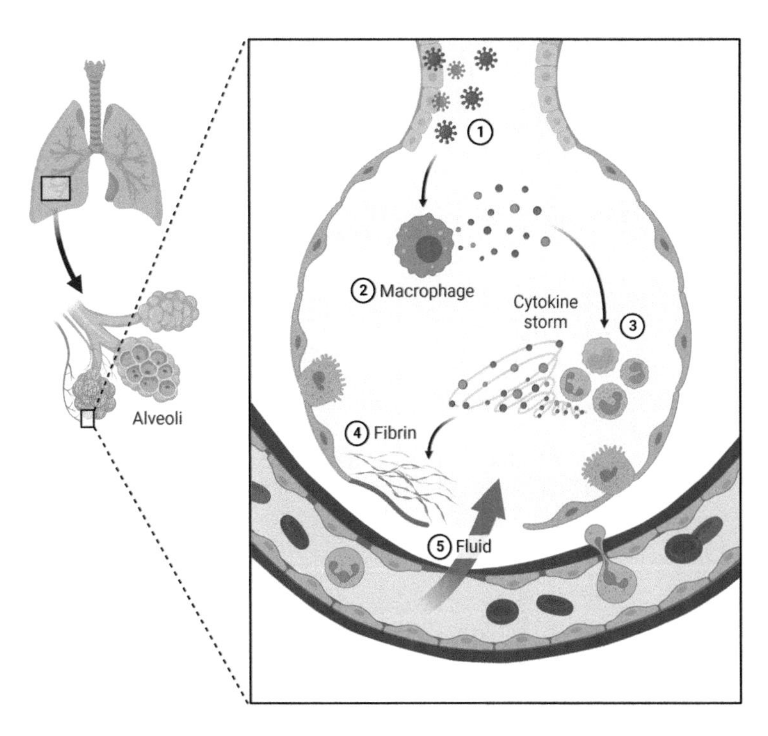

FIGURE 11.2 Schematic diagram of a cytokine storm in COVID-19 infection (created with BioRender.com).

This immune cell dysfunction hampers the body's ability to control the viral infection effectively and contributes to disease progression.

4. **Systemic effects**: The cytokine storm in COVID-19 is not limited to the respiratory system but has systemic effects throughout the body [18]. The excessive release of cytokines can lead to multi-organ dysfunction and contribute to complications in various organs [11,18]. It can also result in a state of systemic inflammation and contribute to the severity of the disease.

Understanding the significance of inflammation and immune dysregulation in COVID-19 pathogenesis is essential for developing targeted therapeutic interventions. Several treatment approaches are being investigated to modulate the immune response and reduce inflammation, such as immunomodulatory drugs and monoclonal antibodies targeting specific cytokines. By dampening the cytokine storm, it is hoped that the severity of COVID-19 and its associated complications can be mitigated. Modulating the immune response and mitigating excessive inflammation are key objectives in managing COVID-19. Strategies that aim to rebalance the immune response, alleviate inflammation, and prevent the cytokine storm hold promise for improving patient outcomes and reducing disease severity.

11.3 EXPLORING THE IMMUNOMODULATORY PROPERTIES OF DIETARY POLYPHENOLS

Polyphenols in various plant-based foods constitute a diverse category of bioactive compounds [24]. They belong to the larger class of natural compounds known as polyphenolic compounds, characterized by multiple phenolic rings derived from the shikimate or polyketide metabolic pathways in plants. Due to their antioxidant, anti-inflammatory, and immunomodulatory properties, these compounds have garnered significant interest in various fields such as medicine, nutrition, and biomedical research [25]. The chemical structures of dietary polyphenols can vary widely, giving rise to their diverse properties and potential health benefits. Some of the major subclasses of nutritional polyphenols are shown in Table 11.1.

Each subclass of dietary polyphenols exhibits distinct chemical structures characterized by variances in the arrangement and quantity of phenolic hydroxyl groups, conjugated double bonds, and other functional groups. These structural differences contribute to variations in their bioavailability, stability, and biological activities. Various factors, such as food processing, metabolism by gut microbiota, and interactions with other dietary components, can affect the bioavailability and potential health impacts of dietary polyphenols, which are commonly found in plant-based foods in complex combinations. The varied chemical compositions of nutritional polyphenols are responsible for their potential health advantages, such as their ability to act as antioxidants and reduce inflammation. They can also influence cellular signalling pathways and interact with enzymes and receptors in different physiological functions. Extensive research has been conducted on these substances to explore their potential in preventing and treating chronic ailments like cardiovascular diseases, neurodegenerative conditions, and specific forms of cancer. Understanding the chemical diversity of dietary polyphenols is crucial for evaluating their potential health benefits, optimizing their extraction and formulation, and exploring

TABLE 11.1
Overview of the Main Subclasses of Dietary Polyphenols, Their Typical Food Sources, and Associated Health Benefits

S. No.	Subclasses of Dietary Polyphenols	Compounds	Food Sources	Health Benefits	References
1.	Flavonoids	Quercetin, kaempferol, catechins, epicatechins, apigenin, luteolin, hesperetin, naringenin, cyanidin, delphinidin	Fruits, vegetables, tea, cocoa, and wine	Antioxidant, anti-inflammatory, immunomodulatory	[26,27]
2.	Phenolic acids	Gallic acid, caffeic acid, ferulic acid	Fruits, vegetables, whole grains, and coffee	Antioxidant, anti-inflammatory, immunomodulatory	[28]
3.	Stilbenes	Resveratrol	Grapes, berries, peanuts, and red wine	Antioxidant, anti-inflammatory	[27,29]
4.	Lignans	(Not specified)	Seeds, whole grains, legumes, and flaxseeds	Health-promoting effects	[30,31]
5.	Tannins	Ellagitannins, gallotannins, proanthocyanidins	Fruits, nuts, tea, and wine	Known for its astringent taste	[32,33]

their therapeutic applications in the context of various diseases and health conditions. Polyphenols have anti-inflammatory and immunomodulatory effects (as given below).

11.3.1 Anti-inflammatory Effects

Polyphenols have been extensively studied for their potent anti-inflammatory properties [34]. Inflammation is a complex physiological response that plays a crucial role in the body's defence against pathogens and injury [35]. Chronic or excessive inflammation can lead to the development and progression of various diseases, such as cardiovascular diseases, neurodegenerative disorders, and cancer. Polyphenols exert their anti-inflammatory effects through several mechanisms:

a. **Inhibition of pro-inflammatory enzymes**: Many polyphenols, such as curcumin, resveratrol, and quercetin, possess inhibitory effects on enzymes involved in the production of pro-inflammatory mediators [36]. They can inhibit the activity of cyclooxygenases (COX) and lipoxygenases (LOX), which are responsible for the synthesis of pro-inflammatory prostaglandins and leukotrienes, respectively [37]. Through the inhibition of these enzymes, polyphenols can decrease the generation of inflammatory mediators and mitigate the occurrence of inflammation.
b. **Modulating inflammatory signalling pathways**: NF-kappa B, MAPKs, and STAT are inflammatory pathways inhibited by polyphenols [38,39]. These signalling pathways regulate pro-inflammatory gene expression and cytokine production. By controlling these pathways, polyphenols can inhibit the expression of inflammatory genes and mitigate the inflammatory response [40].
c. **Antioxidant activity**: Antioxidant polyphenols can efficiently neutralize free radicals and reactive oxygen species (ROS) produced during inflammation, making them beneficial [41]. By neutralizing these harmful molecules, polyphenols can reduce oxidative stress, which is closely associated with chronic inflammation. Polyphenols can also boost the antioxidant enzymes in your body, including superoxide dismutase (SOD) and catalase. This further strengthens their antioxidant capacity [42].
d. **Regulation of immune cell function**: Immune cells involved in inflammatory response can be modulated by polyphenols, including macrophages, dendritic cells, and T cells [43]. They can regulate the formation of inflammatory cytokines by immune cells and promote a shift toward an anti-inflammatory phenotype [44]. Additionally, polyphenols can inhibit the migration of immune cells to the site of inflammation, reducing tissue infiltration and subsequent inflammatory damage [45].

11.3.2 Immunomodulatory Effects

Polyphenols can regulate the activity and function of immune cells, which results in immunomodulatory effects [36,43]. Maintaining homeostasis, defending against pathogens, and regulating immune responses are crucial functions carried out by the immune system [45]. Polyphenols are capable of influencing the immune system through the following mechanisms:

a. **Regulation of immune cell proliferation and differentiation**: Polyphenols can influence the proliferation and differentiation of immune cells, such as T cells, B cells, and natural killer (NK) cells [43]. They can promote the expansion of specific resistant cell populations and enhance their differentiation into effector or regulatory subsets. An instance of this is the ability of particular polyphenols to stimulate the differentiation of regulatory T cells (Tregs). These Tregs are pivotal in preserving immune tolerance and suppressing excessive inflammation [46].
b. **Modulation of cytokine production**: Polyphenols can regulate the production of cytokines by immune cells [43,47]. Polyphenols can hinder the production of pro-inflammatory cytokines, including IL-1β, IL-6, and TNF-α. At the same time, they can stimulate the production of anti-inflammatory cytokines, such as IL-10 and TGF-β [48–50]. Polyphenols can regulate immune responses and promote immune homeostasis by modulating the cytokine balance.
c. **Enhancement of phagocytic activity**: Green tea contains polyphenol epigallocatechin gallate (EGCG), which boosts immune cell phagocytic activity [51,52]. Phagocytosis is an essential immune function that involves engulfing and clearing pathogens and cellular debris. By enhancing phagocytic activity, polyphenols can improve immune defence mechanisms against pathogens.
d. **Modulation of immune signalling pathways**: Polyphenols can modulate various signalling pathways involved in immune cell activation and function [53]. They can suppress the activation of NF-κB, a vital transcription factor in immune and inflammatory responses. By modulating these signalling pathways, polyphenols can regulate immune cell activation, cytokine production, and immune responses.

Polyphenols' anti-inflammatory and immunomodulatory activities are essential for many inflammation or immunological dysregulation diseases. These effects make polyphenols attractive candidates for developing novel therapeutic approaches in preventing and managing conditions such as cardiovascular diseases, neurodegenerative disorders, autoimmune diseases, and cancer. However, the optimal medicinal use of polyphenols requires more study to determine their mechanisms of action.

It is crucial to emphasize that although these polyphenols have demonstrated potential in laboratory settings and certain preliminary studies, additional investigation, including clinical trials, is necessary to comprehensively comprehend their effectiveness, safety, and recommended dosage for viral infections. Moreover, the specific mechanisms by which these polyphenols exert their antiviral effects may vary depending on the virus and the stage of the viral replication cycle.

11.4 POLYPHENOLS FROM HERBAL MEDICINES: POTENTIAL ROLE IN COMBATING COVID-19

The search for effective COVID-19 strategies has led researchers to investigate the potential benefits of plant-derived natural compounds. Polyphenols have garnered significant attention due to their well-documented immunomodulatory, anti-inflammatory, and antiviral properties. Herbal medicines rich in polyphenols have been traditionally used for their therapeutic effects, and their potential in modulating the immune response against COVID-19 is an area of growing interest. Among the diverse compounds that garnered attention in this context, polyphenols from herbal medicines have shown promise due to their well-documented anti-inflammatory, antioxidant, and immunomodulatory properties. These natural bioactive compounds are abundant in various plant-based sources, such as fruits, vegetables, spices, and traditional medicinal herbs.

11.4.1 Quercetin

Quercetin, a widely studied flavonoid, is found in various plant sources such as onions, apples, and citrus fruits. It has been recognized for its anti-inflammatory and antioxidant properties. Quercetin has shown the ability to inhibit viral replication and entry by targeting viral proteins. Studies have demonstrated its potential to reduce viral load and inflammation in respiratory infections [67].

11.4.2 Epigallocatechin Gallate (EGCG)

EGCG, a primary polyphenol in green tea, has exhibited antiviral activity against various viruses, including coronaviruses. It inhibits pro-inflammatory cytokines and regulates immune cell functions to modulate immunity. EGCG's ability to interfere with the viral attachment and entry process makes it a potential candidate for mitigating COVID-19 [54,55].

11.4.3 Resveratrol

Resveratrol, known for its antioxidant and anti-inflammatory properties, is found in red grapes, berries, and nuts. Resveratrol has shown potential in reducing lung inflammation and improving function. Its effect on ACE2 expression, the receptor for SARS-CoV-2, suggests a potential role in limiting viral entry [59–61].

11.4.4 Curcumin

Curcumin, derived from turmeric, is renowned for its anti-inflammatory and immunomodulatory effects. It has been investigated for its potential to mitigate cytokine storms and hyperinflammation, contributing to severe COVID-19 cases. Curcumin's interference with viral replication and its ability to inhibit inflammatory pathways make it a valuable polyphenol for further exploration [62,63].

11.4.5 Andrographolide

Andrographis paniculata, a medicinal herb, contains andrographolide, a polyphenol with notable immunomodulatory and anti-inflammatory properties. Andrographolide has demonstrated antiviral effects against several viruses, and its potential to inhibit viral entry and replication makes it a compelling candidate for COVID-19 intervention [86].

11.4.6 Baicalin

Baicalin, derived from *Scutellaria baicalensis*, has exhibited antiviral and anti-inflammatory effects. It has been shown to inhibit the production of pro-inflammatory cytokines and modulate immune responses. Baicalin's ability to inhibit viral proteases and interfere with viral replication processes underscores its potential relevance in combating COVID-19 [87].

These herbal polyphenols offer a multifaceted approach to addressing the complex immune dysregulation and inflammation associated with COVID-19. Their ability to target various stages of the viral life cycle, modulate immune responses, and attenuate inflammation makes them attractive candidates for adjunctive therapeutic strategies. However, it's important to note that further rigorous research, including clinical trials, is needed to fully establish the efficacy and safety of these polyphenols in the context of COVID-19. The rich repertoire of polyphenols from herbal medicines presents a promising avenue for developing adjunctive therapies against COVID-19. These polyphenols' immunomodulatory, anti-inflammatory, and antiviral properties highlight their potential to manage immune dysregulation and mitigate severe outcomes associated with the disease. Integrating these natural compounds into the arsenal of COVID-19 therapeutic strategies could lead to innovative approaches for addressing the ongoing global health crisis. Here are some specific polyphenols that have been shown in Table 11.2 with potential efficacy against viral infections.

TABLE 11.2
Overview of Select Polyphenols with Potential Antiviral Activity, Their Primary Food Sources, Types of Viruses Affected, and Proposed Mechanisms of Action

S. No.	Polyphenols with Antiviral Activity	Food Source	Viruses Affected	Mechanisms of Action	References
1.	**Epigallocatechin gallate (EGCG)**	Green tea	Influenza viruses, HSV, HIV, hepatitis B and C viruses	Blocking viral entry and attachment, interfering with protein synthesis, and reducing viral enzyme activity	[54,55]
2.	**Quercetin**	Various fruits, vegetables, and herbs	Influenza A and B viruses, RSV, SARS-CoV	Inhibition of viral replication, suppression of viral entry, and modulation of host immune responses	[56–58]
3.	**Resveratrol**	Grapes, berries, red wine	Respiratory viruses, herpesviruses, influenza viruses, HIV	Interference with viral replication, inhibition of viral enzymes, and modulation of host immune responses	[59–61]
4.	**Curcumin**	Turmeric	Influenza viruses, herpesviruses, hepatitis viruses, HIV	Interference with viral entry, replication, and gene expression; modulation of inflammatory responses	[62,63]
5.	**Ellagic acid**	Various fruits, nuts, and berries	Hepatitis B and C viruses, herpesviruses, HPV	Inhibition of viral replication, interference with viral attachment, and modulation of host immune responses	[64,65]

11.5 POTENTIAL BENEFITS OF DIETARY POLYPHENOLS IN COVID-19 MANAGEMENT

The benefits of dietary polyphenols on viral infections have been the subject of numerous investigations, with encouraging results. Here are some critical studies highlighting the effects of polyphenols on various viral infections:

11.5.1 Effects of Polyphenols on Influenza Virus Infections

Using the viral hemagglutinin protein as a target, EGCG, a polyphenol present in green tea, was demonstrated to prevent influenza A and B virus entry into host cells [66]. It has also shown inhibitory effects on viral replication and reduced virus-induced inflammation [66].

11.5.2 Effects of Polyphenols on Herpesvirus Infections

Resveratrol, a polyphenol found in grapes, can prevent the spread of herpes simplex virus (HSV) by inhibiting the production of viral DNA and proteins [68]. It also reduced virus-induced inflammation and protected host cells from viral-induced damage. Ellagic acid, a polyphenol abundant in berries, was found to inhibit the replication of herpesviruses, including herpes simplex virus-1 (HSV-1) and human cytomegalovirus (HCMV), by targeting viral DNA synthesis and viral gene expression [69].

11.5.3 Effects of Polyphenols on Hepatitis Virus Infections

Curcumin, a polyphenol derived from turmeric, demonstrated antiviral effects against hepatitis B virus (HBV) and hepatitis C virus (HCV) in various studies [70]. It inhibited viral replication, reduced viral protein expression, and suppressed virus-induced inflammation [71,72]. Epicatechin gallate (ECG), a polyphenol present in green tea, inhibited HCV entrance in host cells by disrupting viral attachment and fusion [73]. It also suppressed viral replication and reduced the release of infectious viral particles.

11.5.4 Effects of Polyphenols on Human Immunodeficiency Virus (HIV) Infections

Several polyphenols, including EGCG, curcumin, and resveratrol, have been investigated for their effects on HIV infection [74]. They demonstrated inhibitory effects on viral replication, reduced viral protein expression, and modulated host immune responses to suppress viral infection and replication. Quercetin inhibits HIV-1 infection by blocking viral entry and integration into host cells [75,76]. In addition, it exhibited anti-inflammatory properties by decreasing the production of cytokines associated with inflammation.

These studies collectively suggest that dietary polyphenols have the potential to modulate viral infections by targeting different stages of the viral replication cycle, inhibiting viral entry, replication, and gene expression. They also demonstrate anti-inflammatory effects and modulation of host immune responses. However, it is essential to note that further research, including clinical trials, is necessary to validate the efficacy and safety of polyphenols as antiviral agents in humans and to determine the optimal dosage and formulation for specific viral infections. The potential of dietary polyphenols in modulating the immune response to SARS-CoV-2, the virus responsible for COVID-19, has gained significant interest. While research specific to SARS-CoV-2 is still emerging, studies on related viruses and immune modulation provide valuable insights. Here, we take a look at how dietary polyphenols may be able to alter the body's reaction to SARS-CoV-2:

1. **Anti-inflammatory effects**: Dietary polyphenols have been shown to reduce inflammation [77]. In the context of SARS-CoV-2, excessive inflammation, often referred to as a cytokine storm, plays a crucial role in disease severity. Polyphenols possess the ability to inhibit pro-inflammatory signalling pathways, specifically NF-κB and MAPKs, thereby leading to a reduction in the production of pro-inflammatory cytokines such as IL-6, IL-1β, and TNF-α [78,79]. By attenuating the inflammatory response, polyphenols may help prevent the detrimental effects of excessive inflammation in severe COVID-19 cases [80].
2. **Antioxidant activity**: Polyphenols possess potent antioxidant properties, allowing them to scavenge free radicals and reduce oxidative stress [81]. SARS-CoV-2 infection leads to increased oxidative stress due to the release of ROS [82]. Polyphenols can help protect immune cells from damage by reducing oxidative stress, maintaining their functionality, and supporting an appropriate immune response.
3. **Immunomodulation**: Polyphenols can modulate immune responses by regulating the activity and function of immune cells involved in the antiviral defence [83]. They can enhance the phagocytic activity of macrophages and promote the maturation and activation of dendritic cells, leading to improved antigen presentation and activation of adaptive immune responses [43]. Polyphenols can also modulate T-cell responses, differentiating T-helper (Th) cells into specific subsets [46]. As an illustration, particular polyphenols can stimulate the differentiation of regulatory T cells (Tregs), which play a vital role in immune tolerance and controlling excessive inflammation. These immunomodulatory effects may contribute to a balanced and effective immune response to SARS-CoV-2.
4. **Antiviral activity**: While the direct antiviral effects of polyphenols against SARS-CoV-2 are still being explored, studies on related viruses suggest their potential [55,56,62]. Polyphenols have demonstrated antiviral activity against other respiratory viruses, including influenza and coronaviruses [62,64,73]. They can inhibit viral attachment, entry, and replication, disrupting various stages of the viral life cycle. NK cells play an essential part in the body's early antiviral defence, and their activity can be boosted by polyphenols [84].
5. **ACE2 modulation**: SARS-CoV-2 enters host cells through the angiotensin-converting enzyme 2 (ACE2) receptor [10]. Evidence suggests that polyphenols may modulate ACE2 expression and activity [66,85]. Some polyphenols, such as quercetin, have been shown to downregulate ACE2 expression, potentially reducing viral entry into host cells [85]. Modulating ACE2 expression may help limit viral infectivity and subsequent inflammatory responses.

Preclinical and *in vitro* research, as well as investigations on similar viruses, suggest that dietary polyphenols may be able to modulate the immune response to SARS-CoV-2. Further research, including clinical trials, is required to validate these effects specifically for SARS-CoV-2 and determine optimal dosage, combinations, and formulations. Additionally, it is essential to note that dietary polyphenols should not be considered as a substitute for vaccination or standard medical treatments, but rather as potential complementary strategies to support immune function and overall health.

11.6 CONCLUSION AND FUTURE PERSPECTIVES

The dietary polyphenols have shown great potential in modulating immune responses and exerting beneficial effects against viral infections, including SARS-CoV-2. Their anti-inflammatory, antioxidant, and immunomodulatory properties contribute to their ability to regulate immune function, attenuate excessive inflammation, and enhance antiviral defence mechanisms. While research specific to SARS-CoV-2 is still evolving, studies on related viruses provide valuable insights into the potential of dietary polyphenols in combating viral infections. However, it is crucial to note that the current evidence is predominantly based on preclinical and *in vitro* studies, and further research, including well-designed clinical trials, is necessary to verify their efficacy, safety, and optimal dosage in terms of SARS-CoV-2.

Significant opportunities exist for further study into the impact of dietary polyphenols on infectious diseases like SARS-CoV-2. Several avenues of investigation can be pursued to advance our understanding and utilization of polyphenols in combating viral infections. To determine whether or not individual polyphenols or polyphenol-rich therapies effectively prevent and control viral infections like COVID-19, well-designed clinical trials are required. To provide robust evidence, these trials should consider different populations, dosages, and intervention durations. Further mechanistic studies are warranted to elucidate the specific pathways and molecular targets through which polyphenols modulate immune responses and inhibit viral replication. Understanding the underlying mechanisms will aid in developing targeted interventions and potentially synergistic combinations with existing antiviral therapies. Investigating polyphenols' formulation and delivery methods is essential for enhancing their bioavailability, stability, and targeted delivery to specific tissues or cells. Nanotechnology-based approaches and encapsulation techniques can be explored to improve their therapeutic potential. Synergistic effects may be achieved by combining different polyphenols or polyphenols with conventional antiviral drugs or immunomodulators. Investigating these combination approaches can enhance their efficacy and broaden their antiviral spectrum. Considering inter-individual variability in response to polyphenols and viral infections, personalized methods could be developed based on genetic, epigenetic, and metabolic profiling to optimize using polyphenols as adjunct therapies. A thorough evaluation of polyphenol interventions' safety and potential adverse effects is essential. Understanding their pharmacokinetics, possible drug interactions, and long-term impact will ensure their safe and responsible use.

In a nutshell, research into the impact of dietary polyphenols on infectious diseases like SARS-CoV-2 is a developing topic. Polyphenols' anti-inflammatory, immunomodulatory, and antiviral properties hold promise in complementing existing therapeutic strategies and promoting immune health. However, further research is necessary to translate these findings into clinical applications, paving the way for evidence-based interventions against viral infections.

CONFLICT OF INTEREST

The authors declare no conflict of interest, financial or otherwise.

REFERENCES

1. Sohrabi C, Alsafi Z, O'neill N, Khan M, Kerwan A, Al-Jabir A, Iosifidis C, Agha R. World health organization declares global emergency: A review of the 2019 novel coronavirus (COVID-19). *International Journal of Surgery.* 2020 Apr 1; 76:71–6.
2. Nugraha RV, Ridwansyah H, Ghozali M, Khairani AF, Atik N. Traditional herbal medicine candidates as complementary treatments for COVID-19: A review of their mechanisms, pros and cons. *Evidence-Based Complementary and Alternative Medicine.* 2020 Oct 10; 2020:2560645.
3. Islam MT, Sarkar C, El-Kersh DM, Jamaddar S, Uddin SJ, Shilpi JA, Mubarak MS. Natural products and their derivatives against coronavirus: A review of the non-clinical and pre-clinical data. *Phytotherapy Research.* 2020 Oct; 34(10):2471–92.
4. Costagliola G, Spada E, Consolini R. Age-related differences in the immune response could contribute to determining the spectrum of severity of COVID-19. *Immunity, Inflammation and Disease.* 2021 Jun; 9(2):331–9.
5. Qudus MS, Tian M, Sirajuddin S, Liu S, Afaq U, Wali M, Liu J, Pan P, Luo Z, Zhang Q, Yang G. The roles of critical pro-inflammatory cytokines in the drive of cytokine storm during SARS-CoV-2 infection. *Journal of Medical Virology.* 2023 Apr; 95(4):e28751.
6. Lu J, Zeng X, Lu W, Feng J, Yang Y, Wei Y, Chen Y, Zhang J, Pinhu L. Documenting the immune response in patients with COVID-19-induced acute respiratory distress syndrome. *Frontiers in Cell and Developmental Biology.* 2023 Jun 9; 11:1207960.
7. Zhang Z, Li X, Sang S, McClements DJ, Chen L, Long J, Jiao A, Jin Z, Qiu C. Polyphenols as plant-based nutraceuticals: Health effects, encapsulation, nano-delivery, and application. *Foods.* 2022 Jul 23; 11(15):2189.
8. Mitra SS, Nandy S, Dey A. Promising plant-based bioactive natural products in combating SARS-CoV2 novel corona (COVID-19) virus infection. *Medicinal Plants for Lung Diseases: A Pharmacological and Immunological Perspective.* 2021; 2021:497–514.
9. Fathi F, Sami R, Mozafarpoor S, Hafezi H, Motedayyen H, Arefnezhad R, Eskandari N. Immune system changes during COVID-19 recovery play a key role in determining disease severity. *International Journal of Immunopathology and Pharmacology.* 2020 Oct; 34:2058738420966497.
10. Rath S, Perikala V, Jena AB, Dandapat J. Factors regulating dynamics of angiotensin-converting enzyme-2 (ACE2), the gateway of SARS-CoV-2: Epigenetic modifications and therapeutic interventions by epidrugs. *Biomedicine & Pharmacotherapy.* 2021 Nov 1; 143:112095.
11. Somade OT, Ajayi BO, Adeyi OE, Aina BO, David BO, Sodiya ID. Activation of NF-kB mediates up-regulation of cerebellar and hypothalamic pro-inflammatory chemokines (RANTES and MCP-1) and cytokines (TNF-α, IL-1β, IL-6) in acute edible camphor administration. *Scientific African.* 2019 Sep 1; 5:e00114.
12. Tavakolpour S, Rakhshandehroo T, Wei EX, Rashidian M. Lymphopenia during the COVID-19 infection: What it shows and what can be learned. *Immunology Letters.* 2020 Sep; 225:31.
13. Huang W, Berube J, McNamara M, Saksena S, Hartman M, Arshad T, Bornheimer SJ, O'Gorman M. Lymphocyte subset counts in COVID-19 patients: A meta-analysis. *Cytometry Part A.* 2020 Aug; 97(8):772–6.
14. Presti EL, Nuzzo D, Al Mahmood W, Al-Rasadi K, Al-Alawi K, Banach M, Banerjee Y, Ceriello A, Cesur M, Cosentino F, Firenze A. Molecular and pro-inflammatory aspects of COVID-19: The impact on cardiometabolic health. *Biochimica et Biophysica Acta (BBA)-Molecular Basis of Disease.* 2022 Sep 26; 1868:166559.
15. Lazzaroni MG, Piantoni S, Masneri S, Garrafa E, Martini G, Tincani A, Andreoli L, Franceschini F. Coagulation dysfunction in COVID-19: The interplay between inflammation, viral infection and the coagulation system. *Blood Reviews.* 2021 Mar 1; 46:100745.
16. Noris M, Benigni A, Remuzzi G. The case of complement activation in COVID-19 multiorgan impact. *Kidney International.* 2020 Aug 1; 98(2):314–22.
17. Quirch M, Lee J, Rehman S. Hazards of the cytokine storm and cytokine-targeted therapy in patients with COVID-19. *Journal of Medical Internet Research.* 2020 Aug 13; 22(8):e20193.

18. Birman D. Investigation of the effects of covid-19 on different organs of the body. *Eurasian Journal of Chemical, Medicinal and Petroleum Research.* 2023 Jan 1; 2(1):24–36.
19. Huang Q, Wu X, Zheng X, Luo S, Xu S, Weng J. Targeting inflammation and cytokine storm in COVID-19. *Pharmacological Research.* 2020 Sep 1; 159:105051.
20. Mohamud AY, Griffith B, Rehman M, Miller D, Chebl A, Patel SC, Howell B, Kole M, Marin H. Intraluminal carotid artery thrombus in COVID-19: Another danger of cytokine storm?. *American Journal of Neuroradiology.* 2020 Sep 1; 41(9):1677–82.
21. Luo XH, Zhu Y, Mao J, Du RC. T cell immunobiology and cytokine storm of COVID-19. *Scandinavian Journal of Immunology.* 2021 Mar; 93(3):e12989.
22. Toor SM, Saleh R, Sasidharan Nair V, Taha RZ, Elkord E. T-cell responses and therapies against SARS-CoV-2 infection. *Immunology.* 2021 Jan; 162(1):30–43.
23. Adamo S, Chevrier S, Cervia C, Zurbuchen Y, Raeber ME, Yang L, Sivapatham S, Jacobs A, Baechli E, Rudiger A, Stüssi-Helbling M. Profound dysregulation of T cell homeostasis and function in patients with severe COVID-19. *Allergy.* 2021 Sep; 76(9):2866–81.
24. Ketnawa S, Reginio Jr FC, Thuengtung S, Ogawa Y. Changes in bioactive compounds and antioxidant activity of plant-based foods by gastrointestinal digestion: A review. *Critical Reviews in Food Science and Nutrition.* 2022 Jun 23; 62(17):4684–705.
25. Gengatharan A, Abd Rahim MH. The application of cloves as a potential functional component in active food packaging material and model food system: A mini-review. *Applied Food Research.* 2023 Mar 5; 3(1):100283.
26. Cheng N, Bell L, Lamport DJ, Williams CM. Dietary flavonoids and human cognition: A meta-analysis. *Molecular Nutrition & Food Research.* 2022 Nov; 66(21):2100976.
27. Durazzo A, Lucarini M, Souto EB, Cicala C, Caiazzo E, Izzo AA, Novellino E, Santini A. Polyphenols: A concise overview of chemistry, occurrence, and human health. *Phytotherapy Research.* 2019 Sep; 33(9):2221–43.
28. Sun W, Shahrajabian MH. Therapeutic potential of phenolic compounds in medicinal plants-natural health products for human health. *Molecules.* 2023 Feb 15; 28(4):1845.
29. Saad NM, Sekar M, Gan SH, Lum PT, Vaijanathappa J, Ravi S. Resveratrol: Latest scientific evidence of its chemical, biological activities and therapeutic potentials. *Pharmacognosy Journal*, 2020 Nov; 12(6s):1779–91.
30. Nadeem M, Taj Khan I, Khan F, Ajmal Shah M. Lignans and flavonolignans. Silva, AS, Nabavi, SF, Saeedi, M., Nabavi, SM, eds. *Recent Advances in Natural Products Analysis.* 2020, pp. 98–116. Singapore: Springer Nature Singapore.
31. Karimi R, Rashidinejad A. Lignans: Properties, health effects, and applications. In *Handbook of Food Bioactive Ingredients: Properties and Applications.* 2022 Nov 18, pp. 1–26. Cham: Springer International Publishing.
32. Soares S, Brandão E, Guerreiro C, Soares S, Mateus N, De Freitas V. Tannins in food: Insights into the molecular perception of astringency and bitter taste. *Molecules.* 2020 Jun 2; 25(11):2590.
33. de Melo LF, Martins VG, da Silva AP, de Oliveira Rocha HA, Scortecci KC. Biological and pharmacological aspects of tannins and potential biotechnological applications. *Food Chemistry.* 2023 Feb 6; 414:135645.
34. Bucciantini M, Leri M, Nardiello P, Casamenti F, Stefani M. Olive polyphenols: Antioxidant and anti-inflammatory properties. *Antioxidants.* 2021 Jun 29; 10(7):1044.
35. Thierry AR, Roch B. Neutrophil extracellular traps and by-products play a key role in COVID-19: Pathogenesis, risk factors, and therapy. *Journal of Clinical Medicine.* 2020 Sep 11; 9(9):2942.
36. Arya VS, Kanthlal SK, Linda G. The role of dietary polyphenols in inflammatory bowel disease: A possible clue on the molecular mechanisms involved in the prevention of immune and inflammatory reactions. *Journal of Food Biochemistry.* 2020 Nov; 44(11):e13369.
37. Rudrapal M, Eltayeb WA, Rakshit G, El-Arabey AA, Khan J, Aldosari SM, Alshehri B, Abdalla M. Dual synergistic inhibition of COX and LOX by potential chemicals from Indian daily spices investigated through detailed computational studies. *Scientific Reports.* 2023 May 27; 13(1):8656.
38. Hamsalakshmi, Alex AM, Arehally Marappa M, Joghee S, Chidambaram SB. Therapeutic benefits of flavonoids against neuroinflammation: A systematic review. *Inflammopharmacology.* 2022; 30:1–26.
39. Yu C, Wang D, Yang Z, Wang T. Pharmacological effects of polyphenol phytochemicals on the intestinal inflammation via targeting TLR4/NF-κB signaling pathway. *International Journal of Molecular Sciences.* 2022 Jun 22; 23(13):6939.
40. Castejón ML, Montoya T, Ortega-Vidal J, Altarejos J, Alarcón-de-la-Lastra C. Ligstroside aglycon, an extra virgin olive oil secoiridoid, prevents inflammation by regulation of MAPKs, JAK/STAT, NF-κB, Nrf2/HO-1, and NLRP3 inflammasome signaling pathways in LPS-stimulated murine peritoneal macrophages. *Food & Function.* 2022; 13(19):10200–9.
41. Tian M, Chen G, Xu J, Lin Y, Yi Z, Chen X, Li X, Chen S. Epigallocatechin gallate-based nanoparticles with reactive oxygen species scavenging property for effective chronic periodontitis treatment. *Chemical Engineering Journal.* 2022 Apr 1; 433:132197.
42. Jena AB, Samal RR, Bhol NK, Duttaroy AK. Cellular Red-Ox system in health and disease: The latest update. *Biomedicine & Pharmacotherapy.* 2023 Jun 1; 162:114606.
43. Shakoor H, Feehan J, Apostolopoulos V, Platat C, Al Dhaheri AS, Ali HI, Ismail LC, Bosevski M, Stojanovska L. Immunomodulatory effects of dietary polyphenols. *Nutrients.* 2021 Feb 25; 13(3):728.
44. Li R, Zhou Y, Zhang S, Li J, Zheng Y, Fan X. The natural (poly) phenols as modulators of microglia polarization via TLR4/NF-κB pathway exert anti-inflammatory activity in ischemic stroke. *European Journal of Pharmacology.* 2022 Jan 5; 914:174660.
45. Mohammadi A, Blesso CN, Barreto GE, Banach M, Majeed M, Sahebkar A. Macrophage plasticity, polarization and function in response to curcumin, a diet-derived polyphenol, as an immunomodulatory agent. *The Journal of Nutritional Biochemistry.* 2019 Apr 1; 66:1–6.
46. Fujiki T, Shinozaki R, Udono M, Katakura Y. Identification and functional evaluation of polyphenols that induce regulatory T cells. *Nutrients.* 2022 Jul 13; 14(14):2862.
47. Alesci A, Nicosia N, Fumia A, Giorgianni F, Santini A, Cicero N. Resveratrol and immune cells: a link to improve human health. *Molecules.* 2022 Jan 10; 27(2):424.

48. Chen W, Balan P, Popovich DG. The effects of New Zealand grown ginseng fractions on cytokine production from human monocytic THP-1 cells. *Molecules.* 2021 Feb 22; 26(4):1158.
49. Ondua M. Effect of fractions, and compounds from typha capensis in LPS-stimulated raw 264.7 cells. Pro and anti-inflammatory cytokines. *Current Journal of Applied Science and Technology.* 2021 Jun 22; 40(14):20–7.
50. Bocsan IC, Măgureanu DC, Pop RM, Levai AM, Macovei ŞO, Pătraşca IM, Chedea VS, Buzoianu AD. Antioxidant and anti-inflammatory actions of polyphenols from red and white grape pomace in ischemic heart diseases. *Biomedicines.* 2022 Sep 20; 10(10):2337.
51. Sun J, Dong S, Li J, Zhao H. A comprehensive review on the effects of green tea and its components on the immune function. *Food Science and Human Wellness.* 2022 Sep 1; 11(5):1143–55.
52. Hossen I, Kaiqi Z, Hua W, Junsong X, Mingquan H, Yanping C. Epigallocatechin gallate (EGCG) inhibits lipopolysaccharide-induced inflammation in RAW 264.7 macrophage cells via modulating nuclear factor kappa-light-chain enhancer of activated B cells (NF-κ B) signaling pathway. *Food Science & Nutrition.* 2023 May 25; 11(8):4634–50.
53. Jantan I, Haque MA, Arshad L, Harikrishnan H, Septama AW, Mohamed-Hussein ZA. Dietary polyphenols suppress chronic inflammation by modulation of multiple inflammation-associated cell signaling pathways. *The Journal of Nutritional Biochemistry.* 2021 Jul 1; 93:108634.
54. Wang YQ, Li QS, Zheng XQ, Lu JL, Liang YR. Anti-viral effects of green tea EGCG and its potential application against COVID-19. *Molecules.* 2021 Jun 29; 26(13):3962.
55. Li J, Song D, Wang S, Dai Y, Zhou J, Gu J. Anti-viral effect of epigallocatechin gallate via impairing porcine circovirus type 2 attachment to host cell receptor. *Viruses.* 2020 Feb 4; 12(2):176.
56. Venu LN, Austin A. Anti-viral efficacy of medicinal plants against respiratory viruses: Respiratory syncytial virus (RSV) and coronavirus (CoV)/COVID 19. *Journal of Pharmacology.* 2020; 9(4):281–90.
57. Rizky WC, Jihwaprani MC, Mushtaq M. Protective mechanism of quercetin and its derivatives in viral-induced respiratory illnesses. *The Egyptian Journal of Bronchology.* 2022 Dec; 16(1):58.
58. Yang CY, Chen YH, Liu PJ, Hu WC, Lu KC, Tsai KW. The emerging role of miRNAs in the pathogenesis of COVID-19: Protective effects of nutraceutical polyphenolic compounds against SARS-CoV-2 infection. *International Journal of Medical Sciences.* 2022; 19(8):1340.
59. Ricci A, Roviello GN. Exploring the protective effect of food drugs against viral diseases: Interaction of functional food ingredients and SARS-CoV-2, influenza virus, and HSV. *Life.* 2023 Feb 1; 13(2):402.
60. Domi E, Hoxha M, Kolovani E, Tricarico D, Zappacosta B. The importance of nutraceuticals in COVID-19: What's the role of resveratrol?. *Molecules.* 2022 Apr 7; 27(8):2376.
61. Filardo S, Di Pietro M, Mastromarino P, Sessa R. Therapeutic potential of resveratrol against emerging respiratory viral infections. *Pharmacology & Therapeutics.* 2020 Oct 1; 214:107613.
62. Kim DH, Kim JH, Kim DH, Jo JY, Byun S. Functional foods with anti-viral activity. *Food Science and Biotechnology.* 2022 May; 31(5):527–38.
63. Suryawati B, Sari Y, Avicena A, Sukmagautama C, Apriningsih H, Shofiyah L, Novika RG, Wahidah NJ, Rahmawati NY, Ansori AN, Sumarno L. The effect of curcumin and virgin coconut oil towards cytokines levels in COVID-19 patients at Universitas Sebelas Maret Hospital, Surakarta, Indonesia. *Pharmacognosy Journal.* 2022;14(1):216–225.
64. Musarra-Pizzo M, Pennisi R, Ben-Amor I, Mandalari G, Sciortino MT. Anti-viral activity exerted by natural products against human viruses. *Viruses.* 2021 May 4; 13(5):828.
65. Alexova R, Alexandrova S, Dragomanova S, Kalfin R, Solak A, Mehan S, Petralia MC, Fagone P, Mangano K, Nicoletti F, Tancheva L. Anti-COVID-19 potential of ellagic acid and polyphenols of Punica granatum L. *Molecules.* 2023 Apr 27; 28(9):3772.
66. Liu J, Bodnar BH, Meng F, Khan AI, Wang X, Saribas S, Wang T, Lohani SC, Wang P, Wei Z, Luo J. Epigallocatechin gallate from green tea effectively blocks infection of SARS-CoV-2 and new variants by inhibiting spike binding to ACE2 receptor. *Cell & Bioscience.* 2021 Dec; 11(1):1–5.
67. Khazdair MR, Anaeigoudari A, Agbor GA. Anti-viral and anti-inflammatory effects of kaempferol and quercetin and COVID-2019: A scoping review. *Asian Pacific Journal of Tropical Biomedicine.* 2021 Aug 1; 11(8):327.
68. Chen X, Song X, Zhao X, Zhang Y, Wang Y, Jia R, Zou Y, Li L, Yin Z. Insights into the anti-inflammatory and anti-viral mechanisms of resveratrol. *Mediators of Inflammation.* 2022 Aug 12; 2022: 7138756
69. Hassan ST, Šudomová M, Mazurakova A, Kubatka P. Insights into anti-viral properties and molecular mechanisms of non-flavonoid polyphenols against human herpesviruses. *International Journal of Molecular Sciences.* 2022 Nov 11; 23(22):13891.
70. Ardebili A, Pouriayevali MH, Aleshikh S, Zahani M, Ajorloo M, Izanloo A, Siyadatpanah A, Razavi Nikoo H, Wilairatana P, Coutinho HD. Anti-viral therapeutic potential of curcumin: An update. *Molecules.* 2021 Nov 19; 26(22):6994.
71. Liu Z, Ying Y. The inhibitory effect of curcumin on virus-induced cytokine storm and its potential use in the associated severe pneumonia. *Frontiers in Cell and Developmental Biology.* 2020 Jun 12; 8:479.
72. Baranwal M, Gupta Y, Dey P, Majaw S. Anti-inflammatory phytochemicals against virus-induced hyperinflammatory responses: Scope, rationale, application, and limitations. *Phytotherapy Research.* 2021 Nov; 35(11):6148–69.
73. Chojnacka K, Skrzypczak D, Izydorczyk G, Mikula K, Szopa D, Witek-Krowiak A. Anti-viral properties of polyphenols from plants. *Foods.* 2021 Sep 26; 10(10):2277.
74. Rana A, Samtiya M, Dhewa T, Mishra V, Aluko RE. Health benefits of polyphenols: A concise review. *Journal of Food Biochemistry.* 2022 Oct; 46(10):e14264.
75. Mahmud AR, Ema TI, Siddiquee M, Shahriar A, Ahmed H, Mosfeq-Ul-Hasan M, Rahman N, Islam R, Uddin MR, Mizan MF. Natural flavonols: Actions, mechanisms, and potential therapeutic utility for various diseases. *Beni-Suef University Journal of Basic and Applied Sciences.* 2023 Dec; 12(1):1–8.
76. Sun Y, Li C, Li Z, Shangguan A, Jiang J, Zeng W, Zhang S, He Q. Quercetin as an anti-viral agent inhibits the Pseudorabies virus in vitro and in vivo. *Virus Research.* 2021 Nov 1; 305:198556.

77. Arya VS, Kanthlal SK, Linda G. The role of dietary polyphenols in inflammatory bowel disease: A possible clue on the molecular mechanisms involved in the prevention of immune and inflammatory reactions. *Journal of Food Biochemistry.* 2020 Nov; 44(11):e13369.
78. Huang P, Hong J, Mi J, Sun B, Zhang J, Li C, Yang W. Polyphenols extracted from Enteromorpha clathrata alleviates inflammation in lipopolysaccharide-induced RAW 264.7 cells by inhibiting the MAPKs/NF-κB signaling pathways. *Journal of Ethnopharmacology.* 2022 Mar 25; 286:114897.
79. Lee HH, Jang E, Kang SY, Shin JS, Han HS, Kim TW, Lee DH, Lee JH, Jang DS, Lee KT. Anti-inflammatory potential of Patrineolignan B isolated from Patrinia scabra in LPS-stimulated macrophages via inhibition of NF-κB, AP-1, and JAK/STAT pathways. *International Immunopharmacology.* 2020 Sep 1; 86:106726.
80. Milton-Laskibar I, Trepiana J, Macarulla MT, Gómez-Zorita S, Arellano-García L, Fernández-Quintela A, Portillo MP. Potential usefulness of Mediterranean diet polyphenols against COVID-19-induced inflammation: a review of the current knowledge. *Journal of Physiology and Biochemistry.* 2022 Nov 8; 79:1–2.
81. Rudrapal M, Khairnar SJ, Khan J, Dukhyil AB, Ansari MA, Alomary MN, Alshabrmi FM, Palai S, Deb PK, Devi R. Dietary polyphenols and their role in oxidative stress-induced human diseases: Insights into protective effects, antioxidant potentials and mechanism (s) of action. *Frontiers in Pharmacology.* 2022 Feb 14; 13:283.
82. Wieczfinska J, Kleniewska P, Pawliczak R. Oxidative stress-related mechanisms in SARS-CoV-2 infections. *Oxidative Medicine and Cellular Longevity.* 2022 Mar 8; 2022:5589089.
83. Gairola K, Gururani S, Dubey SK. Polyphenols and its effect on the immune system. In *Nutraceuticals and Functional Foods in Immunomodulators.* 2023 Jan 1, pp. 121–140. Singapore: Springer Nature Singapore.
84. Kalathil SG, Thanavala Y. Natural killer cells and T cells in hepatocellular carcinoma and viral hepatitis: Current status and perspectives for future immunotherapeutic approaches. *Cells.* 2021 May 28; 10(6):1332.
85. Junior AG, Tolouei SE, dos Reis Livero FA, Gasparotto F, Boeing T, de Souza P. Natural agents modulating ACE-2: A review of compounds with potential against SARS-CoV-2 infections. *Current Pharmaceutical Design.* 2021 Apr 1; 27(13):1588–96.
86. Intharuksa A, Arunotayanun W, Yooin W, Sirisa-Ard P. A comprehensive review of *Andrographis paniculata* (Burm. f.) Nees and its constituents as potential lead compounds for COVID-19 drug discovery. *Molecules.* 2022 Jul 13; 27(14):4479.
87. Wen Y, Wang Y, Zhao C, Zhao B, Wang J. The pharmacological efficacy of baicalin in inflammatory diseases. *International Journal of Molecular Sciences.* 2023 May 26; 24(11):9317.

12 Molecular Insights into Anti-SARS-CoV-2 Activity of Catechins against Protein Targets for COVID-19 Management

Naina Rajak, Vipendra Kumar Singh, and Neha Garg

ABBREVIATION

ACE2	Angiotensin-converting enzyme 2
3CLpro	3C-like protease
DNMTs	DNA methyltransferase
HCoVs	Human coronaviruses
IL-6	Interleukin-6
MERS-CoV	Middle East Respiratory Syndrome Coronavirus
Mpro	Main protease
PAI-1	Plasminogen activator inhibitor-1
PLpro	Papain-like protease
SAH	S-Adenosylhomocysteine
TMPRSS2	Transmembrane protease, serine 2
TNF	Tumor necrosis factor
TF	Tissue factor

12.1 INTRODUCTION

Coronaviruses are single-strand (+ss) positive RNA viruses with glycosylated spike proteins covering their surface that are investigated to affect various avian and mammalian species [1]. Approximately seven human CoVs (HCoVs) have been revealed to have critical roles in respiratory diseases: three of seven are linked with severe respiratory illnesses with higher mortality, whereas other strains cause limited mortality [2,3]. SARS-CoV was first identified in China in 2002–2003. In contrast, another variant, MERS-CoV, was identified in 2012 following outlines of severe respiratory disease in people of Saudi Arabia. SARS-CoV-2 emerged in China (Wuhan City) as a new strain of coronavirus at the end of 2019, creating an abnormal viral pneumonia outbreak. It is transmitted mainly via respiratory droplets, affects and grows in epithelial cells of the lungs, and primarily affects the respiratory tract, as observed in COVID-19 patients [4]. Due to the high rate of transmissible potential, this new strain of virus, also called COVID-19, has propagated worldwide in very little time [5,6]. It has intensely exceeded MERS and SARS in the number of affected humans and the geographical limitation of a pandemic. The continual outburst of COVID-19 has created a remarkable risk to public health worldwide [7,8].

In brief, SARS-CoV-2 utilizes its receptor-binding domain surrounded by the spike protein (S1 domain) on the virus's surface for thė interaction with the ACE2 receptor and affecting lung cells. Post interaction to ACE2, the S protein is fixed by cell membrane-bound serine protease TMPRSS2 in the late stage of endosomes into their two subunits, S1 and S2. In the S2 domain, a fusion peptide activates the combination of the envelope with the membrane, delivering the viral RNA into the cytoplasm [9]. After entering the cytoplasm, the 5¢ domain is converted into two subunits, namely pplab and ppla. Thus, these polyproteins are further fixed by two major proteases, PL^{pro} and $3CL^{pro}$ or M^{pro}, which are a component of polyproteins. ppla and pplab generate 16 non-structural proteins with diverse roles in virus replication. The rest of the viral genome is divided into non-structural and structural proteins. Further, the structural portion like nucleocapsid (N), envelope (E), membrane (M), and spike (S) proteins construct viral RNA, subsequently replicating the genome to form novel viral fragments that are free from infected host cells to induce infection to other healthy cells [10].

A cohort study on 1099 COVID-19 patients revealed general symptoms such as myalgias (14.9%), dry cough (67%), fever (88%), dyspnea (18.7%), and fatigue (38%) at the onset of disease. Pneumonia was established as one of the most critical outcomes of the infection in most patients of COVID-19. Most patients with pneumonia have difficulty breathing after 5 days of infection [11]. Some patients have shown unreliable and irregular results at clinical levels, ranging from a subclinical introduction to failure of multiple organs and even death [12,13].

As per the epidemiological data of a large set of 72,314 patients, most cases (approximately 81%) showed moderate or mild symptoms, whereas 14% of patients sooner or later showed severe pneumonia, which needed ventilation in an intensive care unit. Some patients (5%) had severe manifestations, including COVID-19, septic shock, respiratory damage, and multiple vital organ failure [14,15]. Various studies have demonstrated that COVID-19 patients can establish a post-chronic or chronic disease syndrome that covers abnormalities and symptoms after 3 months of

DOI: 10.1201/9781003433200-12

onset. The post-COVID-19 difficulty in patients may be directly associated with impaired functions of the heart, liver, brain, lung, kidney, and coagulation system [16,17]. The older population (such as aged ≥65) is prone to various types of disease with the development of COVID-19, like obesity, chronic kidney disease, immunodeficiencies, cancer, diabetes, chronic pulmonary disease, cardiovascular disease, hypertension, and chronic liver disease [18–23]. In this framework, plant-derived compounds that showed minimal side effects and more significant antiviral potential in numerous preclinical models against SARS-CoV-2 would be novel candidates in COVID-19 patients.

Catechins are bioactive polyphenolic molecules from various vegetables, fruits, beverages, etc. The most common source of catechins is green tea extract that holds a sufficient amount of (–)-epigallocatechin-3-gallate (EGCG), (–)-epigallocatechin (EGC), (–)-epicatechin (EC), and (–)-epicatechin-3-gallate (ECG) [24,25]. To date, a wide range of pharmacological potentials has been described for catechins, such as anticancer, anti-diabetic, and antiviral properties, which are directly linked with their antioxidant and anti-inflammatory characteristics [26–29]. In the present chapter, we discuss the SARS-CoV-2 structure and epidemiology study of COVID-19, as well as provide preclinical and clinical evidence that catechins could be utilized as antiviral agents in managing COVID-19.

12.2 ARRANGEMENT OF SARS-COV-2 STRUCTURE

SARS-CoV, SARS-CoV-2, and MERS-CoV are a member of the same family and genus; SARS-CoV-2 shares a substantial amount of genetic similarity with SARS-CoV (79%), and MERS-CoV (50%), respectively. Some studies have reported that SARS-CoV and SARS-CoV-2 contain 80% sequence equality and 89% nucleotide similarity. Some genomic studies also show that Bat-CoV-RaTG13 (bat coronavirus) and SARS-CoV-2 contain 96.2% similar sequence identity [30]. Along with 16 non-structural proteins (Nsps 1–16), the virions contain four vital structural proteins: an envelope, spike, membrane, and nucleocapsid protein, as shown in Figure 12.1. The external area of SARS-CoV-2 is covered with the surface of spike glycoprotein. The SARS-CoV-2 entry into the host cell via binding to the ACE2 (angiotensin-converting enzyme) is mediated by spike glycoprotein. Spike glycoprotein contains two functional subunits, S1 and S2. The functional subunit S1 involves two domain-first N-terminal domains and receptor-binding domains that help to interact via the host cell receptor, i.e., ACE2. The S2 subunit is known for the membrane fusion between the host and virus cells [31]. The outer surface of the virions is covered by envelope protein and membrane protein.

12.3 CATECHINS

Catechins, or flavan-3-ol, are the most commonly studied phenolic compounds known as flavonoids. It is a significant antioxidant and secondary metabolite found in particular food and medicinal herbs, including tea, cocoa, berries, and legumes. Catechins contain a dihydropyran heterocycle (ring C) with a hydroxyl group attached to carbon 3 and two benzene rings known as A and B. Various types of catechins have been reported, including epicatechin gallate (ECG), epicatechin (EC), catechin, epigallocatechin (EGC) and its enantiomers gallocatechin (GC), EGCG and its enantiomers gallocatechin gallate (GCG), which are known as significant types of catechins (as shown in Figure 12.2) [27]. Several epidemiological reports have suggested that catechins play an essential role against human diseases like rectal, colon, and breast cancer by inhibiting DNA methylation via suppressing DNMTs and enhancing the SAH level, preventing tumor formation and cancer cell invasion [32,33]. Catechins also contain bactericidal, antimicrobial,

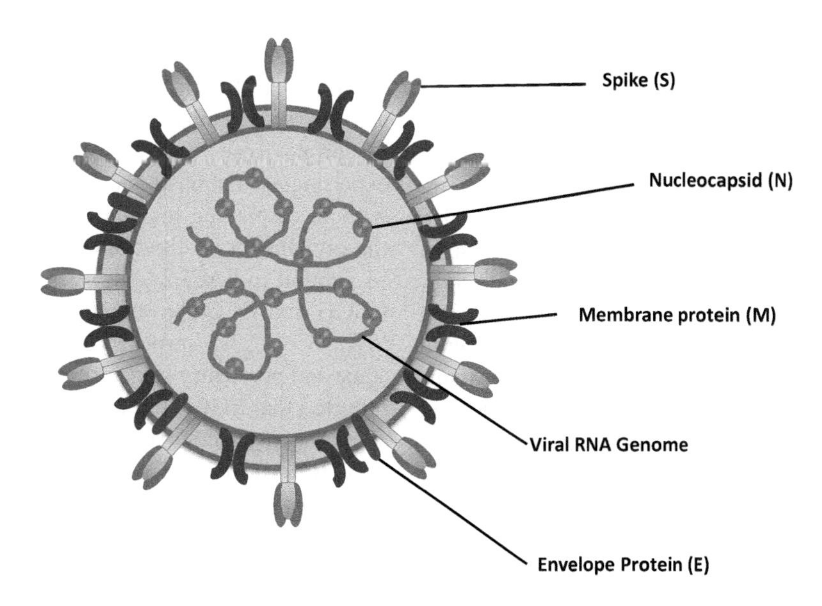

FIGURE 12.1 Structure of SARS-CoV-2.

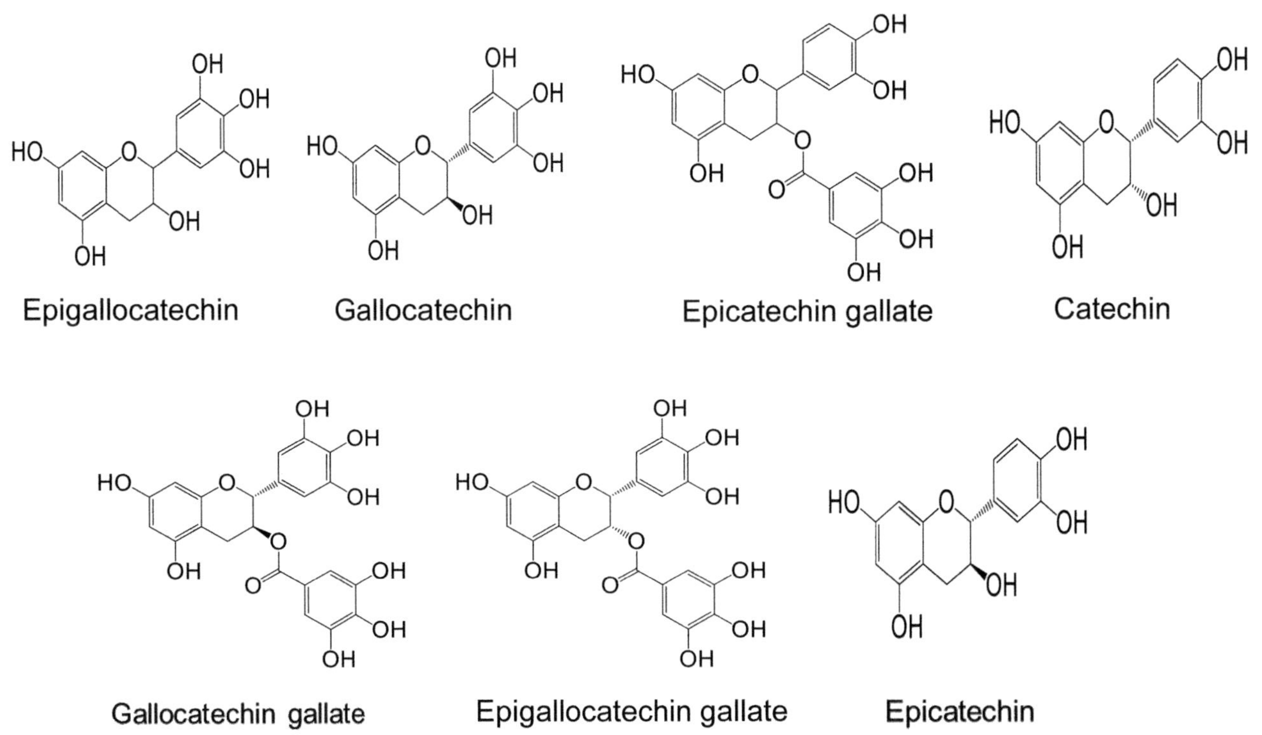

Epigallocatechin Gallocatechin Epicatechin gallate Catechin

Gallocatechin gallate Epigallocatechin gallate Epicatechin

FIGURE 12.2 The structure of some significant types of catechins.

and antiviral properties [34,35]. Various studies have reported that oral intake of catechin, which is present in foods, can prevent neurodegenerative and chronic diseases, prevent excessive oxidative stress, and enhance the activation of antioxidative molecules such as glutathione peroxidases and glutathione [36]. Some recent studies showed that catechins have inhibitory activities against many viruses, such as Zika, Chikungunya, Dengue, Ebola, H1N1, human immunodeficiency, and influenza A. The significant antiviral effects of catechins may serve as a future therapeutic use to treat COVID-19 infection [37].

12.4 CATECHINS AS AN ALTERNATIVE MEDICINE AGAINST SARS-COV-2

Due to the frequent mutations in the coronavirus genome, no specific antiviral drug for SARS-CoV-2 is presently accessible. It significantly affects the respiratory system, and multi-organ failure, especially kidneys and heart, also causes cytokine storms; as a result, there is an increase in mortality rate. Some antiviral drugs, such as Remdesivir, Favipiravir, and Lopinavir, have an inhibitory effect on HIV, Ebola virus, and influenza virus and are considered for treating COVID-19. Various plant-based phytochemicals such as quercetin, catechins, kaempferol, and baicalein are the most thoroughly studied flavonoids with antiviral properties that can be used for the better management of SARS-CoV-2 infection [38,39]. EGCG is a significant type of catechin extracted from green tea that has numerous health benefits with broad-spectrum antiviral activity toward a variety of human RNA viruses and is also used against several diseases, including anti-mutagenic [40], hypoglycemic [41], cardioprotective, antioxidative, anti-inflammatory, and immunomodulatory properties [42].

12.5 CATECHINS PREVENT VIRION ENTRY INTO THE HOST CELLS

The epithelial cells of the salivary glands and oral mucosa are strongly expressed SARS-CoV-2 entry factors, such as TMPRSS2 and ACE2 proteases, and SARS-CoV-2 accumulation in these cells has been detected in COVID-19-infected people [43]. According to Gupta et al., SARS-CoV-2 entry points showed viral RNA in the gingival crevicular fluid of 63.64% of COVID-19-infected patients [44]. Another study by Henss et al. reported that EGCG has a broad antiviral activity that inhibits virus infection by blocking the entry of coronaviruses pseudotyped lentiviral vectors. *In vitro* experiments demonstrated that EGCG inhibits the virus without displaying toxicity to the target cells; it prevents the attachment of the SARS-CoV-2 RBD present in the S1 subunit to hACE2 [45]. According to Ngwe Tun et al., EGCG and ECG showed significant inhibition effects on SARS-CoV-2 3CL-protease activity by preventing the entry stage and post-entry stage of the virus life cycle where they reported that EGCG has a higher affinity for SARS-CoV-2 suppression as compared to ECG, as confirmed by the viral RNA level and infection titer [46]. At the time of SARS-CoV-2 infection in hospital patients, catechins interfere with viral enzymes and endosome acidification, activating virus particles and targeting cells to inhibit fusion [47]. The primary step for coronaviruses is to first interact with the ACE2 receptor of host cells through the virus spike protein and form a complex of RBD–ACE2 by which the virus enters the host cell, where it starts replication and causes infection. Some mechanistic studies demonstrated that EGCG has a tremendous binding affinity for spike protein. It interacts with amino acids near the receptor-binding domain of spike proteins; as a result, fluctuation is generated between the alpha and beta helices of the RBD/ACE2 complex, thus inhibiting complex formation between RBD/ACE2 and preventing the viral infection [48]. Another study suggests that EGCG can be a potent inhibitor against SARS-CoV-1, HCoV-OC43 (human coronavirus OC43), SARS-CoV-2, and MERS-CoV by preventing the entry of these viruses in the host cells [49,50]. Further, Hong et al. revealed that EGCG strongly inhibits the Nsp 15 enzyme of SARS-CoV-2. Previously, they screened the EGCG, epigallocatechin, and epicatechin gallate from green tea extract. To analyze which compound has more potent inhibitory activity against Nsp 15, they tested EGCG using Nsp15 endoribonuclease assay. By measuring the IC_{50} value of green tea extract and EGCG, they showed that pure EGCG has three times more potent inhibitory activity against Nsp 15 than the other green tea extract. This study demonstrated that IC_{50} of green tea extract had a higher effect than EGCG. Furthermore, via computational study, they prove that numerous −OH groups of EGCG correspond with the binding site of Nsp 15. This study showed the significant importance of gallate moiety in EGCG against the Nsp 15 from SARS-CoV-2 [51].

12.6 CATECHINS TARGETING PROTEASE INHIBITOR AGAINST SARS-COV-2

Catechins significantly inhibit the $3CL^{pro}$ of SARS-CoV-2 activity with an IC_{50} value of 0.847 µM, as compared to luteolin and quercetin with an IC_{50} score of 89.670 and 97.460 µM in a FRET (fluorescence resonance energy transfer) protease assay. The binding potential of $3CL^{pro}$ of SARS-CoV-2 to these compounds is as follows: quercetin 1.24 µM, luteolin 1.63 µM, and EGCG 6.17 µM. The molecular docking result suggests that EGCG demonstrates an excellent binding affinity with $3CL^{pro}$ of SARS-CoV-2 with a calculated docking score of –7.9 kcal/mol and forms H_2 (hydrogen) bonds with Cys145 and His41, as well as different non-covalent interactions at the binding site (active site) of $3CL^{pro}$ of SARS-CoV-2 [52]. In a recent study, Liu et al. reported that all catechins can potentially inhibit the activity of SARS-CoV-2 M^{pro}. Their molecular docking study revealed the complex nature of SARS-CoV-2 M^{pro}-catechins interacting with central amino acid residues, which are His163, Asn142, Gln189, Gly143, Thr24, and Thr26. Further, the intracellular inhibition assay validated the docking result with major catechins such as EGCG, GCG, CG, and ECG. These catechins showed dose-dependent inhibition against SARS-CoV-2 M^{pro}. The GCG and ECG were reported as more potent intracellular inhibitors against SARS-CoV-2 M^{pro}; while EGCG showed poorer SARS-CoV-2 M^{pro} inhibition activity, EGCG has been known to prevent SARS-CoV-2 infections via other mechanisms [53] (Figure 12.3).

Various studies on SARS-hCoV have reported that PL^{pro} (papain-like protease) is necessary for SARS-CoV-2 replication [54,55]. PL^{pro} shows deubiquitinase and proteolytic enzymatic activity that discomfits the host's antiviral immune reactions against the SARS-CoV-2 virus. A recent study has revealed that catechins prevent PL^{pro} proteolytic and deubiquitinase enzymatic activity with IC_{50} values of 14.2 µM and 44.7 µM, respectively. The fluorescence intensity assay analysis revealed that at 50, 100, and 200 µM, catechins inhibit the deubiquitinase activity by 30%, 40%, and 50%, while EGCG by 70%, 90%, and 95%, respectively. The cytotoxic study also demonstrated that EGCG inhibits the proteolytic and deubiquitinase activity of PL^{pro}. The computational study demonstrates the complex structure

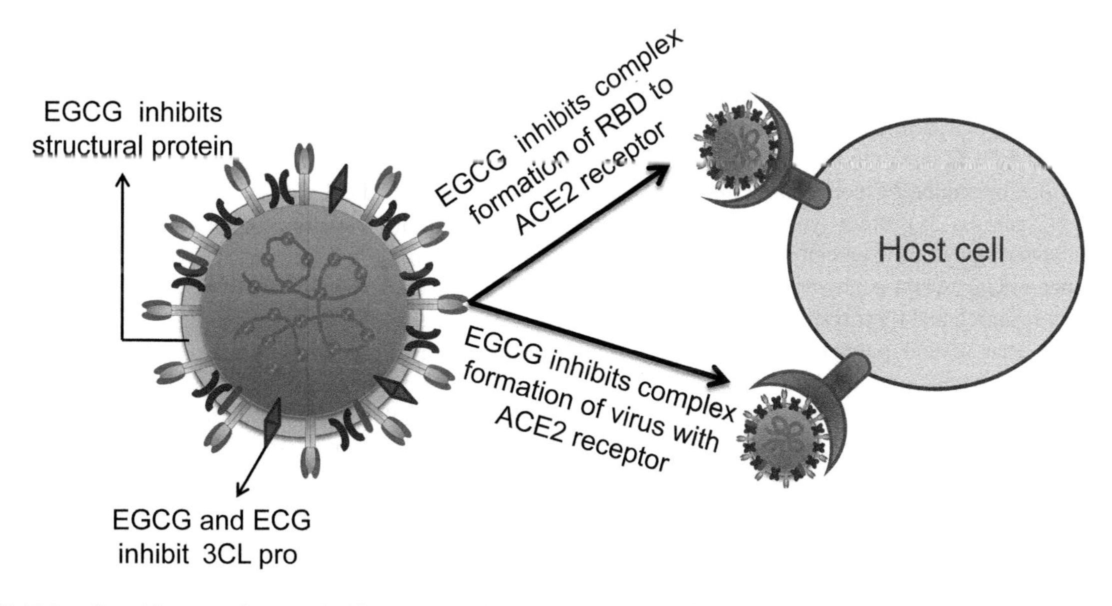

FIGURE 12.3 Catechins targeting papain-like protease (PL^{pro}) against SARS-CoV-2 activity.

of EGCG-PL^{pro}s, suggesting that the residues Glu167 and Tyr268 are the primary amino acids for the interaction between the PL^{pro} and EGCG. This study shows that catechins can be a robust potential inhibitor for SARS-CoV-2 by preventing its replication in host cells and regulating cytokine storms in prevailing coronavirus-infected patients [56].

12.7 CATECHINS PROTECT FROM THROMBOTIC, THROMBOEMBOLIC, AND LUNG FIBROSIS COMPLICATIONS IN SEVERE COVID-19 PATIENTS

An up-to-date study on COVID-19 infection demonstrated that infected patients have been allied with thrombotic and thromboembolic complications because exceeding inflammation, endothelial dysfunction, and cell and platelet activation cause significant mortality risk. Excess inflammatory cytokines such as TNF and IL-6 and platelet surface P selectin protein, released by endothelial dysfunction, encourage thrombus formation. The decreasing level of ACE2 leads to enhanced expression of Ang II, which stimulates the level of PAI-1 in several cells, such as endothelial, adipocytes, and muscle cells. The high levels of PAI-1 stimulate hypofibrinolysis; as a result, an increased risk of thrombosis has been observed [57]. Holy et al., in their *in vivo* study, have documented that EGCG, extracted from tea, can inhibit thrombosis of carotid arteries through the prevention of tissue factor activity. The molecular level experiment results demonstrated that EGCG activates the laminin receptor (endothelial cell surface membrane protein) to prevent the phosphorylation of JNK1/2 and repression of JNK-induced TF level in aortic endothelial cells. The increasing expression of TF is detected in atherosclerotic plaques, and thereby, repression of TF levels in vascular cells is a probable strategy to treat and prevent arterial thrombosis [58]. According to another *in vivo* study, EGCG protects mice against dying from pulmonary thrombosis by inhibiting platelet aggregation when given orally in doses of 10 or 50 mg/kg before injecting adrenaline and collagen into their tail veins. The molecular mechanism study demonstrated that EGCG prevents the formation of thrombus in lung tissues by inhibiting collagen-mediated PL (phospholipase) C-gamma 2, protein tyrosine phosphorylation, as well as lowering cytosolic calcium mobilization through maintenance of Ca^{2+} ATPase activity [59–61].

Growing research and their results on COVID-19 infection indicate that several patients experience respiratory symptoms after being infected by SARS-CoV-2. These studies also suggest that COVID-19-infected patients have pulmonary fibrosis and various symptoms, including dry cough, dyspnea, fatigue, and long-term disability, which may affect the patient's quality of life. Hua et al., in their *in vivo* study, reported that the most abundant catechin EGCG can significantly decrease the subsequent fibrosis by inhibiting neutrophils, ROS, and apoptosis by activating neutrophils and increasing the regression of pulmonary inflammation [62]. Several studies also demonstrated that catechin can serve as a potent anti-fibrotic drug for treating pulmonary fibrosis as it reduces inflammation and alveolar damage, prevents interstitial and peribronchial fibrosis, and decreases perivascular fibrosis [63,64].

12.8 CONCLUSION

The pandemic of COVID-19 not only caused a crisis in human health, but also had a negative impact on the world economy. Even though the epidemic has been considerably reduced worldwide, the COVID-19 virus continues to coexist in the environment and persistently mutate. In the last 3–4 years, monoclonal antibodies and few vaccines have effectively progressed to fight COVID-19 disease. The lack of fruitful therapy against SARS-CoV-2 has shifted some concerns toward plant-derived novel molecules. So, there is an urgent need to identify anti-COVID-19 from plants. Catechins are a rich source of antioxidants and are effective against various antiviral infections. Due to their greater bioavailability, easy availability, and economic feasibility, catechins-derived molecules describe a promising potential to find a novel treatment against COVID-19 infections. The catechin-derived bioactive molecules are ideally well-tolerated with low toxicity and are relatively safe and well-absorbed by the human body. Catechin constitutes a significantly underutilized origin of bioactive molecules with a wide range of antiviral potential. Various *in silico* and preclinical studies have recently identified catechin-derived molecules with auspicious anti-SARS-CoV-2 potential. Catechins can soon develop novel plant-based remedies for treating COVID-19.

12.9 FUTURE PERSPECTIVES

The COVID pandemic has affected human life worldwide for the last few years. It is assumed that COVID-19 will not be eliminated. So, there is an immediate requirement to develop novel, safe, effective, and low-toxic plant-derived alternatives for the cure of COVID-19. The speedy progression of vaccines against COVID-19 has a widespread significant step-up in lowering the impact of COVID-19, depicting a medical victory. Various persons, though completely immunized, drop their immunity eventually. The issue is mainly notable with the most recent strain, such as omicron. Thus, there is still a need for alternative drugs that can allow excellent protection against current and mutant variants of future coronaviruses.

Regarding this, catechins may be a better plant-derived bioactive molecule to fight the pandemic of COVID-19. Plant-derived biomolecules are a new juncture for use as drugs and other natural formulations for the better management of COVID-19 and other viral diseases in the future due to their therapeutic characteristics, such as safety, easy accessibility, and low cost. The significant advantage of plant-derived products is their edible nature, high thermostability, and ease of administration. On the other hand, catechins contribute a relevant chance for searching novel and fruitful anti-COVID-19 drugs. Catechins are a strong and safe candidate for combating SARS-CoV-2 as they can mess the life span of viruses, such as replication, virus-specific host targets, assembly, and viral entrance. Albeit compared to synthetic drugs, catechins are relatively less toxic due to their natural source and conventional application as folk medicine. Still, catechins may also have side effects at higher concentrations. Hence, more studies, especially *in vitro* and *in vivo*, are required to calculate the effective therapeutic dose of various catechins before use in clinical settings. Lastly, to enhance the application of catechins, combination usages, such as the treatments with other FDA-authorized anti-SARS-CoV-2 medicines or with the cooperation of advanced technology such as nanotechnology, may be a positive approach to progress catechin which has shown excellent additive/synergistic impacts against COVID-19.

ACKNOWLEDGMENT

Ms. Naina Rajak acknowledges the Council of Scientific and Industrial Research-Senior Research Fellowship (CSIR-SRF), New Delhi, India. Dr. Vipendra Kumar Singh acknowledges the Indian Council of Medical Research-Associateship Fellowship, New Delhi, India. Dr. Neha Garg would like to acknowledge the seed grant under the Institute of Eminence Scheme, BHU, for financial support.

REFERENCES

[1] Mercurio I, Tragni V, Busto F, De Grassi A, Pierri CL. Protein structure analysis of the interactions between SARS-CoV-2 spike protein and the human ACE2 receptor: From conformational changes to novel neutralizing antibodies. *Cell Mol Life Sci* 2021;78:1501–22. https://doi.org/10.1007/s00018-020-03580-1.

[2] Lim YX, Ng YL, Tam JP, Liu DX. Human coronaviruses: A review of virus-host interactions. *Dis (Basel, Switzerland)* 2016;4:26. https://doi.org/10.3390/diseases4030026.

[3] Zhu N, Zhang D, Wang W, Li X, Yang B, Song J, et al. A novel coronavirus from patients with pneumonia in China, 2019. *N Engl J Med* 2020;382:727–33. https://doi.org/10.1056/NEJMoa2001017.

[4] Li B, Li X, Wang Y, Han Y, Wang Y, Wang C, et al. Diagnostic value and key features of computed tomography in Coronavirus Disease 2019. *Emerg Microbes Infect* 2020;9:787–93. https://doi.org/10.1080/22221751.2020.1750307.

[5] Wu JT, Leung K, Leung GM. Nowcasting and forecasting the potential domestic and international spread of the 2019-nCoV outbreak originating in Wuhan, China: A modeling study. *Lancet (London, England)* 2020;395:689–97. https://doi.org/10.1016/S0140-6736(20)30260-9.

[6] Hui DS, Azhar E, Madani TA, Ntoumi F, Kock R, Dar O, et al. The continuing 2019-nCoV epidemic threat of novel coronaviruses to global health - The latest 2019 novel coronavirus outbreak in Wuhan, China. *Int J Infect Dis* 2020;91:264–6. https://doi.org/10.1016/j.ijid.2020.01.009.

[7] Deng S-Q, Peng H-J. Characteristics of and public health responses to the coronavirus disease 2019 outbreak in China. *J Clin Med* 2020;9:575. https://doi.org/10.3390/jcm9020575.

[8] Han Q, Lin Q, Jin S, You L. Coronavirus 2019-nCoV: A brief perspective from the front line. *J Infect* 2020;80:373–7. https://doi.org/10.1016/j.jinf.2020.02.010.

[9] Hoffmann M, Kleine-Weber H, Schroeder S, Krüger N, Herrler T, Erichsen S, et al. SARS-CoV-2 cell entry depends on ACE2 and TMPRSS2 and is blocked by a clinically proven protease inhibitor. *Cell* 2020;181:271–80. https://doi.org/10.1016/j.cell.2020.02.052.

[10] Wu A, Peng Y, Huang B, Ding X, Wang X, Niu P, et al. Genome composition and divergence of the novel coronavirus (2019-nCoV) originating in China. *Cell Host Microbe* 2020;27:325–8. https://doi.org/10.1016/j.chom.2020.02.001.

[11] Guan W-J, Ni Z-Y, Hu Y, Liang W-H, Ou C-Q, He J-X, et al. Clinical characteristics of coronavirus disease 2019 in China. *N Engl J Med* 2020;382:1708–20. https://doi.org/10.1056/NEJMoa2002032.

[12] van Steenbrugge GJ, Schröder FH. Androgen-dependent human prostate cancer in nude mice. The PC-82 tumor model. *Am J Clin Oncol* 1988;11(Suppl 2):S8–12. https://doi.org/10.1097/00000421-198801102-00003.

[13] Yao R, Martin CB, Haase VS, Tse BC, Nishino M, Gheorghe C, et al. Initial clinical characteristics of gravid severe acute respiratory syndrome coronavirus 2-positive patients and the risk of progression to severe coronavirus disease 2019. *Am J Obstet Gynecol MFM* 2021;3:100365. https://doi.org/10.1016/j.ajogmf.2021.100365.

[14] Wu Z, McGoogan JM. Characteristics of and important lessons from the coronavirus disease 2019 (COVID-19) outbreak in China: Summary of a report of 72,314 cases from the Chinese center for disease control and prevention. *JAMA* 2020;323:1239–42. https://doi.org/10.1001/jama.2020.2648.

[15] Chen T, Wu D, Chen H, Yan W, Yang D, Chen G, et al. Clinical characteristics of 113 deceased patients with coronavirus disease 2019: Retrospective study. *BMJ* 2020;368:m1091. https://doi.org/10.1136/bmj.m1091.

[16] Wiersinga WJ, Rhodes A, Cheng AC, Peacock SJ, Prescott HC. Pathophysiology, transmission, diagnosis, and treatment of coronavirus disease 2019 (COVID-19): A review. *JAMA* 2020;324:782–93. https://doi.org/10.1001/jama.2020.12839.

[17] Nalbandian A, Sehgal K, Gupta A, Madhavan MV, McGroder C, Stevens JS, et al. Post-acute COVID-19 syndrome. *Nat Med* 2021;27:601–15. https://doi.org/10.1038/s41591-021-01283-z.

[18] Du R-H, Liang L-R, Yang C-Q, Wang W, Cao T-Z, Li M, et al. Predictors of mortality for patients with COVID-19 pneumonia caused by SARS-CoV-2: A prospective cohort study. *Eur Respir J* 2020;55:2000524. https://doi.org/10.1183/13993003.00524-2020.

[19] Liang W, Guan W, Chen R, Wang W, Li J, Xu K, et al. Cancer patients in SARS-CoV-2 infection: A nationwide analysis in China. *Lancet Oncol* 2020;21:335–7. https://doi.org/10.1016/S1470-2045(20)30096-6.

[20] Garg S, Kim L, Whitaker M, O'Halloran A, Cummings C, Holstein R, et al. Hospitalization rates and characteristics of patients hospitalized with laboratory-confirmed coronavirus disease 2019- COVID-NET, 14 states, March 1–30, 2020. *MMWR Morb Mortal Wkly Rep* 2020;69:458–64. https://doi.org/10.15585/mmwr.mm6915e3.
[21] Richardson S, Hirsch JS, Narasimhan M, Crawford JM, McGinn T, Davidson KW, et al. Presenting characteristics, comorbidities, and outcomes among 5,700 patients hospitalized with COVID-19 in the New York City area. *JAMA* 2020;323:2052–9. https://doi.org/10.1001/jama.2020.6775.
[22] Grasselli G, Zangrillo A, Zanella A, Antonelli M, Cabrini L, Castelli A, et al. Baseline characteristics and outcomes of 1,591 patients infected with SARS-CoV-2 admitted to ICUs of the Lombardy Region, Italy. *JAMA* 2020;323:1574–81. https://doi.org/10.1001/jama.2020.5394.
[23] Petrilli CM, Jones SA, Yang J, Rajagopalan H, O'Donnell L, Chernyak Y, et al. Factors associated with hospital admission and critical illness among 5,279 people with coronavirus disease 2019 in New York City: Prospective cohort study. *BMJ* 2020;369:m1966. https://doi.org/10.1136/bmj.m1966.
[24] Ramdani D, Chaudhry AS, Seal CJ. Chemical composition, plant secondary metabolites, and minerals of green and black teas and the effect of different tea-to-water ratios during their extraction on the composition of their spent leaves as potential additives for ruminants. *J Agric Food Chem* 2013;61:4961–7. https://doi.org/10.1021/jf4002439.
[25] Shii T, Tanaka T, Watarumi S, Matsuo Y, Miyata Y, Tamaya K, et al. Polyphenol composition of a functional fermented tea obtained by tea-rolling processing of green tea and loquat leaves. *J Agric Food Chem* 2011;59:7253–60. https://doi.org/10.1021/jf201499n.
[26] Singh VK, Arora D, Ansari MI, Sharma PK. Phytochemicals-based chemopreventive and chemotherapeutic strategies and modern technologies to overcome limitations for better clinical applications. *Phytother Res* 2019;33:3064–89. https://doi.org/10.1002/ptr.6508.
[27] Fan F-Y, Sang L-X, Jiang M. Catechins and their therapeutic benefits to inflammatory bowel disease. *Molecules* 2017;22:484. https://doi.org/10.3390/molecules22030484.
[28] Chen D, Chen G, Sun Y, Zeng X, Ye H. Physiological genetics, chemical composition, health benefits and toxicology of tea (Camellia sinensis L.) flower: A review. *Food Res Int* 2020;137:109584. https://doi.org/10.1016/j.foodres.2020.109584.
[29] Kochman J, Jakubczyk K, Antoniewicz J, Mruk H, Janda K. Health benefits and chemical composition of matcha green tea: A review. *Molecules* 2020;26:85. https://doi.org/10.3390/molecules26010085.
[30] Abdelrahman Z, Li M, Wang X. Comparative review of SARS-CoV-2, SARS-CoV, MERS-CoV, and influenza A respiratory viruses. *Front Immunol* 2020;11:552909. https://doi.org/10.3389/fimmu.2020.552909.
[31] Wang M-Y, Zhao R, Gao L-J, Gao X-F, Wang D-P, Cao J-M. SARS-CoV-2: Structure, biology, and structure-based therapeutics development. *Front Cell Infect Microbiol* 2020;10:587269. https://doi.org/10.3389/fcimb.2020.587269.
[32] Montgomery M, Srinivasan A. Epigenetic gene regulation by dietary compounds in cancer prevention. *Adv Nutr* 2019;10:1012–28. https://doi.org/10.1093/advances/nmz046.
[33] Lambert JD, Hong J, Yang G-Y, Liao J, Yang CS. Inhibition of carcinogenesis by polyphenols: Evidence from laboratory investigations. *Am J Clin Nutr* 2005;81:284S–91S. https://doi.org/10.1093/ajcn/81.1.284S.
[34] Fathima A, Rao JR. Selective toxicity of Catechin-a natural flavonoid towards bacteria. *Appl Microbiol Biotechnol* 2016;100:6395–402. https://doi.org/10.1007/s00253-016-7492-x.
[35] Daglia M. Polyphenols as antimicrobial agents. *Curr Opin Biotechnol* 2012;23:174–81. https://doi.org/10.1016/j.copbio.2011.08.007.
[36] Pham-Huy LA, He H, Pham-Huy C. Free radicals, antioxidants in disease and health. *Int J Biomed Sci* 2008;4:89–96.
[37] Xu J, Xu Z, Zheng W. A review of the antiviral role of green tea catechins. *Molecules* 2017;22:1337. https://doi.org/10.3390/molecules22081337.
[38] Sissoko D, Laouenan C, Folkesson E, M'Lebing A-B, Beavogui A-H, Baize S, et al. Experimental treatment with favipiravir for Ebola virus disease (the JIKI trial): A historically controlled, single-arm proof-of-concept trial in guinea. *PLoS Med* 2016;13:e1001967. https://doi.org/10.1371/journal.pmed.1001967.
[39] Khazeei Tabari MA, Iranpanah A, Bahramsoltani R, Rahimi R. Flavonoids as promising antiviral agents against SARS-CoV-2 infection: A mechanistic review. *Molecules* 2021;26:3900. https://doi.org/10.3390/molecules26133900.
[40] Ellis LZ, Liu W, Luo Y, Okamoto M, Qu D, Dunn JH, et al. Green tea polyphenol epigallocatechin-3-gallate suppresses melanoma growth by inhibiting inflammasome and IL-1β secretion. *Biochem Biophys Res Commun* 2011;414:551–6. https://doi.org/10.1016/j.bbrc.2011.09.115.
[41] Tsuneki H, Ishizuka M, Terasawa M, Wu J-B, Sasaoka T, Kimura I. Effect of green tea on blood glucose levels and serum proteomic patterns in diabetic (db/db) mice and on glucose metabolism in healthy humans. *BMC Pharmacol* 2004;4:18. https://doi.org/10.1186/1471-2210-4-18.
[42] Kuo C-L, Chen T-S, Liou S-Y, Hsieh C-C. Immunomodulatory effects of EGCG fraction of green tea extract in innate and adaptive immunity via T regulatory cells in murine model. *Immunopharmacol Immunotoxicol* 2014;36:364–70. https://doi.org/10.3109/08923973.2014.953637.
[43] Dinda B, Dinda S, Dinda M. Therapeutic potential of green tea catechin, (–)-epigallocatechin-3-O-gallate (EGCG) in SARS-CoV-2 infection: Major interactions with host/virus proteases. *Phytomedicine plus Int J Phyther Phytopharm* 2023;3:100402. https://doi.org/10.1016/j.phyplu.2022.100402.
[44] Gupta S, Mohindra R, Chauhan PK, Singla V, Goyal K, Sahni V, et al. SARS-CoV-2 detection in gingival crevicular fluid. *J Dent Res* 2021;100:187–93. https://doi.org/10.1177/0022034520970536.
[45] Henss L, Auste A, Schürmann C, Schmidt C, von Rhein C, Mühlebach MD, et al. The green tea catechin epigallocatechin gallate inhibits SARS-CoV-2 infection. *J Gen Virol* 2021;102:001574. https://doi.org/10.1099/jgv.0.001574.
[46] Chiou W-C, Chen J-C, Chen Y-T, Yang J-M, Hwang L-H, Lyu Y-S, et al. The inhibitory effects of PGG and EGCG against the SARS-CoV-2 3C-like protease. *Biochem Biophys Res Commun* 2022;591:130–6. https://doi.org/10.1016/j.bbrc.2020.12.106.

[47] Ngwe Tun MM, Luvai E, Nwe KM, Toume K, Mizukami S, Hirayama K, et al. Anti-SARS-CoV-2 activity of various PET-bottled Japanese green teas and tea compounds in vitro. *Arch Virol* 2022;167:1547–57. https://doi.org/10.1007/s00705-022-05483-x.

[48] Jena AB, Kanungo N, Nayak V, Chainy GBN, Dandapat J. Catechin and curcumin interact with S protein of SARS-CoV-2 and ACE2 of human cell membrane: Insights from computational studies. *Sci Rep* 2021;11:2043. https://doi.org/10.1038/s41598-021-81462-7.

[49] Mhatre S, Gurav N, Shah M, Patravale V. Entry-inhibitory role of catechins against SARS-CoV-2 and its UK variant. *Comput Biol Med* 2021;135:104560. https://doi.org/10.1016/j.compbiomed.2021.104560.

[50] Liu J, Bodnar BH, Meng F, Khan AI, Wang X, Saribas S, et al. Epigallocatechin gallate from green tea effectively blocks infection of SARS-CoV-2 and new variants by inhibiting spike binding to ACE2 receptor. *Cell Biosci* 2021;11:168. https://doi.org/10.1186/s13578-021-00680-8.

[51] Hong S, Seo SH, Woo S-J, Kwon Y, Song M, Ha N-C. Epigallocatechin gallate inhibits the uridylate-specific endoribonuclease Nsp15 and efficiently neutralizes the SARS-CoV-2 strain. *J Agric Food Chem* 2021;69:5948–54. https://doi.org/10.1021/acs.jafc.1c02050.

[52] Du A, Zheng R, Disoma C, Li S, Chen Z, Li S, et al. Epigallocatechin-3-gallate, an active ingredient of traditional Chinese medicines, inhibits the 3CLpro activity of SARS-CoV-2. *Int J Biol Macromol* 2021;176:1–12. https://doi.org/10.1016/j.ijbiomac.2021.02.012.

[53] Liu S-Y, Wang W, Ke J-P, Zhang P, Chu G-X, Bao G-H. Discovery of Camellia sinensis catechins as SARS-CoV-2 3CL protease inhibitors through molecular docking, intra and extra cellular assays. *Phytomedicine* 2022;96:153853. https://doi.org/10.1016/j.phymed.2021.153853.

[54] Osipiuk J, Azizi S-A, Dvorkin S, Endres M, Jedrzejczak R, Jones KA, et al. Structure of papain-like protease from SARS-CoV-2 and its complexes with non-covalent inhibitors. *Nat Commun* 2021;12:743. https://doi.org/10.1038/s41467-021-21060-3.

[55] Ershov PV, Yablokov EO, Mezentsev YV, Chuev GN, Fedotova MV, Kruchinin SE, et al. SARS-COV-2 coronavirus papain-like protease PLpro as an antiviral target for inhibitors of active site and protein-protein interactions. *Biophysics (Oxf)* 2022;67:902–12. https://doi.org/10.1134/S0006350922060082.

[56] Kawall A, Lewis DSM, Sharma A, Chavada K, Deshmukh R, Rayalam S, et al. Inhibitory effect of phytochemicals towards SARS-CoV-2 papain like protease (PLpro) proteolytic and deubiquitinase activity. *Front Chem* 2023;10:1100460. https://doi.org/10.3389/fchem.2022.1100460.

[57] Puhm F, Allaeys I, Lacasse E, Dubuc I, Galipeau Y, Zaid Y, et al. Platelet activation by SARS-CoV-2 implicates the release of active tissue factor by infected cells. *Blood Adv* 2022;6:3593–605. https://doi.org/10.1182/bloodadvances.2022007444.

[58] Holy EW, Stämpfli SF, Akhmedov A, Holm N, Camici GG, Lüscher TF, et al. Laminin receptor activation inhibits endothelial tissue factor expression. *J Mol Cell Cardiol* 2010;48:1138–45. https://doi.org/10.1016/j.yjmcc.2009.08.012.

[59] Kang WS, Lim IH, Yuk DY, Chung KH, Park JB, Yoo HS, et al. Antithrombotic activities of green tea catechins and (−)-epigallocatechin gallate. *Thromb Res* 1999;96:229–37. https://doi.org/10.1016/s0049-3848(99)00104-8.

[60] Jin Y-R, Im J-H, Park E-S, Cho M-R, Han X-H, Lee J-J, et al. Antiplatelet activity of epigallocatechin gallate is mediated by the inhibition of PLCgamma2 phosphorylation, elevation of PGD2 production, and maintaining calcium-ATPase activity. *J Cardiovasc Pharmacol* 2008;51:45–54. https://doi.org/10.1097/FJC.0b013e31815ab4b6.

[61] Zhang Z, Zhang X, Bi K, He Y, Yan W, Yang CS, et al. Potential protective mechanisms of green tea polyphenol EGCG against COVID-19. *Trends Food Sci Technol* 2021;114:11–24. https://doi.org/10.1016/j.tifs.2021.05.023.

[62] You H, Wei L, Sun W-L, Wang L, Yang Z-L, Liu Y, et al. The green tea extract epigallocatechin-3-gallate inhibits irradiation-induced pulmonary fibrosis in adult rats. *Int J Mol Med* 2014;34:92–102. https://doi.org/10.3892/ijmm.2014.1745.

[63] Sriram N, Kalayarasan S, Sudhandiran G. Epigallocatechin-3-gallate exhibits anti-fibrotic effect by attenuating bleomycin-induced glycoconjugates, lysosomal hydrolases and ultrastructural changes in rat model pulmonary fibrosis. *Chem Biol Interact* 2009;180:271–80. https://doi.org/10.1016/j.cbi.2009.02.017.

[64] Donà M, Dell'Aica I, Calabrese F, Benelli R, Morini M, Albini A, et al. Neutrophil restraint by green tea: Inhibition of inflammation, associated angiogenesis, and pulmonary fibrosis. *J Immunol* 2003;170:4335–41. https://doi.org/10.4049/jimmunol.170.8.4335.

13 Role of Quercetin in Managing COVID-19

Lin Ang, Yixian Quah, and Eunhye Song

13.1 INTRODUCTION

COVID-19 has infected more than 676 million people worldwide as of March 10, 2023 (1). The primary clinical manifestations commonly observed are cough, fever, headache, fatigue, muscle and joint pain, and diarrhea (2,3). Some patients with COVID-19 experience a significant deterioration in their health, of which severe cases of COVID-19 often progress to respiratory failure or even life-threatening complications such as pneumonia, multi-organ failure, and death (4). Besides, a considerable number of patients with COVID-19 have reported experiencing long-term COVID-19 symptoms, including shortness of breath, cough, dizziness, brain fog, pins-and-needles feelings, and insomnia, which persist for several weeks, months, or longer (5).

The outbreak of COVID-19 has led to an increase in the public's interest in immunity and such awareness of enhancing the immune system results in an increased demand for dietary supplementation (6). At present, the available evidence is inadequate to support conclusive recommendations either in favor or against the utilization of various dietary supplements such as herbs, vitamins, minerals, fatty acids, or other supplements in treating or preventing COVID-19 (7). According to regulatory regulations in most countries, dietary supplements are restricted from being advertised as a means of treating, preventing, or curing any diseases, as only approved drugs are permitted to make such marketing legal (8). Nonetheless, there was a notable rise in the sales of dietary supplements, specifically those promoted for enhancing immune health following the onset of the COVID-19 pandemic, as many individuals believed that these products could offer some level of defense against COVID-19 infection or potentially reduce the disease severity (9).

13.1.1 Dietary Supplementation for Managing COVID-19

Many studies have investigated the role of dietary supplementation, including vitamins C and D, zinc, calcium, probiotics, fatty acids, and quercetin, in the management of COVID-19. A systematic review reported that the use of vitamin C in randomized controlled trials showed a significant decrease ($p=0.003$) in in-hospital mortality among patients with COVID-19 (10). The administration of vitamin D supplements demonstrated a significant decrease of 33% in mortality rates occurring within a 30-day timeframe of COVID-19 infection (11). An umbrella review investigating the association between micronutrients reported that zinc deficiencies increased the risk of COVID-19 infection by 1.53-fold and calcium deficiencies increased ICU admission by 4.09-fold (12).

A 445,850 subscribers app-based community survey launched to investigate the use of dietary supplements in the general population of the United Kingdom, United States, and Sweden showed that there is a significant association between the use of vitamin D, multivitamins, probiotics, or omega-3 fatty acid supplements and lower risk of testing positive COVID-19 (13). A review investigating the consumption of dietary supplements among populations ($n=22{,}518$) from Asia, Europe, and the Middle East during the first 2 years of the COVID-19 pandemic reported that the most commonly used dietary supplements were vitamin C, vitamin D, multivitamins, and zinc to enhance immunity and prevent COVID-19 infection (7). Nevertheless, the role of dietary supplements in promoting immune function has been shown. However, the association between specific dietary supplements and their potential to decrease the risk of COVID-19 infection remains uncertain.

13.2 QUERCETIN

Quercetin contains five hydroxyl groups at 3,5,7,3¢ and 4¢ of the basic skeleton of flavonol. Several hydroxyl groups in quercetin can undergo glycosylation, leading to the formation of a variety of quercetin glycosides. These glycosides are considered the main derivatives of quercetin (14). These derivatives of quercetin have often been subject to modifications aimed at enhancing their diversity and thereby yielding a range of distinct biological activities. Considerable research has also been dedicated to explore the relationship between the structure of quercetin and its derivatives and their corresponding activities, yielding valuable insights into this area of study. Extensive scientific investigations have been conducted at the molecular level to explore the properties and effects of quercetin and its derivatives, with a specific emphasis on their bioactivities that are pertinent to the infection caused by the COVID-19 virus. These features encompass antioxidative, anti-inflammatory, and antiviral characteristics.

DOI: 10.1201/9781003433200-13

13.2.1 Structure–Activity Relationships (SARs) of Quercetin

13.2.1.1 Anti-oxidative and Anti-inflammation Activities

Quercetin has been documented to display stronger radical scavenging activity in comparison to its derivatives. The efficient radical scavenging ability of quercetin can be attributed to its fulfillment of three fundamental conditions. First, quercetin possesses an o-dihydroxy structure within its B ring, which serves to stabilize the radical state and participates in the dispersion of electrons. Second, quercetin possesses a 2,3 double bond in conjunction with a 4-oxo functional group located in the C ring. This relationship between the antioxidative potential and the molecular structure in terms of electron dispersion within the aromatic nucleus is significant. When quercetin reacts with free radicals, the phenoxyl radicals produced are stabilized by the resonance effect of the aromatic nucleus. The third criterion is that quercetin contains 3- and 5-OH groups with 4-oxo function in A and C rings; these specific structural features are essential in order to achieve the highest level of radical scavenging capacity (15). A study comparing the antioxidant activity of quercetin and its derivatives concluded that the derivatization of quercetin's hydroxyl groups decreased its antioxidant potency (16). The author proposed a connection between the antioxidant potential and the quantity of available hydroxyl groups within quercetin (16). Conversely, the modification of quercetin resulted in enhanced anti-inflammatory effects, which are distinct from its antioxidant activity.

In contrast, another study indicated a correlation between quercetin's anti-inflammatory effectiveness and its inherent structural attributes compared to its derivatives. In particular, quercetin exhibited a remarkable ability to suppress elastase release from neutrophils, achieving a reduction of over 90% at a concentration of 30 μM and displaying an IC_{50} of 6.25 μM, which makes it the strongest elastase inhibitor among derivatives with bulkier functional groups (17). In accordance with this observation, the anti-inflammatory potency of quercetin, as evaluated through its capacity to inhibit elastase release, may be associated with the presence of hydroxyl groups.

13.2.1.2 Antiviral Activity

Quercetin and its derivatives have shown antiviral activity toward a wide range of viruses, including influenza A virus (18), hepatitis C virus (HCV) (19), Ebola virus (20), Mayaro virus (21), and Dengue virus (22). An investigation employing molecular docking analysis was conducted to elucidate specific structure–anti-influenza activity relationships associated with quercetin. In one study, both the general avian influenza A (H7N9) virus neuraminidase and the neuraminidase of an oseltamivir-resistant influenza virus, a key virulence factor, were employed as the docking analysis targets. Given the critical role of hydrogen bonds in mediating interactions and bindings between proteins and ligands, it's noteworthy that the researchers highlighted quercetin's possession of five hydrogen bond donors and seven hydrogen bond acceptors. Notably, their findings underscored that quercetin engages with the neuraminidase by forming at least 11 hydrogen bonds, yielding a binding energy of –9.41 kcal/mol (23).

A comprehensive examination employed molecular docking analysis to explore the interplay between quercetin and various receptors originating from diverse viruses, such as HCV, dengue type 2 virus, Ebola virus, and influenza A virus. Their observations suggest that the specific amino acids within the viral protein that exhibit a higher tendency for bonding or interaction with quercetin remain ambiguous; nevertheless, a slightly heightened frequency of interaction with quercetin is evident for cysteine compared to other amino acid residues (24). Another docking simulation provided insights into the probable inhibitory mechanism of quercetin against Dengue infection by delving into its interaction with the viral non-structural protein 3 (NS3). Their predictive analysis effectively identified that quercetin impedes viral replication by attaching itself to the JNK phosphorylation site in proximity to the amino acid residue SER137 on NS3. This discovery underscores the potential for forthcoming advancements in antiviral drug development, centered on compounds akin to quercetin that focus on this specific protein site to impede viral replication within the flavivirus family (25).

Coronavirus belongs to the Coronaviridae family which is a family of enveloped and positive-strand RNA viruses. The viral lipid envelope protects the virus and facilitates its interaction with the host, whereas the positive-strand RNA acts as a template for the synthesis of proteins (26). Severe Acute Respiratory Syndrome Coronaviruses (SARS-CoV-1 and 2) and Middle East Respiratory Syndrome Coronavirus (MERS-CoV) are highly pathogenic viruses and they use human angiotensin-converting enzyme 2 (ACE2) and dipeptidyl peptidase-4 (DPP4) for binding to their host, respectively. Specifically, to facilitate fusion, the transmembrane S protein of SARS-CoV-2 interacts with the human ACE2 receptor with its S1 subunit containing the receptor-binding domain (RBD), while the S2 subunit, upon fusion, was cleaved by host proteases leading to the activation of the glycoprotein (27). *In silico* molecular docking analyses evaluated the binding of quercetin with the S protein of SARS-CoV-2, particularly at the S2 domain with the binding energy of –8.5 kcal/mol, which is the lowest among the selected inhibitors of SARS-CoV-2 S protein (Figure 13.1). From their finding, the amino acid residues that interacted with quercetin were ILE 870, ASP 867, ALA 1056, PRO 1057, GLY 1059, HIS 1058, SER 730, MET 730, MET 731, LYS 733, VAL 860, LEU 861, and PRO 863. Quercetin forms hydrogen bonds with 8 of the amino acid residues, while the rest were through hydrophobic interaction (28).

The S protein initially exists as a protein precursor and requires proteolytic cleavage at the juncture between its S1 and S2 subunits to enable effective receptor binding. Orchestrating this cleavage process is the

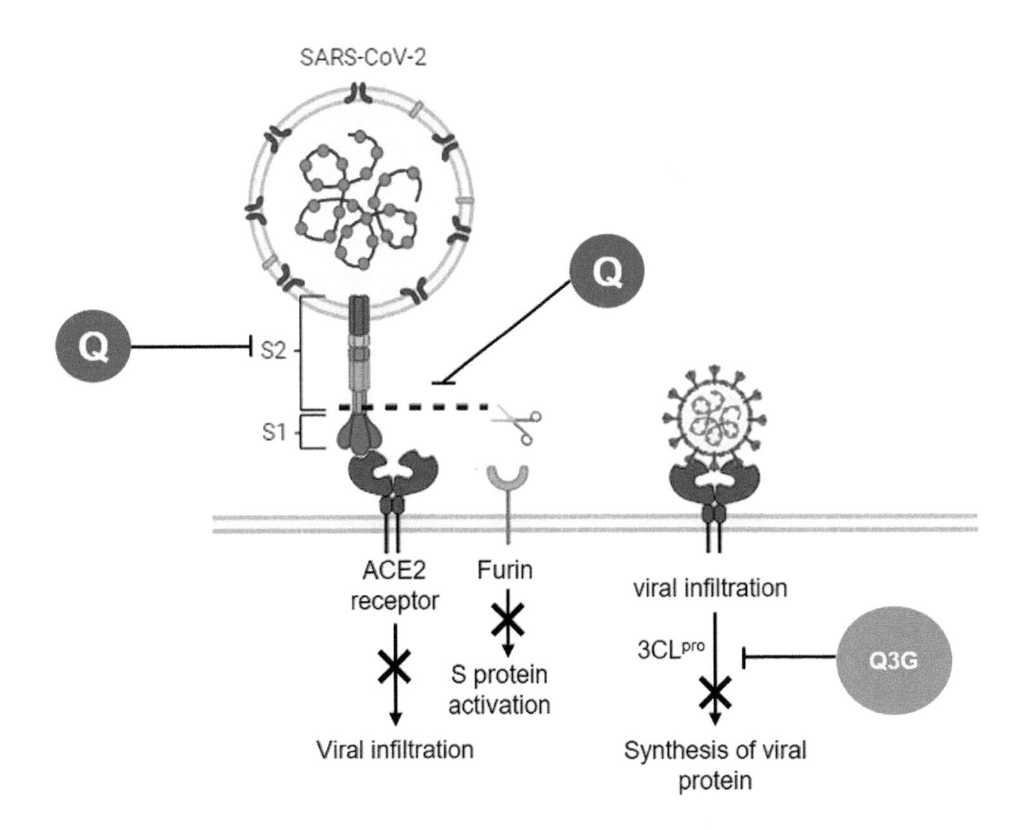

FIGURE 13.1 Graphical presentation of the antiviral mechanism of actions of quercetin against SARS-CoV-2 virus. Quercetin binds to S2 protein and furin, interrupting their interactions to cause viral infiltration. Q3G interacted with 3CL^{pro}, inhibiting its enzymatic function in viral protein synthesis. Q, quercetin; Q3G, quercetin-3-β-galactoside; S1/S2, spike proteins.

furin protease, a proprotein convertase important for the S protein's activation and subsequent enhancement of viral infectivity (29). A study examined the interaction between furin and two compounds, quercetin and its metabolite 2-(3,4-dihydroxybenzoyl)-2,4,6-trihydroxy-3(2H) benzofuranone (BZF), using molecular docking analysis (30). The simulation results indicated that the furin–quercetin interaction had lower binding free energy and inhibition constants (Ki) compared to the furin–BZF interaction. Remarkably, quercetin displayed a notable binding affinity with furin, showing a binding free energy of −7.77 kcal/mol (Figure 13.1). Notably, both quercetin and BZF displayed competitive and superior inhibitory activity when compared to established antiviral drugs like chloroquine and hydroxychloroquine. However, their inhibitory effects on the S proteins were less pronounced in comparison to their effects on furin. This discrepancy may be attributed to the higher abundance of positively charged amino acid residues in the S proteins (30).

The main protease, denoted as M^{pro}, comprising 3C-like protease (3CL^{pro}) and papain-like proteases (PL^{pro}), is considered a highly promising candidate for developing effective pharmaceutics against SARS-CoV-2 and potential future strains of coronaviruses due to various factors. Following infection, the viral RNA initiates the synthesis of viral polyproteins, which subsequently undergo processing by M^{pro} to render these proteins biologically functional and active. Quercetin-3-β-galactoside (Q3G) was identified as a potent inhibitor of SARS-CoV 3CL^{pro} through molecular docking analysis and enzymatic inhibition assays. Q3G exhibited notable competitive inhibition against SARS-CoV 3CL^{pro} (Figure 13.1). Molecular simulations further emphasized the importance of residue Gln189 in mediating the binding interaction between Q3G and SARS-CoV 3CL^{pro} (31). The same study also suggested a strong correlation between the inhibitory activity of Q3G and the presence of four hydroxyl groups on the quercetin structure. This is supported by the notable decrease in inhibitory activity when these hydroxyl groups are eliminated, as well as the observed tolerance when a substantial sugar substituent is introduced to the 7-hydroxyl position of quercetin (31). Furthermore, the substitution of the galactose moiety with alternative sugars such as fucose, arabinose, and glucose exhibited no discernible impact on the inhibitory efficacy of the compound, thereby emphasizing the crucial role played by the hydroxyl groups of the quercetin moiety (31). Docking analysis revealed that the Q3G interacted strongly with 3CL^{pro} at Gln189 residue by forming 4 hydrogen bonds with O and N atoms. Gln189 residue is one of the residues located near the pharmacophore spheres in 3CL^{pro}. Besides, Q3G also interacted with 3CL^{pro} at Leu141, Asn142, Met165, and Glu166 residues via hydrophobic interactions (32). Another study reported that the increased number of hydroxyl groups in the flavone backbone in quercetin allows it to exhibit higher inhibitory activity against PL^{pro} compared to that of kaempferol (33).

To summarize, the hydroxyl groups present within quercetin play a pivotal role in mediating the interactions between this compound and viral proteins. Quercetin demonstrates an affinity for key coronavirus proteins, including the S protein, furin protease, and M^{pro}. The primary modes of interaction observed between quercetin and its derivative, Q3G, with viral proteins primarily involve hydrogen bonding and hydrophobic interactions.

13.3 ROLES OF QUERCETIN IN MANAGING COVID-19

13.3.1 Anti-hypertension and Cardioprotective Activities of Quercetin

There have been reports highlighting the connection between angiotensin-converting enzyme 2 (ACE2) and an increased likelihood of earlier onset and more severe cases of hypertension, as well as heightened COVID-19 severity (34,35). Human ACE2 is found in the epithelial cells, particularly in the respiratory tract and pulmonary system. It serves as a pivotal function in the regulation of blood pressure and is also utilized by SARS-CoV-2 as the point of infiltration to human host cells. The S protein interacts with ACE2 receptors in human cells, facilitating viral entry. According to existing research, it has been proposed that quercetin may possess the ability to impede the interaction between the spike protein and ACE2 receptors (36). This potential inhibitory effect could potentially lead to a decrease in viral entry and subsequent infection.

Quercetin demonstrates significant inhibitory effects on recombinant human angiotensin-converting enzyme 2 (rhACE2) at concentrations that are relevant to physiological conditions *in vitro*, with an IC_{50} value of 4.48 μM (37). In humans, quercetin intake results in C_{max} values of ≤10 μM (38). At a concentration of 10 μM, quercetin demonstrated inhibitory effects on the activity of rhACE2 by decreasing its binding affinity for the Mca-APK (Dnp) substrate (37). Another study examined the effects of quercetin and Q3G on the protein expression of ACE in Human umbilical vein endothelial cells (HUVECs). They found that these two compounds could almost completely suppress ACE protein levels induced by TNF-α in the endothelial cells (Figure 13.2). The findings presented by the authors imply that quercetin and Q3G possess vasodilatory effects.

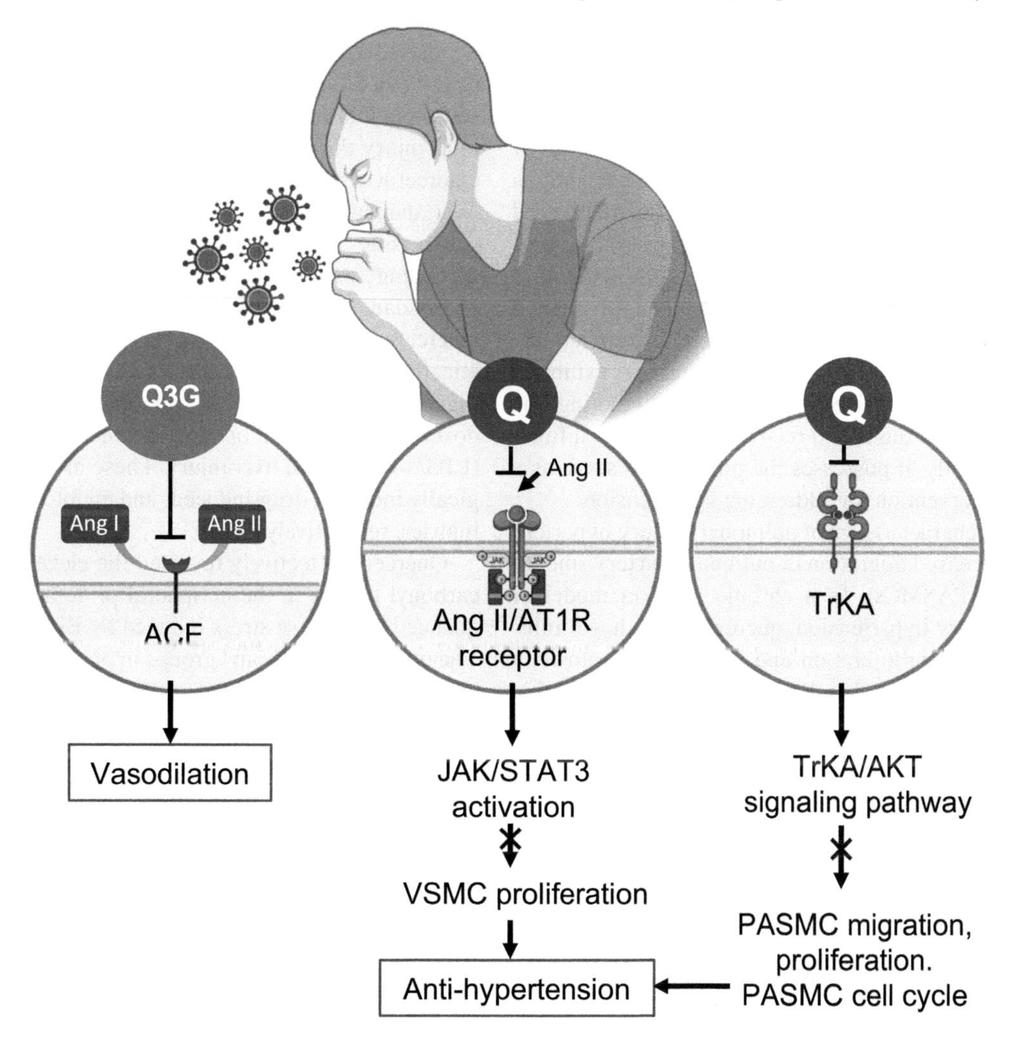

FIGURE 13.2 Graphical presentation of the anti-hypertension mechanisms of quercetin. Quercetin suppressed ACE protein levels and inhibited the JAK/STAT3 and TrKA/AKT signaling pathways, resulting in anti-hypertension effects. Q, quercetin; Q3G, quercetin-3-β-galactoside; ACE2, angiotensin-converting enzyme 2; PASMC, pulmonary artery smooth muscle cells; VSMC, vascular smooth muscle cells.

Hypertension and cardiovascular diseases have been identified as significant risk factors associated with adverse outcomes in individuals diagnosed with COVID-19 (39,40). Several studies investigated the anti-hypertension mechanism of quercetin, at a physiologically attainable concentration, in modulating the effects of angiotensin II (Ang II) induction on vascular smooth muscle cells (VSMCs) (41), vein endothelial cells (42), and cardiomyocytes (43). ACE2 converts Ang II into Ang-(1–7), and the ACE2/Ang-(1–7)/MAS axis stabilizes the negative effects of the renin–angiotensin system. Ang II plays a pivotal role in elevating blood pressure, prompting renal tubules to retain sodium and water, and stimulating aldosterone release. This potent vasoconstrictor also exhibits proliferative, pro-inflammatory, and pro-fibrotic activities. At the cellular level, Ang II induces VSMC growth, increases collagen deposition, induces inflammation, and alters vascular contractility and dilation (44–46).

A network pharmacology investigation with a target-based approach has revealed that quercetin demonstrates bioactivity against essential hypertension targets by engaging several signaling pathways. The effect of quercetin on isolated Wistar rat VSMCs was then investigated. VSMC proliferation induced by Ang II was inhibited by the presence of quercetin. In addition, it inhibits the activation of JAK2/STAT3 signaling by reducing its phosphorylation in Ang II and substantially reduces the expression level of the proliferative marker, proliferating cell nuclear antigen (PCNA), induced by Ang II (41). Ang II has been observed to reduce nitric oxide levels and enhance superoxide production in HUVECs. Nevertheless, when administered at a concentration of 3 μM, quercetin effectively counteracted the negative effects induced by Ang II on nitric oxide and superoxide levels (42). In brief, quercetin has exhibited favorable anti-hypertensive effects by counteracting the impacts caused by Ang II and recovering endothelial function. Consequently, it possesses the potential to serve as a therapeutic intervention for addressing hypertension.

One of the characteristics of pulmonary artery hypertension is the increased migration of pulmonary artery smooth muscle cells (PASMCs). In a chronic hypoxia model of pulmonary artery hypertension, quercetin has shown inhibition in PASMC proliferation and increased apoptosis in PASMCs (47). At the molecular level, it was observed that quercetin caused cell cycle arrest in the G1/G0 phase and hindered cell migration in PASMCs by inhibiting the TrkA/AKT signaling cascade, which is activated in hypoxic conditions (Figure 13.2). *In vitro* study revealed that quercetin treatment resulted in the dilation of the pulmonary artery, suppression of proliferation in smooth muscle cells of the pulmonary artery, and initiation of apoptosis in the isolated pulmonary arteries (48).

The known anti-ACE activity of quercetin has led to concerns regarding the potential increased risk of obtaining COVID-19 when utilizing ACE inhibitors (ACEI) and angiotensin-receptor blockers (ARB). The administration of ACEI and ARB has been found to upregulate the expression of ACE2, hence potentially enhancing the patient's vulnerability to COVID-19 infection (35). However, the available evidence suggests that individuals on long-term therapy with ACE inhibitors or ARBs may not have a significantly higher risk of severe COVID-19 outcomes, but the certainty of this conclusion is low due to limitations in the data (49).

13.3.2 Hepatoprotective Activities of Quercetin

Respiratory symptoms are commonly observed in individuals afflicted with COVID-19, but it is noteworthy to acknowledge that liver injury is not an uncommon indication. Studies have consistently shown that individuals with COVID-19 infection often experience varying degrees of liver injury. The liver damage, as revealed by numerous investigations, can play a role in the overall severity of the disease (50–52). Direct liver damage might be caused by direct viral invasion as ACE2 was found to be expressed in 2.6% of hepatocytes and 59.7% of cholangiocytes (53). Indirect hepatotoxicity arises from the interaction between antiviral agents and hepatic transporters, ATP Binding Cassette Subfamily C member 2/Multidrug Resistance-Associated Protein 2 (ABVV2/MRP2) (54). Oxidative stress is a significant factor in the occurrence of drug-induced liver injury. This is due to the initiation of a self-perpetuating cycle of mitochondrial damage and cellular injury through covalent modification of proteins (55). Quercetin has the potential to ameliorate drug-induced liver damage mainly by mitigating oxidative stress and suppressing inflammation (56–59), as elaborated in the preceding section where its mechanisms of action as an antioxidant and anti-inflammatory agent were discussed. Quercetin's hepatoprotective potential has been systematically assessed in various *in vivo* models encompassing liver injury induced by bile duct ligation (BDL) (56), doxorubicin (57), and lipopolysaccharides/d-galactosamine (LPS/d-GalN) (58) liver injury. These models represent surgically induced, drug-induced, and endotoxin-induced liver injuries, respectively.

Quercetin effectively reversed the elevated formation of carbonyl groups in the peripheral protein chains, a consequence of oxidative stress induced by BDL in the rats (56). The presence of carbonyl groups in proteins signifies significant oxidative damage and impaired protein function associated with diseases, which can be exacerbated by reactive oxygen species (ROS) (60). Furthermore, the same study also showed that quercetin managed to suppress the expression level of several liver fibrosis markers while increasing antioxidant protection. In doxorubicin-induced liver-injured rats, quercetin inhibits inflammation by reducing the elevated TNF-α level in the liver (57). Similarly, quercetin was found to inhibit inflammation in mice with acute liver injury by suppressing inflammatory cytokines via the IKK/NF-κB and MAPK signaling pathways (58). Quercetin exhibited a significant reduction in the elevated levels of lipid peroxidation observed in the liver exposed to doxorubicin, as evidenced by the decrease in malondialdehyde

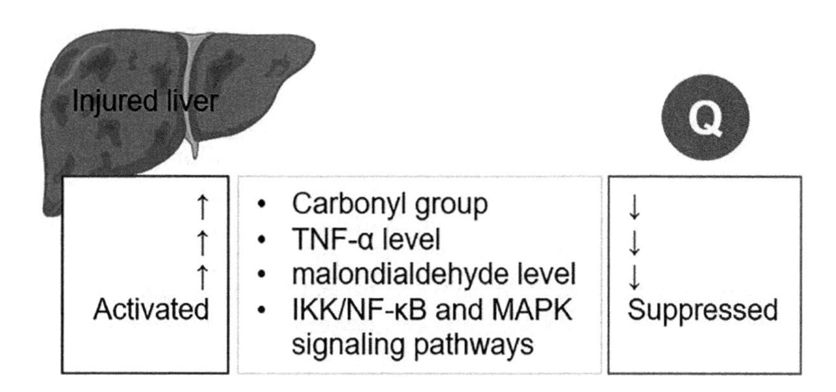

FIGURE 13.3 The summary of the hepatoprotective activities of quercetin in liver injury models.

(MDA) levels, a well-established biomarker of lipid peroxidation (57). Similarly, the level of MDA was observed to increase following treatment with LPS/d-GalN. However, the administration of quercetin significantly reduced these elevated levels (58). Quercetin demonstrates its hepatoprotective potential by effectively reducing oxidative stress, suppressing inflammation, and mitigating lipid peroxidation across various *in vivo* liver injury models (Figure 13.3). This exemplifies the potential of quercetin as a hepatoprotective agent, highlighting its diverse mechanisms of action in the management of COVID-19 infection.

13.3.3 Anti-coagulation Activities of Quercetin

Individuals infected with COVID-19 often exhibit a higher incidence of severe coagulation problems when compared to other respiratory viral infections. Blood coagulation abnormalities in patients with COVID-19 are often accompanied by increased levels of D-dimer, C-reactive protein, lactate dehydrogenase, P-selectin, fibrinogen, and Von Willebrand factor antigen. D-dimer demonstrates a direct correlation with diverse coagulation abnormalities and indicators, providing significant insights into the continuum of disease severity (61, 62). In essence, COVID-19 infection elicits the production of proinflammatory cytokines, thereby initiating the activation of tissue factor (TF). TF plays a crucial role as the principal initiator of the extrinsic coagulation pathway, thereby significantly contributing to the hypercoagulable state (63).

The connection between diabetes and COVID-19 infection is bidirectional. From a physiological standpoint, individuals with diabetes are more likely to have underlying health conditions such as hypertension, proinflammatory, and hypercoagulable states. Conversely, the treatments used for COVID-19 may exacerbate the underlying diabetes condition (64). In an induced diabetic mice model, quercetin demonstrated a notable improvement in blood flow in the carotid artery within just 5 minutes of administration at a dosage of 6 mg/kg (65). Furthermore, quercetin was observed to sustain this improved blood flow in diabetic mice even 30 minutes after the injury induced by FeCl3. An examination of platelets obtained from diabetic mice revealed that after a 7-day treatment, quercetin effectively mitigated the platelet hyper-aggregability in these diabetic mice. Additionally, the study also illustrated quercetin's capacity to delay thrombus formation in the mice.

Quercetin was studied for its antihemolytic effects and platelet aggregation inhibitory effects in human blood (66). At the physiologically attainable concentration, 10 μM, the study showed that quercetin effectively inhibited hemolysis induced by 2,2′-azobis (2-amidinopropane) dihydrochloride, a substance used to generate free radicals in human blood. They also showed that pre-incubation with quercetin was able to inhibit the aggregation of platelets which is activated by adenosine diphosphate. To ascertain whether quercetin's antihemolytic effect arises solely from its antioxidant properties or if there are different mechanisms, the author conducted further investigations by examining hemolysis induced by phospholipase C (PLC), an enzyme with known hemolytic properties that hydrolyzes phospholipids at the cellular membrane. Their findings indicate that these effects are probably facilitated by quercetin interacting with cellular membranes rather than solely through the scavenging of free radicals.

In summary, quercetin demonstrates considerable potential as a viable candidate for addressing coagulation problems associated with COVID-19 infection. Quercetin demonstrates its capacity to enhance blood circulation, reduce excessive platelet aggregation, and delay the formation of blood clots in mice with induced diabetes. Furthermore, the antihemolytic properties of quercetin and its ability to inhibit platelet aggregation in human blood, particularly through interactions with cellular membranes, demonstrate its diverse mechanisms of action. These findings highlight the potential of quercetin as a valuable therapeutic agent with diverse advantages in the management of coagulation disorders and complications related to COVID-19 infection. Additional research and clinical investigations have the potential to reveal the wider range of applications of quercetin in these crucial health scenarios.

13.4 CLINICAL STUDIES OF QUERCETIN IN MANAGING COVID-19

Many studies have demonstrated how quercetin presents a wide spectrum of pharmacological activities on viral-induced respiratory illnesses (67). A clinical study, including a large community sample, compared doses of 500 mg and 1,000 mg of quercetin to placebo found that subjects above 40 years of age taking 1,000 mg of oral quercetin supplement showed lower disease severity ($p=0.020$) and fewer sick days ($p=0.048$) for upper respiratory tract infection (68). Due to quercetin's pharmacological properties and its potential effectiveness in decreasing the risk and duration of respiratory infections, some researchers recommend studying quercetin use in managing COVID-19. Several clinical studies have examined the use of quercetin supplementation in COVID-19 patients (Table 13.1).

In an open-label randomized controlled trial (RCT) in Pakistan (69), 152 individuals who were diagnosed with COVID-19 and exhibited mild to moderate symptoms without requiring hospitalization, were enrolled and randomized into two groups. The intervention group was given daily Quevir (a supplement containing quercetin with sunflower lecithin) with standard care, while the control group was given only standard care. The duration of hospitalization and the requirement for invasive oxygen therapy were found to be significantly reduced among patients who were receiving quercetin supplements. Quercetin supplementation also reduced the progression to intensive care units and the number of deaths.

In a second RCT conducted at a similar period by the same primary investigator in Pakistan (70), 42 COVID-19 outpatients were enrolled and randomized into a quercetin supplementation group and standard care group. Of 21 patients in the quercetin supplementation group, 16 patients tested negative for COVID-19, while only 2 out of 21 patients in the standard care group tested negative after a week. After 2 weeks, all patients in the quercetin supplementation group tested negative, while 17 out of 21 patients in the standard care group tested negative. The quercetin supplementation group also showed reduced levels of lactate dehydrogenase (–35.5%), ferritin (–40%), C-reactive protein (–54.8%), and D-dimer (–11.9%).

One RCT (71) in Türkiye investigated whether co-supplementation of quercetin, vitamin C, and bromelain as an adjunctive treatment had a curative role in treating COVID-19, compared to standard care. The co-supplementation group showed a significant reduction in C-reactive protein and ferritin levels ($p<0.05$), as well as significantly increased platelet and lymphocyte counts ($p<0.05$). However, the co-supplementation group did not significantly reduce the risk of COVID-19 infections.

Another RCT (72) investigating intravenous drip of quercetin supplementation on COVID-19 patients with pneumonia found that quercetin improves pulmonary gas exchange and increases lung recovery. Quercetin supplementation also contributes to the stabilization of D-dimer levels in COVID-19 patients.

An open-label RCT (73) evaluated co-supplementation of curcumin, quercetin, and vitamin D3 in patients with mild to moderate symptoms of COVID-19 and found that co-supplementation reduced the duration of hospitalization and the serum levels of alkaline phosphatase, quantitative c-reactive protein, and lactate dehydrogenase. Co-supplementation of curcumin, quercetin, and vitamin D3 also significantly increased hemoglobin levels and respiratory rate in COVID-19 patients. However, no significant differences were found for mortality, duration, and number of intensive care unit admissions in the co-supplementation group as compared to standard care.

Another RCT in Iran (74) examining 1,000 mg of quercetin supplementation in severe COVID-19 patients reported that quercetin supplementation significantly reduced the duration of hospitalization ($p<0.039$). The serum levels of alkaline phosphatase, quantitative c-reactive protein, and lactate dehydrogenase were also significantly reduced in patients who received quercetin supplementation.

A RCT (75) exploring the possible therapeutic effect of quercetin in non-severe COVID-19 patients found that 34 out of 50 patients in the quercetin supplementation group tested negative for COVID-19, while only 12 out of 50 patients of the standard care group tested negative after a week. The quercetin supplementation group also showed a significant fall in the serum lactate dehydrogenase ($p=0.001$).

13.5 SAFETY ASPECTS OF QUERCETIN

The primary dietary sources of quercetin in the human diet are onions and apple peels (76). Quercetin exists naturally as quercetin glucoside, a compound in which quercetin is bound to glucose. This particular form of quercetin undergoes a process of deglycosylation within the gut by enterobacteria, resulting in the conversion of quercetin glucoside to quercetin aglycone, which is the unbound form of quercetin and is conducive to absorption (77). The estimated daily intake of quercetin from the dietary source has been observed in the range of 30–40 mg aglycone equivalent in Western diets (78). In dietary supplements, quercetin aglycone is commonly recommended at daily doses of 500–1,000 mg, and it was granted a 'Generally Recognized as Safe (GRAS)' by the USFDA up to 500 mg per serving (79). However, concern has been raised regarding the toxicity of quercetin. Several chronic toxicity studies in rats have revealed adverse effects including reduced body weights, increased incidence of non-neoplastic hyperplastic polyps of the cecum, and parathyroid hyperplasia. Abnormal pigmentation in the glandular stomach and the small intestine was also observed in the animals administered with high quercetin doses (80–82). In contrast, a study on the sub-chronic toxicity of quercetin was conducted by administering varying doses of quercetin (62, 125, and 250 mg/kg of diet) to CD2F1 mice. The study found no observable impact on body composition, organ function, behavior, or metabolism (83).

One of the bioactivities exhibited by quercetin is its ability to inhibit COX-2. Multiple studies have documented the capacity of quercetin to modulate the expression of COX-2,

TABLE 13.1
Characteristics of Clinical Studies Investigating Quercetin in Managing COVID-19

Study Population (*n*)	Country	Study Design, Duration	Intervention (Regimen, *n*)	Control (Regimen, *n*)	Findings	References
Non-severe COVID-19 outpatients, *n*=152	Pakistan	Open-label, 30 days	400 mg quercetin supplementation (Quevir®—quercetin formulated with sunflower lecithin, 2 tablets per day, *n*=76), plus standard care	Standard care (Analgesics/ antipyretic, oral steroids, and antibiotics—according to hospital guidelines, *n*=76)	Quercetin supplementation reduced the frequency and length of hospitalization, the need for non-invasive oxygen therapy, the progression to intensive care units, and the number of deaths	69
Non-severe COVID-19 outpatients, *n*=42	Pakistan	Open-label, 14 days	400–600 mg quercetin supplementation (Quevir®—quercetin formulated with sunflower lecithin, 2–3 tablets per day, *n*=21), plus standard care	Standard care (Analgesics/ antipyretics, oral steroids, and antibiotics—according to hospital guidelines, *n*=21)	Quercetin supplementation reduced the conversion time from positive to negative and symptom severity, as well as improved laboratory indicators	70
Hospitalized COVID-19 patients with at least one chronic disease and moderate-to-severe respiratory symptoms, *n*=434	Türkiye	Open-label, 10 days	QCB (quercetin, vitamin C, bromelain) supplementation (1,000 mg quercetin, 1,000 mg vitamin C, and 100 mg bromelain, *n*=52), plus standard care	Standard care (hydroxychloroquine, 400 mg daily for another 5 days, and favipiravir, 2×600 mg for 4 days following a 2×1,600 mg loading dose on day one, *n*=382)	Quercetin supplementation improved laboratory indicators	71
COVID-19 patients with at least 4 symptoms, *n*=200	Ukraine	Open-label, 20 days	Quercetin supplementation (Intravenous drip of Quercetin/ Polyvinylpyrrolidone for 10 days followed by oral administration of Quercetin/Pectin for 10 days, *n*=99), plus standard care	Standard care (antibiotics, antipyretic, etc., *n*=101)	Quercetin supplementation improved symptom recovery and laboratory indicators	72

(*Continued*)

TABLE 13.1 (*Continued*)
Characteristics of Clinical Studies Investigating Quercetin in Managing COVID-19

Study Population (*n*)	Country	Study Design, Duration	Intervention (Regimen, *n*)	Control (Regimen, *n*)	Findings	References
Non-severe COVID-19 outpatients, $n=56$	Pakistan	Open-label, 14 days	Quercetin-included supplementation (single oral supplement formulation containing 42 mg of specific curcumin extract, 65 mg specific quercetin extract, and 2.25 μg (90 IU) cholecalciferol, 2 capsules per day, $n=28$), plus standard care	Standard care (500 mg paracetamol with or without oral azithromycin—according to hospital guidelines, $n=28$)	Quercetin supplementation reduced the conversion time from positive to negative, improved acute symptoms, and modulated hyperinflammatory response	73
Severe COVID-19 inpatients, $n=60$	Iran	Open-label, 7 days	Quercetin supplementation (1,000 mg of quercetin, $n=30$), plus standard care	Standard care (100–200 mg Remdesivir, 1,200–3,200 mg Favipiravir, 1,000 IU vitamin D, 250 mg $MgSO_4$, 40 mg Famotidine, 30 mg Zinc sulfate, 500–1,000 mg vitamin C, 8 mg daily IV injection Dexamethasone, $n=30$)	Quercetin supplementation reduced disease severity and improved laboratory indicators	74
Non-severe COVID-19 outpatients, $n=108$	Pakistan	Open-label, 14 days	500 mg quercetin supplementation (Quercetin Phytosome® Quevir® tablet-200 mg pure quercetin formulated with sunflower lecithin, $n=54$), plus standard care	Standard care (Analgesics/ antipyretic, and antibiotics-according to hospital guidelines, $n=54$)	Quercetin supplementation reduced the conversion time from positive to negative as well as improved acute symptoms and laboratory indicators	75

thereby contributing to the anti-inflammatory properties of this flavonoid (84–86). The COX-2 enzyme not only serves as a significant mediator in the process of inflammation but also plays a crucial role in the promotion of tumor growth. Quercetin has demonstrated the ability to inhibit the expression of COX-2 and angiogenesis by suppressing p300 signaling and preventing the binding of various transactivators to the COX-2 promoter (84). While a structural docking study suggested that quercetin might function as a potential selective COX-2 inhibitor (87), it is worth noting that, regardless of its selectivity, an increased risk of developmental toxicity in rats associated with both selective and non-selective COX-2 inhibitors was reported in another previous study (88). While the inhibition of COX-2 leads to the disruption of the angiogenesis pathway, which is essential in suppressing tumor growth, this mechanism of action can have both positive and negative consequences. On one hand, quercetin's anti-angiogenic effect proves advantageous in impeding tumor or cancer progression (89), as it impedes the formation of blood vessels essential for supplying nutrients and growth factors necessary for tumor development (90). Conversely, this anti-angiogenic action can yield detrimental effect for embryonic or fetal growth, as the normal formation of blood vessels is indispensable for sustaining embryonic development (91). However, signs of maternal or fetal toxicity were not observed in repeated oral quercetin doses up to 2,000 mg/kg body weight administered in female rats during pregnancy (92,93). A separate investigation was conducted to assess the developmental toxicity of quercetin and its derivatives using the chicken embryonic assay. The results indicated that quercetin exhibited a lower incidence of chicken embryonic mortality and deformities in comparison to its derivatives (94).

Quercetin, classified as a phytoestrogen, has the potential to influence hormone production and the reproductive systems of both males and females upon prolonged exposure (95). While previous study has reported adverse effects of quercetin on sperm quality in rats (96), a majority of research underscores the positive impact of quercetin on the reproductive system. Specifically, investigations have shown that quercetin can enhance progesterone synthesis in MA-10 cells, a mouse Leydig tumor cell line, implying a stimulatory role in steroidogenesis (97). Furthermore, research has demonstrated an increase in Cyp11a1 promoter activity in MA-10 cells when exposed to quercetin (98). Cyp11a1 is a key steroidogenic enzyme responsible for metabolizing cholesterol into pregnenolone, a precursor to testosterone. This elevation in testosterone levels is commonly associated with enhanced spermatogenesis (99). Quercetin exhibits a protective role in shielding preimplantation embryos from the deleterious effects of diabetes. The mitigation of disturbances in sex hormones during early pregnancy holds promise for addressing reproductive disorders in diabetic women. Therefore, quercetin emerges as a novel and viable solution for enhancing reproductive disorders in diabetic females (100).

Based on the results of the Ames assay, quercetin demonstrated mutagenicity at a concentration of 5×10^{-3} pmol/plate (101). The Ames assay employs *Salmonella typhimurium* as a means of assessing DNA mutations induced by the test compound. Quercetin acted directly and its mutagenicity increased with metabolic activation. The mutagenic properties of quercetin were attributed to the structural features. The fundamental structural prerequisites for eliciting mutagenicity encompass a flavonoid ring structure that includes a free hydroxyl group at the 3-position, a double bond at the 2,3-position, and a keto-group positioned at the 4-position allowing the proton of the hydroxyl group at 3-position to convert into a 3-keto moiety (102). Notably, quercetin exhibits these distinctive characteristics. In addition to its mutagenic potential, DataWarrior analysis predicted mutagenic and tumorigenic properties of quercetin (103). Flavonols, such as quercetin, are highly susceptible to oxidation by atmospheric oxygen (104). Attempts have been made to study the oxidation of quercetin (104,105). The most common natural process is oxidation by atmospheric oxygen. When high-purity quercetin is subjected to minimal oxidation, it undergoes oxidative degradation resulting in the formation of protocatechuic acid, phloroglucinic acid, and phloroglucinol. These products arise due to microbial degradation of unabsorbed quercetin in the colon. Fortunately, they are not known to have any adverse effects on human health (106,107).

In terms of COVID-19-related clinical trials, no serious adverse effects have been reported. Only one clinical study that used 200 mg of quercetin twice a day for 30 days reported adverse effects such as sleep disorder, gastric pain, flatulence, diarrhea, constipation, and reflux; however, the treatment and control groups had these side effects at similar rates (69).

13.6 FUTURE IMPLICATIONS

The health benefits of quercetin have been extensively explored for diverse diseases and health conditions. Such information could enable COVID-19 patients to consume quercetin as a dietary supplement since COVID-19 has transitioned into an endemic. Although the number of clinical studies is limited and the clinical settings or populations were restricted to certain Asian countries, the findings suggest the effectiveness of quercetin for the treatment of COVID-19. Further randomized clinical trials involving worldwide populations could support the generalizability of the results. Current clinical evidence showed promising results of quercetin in managing COVID-19, but many of these clinical trials had small sample sizes, open-labeled designed, and measured multiple outcomes which increases the risk of bias in the study. Large well-designed randomized clinical trials of quercetin are still needed for the validation of the effectiveness of quercetin in managing COVID-19.

13.7 CONCLUSION

Quercetin, as an antioxidant and flavonoid found in herbs, vegetables, and fruits, has antioxidative, anti-inflammatory, and antiviral features without any major adverse effects. With such features, quercetin has shown its potential in the treatment of COVID-19 and infectious diseases in general. Quercetin alone or in combination with other treatments is beneficial in the treatment and prevention of COVID-19, as the mechanisms of quercetin and the clinical evidence support its effectiveness against COVID-19. Quercetin has not been granted current regulatory approval as a novel pharmaceutical intervention for either mitigating symptoms associated with COVID-19 or preventing infections caused by the SARS-CoV-2 virus. It is imperative to acknowledge that the dissemination of inaccurate information pertaining to the effectiveness of quercetin in combating COVID-19 has the potential to dissuade individuals from seeking proper medical care upon infection and undermine compliance with preventive measures for subsequent viral epidemics.

REFERENCES

1. World Health Organization (WHO). WHO Coronavirus (COVID-19) Dashboard: World Health Organization; 2023 [Available from: https://covid19.who.int/].
2. Eccles R, Boivin G, Cowling BJ, Pavia A, Selvarangan R. Treatment of COVID-19 symptoms with over the counter (OTC) medicines used for treatment of common cold and flu. *Clinical Infection in Practice.* 2023;19:100230.
3. Ang L, Lee HW, Choi JY, Zhang J, Lee MS. Herbal medicine and pattern identification for treating COVID-19: a rapid review of guidelines. *Integrative Medicine Research.* 2020;9(2):100407.
4. Osuchowski MF, Winkler MS, Skirecki T, Cajander S, Shankar-Hari M, Lachmann G, et al. The COVID-19 puzzle: deciphering pathophysiology and phenotypes of a new disease entity. *The Lancet Respiratory Medicine.* 2021;9(6):622–42.
5. Davis HE, McCorkell L, Vogel JM, Topol EJ. Long COVID: major findings, mechanisms and recommendations. *Nature Reviews Microbiology.* 2023;21(3):133–46.
6. Iddir M, Brito A, Dingeo G, Fernandez Del Campo SS, Samouda H, La Frano MR, Bohn T. Strengthening the immune system and reducing inflammation and oxidative stress through diet and nutrition: considerations during the COVID-19 crisis. *Nutrients [Internet].* 2020; 12(6):1562.
7. Arora I, White S, Mathews R. Global dietary and herbal supplement use during COVID-19-a scoping review. *Nutrients.* 2023;15(3):771.
8. Djaoudene O, Romano A, Bradai YD, Zebiri F, Ouchene A, Yousfi Y, et al. A global overview of dietary supplements: regulation, market trends, usage during the COVID-19 pandemic, and health effects. *Nutrients.* 2023;15(15):3320.
9. Mullin GE, Limektkai B, Wang L, Hanaway P, Marks L, Giovannucci E. Dietary supplements for COVID-19. In: Rezaei N, editor. *Coronavirus Disease - COVID-19.* Cham: Springer International Publishing; 2021, pp. 499–515.
10. Olczak-Pruc M, Swieczkowski D, Ladny JR, Pruc M, Juarez-Vela R, Rafique Z, et al. Vitamin C supplementation for the treatment of COVID-19: a systematic review and meta-analysis. *Nutrients.* 2022;14(19):4217.
11. Gibbons JB, Norton EC, McCullough JS, Meltzer DO, Lavigne J, Fiedler VC, Gibbons RD. Association between vitamin D supplementation and COVID-19 infection and mortality. *Scientific Reports.* 2022;12(1):19397.
12. Xie Y, Xu J, Zhou D, Guo M, Zhang M, Gao Y, et al. Micronutrient perspective on COVID-19: umbrella review and reanalysis of meta-analyses. *Critical Reviews in Food Science and Nutrition.* 2023:1–19.
13. Panayiotis L, Benjamin M, Kerstin K, Mark SG, Mohsen M, Emily RL, et al. Modest effects of dietary supplements during the COVID-19 pandemic: insights from 445 850 users of the COVID-19 Symptom Study app. *BMJ Nutrition, Prevention & Health.* 2021;4(1):149.
14. Magar RT, Sohng JK. A review on structure, modifications and structure-activity relation of quercetin and its derivatives. *Journal of Microbiolog and Biotechnology.* 2020;30(1):11–20.
15. Rice-Evans CA, Miller NJ, Paganga G. Structure-antioxidant activity relationships of flavonoids and phenolic acids. *Free Radical Biology and Medicine.* 1996;20(7):933–56.
16. Lesjak M, Beara I, Simin N, Pintać D, Majkić T, Bekvalac K, et al. Antioxidant and anti-inflammatory activities of quercetin and its derivatives. *Journal of Functional Foods.* 2018;40:68–75.
17. Lin C-F, Leu Y-L, Al-Suwayeh SA, Ku M-C, Hwang T-L, Fang J-Y. Anti-inflammatory activity and percutaneous absorption of quercetin and its polymethoxylated compound and glycosides: the relationships to chemical structures. *European Journal of Pharmaceutical Sciences.* 2012;47(5):857–64.
18. Kim Y, Narayanan S, Chang K-O. Inhibition of influenza virus replication by plant-derived isoquercetin. *Antiviral Research.* 2010;88(2):227–35.
19. Khachatoorian R, Arumugaswami V, Raychaudhuri S, Yeh GK, Maloney EM, Wang J, et al. Divergent antiviral effects of bioflavonoids on the hepatitis C virus life cycle. *Virology.* 2012;433(2):346–55.
20. Fanunza E, Iampietro M, Distinto S, Corona A, Quartu M, Maccioni E, et al. Quercetin blocks Ebola virus infection by counteracting the VP24 interferon-inhibitory function. *Antimicrobial Agents and Chemotherapy.* 2020;64(7):10.
21. dos Santos AE, Kuster RM, Yamamoto KA, Salles TS, Campos R, de Meneses MDF, et al. Quercetin and quercetin 3-O-glycosides from Bauhinia longifolia (Bong.) Steud. show anti-Mayaro virus activity. *Parasites & Vectors.* 2014;7(1):130.
22. Zandi K, Teoh B-T, Sam S-S, Wong P-F, Mustafa MR, AbuBakar S. Antiviral activity of four types of bioflavonoid against dengue virus type-2. *Virology Journal.* 2011;8(1):560.
23. Liu Z, Zhao J, Li W, Wang X, Xu J, Xie J, et al. Molecular docking of potential inhibitors for influenza H7N9. *Computational and Mathematical Methods in Medicine.* 2015;2015:480764.
24. Rahman MA, Shorobi FM, Uddin MN, Saha S, Hossain MA. Quercetin attenuates viral infections by interacting with target proteins and linked genes in chemicobiological models. *In Silico Pharmacology.* 2022;10(1):17.
25. Alomair L, Almsned F, Ullah A, Jafri MS. In silico prediction of the phosphorylation of NS3 as an essential mechanism for dengue virus replication and the antiviral activity of quercetin. *Biology (Basel).* 2021;10(10):1067.
26. Wong NA, Saier MH, Jr. The SARS-coronavirus infection cycle: a survey of viral membrane proteins, their functional interactions and pathogenesis. *International Journal of Molecular Sciences.* 2021;22(3):1308.
27. Yang J, Petitjean SJL, Koehler M, Zhang Q, Dumitru AC, Chen W, et al. Molecular interaction and inhibition of SARS-CoV-2 binding to the ACE2 receptor. *Nature Communications.* 2020;11(1):4541.
28. Pandey P, Rane JS, Chatterjee A, Kumar A, Khan R, Prakash A, Ray S. Targeting SARS-CoV-2 spike protein of COVID-19 with naturally occurring phytochemicals: an in silico study for drug development. *Journal of Biomolecular Structure and Dynamics.* 2021;39(16):6306–16.
29. Villoutreix BO, Badiola I, Khatib A-M. Furin and COVID-19: structure, function and chemoinformatic analysis of representative active site inhibitors. *Frontiers in Drug Discovery.* 2022;2:899239.
30. Milanović Ž B, Antonijević MR, Amić AD, Avdović EH, Dimić DS, Milenković DA, Marković ZS. Inhibitory activity of quercetin, its metabolite, and standard antiviral drugs towards enzymes essential for SARS-CoV-2: the role of acid-base equilibria. *RSC Advances.* 2021;11(5): 2838–47.
31. Chen L, Li J, Luo C, Liu H, Xu W, Chen G, et al. Binding interaction of quercetin-3-beta-galactoside and its synthetic derivatives with SARS-CoV 3CL(pro): structure-activity

relationship studies reveal salient pharmacophore features. *Bioorganic & Medicinal Chemistry.* 2006;14(24): 8295–306.
32. Chen L, Li J, Luo C, Liu H, Xu W, Chen G, et al. Binding interaction of quercetin-3-β-galactoside and its synthetic derivatives with SARS-CoV 3CLpro: structure-activity relationship studies reveal salient pharmacophore features. *Bioorganic & Medicinal Chemistry.* 2006;14(24):8295–306.
33. Park J-Y, Yuk HJ, Ryu HW, Lim SH, Kim KS, Park KH, et al. Evaluation of polyphenols from Broussonetia papyrifera as coronavirus protease inhibitors. *Journal of Enzyme Inhibition and Medicinal Chemistry.* 2017;32(1):504–12.
34. Hamet P, Pausova Z, Attaoua R, Hishmih C, Haloui M, Shin J, et al. SARS-CoV-2 receptor ACE2 gene is associated with hypertension and severity of COVID 19: interaction with sex, obesity, and smoking. *American Journal of Hypertension.* 2021;34(4):367–76.
35. Bosso M, Thanaraj TA, Abu-Farha M, Alanbaei M, Abubaker J, Al-Mulla F. The two faces of ACE2: the role of ACE2 receptor and its polymorphisms in hypertension and COVID-19. *Molecular Therapy – Methods & Clinical Development.* 2020;18:321–7.
36. Gasmi A, Mujawdiya PK, Lysiuk R, Shanaida M, Peana M, Gasmi Benahmed A, et al. Quercetin in the prevention and treatment of coronavirus infections: a focus on SARS-CoV-2. *Pharmaceuticals (Basel).* 2022;15(9):1049.
37. Liu X, Raghuvanshi R, Ceylan FD, Bolling BW. Quercetin and its metabolites inhibit recombinant human angiotensin-converting enzyme 2 (ACE2) activity. *Journal of Agricultural and Food Chemistry.* 2020;68(47):13982–9.
38. Terao J. Factors modulating bioavailability of quercetin-related flavonoids and the consequences of their vascular function. *Biochemical Pharmacology.* 2017;139:15–23.
39. Tadic M, Saeed S, Grassi G, Taddei S, Mancia G, Cuspidi C. Hypertension and COVID-19: ongoing controversies. *Frontiers in Cardiovascular Medicine.* 2021;8:639222.
40. Gallo G, Calvez V, Savoia C. Hypertension and COVID-19: current evidence and perspectives. *High Blood Pressure & Cardiovascular Prevention.* 2022;29(2):115–23.
41. Wang D, Ali F, Liu H, Cheng Y, Wu M, Saleem MZ, et al. Quercetin inhibits angiotensin II-induced vascular smooth muscle cell proliferation and activation of JAK2/STAT3 pathway: a target based networking pharmacology approach. *Frontiers in Pharmacology.* 2022;13:1002363.
42. Jones HS, Gordon A, Magwenzi SG, Naseem K, Atkin SL, Courts FL. The dietary flavonol quercetin ameliorates angiotensin II-induced redox signaling imbalance in a human umbilical vein endothelial cell model of endothelial dysfunction via ablation of p47phox expression. *Molecular Nutrition & Food Research.* 2016;60(4):787–97.
43. Yan L, Zhang JD, Wang B, Lv YJ, Jiang H, Liu GL, et al. Quercetin inhibits left ventricular hypertrophy in spontaneously hypertensive rats and inhibits angiotensin II-induced H9C2 cells hypertrophy by enhancing PPAR-γ expression and suppressing AP-1 activity. *PLoS One.* 2013;8(9):e72548.
44. Ni W, Yang X, Yang D, Bao J, Li R, Xiao Y, et al. Role of angiotensin-converting enzyme 2 (ACE2) in COVID-19. *Critical Care.* 2020;24(1):422.
45. Benigni A, Cassis P, Remuzzi G. Angiotensin II revisited: new roles in inflammation, immunology and aging. *EMBO Molecular Medicine.* 2010;2(7):247–57.
46. Touyz RM. The role of angiotensin II in regulating vascular structural and functional changes in hypertension. *Current Hypertension Reports.* 2003;5(2):155–64.
47. He Y, Cao X, Liu X, Li X, Xu Y, Liu J, Shi J. Quercetin reverses experimental pulmonary arterial hypertension by modulating the TrkA pathway. *Experimental Cell Research.* 2015;339(1):122–34.
48. Morales-Cano D, Menendez C, Moreno E, Moral-Sanz J, Barreira B, Galindo P, et al. The flavonoid quercetin reverses pulmonary hypertension in rats. *PLoS One.* 2014;9(12):e114492.
49. Patel AB, Verma A. COVID-19 and angiotensin-converting enzyme inhibitors and angiotensin receptor blockers: what is the evidence? *JAMA.* 2020;323(18):1769–70.
50. Sadeghi Dousari A, Hosseininasab SS, Sadeghi Dousari F, Fuladvandi M, Satarzadeh N. The impact of COVID-19 on liver injury in various age. *World Journal of Virology.* 2023;12(2):91–9.
51. Saha L, Vij S, Rawat K. Liver injury induced by COVID 19 treatment - what do we know? *World Journal of Gastroenterology.* 2022;28(45):6314–27.
52. Yu D, Du Q, Yan S, Guo X-G, He Y, Zhu G, et al. Liver injury in COVID-19: clinical features and treatment management. *Virology Journal.* 2021;18(1):121.
53. Chai X, Hu L, Zhang Y, Han W, Lu Z, Ke A, et al. Specific ACE2 expression in cholangiocytes may cause liver damage after 2019-nCoV infection. *bioRxiv*; 2020:931766.
54. Li X, Wang W, Yan S, Zhao W, Xiong H, Bao C, et al. Drug-induced liver injury in COVID-19 treatment: incidence, mechanisms and clinical management. *Frontiers in Pharmacology.* 2022;13:1019487.
55. McGill MR, Ramachandran A, Jaeschke H. Oxidant stress and drug-induced hepatotoxicity. In: Laher I, editor. *Systems Biology of Free Radicals and Antioxidants.* Berlin, Heidelberg: Springer; 2014, pp. 1757–85.
56. Kabirifar R, Ghoreshi Z-A-S, Safari F, Karimollah A, Moradi A, Eskandari-nasab E. Quercetin protects liver injury induced by bile duct ligation via attenuation of Rac1 and NADPH oxidase1 expression in rats. *Hepatobiliary & Pancreatic Diseases International.* 2017;16(1):88–95.
57. Ahmed OM, Elkomy MH, Fahim HI, Ashour MB, Naguib IA, Alghamdi BS, et al. Rutin and quercetin counter doxorubicin-induced liver toxicity in wistar rats via their modulatory effects on inflammation, oxidative stress, apoptosis, and Nrf2. *Oxidative Medicine and Cellular Longevity.* 2022;2022:2710607.
58. Peng Z, Gong X, Yang Y, Huang L, Zhang Q, Zhang P, et al. Hepatoprotective effect of quercetin against LPS/d-GalN induced acute liver injury in mice by inhibiting the IKK/NF-κB and MAPK signal pathways. *International Immunopharmacology.* 2017;52:281–9.
59. Abo-Salem OM, Abd-Ellah MF, Ghonaim MM. Hepatoprotective activity of quercetin against acrylonitrile-induced hepatotoxicity in rats. *Journal of Biochemical and Molecular Toxicology.* 2011;25(6):386–92.
60. Fedorova M, Bollineni RC, Hoffmann R. Protein carbonylation as a major hallmark of oxidative damage: update of analytical strategies. *Mass Spectrometry Reviews.* 2014;33(2):79–97.
61. Conway EM, Mackman N, Warren RQ, Wolberg AS, Mosnier LO, Campbell RA, et al. Understanding COVID-19-associated coagulopathy. *Nature Reviews Immunology.* 2022;22(10): 639–49.
62. Teimury A, Khameneh MT, Khaledi EM. Major coagulation disorders and parameters in COVID-19 patients. *European Journal of Medical Research.* 2022;27(1):25.

63. Kohansal Vajari M, Shirin M, Pourbagheri-Sigaroodi A, Akbari ME, Abolghasemi H, Bashash D. COVID-19-related coagulopathy: a review of pathophysiology and pharmaceutical management. *Cell Biology International.* 2021;45(9):1832–50.
64. Landstra CP, de Koning EJP. COVID-19 and diabetes: understanding the interrelationship and risks for a severe course. *Frontiers in Endocrinology.* 2021;12:649525.
65. Mosawy S, Jackson DE, Woodman OL, Linden MD. The flavonols quercetin and 3′,4′-dihydroxyflavonol reduce platelet function and delay thrombus formation in a model of type 1 diabetes. *Diabetes and Vascular Disease Research.* 2014;11(3):174–81.
66. Chen Y, Deuster P. Comparison of quercetin and dihydroquercetin: antioxidant-independent actions on erythrocyte and platelet membrane. *Chemico-Biological Interactions.* 2009;182(1):7–12.
67. Rizky WC, Jihwaprani MC, Mushtaq M. Protective mechanism of quercetin and its derivatives in viral-induced respiratory illnesses. *The Egyptian Journal of Bronchology.* 2022;16(1):58.
68. Heinz SA, Henson DA, Austin MD, Jin F, Nieman DC. Quercetin supplementation and upper respiratory tract infection: a randomized community clinical trial. *Pharmacological Research.* 2010;62(3):237–42.
69. Di Pierro F, Derosa G, Maffioli P, Bertuccioli A, Togni S, Riva A, et al. Possible therapeutic effects of adjuvant quercetin supplementation against early-stage COVID-19 infection: a prospective, randomized, controlled, and open-label study. *International Journal of General Medicine.* 2021;14: 2359–66.
70. Di Pierro F, Iqtadar S, Khan A, Ullah Mumtaz S, Masud Chaudhry M, Bertuccioli A, et al. Potential clinical benefits of quercetin in the early stage of COVID-19: results of a second, pilot, randomized, controlled and open-label clinical trial. *International Journal of General Medicine.* 2021;14:2807–16.
71. Önal H, Arslan B, Üçüncü Ergun N, Topuz Ş, Yilmaz Semerci S, Kurnaz ME, et al. Treatment of COVID-19 patients with quercetin: a prospective, single center, randomized, controlled trial. *Turkish Journal of Biology.* 2021;45(4):518–29.
72. Zupanets IA, Holubovska OA, Tarasenko OO, Bezuhla NP, Pasichnyk MF, Karabynosh SO, et al. Quercetin effectiveness in patients with COVID-19 associated pneumonia. *Zaporozhye Medical Journal.* 2021;23(5):636–43.
73. Khan A, Iqtadar S, Mumtaz SU, Heinrich M, Pascual-Figal DA, Livingstone S, Abaidullah S. Oral co-supplementation of curcumin, quercetin, and vitamin D3 as an adjuvant therapy for mild to moderate symptoms of COVID-19-results from a pilot open-label, randomized controlled trial. *Frontiers in Pharmacology.* 2022;13:898062.
74. Shohan M, Nashibi R, Mahmoudian-Sani MR, Abolnezhadian F, Ghafourian M, Alavi SM, et al. The therapeutic efficacy of quercetin in combination with antiviral drugs in hospitalized COVID-19 patients: a randomized controlled trial. *European Journal of Pharmacology.* 2022;914:174615.
75. Di Pierro F, Khan A, Iqtadar S, Mumtaz SU, Chaudhry MNA, Bertuccioli A, et al. Quercetin as a possible complementary agent for early-stage COVID-19: concluding results of a randomized clinical trial. *Frontiers in Pharmacology.* 2022;13:1096853.
76. Thilakarathna SH, Rupasinghe HP. Flavonoid bioavailability and attempts for bioavailability enhancement. *Nutrients.* 2013;5(9):3367–87.
77. Li Y, Yao J, Han C, Yang J, Chaudhry MT, Wang S, et al. Quercetin, inflammation and immunity. *Nutrients.* 2016;8(3):167.
78. Andres S, Pevny S, Ziegenhagen R, Bakhiya N, Schäfer B, Hirsch-Ernst KI, Lampen A. Safety aspects of the use of quercetin as a dietary supplement. *Molecular Nutrition & Food Research.* 2018;62(1):1700447.
79. Dagher O, Mury P, Thorin-Trescases N, Noly PE, Thorin E, Carrier M. Therapeutic potential of quercetin to alleviate endothelial dysfunction in age-related cardiovascular diseases. *Frontiers in Cardiovascular Medicine.* 2021;8:658400.
80. Dunnick JK, Hailey JR. Toxicity and carcinogenicity studies of quercetin, a natural component of foods. *Fundamental and Applied Toxicology.* 1992;19(3):423–31.
81. Ito N, Hagiwara A, Tamano S, Kagawa M, Shibata M, Kurata Y, Fukushima S. Lack of carcinogenicity of quercetin in F344/DuCrj rats. *Japanese Journal of Cancer Research.* 1989;80(4):317–25.
82. National Toxicology Program. Toxicology and carcinogenesis studies of quercetin (CAS No. 117-39-5) in F344 rats (feed studies). *National Toxicology Program Technical Report Series.* 1992;409:1–171.
83. Cunningham P, Patton E, VanderVeen BN, Unger C, Aladhami A, Enos RT, et al. Sub-chronic oral toxicity screening of quercetin in mice. *BMC Complementary Medicine and Therapies.* 2022;22(1):279.
84. Xiao X, Shi D, Liu L, Wang J, Xie X, Kang T, Deng W. Quercetin suppresses cyclooxygenase-2 expression and angiogenesis through inactivation of P300 signaling. *PLoS One.* 2011;6(8):e22934.
85. García-Mediavilla V, Crespo I, Collado PS, Esteller A, Sánchez-Campos S, Tuñón MJ, González-Gallego J. The anti-inflammatory flavones quercetin and kaempferol cause inhibition of inducible nitric oxide synthase, cyclooxygenase-2 and reactive C-protein, and down-regulation of the nuclear factor kappaB pathway in Chang liver cells. *European Journal of Pharmacology.* 2007;557(2):221–9.
86. Lee KM, Hwang MK, Lee DE, Lee KW, Lee HJ. protective effect of quercetin against arsenite-induced COX-2 expression by targeting PI3K in rat liver epithelial cells. *Journal of Agricultural and Food Chemistry.* 2010;58(9):5815–20.
87. Mandour Y, Handoussa H, Swilam N, Hanafi R, Mahran L. Structural docking studies of COX-II inhibitory activity for metabolites derived from corchorus olitorius and vitis vinifera. *International Journal of Food Properties.* 2016;19(10):2377–84.
88. Burdan F. Comparison of developmental toxicity of selective and non-selective cyclooxygenase-2 inhibitors in CRL:(WI)WUBR Wistar rats - DFU and piroxicam study. *Toxicology.* 2005;211(1):12–25.
89. Pratheeshkumar P, Budhraja A, Son YO, Wang X, Zhang Z, Ding S, et al. Quercetin inhibits angiogenesis mediated human prostate tumor growth by targeting VEGFR-2 regulated AKT/mTOR/P70S6K signaling pathways. *PLoS One.* 2012;7(10):e47516.

90. Gately S, Li WW. Multiple roles of COX-2 in tumor angiogenesis: a target for antiangiogenic therapy. *Seminars in Oncology*. 2004;31(2 Suppl 7):2–11.
91. Huang Z, Huang S, Song T, Yin Y, Tan C. Placental angiogenesis in mammals: a review of the regulatory effects of signaling pathways and functional nutrients. *Advances in Nutrition*. 2021;12(6):2415–34.
92. Beazley KE, Nurminskaya M. Effects of dietary quercetin on female fertility in mice: implication of transglutaminase 2. *Reproduction, Fertility and Development*. 2016;28(7):974–81.
93. Willhite CC. Teratogenic potential of quercetin in the rat. *Food and Chemical Toxicology*. 1982;20(1):75–9.
94. Zhang X, Wu C. In silico, in vitro, and in vivo evaluation of the developmental toxicity, estrogenic activity, and mutagenicity of four natural phenolic flavonoids at low exposure levels. *ACS Omega*. 2022;7(6):4757–68.
95. Cao Y, Zhuang M-F, Yang Y, Xie S-W, Cui J-G, Cao L, et al. Preliminary study of quercetin affecting the hypothalamic-pituitary-gonadal axis on rat endometriosis model. *Evidence-Based Complementary and Alternative Medicine*. 2014;2014:781684.
96. Farombi EO, Abarikwu SO, Adesiyan AC, Oyejola TO. Quercetin exacerbates the effects of subacute treatment of atrazine on reproductive tissue antioxidant defence system, lipid peroxidation and sperm quality in rats. *Andrologia*. 2013;45(4):256–65.
97. Chen Y-C, Nagpal ML, Stocco DM, Lin T. Effects of genistein, resveratrol, and quercetin on steroidogenesis and proliferation of MA-10 mouse Leydig tumor cells. *Journal of Endocrinology*. 2007;192(3):527–37.
98. Cormier M, Ghouili F, Roumaud P, Martin LJ, Touaibia M. Influence of flavonols and quercetin derivative compounds on MA-10 Leydig cells steroidogenic genes expressions. *Toxicology in Vitro*. 2017;44:111–21.
99. Grande G, Barrachina F, Soler-Ventura A, Jodar M, Mancini F, Marana R, et al. The role of testosterone in spermatogenesis: lessons from proteome profiling of human spermatozoa in testosterone deficiency. *Frontiers in Endocrinology*. 2022;13:852661.
100. Bolouki A, Zal F, Alaee S. Ameliorative effects of quercetin on the preimplantation embryos development in diabetic pregnant mice. *Journal of Obstetrics and Gynaecology Research*. 2020;46(5):736–44.
101. Resende FA, Vilegas W, Dos Santos LC, Varanda EA. Mutagenicity of flavonoids assayed by bacterial reverse mutation (AMES) test. *Molecules [Internet]*. 2012;17(5):5255–68.
102. Rietjens IMCM, Boersma MG, van der Woude H, Jeurissen SMF, Schutte ME, Alink GM. Flavonoids and alkenylbenzenes: mechanisms of mutagenic action and carcinogenic risk. *Mutation Research/Fundamental and Molecular Mechanisms of Mutagenesis*. 2005;574(1):124–38.
103. Gogoi N, Chowdhury P, Goswami AK, Das A, Chetia D, Gogoi B. Computational guided identification of a citrus flavonoid as potential inhibitor of SARS-CoV-2 main protease. *Molecular Diversity*. 2021;25(3):1745–59.
104. Zenkevich IG, Eshchenko AY, Makarova SV, Vitenberg AG, Dobryakov YG, Utsal VA. Identification of the products of oxidation of quercetin by air oxygen at ambient temperature. *Molecules*. 2007;12(3):654–72.
105. Kubo I, Nihei K, Shimizu K. Oxidation products of quercetin catalyzed by mushroom tyrosinase. *Bioorganic & Medicinal Chemistry*. 2004;12(20):5343–7.
106. Kakkar S, Bais S. A review on protocatechuic acid and its pharmacological potential. *ISRN Pharmacology*. 2014;2014:952943.
107. Yuan S, Gao F, Xin Z, Guo H, Shi S, Shi L, et al. Comparison of the efficacy and safety of phloroglucinol and magnesium sulfate in the treatment of threatened abortion: a meta-analysis of randomized controlled trials. *Medicine (Baltimore)*. 2019;98(24):e16026.

14 Molecular Aspects in Pharmacological Actions of Quercetin Targeting Life Cycle of SARS-CoV-2

R. Manikandan, B. Balamuralikrishnan, P. Gajalakshmi, T. Karpagam, A. Shanmugapriya, M. Abirami, R. Ramya, and A. Vijaya Anand

LIST OF ABBREVIATIONS

ACE2	Angiotensin-converting enzyme 2
ADMET	Absorption, distribution, metabolism, excretion, and toxicity
3CLpro	3-chymotrypsin-like protease
COVID-19	Coronavirus disease-19
EC50	Effective concentration-50
HIV	Human immunodeficiency virus
IC50	Inhibitory concentration-50
MERS	Middle East Respiratory Syndrome
MERS-CoV	Middle East Respiratory Syndrome Coronavirus
nCoV	Novel coronavirus
PLpro	Papain-like protease
RdRp	RNA-dependent RNA polymerase
SARS	Severe Acute Respiratory Syndrome
SARS-CoV-1	Severe Acute Respiratory Syndrome Coronavirus 1
TMPRSS2	Transmembrane protease serine 2
WHO	World Health Organization

14.1 MEDICINAL PLANTS

Medical plants are becoming recognized as a vital resource for the prevention, management, and treatment of many diseases (Rakotoarivelo et al., 2015). Every plant has several significant components that are useful in the medical industry and can be utilized to create various types of medications (Yuan et al., 2016; Anand et al., 2021). Many underdeveloped nations, as well as many industrialized nations, use herbal medicine to cure specific diseases like coughs and to preserve overall health. Echinacea, garlic, ginger, ginkgo, ginseng, and other herbs are among them (Rakotoarivelo et al., 2015; Abidharini et al., 2023; Kaviya et al., 2023).

14.2 IMPORTANCE OF THERAPEUTIC PLANTS

Alternative medicine refers to the use of plants for therapeutic purposes. Almost all cultures have employed alternative medicine, especially Western and Asian ones. Unfortunately, the majority of people still hold the view that the only reliable and efficient medication available today comes in a dose form, such as tablets and capsules. Even though numerous tablets and capsules, including aspirin, digoxin, and paclitaxel, are used regularly, they are derived from plants. Our antique antecedents used plant materials in the past to flavor and preserve foods, alleviate headaches, lessen discomfort, and even stop diseases and epidemics. Over the ages, human groups have shared knowledge of their therapeutic qualities. The biological properties of plant species that are utilized across the world for several purposes, such as the treatment of infectious diseases, are usually linked to active compounds produced during secondary metabolism (Singh, 2015; Poochi et al., 2020; Ajith et al., 2023). People are currently being cautioned by numerous studies about the dangers and risks posed by pathogenic microorganisms that have developed resistance to newly developed antibiotics (Munita and Arias, 2016). As a result, a great deal of research is required to characterize the chemical nature of the plants and their antimicrobial potential and the mechanisms underlying the inhibition of microbial growth, either in isolation or in combination with traditional antimicrobials, since data regarding the antimicrobial activity of many plants that were previously believed to be empirical have been verified by science.

14.3 FUTURE OF MEDICAL PLANTS

Herbal remedies are the mainstay of the traditional medical system. Most of the medicinal plants in India are gathered from forests. About a hundred new plant-based drugs, such as deserpidine, vincristine, and resin amine, which are found in higher plants, were introduced to the US drug market between 1950 and 1970. Medicinal plants have provided humanity with a vast array of efficient treatments to lessen or eradicate illnesses and the pain associated with disease, even in the face of the invention of synthetic drugs. Even now, several of these drugs are still important and relevant.

Worldwide, the benefits of plant-based medicines are rising (Bhat, 1995). There have been remarkable advances in the pharmacological investigation of various plants used in traditional medical systems. The therapeutic potency of medicinal plants, especially their antimicrobial activities, is determined by a comprehensive range of secondary

DOI: 10.1201/9781003433200-14

metabolites or bioactive constituents found in the plants, including alkaloids, terpenoids, flavonoids, and tannins (Evans and Trease, 2002). The benefit of plant-derived materials as an indigenous therapy in conventional systems of medicine has been linked to the introduction of plant-based drugs in modern medication (Igoli et al., 2003). Effective antibacterial, anti-inflammatory, antifungal, antidiuretic, anti-diabetic, and anticancer properties have been found in many plants (Sule et al., 2010; Timothy et al., 2012; Bharathi et al., 2023).

Plant-based medications are employed to treat disorders associated with skin, mental illness, diabetes, tuberculosis, cancer, and jaundice. It is often known that the majority of developing nations maintain good health through the use of conventional therapy and medicinal plants as a standardizing basis. Oral illnesses are major health problems, and two of the most prevalent communicable diseases in the industrialized world that can be prevented are dental caries and periodontal infections. Dental health affects one's whole quality of life, and chronic illnesses and systemic diseases are associated with poor dental health. Many individuals worldwide die every day due to diseases that may be controlled or treated because they lack access to even the most basic medical treatment. In many countries, diseases are usually related to malnutrition.

Consequently, the significance of plants in conventional treatment and the pharmaceutical industry's natural raw materials cannot be magnified. Herbal medicine is utilized to cure illnesses in almost all non-industrialized countries. Among the medications that are nowadays known to doctors that have a long history of usefulness as herbal cures are quinine, digitalis, opium, etc. The use of therapeutic plants is growing worldwide as a result of the rapid development of conventional treatment and the growing acceptance of herbal remedies (Chirumbolo et al., 2018; Kathirvel Bharathi et al., 2023). Eighty percent of the world's population gets their primary healthcare from traditional medicine, according to the World Health Organization (WHO). Plant extracts and their active components are used in most of this therapy (Joshi et al., 2011).

In medicine, plants are used to prevent disease, treat it, and maintain and improve overall health on all levels—physical, mental, and spiritual. Conventional medicines that have been changed are known as "complementary" or "alternative" therapies in the world. The use of conventional treatment has remained widespread all over the world (Kathirvel Bharathi et al., 2023). Currently, nearly 100,000 plants are either unidentified or have not yet had their medicinal properties thoroughly researched and examined. Studies both now and in the future should examine the medicinal efficacy of plants and herbs because it is anticipated that they will be crucial to medicine, particularly in the treatment of serious illnesses like cancer.

Medicinal plants have many features when employed as a remedy, as follows (Rasool Hassan, 2012):

- **Synergic medicine**: Each plant contains a variety of substances that may interact eventually to either enhance or impair the processes of one another or to counteract any potential negative effects.
- **Support of official medicine**: To achieve the desired result, chemical products and plant-based ingredients can be combined.
- **Preventive medicine**: Some plant-based ingredients are effective in preventing or lowering the risk of contracting specific diseases (such as the flu), which can help ease the burden and expense of using chemical treatments.

14.4 FLAVONOIDS

Flavonoids are phytochemical substances with potential medical uses that are highly found in a wide variety of plants, vegetables, fruits, and leaves. Flavonoids have a variety of health advantages, including antioxidant, anti-inflammatory, anticancer, and antiviral effects. The primary component of flavonoids, which are secondary metabolites, is a benzopyrone ring carrying phenolic or polyphenolic groups at various locations (Jaganathan and Mandal, 2009).

The fact that plant-derived compounds have a broad range of biological actions has been generally acknowledged for millennia. The isolation of flavonoids, their identification, characterization, and functions, as well as their applications for medicinal benefits, are the recent trends in flavonoid research and development efforts. Bioinformatics tools and molecular docking expertise are also used to signify possible industrial manufacture and application. In the current study, efforts have been made to examine the mode of action of flavonoids, their roles, and various applications. Additionally, it has been predicted that flavonoids may one day be used as medications to prevent chronic diseases (Arthi Boro et al., 2023).

They are members of a group of phenolic compounds with low molecular weight that are seen in various plants. They belong to one of the most distinctive kinds of chemical constituents found in higher plants. In most angiosperm families, a large number of flavonoids are easily identified in floral pigments. They are present throughout all of a plant's sections, not just the flowers. Flavonoids are called dietary flavonoids since they can be found in large quantities in a variety of foods and drinks that come from plants, including vegetables, fruits, chocolate, and tea. Flavonols, flavones, isoflavones, and chalcones are only a few of the subgroups that make up the class of flavonoids. These groups' principal sources are distinctively their own. The main dietary origins of flavonols, for instance, are tea and onions (Bjorklund et al., 2017; Ozarowski and Karpinski, 2021; Arthi Boro et al., 2023).

14.4.1 Structure of Flavonoids

The carbon on which the B ring is connected, the capacity of unsaturation and oxidation of the C ring, and the carbon of the C ring to which the B ring is joined allow flavonoids to be divided into numerous subgroups. There is a link between the B ring and the third position of the C ring in a subgroup of flavonoids called isoflavones. Neoflavonoids have the B ring coupled in position 4, and subgroups for those with the B ring coupled in place 2 can be created using the C ring's structural characteristics. A subclass of flavonoids known as isoflavones connects the B ring and the third place of the C ring.

In neoflavonoids, the B ring is connected in place 4, and subgroups can be created for those in which the B ring is linked in place 2 established on the structural features of the C ring. Unlike flavones, which only have two hydroxyl groups on the C ring, flavonols have a third hydroxyl group that can also be glycosylated. Like flavones, flavonols show a wide range of methylation and hydroxylation patterns due to the numerous glycosylation types. Additionally, they are probably the most numerous and substantial subclass of flavonoids found in vegetables and fruits. For instance, quercetin is found in numerous plant foods.

14.5 QUERCETIN

The flavonoid quercetin has acquired a lot of attention recently because it is thought to be the most prevalent in fruits and vegetables. This substance offers numerous positive health effects, including anti-inflammatory, immune system-regulating, cardioprotective, and neuroprotective properties. The antioxidant properties of quercetin and the removal of free radicals are primarily responsible for its neuroprotective benefits. In turn, quercetin supplementation influences the generation of free radicals and faulty mitochondria, as well as energy production, electron chain performance, and mitochondrial biogenesis. In addition, quercetin can pass the blood–brain barrier (Filipa Almeida et al., 2018). Figure 14.1 shows the structure of quercetin.

OH O OH OH HO O OH

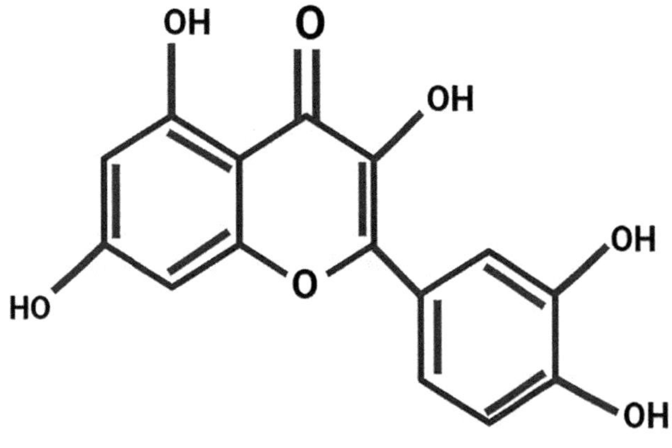

FIGURE 14.1 Structure of quercetin. *Source*: PubChem (CID 5280343).

14.5.1 Sources and Properties of Quercetin

The fact that naturally occurring plant phytochemicals are typically less costly and have rare side effects than manufactured medications has shown an increase in curiosity in them in recent years for treating various illnesses. In addition to acting as antioxidants and free radical scavengers, quercetin and several other natural polyphenols also activate phase II detoxification enzymes (Li et al., 2016; Stanek et al., 2019). A practical confirmation has been conducted on quercetin, a hydrophobic citron-yellow crystal produced from different parts of the plants, to evaluate its characteristics and medicinal effects. One of the most well-known flavonoids is quercetin. The occurrence of five hydroxyl groups with electron-donating capability at positions 3, 5, 7, 3′, and 4′ distinguishes quercetin from other molecules. The threat of disorders associated with metabolism, cardiovascular diseases, and some kinds of cancer can be reduced by its capacity to interfere with free radicals, which are the reason for oxidative stress (Saeedi Boroujeni and Mahmoudian Sani, 2021). Table 14.1 shows that plants, vegetables, and fruits contain rich amounts of quercetin.

In plants, this flavonol is found throughout, predominantly as water-soluble quercetin glycosides. Under the influence of gastric acid, quercetin and its by-products are stable in the stomach. Glucosides are hydrolyzed to the aglycone form and subsequently absorbed in the small intestine with the help of enzymes beta-glucosidase and lactase phlorizin hydrolase. Sugars must therefore be taken out of the molecule before absorption into the enterocyte (Day et al., 2000).

14.6 SEVERE ACUTE RESPIRATORY SYNDROME (SARS) CORONAVIRUS

Viral infections persist to pose a major risk to general health. A few of the viral epidemics the world has seen over the past 20 years include the Middle East Respiratory Syndrome Coronavirus (MERS-CoV) in 2012, the H1N1 influenza pandemic in 2009, and the Severe Acute Respiratory Syndrome Coronavirus 1 (SARS-CoV-1) in 2003. In December 2019, Wuhan City, Province of Hubei (China), declared an outburst of pneumonia with an unidentified etiology to the WHO's China Country Office. China's National Health Commission claims that exposure in a single Wuhan City seafood market caused the incident.

The novel coronavirus (2019-nCoV), which is the etiological cause of atypical pneumonia, was discovered by the authorities of the Chinese on January 7, 2020. The significance of January 22, 2020 is that this is when the virus is considered to have first emerged in Europe. Then, the virus was encountered in a German individual infected by a person from Shanghai (China), in Bavaria (Germany). Afterward, it was identified in Italy on January 25, 2020. It is believed that the unaware victims who were infected with the virus were able to move around in Basso Lodigiano with or without symptoms that were misunderstood as influenza

TABLE 14.1
Source of quercetin in plants, fruits and vegetables

S. No.	Common Name	Scientific Name	Concentration of Quercetin (mg//100g)	Medicinal Uses	Reference
1	Dill weed	*Anethum graveolens*	79.0	Fever, nerve discomfort, genital ulcer, pneumonia, cold, and menstruation cramps are all treated	Wijdan Dabeek et al. (2019)
2	Fennel leaves	*Foeniculum vulgare*	48.80	Traditional medicine for digestive, reproductive, respiratory, endocrine, and respiratory system ailments	
3	Red leaf lettuce	*Lactuca sativa.* var. *crispa*	30.60	It aids in the reduction of high blood pressure, the relaxation of cardiac muscles, and the maintenance of a healthy heartbeat	Gasmi et al. (2022)
4	Kale	*Brassica oleracea*	22.58	It aids in blood pressure regulation, immune system support, and may reduce the risk of several types of cancer	Silvia Tsanova-Savova et al. (2016)
5	Romaine lettuce	*Lactuca sativa* var. *longifolia*	12.00	It provides necessary nitrate for heart health. It is a combination of beta-carotene and vitamin C that protects it from cholesterol accumulation	Gasmi et al. (2022)
6	Juniper berry	*Juniperus communis*	46.60	It has traditionally been used as diuretic, anti-arthritis, anti-diabetes, antiseptic as well as for the treatment of gastrointestinal and autoimmune disorders	Wijdan Dabeek et al. (2019)
7	Cranberry	*Vaccinium* subg. *Oxycoccus*	25.00	Cranberry fruits and leaves have traditionally been used to treat bladder, stomach, and liver illnesses, as well as diabetes, wounds, and other ailments	Silvia Tsanova Savova et al. (2016)
8	Blueberry	*Vaccinium* sect. *cyanococcus*	14.60	Berries' antioxidants may aid in the prevention of heart disease and high blood pressure	
9	Apple	*Malus pumila*	4.01	Apples have been shown to have extremely high antioxidant property and prevent cell proliferation in cancer	Silvia Tsanova Savova et al. (2016)
10	Dark colored Grapes	*Vitis vinifera*	3.00	It functions as an antioxidant. The chemical compound protects cells from cancer, diabetes, neurodegenerative disorder, and cardiovascular disease	Silvia Tsanova Savova et al. (2016)
11	Okra	*Abelmoschus esculentus*	20.03	Okra is low in calories yet high in nutrients. Okra contains vitamin C, which promotes healthy immunological function	Silvia Tsanova Savova et al. (2016)
12	Onion	*Allium cepa*	11.00	It has been used to treat a wide range of conditions, including headaches, fevers, toothaches, coughs, sore throats, flu, baldness, epilepsy, rashes, jaundice, constipation, flatulence, intestinal worms, low sexual strength, rheumatism, bodily pain and muscular cramps, high blood pressure, and diabetes	Silvia Tsanova Savova et al. (2016)
13	Green pepper	*Capsicum annuum*	10.27	Green peppers are high in iron, but they are also strong in vitamin C, which can aid in iron absorption Green peppers are a superfood when it comes to preventing and treating iron deficiency (anemia) because of this combination	Silvia Tsanova Savova et al. (2016)
14	Broccoli	*Brassica oleracea* var. *italica*	13.70	It may aid in the prevention of the type of cell damage that leads to cancer. It aids in the improvement of immunological health and skin health It lowers the diabetes risk	Wijdan Dabeek et al. (2019)
15	Tomato	*Solanum lycopersicum*	0.70–4.40	Tomato is used to prevent breast, bladder, cervix, colon and rectum cancer, as well as stomach, lung, ovaries, pancreatic, and prostate cancer. It is also used to prevent diabetes, heart and blood vessel disease, cataracts, and asthma	Silvia Tsanova-Savova et al. (2016)

(Arumugam et al., 2020; Shanmugam et al., 2020; Pushpara et al., 2022; Chandra Manivannan et al., 2022). On February 11, 2020, the WHO declared that the disease brought on by a new virus would be known as "COVID-19 (coronavirus disease-2019)". On January 30, 2020, the WHO announced the situation as a general health crisis international problem.

The first symptomatic COVID-19 individual in Italy was discovered on February 20, 2020, at the Codogno Hospital Emergency Department in Basso Lodigiano, Province of Lodi. The WHO Director-General declared on February 26, 2020, that for the first time, there have been more newly reported instances of the viral illness, currently and officially known as COVID-19, documented outside of China (Arumugam et al., 2020). The WHO labeled this viral illness a pandemic on March 11, 2020.

Numerous research teams from all around the world have been investigating different natural compounds for COVID-19 treatment (Meyyazhagan et al., 2022). Antiviral quercetin has been proven useful in the treatment of MERS and SARS. This chapter aims to better comprehend the many biochemical, pharmacological, and immunomodulatory characteristics of quercetin that may help with COVID-19 prevention and therapy. The synergistic effects of quercetin in combination with vitamins and minerals also cover the application of bioinformatics tools in identifying the ability of quercetin to prevent, control, or treat COVID-19.

14.6.1 Etiology

A substantial family of viruses is the coronaviruses. They are usually single-stranded RNA viruses that, when evaluated under an electron microscope, resemble crowns. The mid-1960s saw the discovery of coronaviruses, which are comprehended to infect birds, animals, and humans. The epithelial compartments of the digestive and respiratory systems are the main targeted cells.

Despite frequently existing in a pleomorphic state, SARS-CoV-2 is a single-stranded, RNA virus, with a diameter range from 60 to 140 nm and a round or elliptical shape. The RNA genome of the virus has 29,891 nucleotides and 9,860 amino acid codons, and it shares 99.9% sequence uniqueness with the bat genome, indicating that humans are its most recent host. It is heat- and UV-sensitive, just like other CoVs. Additionally, lipid solvents including ether (75%), ethanol, chloroform, and disinfectants, which contain chlorine, can successfully inactivate these viruses. Chlorhexidine does not influence this virus and makes it inactive (Bhotla et al., 2020).

14.7 ANTI-SARS-COV-2 ACTIVITY

SARS-CoV-2 is an emerging international jeopardy that is putting a strain on the ability of healthcare systems around the world. More than 6,400,000 people had died from the SARS-CoV-2 virus as of August 18, with 1,060,000 of those deaths occurring in the United States alone. Traditional Chinese medicine's primary effective therapeutic component, quercetin, may effectively treat and prevent SARS-CoV-2 (Yang et al., 2020; Agrawal et al., 2020; Bhotla et al., 2021). Infectious disease epidemics have historically been a substantial obstacle to the advancement of humans. Specifically, the SARS-CoV-2 pandemic poses a significant dispute with many developing nations that lack access to specialized laboratories and equipment (Bhotla et al., 2021). It is critical to develop biosensors with quick, simple, and non-device-dependent detection to address this crisis. Additionally, it is crucial to develop medicines for clinical treatment. Theoretically, quercetin may hinder the replication of SARS-CoV-2 and lessen the inflammation and harmful effects of COVID-19 vaccinations (Bastaminejad and Bakhtiyari, 2020; Derosa et al., 2021; Boretti, 2021; Kuchi Bhotla et al., 2021).

Additionally, quercetin lessens the severity of COVID-19 symptoms and its negative predictors while also speeding up the conversion of positive molecular tests to negative ones (Di Pierro et al., 2021). COVID-19-positive (individuals with signs, symptoms, and outpatients) were split into two categories in a recent clinical study, with one group acquiring standard care treatment and the other receiving quercetin as an extra treatment. According to blood markers, the patient's blood contained lower levels of C-reactive protein, lactate dehydrogenase, ferritin, and D-dimer after taking quercetin supplements. Supplementing with quercetin not only speeds up the process by which molecular experiments go from positive to negative, but also lessens the stringency of COVID-19 symptoms (Di Pierro et al., 2021). Individuals with COVID-19 can be administered a combination of quercetin and vitamins (Colunga Biancatelli et al., 2020). More research is required, but quercetin may be an effective intervention to reduce the severity of infections in the respiratory tract (Zheng et al., 2017; Aucoin et al., 2020).

SARS-CoV-2 is very similar to quercetin, which is an effective constituent against the SARS virus. Because quercetin inhibits multiple steps of the life cycle of the virus, it may have therapeutic value against SARS-CoV-2 (Derosa et al., 2021). Specifically, quercetin may modify the expression of 30% of genes that code for SARS-CoV-2 protein targets in humans, which may interfere with the functions of 85% of these proteins (Glinsky and Tripartite, 2020). $3CL^{pro}$ (3-chymotrypsin-like protease), PL^{pro} (papain-like protease), RNA-dependent RNA polymerase (RdRp), and viral spike glycoproteins are all potential medicinal targets in the virus. Transmembrane protease serine 2 (TMPRSS2), angiotensin-converting enzyme (ACE2), and the angiotensin AT2 receptor, in contrast, are drug targets that are made available to the host (Liu et al., 2020). A potential antiviral therapeutic method is to prevent the relations between the SARS-CoV-2 glycoprotein spike and ACE2. With an IC_{50} of 4.48 M, quercetin is a powerful inhibitor of recombinant hACE2 *in vitro* (Liu et al., 2020).

Tiny molecules that interact with the SARS virus's spike proteins may be able to stop it from infecting host cells. By preventing its entry, Yi et al. (2004) confirmed that quercetin had antiviral activity as opposed to HIV-luc/SARS (EC_{50}: 83.4 M). Additionally, quercetin showed very little cytotoxicity toward normal cells creating a potential small molecule for treating the SARS virus (Yi et al., 2004). By preventing the SARS virus from functioning, quercetin prevents viral replication. This has been pointed out that a potential treatment for SARS involves targeting the $3CL^{pro}$, which is highly essential for the replication of SARS-CoV. Previous research by Chen et al. (2006) showed that quercetin derivatives had inhibitory influences on SARS-CoV $3CL^{pro}$.

Nguyen et al. (2012) indicated that quercetin inhibits $3CL^{pro}$ with an IC_{50} value of 73 M, expressed in *Pichia pastoris*. The major protease, or $3CL^{pro}$, is a possible target for COVID-19 drugs. Zhang et al. (2020) crystallized SARS-CoV-2 $3CL^{pro}$ using α-ketoamide inhibitor. This is an important point to note: with a binding energy of approximately 8.58 kcal/mol, quercetin binds to the $3CL^{pro}$ of the SARS-CoV-2 (Khaerunnisa et al., 2020). The quercetin derivatives 3-D-galactoside and quercetin 3-β-D-glucosidequercetin demonstrated a few possibilities against SARS-CoV-2 in molecular docking (Adem et al., 2022). The interaction between MERS-CoV and flavonoids depends heavily on S1 and S2 of the $3CL^{pro}$ site. Some hydrophobic and carbohydrate-attached quercetin derivatives may be possible drug targets for the inhibition of $3CL^{pro}$. Using a flavonoid library, Jo et al. (2019) discovered compounds that inhibited the $3CL^{pro}$ of MERS-CoV, a coronavirus that has an extremely high mortality rate of approximately 35%. It was found that $3CL^{pro}$'s activity was inhibited by the quercetin derivative quercetin-3-β-D-glucoside. The tryptophan-based fluorescence technique was utilized to verify the binding between the enzyme and the quercetin derivative (Jo et al., 2019).

Another molecular docking analysis suggested that the SARS-CoV-2 major protease (M^{pro}) protein (6flu7) may be inhibited by quercetin. The ligand binding affinity of quercetin was 7.9 Kcal/mol (Ubani et al., 2020). Munafo et al. (2022) assessed the inhibitory potency of quercetin against the RdRp, which is in charge of replicating viral genomes. In the biochemical assay, the IC_{50} value was $6.9 \pm 1.0\,\mu M$ (Munafo et al., 2022). RdRp could therefore be pharmacologically targeted to inhibit the replication of SARS-CoV-2.

Quercetin may be useful in preventing SARS virus infection and its associated side effects in humans, according to research combining *in vitro* and computational methods (Yi et al., 2004; Chen et al., 2006; Ryu et al., 2010). It is interesting to learn that the human SARS-CoV-2 infection is mediated by the ACE2 receptor, which is expressed in organs including the lung, heart, endothelium cells, intestine, and kidney (Ferrario et al., 2005). By inhibiting spike–ACE2 binding and interacting with PL^{pro}, quercetin can interfere with the expression of ACE2 (Glinsky, 2020). This gene expression is needed for SARS-CoV-2 penetration into the cells of the human (Figure 14.2). Overall, quercetin activity reduces the SARS-CoV-2 possible effectiveness significantly (Huang et al., 2020). According to Shin et al. (2020), PL^{pro} is an essential CoV enzyme involved in both immune evasion and viral dissemination. Using an HIV-luc/SARS

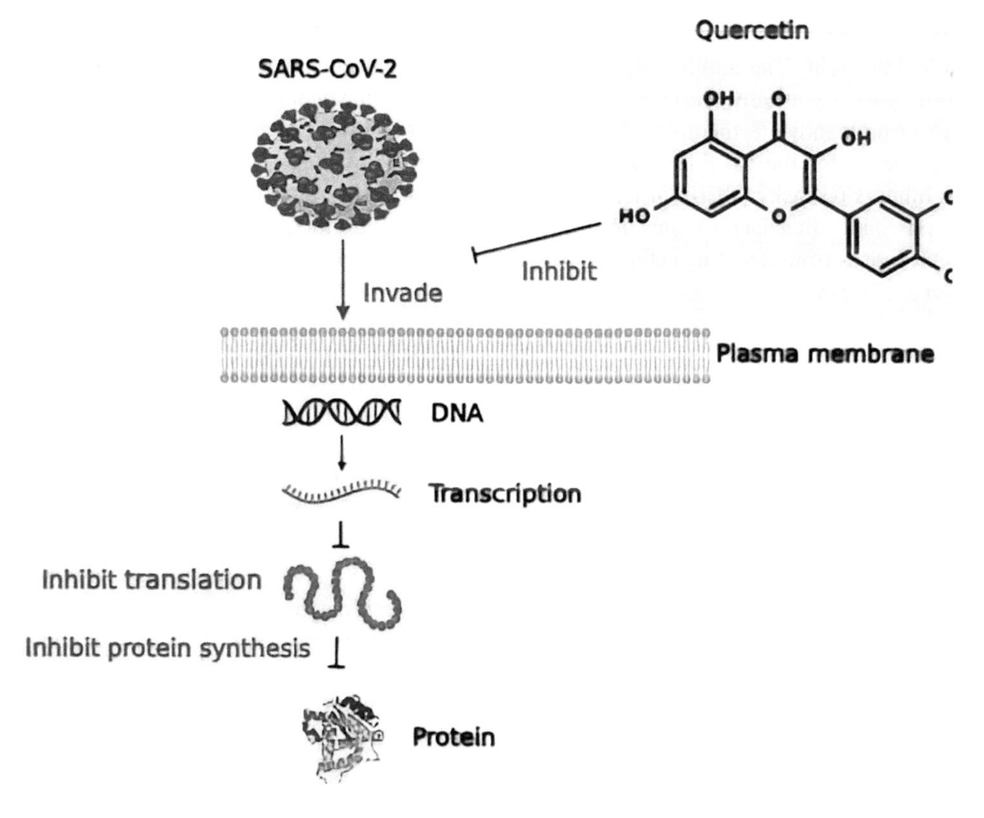

FIGURE 14.2 Possible effect of quercetin on SARS-CoV-2 virus life cycle.

pseudotyped virus, Yi et al. (2004) performed an *in vitro* virus entryway inhibition test. Quercetin and luteolin, two other similar flavonoids, were successful in preventing SARS-CoV from entering the host cell.

It has been found that the action of the proteases linked to SARS-CoV is suppressed by quercetin and its derivatives, luteolin, epigallocatechin gallate, and quercetin-3-β-galactoside. Ryu et al. (2010) discovered that $3CL^{pro}$, also known as the M^{pro} of SARS-CoV and essential to the virus's replication and infection processes, was inhibited by both luteolin and quercetin in an *in vitro* study. Nevertheless, their inhibitory efficacy was lower than that of amentoflavone, a flavonoid obtained from the plant *Torreya nucifera* (Ryu et al., 2010). Nguyen et al. (2012) report that quercetin and epigallocatechin gallate are interfered with $3CL^{pro}$ with an IC_{50} of 73 μM.

Park et al. (2017) found that quercetin inhibited the SARS-CoV proteases with the following IC_{50} values ($3CL^{pro}$—52.7 μM) and (PL^{pro}—8.6 μM), in addition to the MERS-CoV protease ($3CL^{pro}$—34.8 μM). After screening a chemical library containing over 150 chemicals, quercetin was discovered to be a potent inhibitor of SARS-CoV-2 $3CL^{pro}$ (Ki~7 μM; Abian et al., 2020). Quercetin-3-β-galactoside was found to be an inhibitor of SARS-CoV $3CL^{pro}$ by employing mutagenesis analyses, docking studies, fluorescence resonance energy transfer bioassays, and surface plasmon resonance (Chen et al., 2006).

Quercetin is helpful not only in controlling SARS-CoV but also against SARS-CoV-2, according to a provocative prognosis study that proposes the combination of vitamin D, estradiol, and quercetin can be a presumed COVID-19 mitigating agent. The authors identified quercetin and luteolin, two structurally similar flavonoids, as among the top-scoring candidate therapies that could be used as building blocks for the development of potential SARS-CoV-2 inhibitors (Glinsky, 2020). *In vitro* evidence reinforcing the possibility that luteolin and quercetin can both stop the SARS virus from infecting cells supports this conclusion (Yi et al., 2004).

14.8 QUERCETIN: POSSIBLE SYNERGISTIC IMPACTS OF MICRONUTRIENTS IN CONTROL OF COVID-19

Because of their numerous physiological and biological effects, several polyphenols, flavonoids, minerals, and vitamins are helpful in the fight against viral infections (Gasmi et al., 2020). Vitamins, for instance, have been shown to increase the medicinal efficacy of phytoconstituents and strengthen immunity against viruses. In a similar vein, flavonoids lower inflammation, oxidative stress, and the body's capacity to fight disease (Mora et al., 2008).

As a result, it is anticipated that the combination of quercetin with phenolic compounds, minerals, and vitamins will have a multiple effect and aid in the swift removal of the virus. Combining quercetin with phenolic compounds is used to fight viral disease by hitting multiple targets at once (Moran-Santibanez et al., 2018; Shanmugam Ramya et al., 2023). This combination also aids in lowering polyphenol and flavonoid dosages, which lessens the growth of viruses resistant to medications or other natural substances (Moran-Santibanez et al., 2018). Aslam et al.'s recent study, for instance, revealed that a blend of different plant extracts high in polyphenols and flavonoids exhibited synergism, better antioxidant capacity, and the ability to scavenge free radicals (Aslam et al., 2017).

Additionally, it has been claimed that a combination of plants can treat viral infections of the nasal cavity, throat, skin, and respiratory tract. When combined with phenolic compounds, minerals and vitamins, quercetin may be a more effective medicinal option than when taken alone (Gasmi et al., 2020). Utilizing a particular and thoughtfully formulated mixture of phenols has the benefit of having high protection without producing noticeable harmful effects (Mehany et al., 2021).

After screening thousands of possible antiviral bioactive compounds in a therapeutic plant database, the following polyphenols were judged to be the majority of hopeful compounds for suppressing SARS-CoV-2 3CL virus reproduction (Ul Qamar et al., 2020): (2S)7 -O-(6″-O-galloyl)-eriodictyol-beta-D-glucopyranos ide, 5,7,30,40-tetrahydroxy-20-(3,3-dimethylallyl) isoflavone, calceolarioside, methyl rosmarinate, myricetin 3-O-beta-D-glucopyranoside, myricitrin, and flavanone-3-O-beta-D-glucopyranoside B. Lin et al. (2005) found that a phenolic aloe-emodin molecule produced from *Isatis indigotica* roots reduced $3CL^{pro}$ cleavage activity in a dose-dependent manner (IC_{50}=366 mM).

The flavonoids myricetin and scutellarein were discovered by Yu et al. (2012) to be capable of hindering the SARS-CoV helicase protein. Flavonol kaempferol glycosides can be effective antiviral agents against coronaviruses that bind to their 3a channel protein (Schwarz et al., 2014). The PL^{pro} of SARS-CoV was inhibited by the geranylated flavonoids tomentins, which were obtained from the fruit extract of *Paulownia tomentosa* (Cho et al., 2013). Seeds of *Psoralea corylifolia* contain isobavachalcone and psoralidin, which lessen the SARS-CoV activity (Kim et al., 2014).

A study found that the most effective inhibitor of PL coronavirus cysteine proteases (IC_{50}=3.7 μM) was papyriflavonol derived from *Broussonetia papyrifera* extract (Park et al., 2017). Because of the yield of various combinations, butanol extracts of *Cinnamomi cortex* showed anti-SARS-CoV actions via a variety of mechanisms. Procyanidins, however, were considered to be the primary components possessing these qualities (Zhuang et al., 2009).

Among the polyphenols in green tea, only epigallocatechin-3-gallate and their lipophilic by-products were identified as unique compounds (Hsu, 2015). Since fatty acid monoesters of epigallocatechin-3-O-gallate have an attraction for both viruses and cell membranes, they have demonstrated an antiviral impact against different kinds of viruses (Kaihatsu et al., 2018). The polyphenol theaflavin-3,30-digallate exhibits this action, according to research on the 3CL^{pro} inhibitory impact of extracts from seven different teas (Chen et al., 2005).

Along with quercetin, effective inhibitors of the SARS-CoV 3CL^{pro} enzymatic action included the flavonoids apigenin, epigallocatechin, luteolin, amentoflavone, kaempferol, and epigallocatechin gallate (Jo et al., 2020). Among the many polyphenols experimented with *in silico* reinforced by *in vitro* analyses, ellagic acid turned out to be the most effective 3CL^{pro} inhibitor (Bahun et al., 2022). In a rat model, inflammation and oxidative stress were reduced when quercetin and curcumin (both 50 mg/kg) were combined (Heeba et al., 2014). The combination demonstrated greater efficacy in lowering nitric oxide levels, malondialdehyde (a sign of lipid peroxidation), and the inflammatory cytokine Tumor Necrosis Factor-alpha. Additionally, the mixture raised heme oxygenase-1 levels and increased glutathione levels, showing lessened oxidative stress (Heeba et al., 2014).

14.9 COMBINING QUERCETIN WITH TRACE ELEMENTS AND VITAMINS

Vitamins strengthen immunity and are necessary for a number of physiological functions in the body. When combined with quercetin, vitamins can help reduce the signs and symptoms of COVID-19. Because of its strong antioxidant properties, vitamin C may help lessen coronavirus symptoms. Vitamin C has been employed to treat COVID-19-infected individuals in several recent clinical trials.

The RdRp of SARS-CoV-20 is reportedly inhibited by an increase in intracellular zinc concentration, which raises intracellular pH and damages the virus' replication mechanism (Hecel et al., 2020). Zinc influx into cells is aided by quercetin acting as a zinc ionophore. In Vero E6 cells, it has been demonstrated that the supplementation of dietary bromelain from the pineapple plant reduces the ACE2 and TMPRSS2 expression and also prevents SARSCoV-2 infection (Sagar et al., 2020). In COVID-19-positive individuals, SARS-CoV-2-induced sepsis results in an increase in proinflammatory cytokines, which causes neutrophils to accumulate more in the lungs and additionally damage the capillaries alveolus. In addition to preventing this accumulation, vitamin C can also quercetin increase alveolar fluid.

Additionally, vitamin C slows down the oxidative breakdown of flavonols, maintaining higher plasma concentrations of flavonoids like quercetin (Kinker et al., 2014). One important immunomodulatory vitamin that regulates the immune system is vitamin D. According to studies, vitamin D protects against respiratory distress by controlling immune system function and getting rid of viral pathogens. It aids in pathogen removal by downregulating the overproduction of cytokines during viral infection (Yamshchikov et al., 2009). Combining quercetin and these vitamin D properties may be advantageous. Additionally, quercetin works in concert with vitamin E to counteract oxidative damage. Vitamin E and quercetin together improve cellular defense against reactive oxygen species (ROS) and lessen free radical damage (Mostafavi-Pour et al., 2008).

A further observation revealed that the combination of quercetin and vitamin E significantly decreased metal intoxication, especially about non-essential cadmium (Milton Prabu et al., 2010, Zoroddu et al., 2019). According to numerous studies, combining vitamins and quercetin can reduce inflammation, boost antioxidant capacity, and get rid of viruses. These characteristics help treat COVID-19, a condition characterized by symptoms of oxidative stress, inflammation, and respiratory distress.

From viral entry to replication, quercetin demonstrated inhibition at multiple stages of the life cycle of the virus. Particularly, quercetin may instantly bind the glycoprotein spike and block the activity of ACE2, severing the interface between the virus and the host and intercepting the entry of coronavirus. Quercetin may interfere with the normal functions of the proteins present in the virus and in human cells by changing the expression of multiple human genes that code for virus protein targets. By obstructing the actions of PL^{pro}, 3CL^{pro}, and RdRp, quercetin prevents the spread of viruses. Furthermore, quercetin has a variety of antioxidant, immunomodulatory, and anti-inflammatory properties that help to lessen the effects of the disease.

In addition to quercetin, phenols can treat COVID-19 infection. Because quercetin works well with these polyphenols and benefits from vitamin C, vitamin D, vitamin E, and zinc, their effectiveness can be increased. Numerous clinical trials have been initiated concerning the single administration of quercetin and/or along with zinc, vitamin D3, curcumin, vitamin C, and medications like azithromycin, masitinib, hydroxychloroquine, and ivermectin. The effectiveness of some varieties tested for the deterrence of COVID-19 is confirmed by the results that are currently available (Imran et al., 2022).

The molecular docking results and dynamics simulations demonstrate that quercetin's binding sites are especially abundant around a particular "cavity" at the Spike/ACE2 complex interface, suggesting an area that is conducive to quercetin interfering with Spike/ACE2 interaction. Quercetin lowers the contagious of all six tried SARS-CoV-2-spike mutants in addition to attenuating the infectivity of wild-type viruses, as shown by the virus infection assay. As a result, quercetin is a prospective therapeutic candidate that can be used to treat both the wild-type and maybe future highly contagious strains of SARS-CoV-2 (Pan et al., 2023).

Of the most powerful biological inhibitors of SARS-CoV-2 RdRp, eight substances were found. These are the following compounds: asphodelin A 4¢-O-β-D-glucopyranoside, quercetin-O-β-D-3-glucopy ranoside, luteolin 7-O-β-D-glucopyranoside, naringenin, 1-methoxy-3-indolylmethyl glucosinolate, and kaempferol

3-galactoside. The study's findings offer important insights for the creation of COVID-19 medication based on natural products (Elkaeed et al., 2023). Yan et al. (2023) from a series of *in vitro* tests suggest that quercetin is a promiscuous M^{pro} inhibitor. These findings imply that the identification of M^{pro} inhibitors in the future will require a rigorous *in vitro* validation process using a variety of biochemical tests.

It was discovered that quercetin and gallic acid, with IC_{50} values below 10 µM, potently inhibited the SARS-CoV-2 M^{pro} in concentration-dependent ways. Surface plasmon resonance, covalent docking, and fluorescence resonance energy transfer simulations were used to further characterize the make-inoperative kinetics, binding affinity, and mechanism of quercetin and gallic acid. Overall, by combining target-based high-throughput screening, this study established a workable strategy for effectively identifying the covalent inhibitors of M^{pro} from therapeutic plants. This would greatly aid in the identification of important antiviral constituents from therapeutic plants (Zhang et al., 2023). Hussein et al. (2023) highlighted the possibility that recently synthesized compounds based on quercetin could interfere with the SARS-CoV-2 protein. In the ADMET investigation, all of the drugs under study demonstrated good absorption and solubility characteristics together with modest toxicities. The remarkable affinity for binding to SARS-CoV-2 major protease was shown by the docking data.

14.10 CONCLUSION

Numerous health benefits, such as antioxidant, antimicrobial, antifungal, anti-diabetic, hypolipidemic, anti-inflammatory, and anticancer properties, are attributed to quercetin. Its antiviral properties mostly target inhibiting different phases of viral infections, such as blocking the virus's entrance, replication, and translation-related protein synthesis.

ACKNOWLEDGMENT

All the authors are thankful to their respective University and Institutions for their valuable support.

DATA AVAILABILITY

The authors confirm that the data supporting the findings of this study are available within the book/chapter.

REFERENCES

Abian, O., Ortega-Alarcon, D., Jimenez-Alesanco, A., Ceballos-Laita, L., Vega, S., Reyburn, H. T., & Velazquez-Campoy, A., (2020). Structural stability of SARS-CoV-2 3CLpro and identification of quercetin as an inhibitor by experimental screening. *International Journal of Biological* Macromolecules. *164*, 1693–1703. https://doi.org/10.1016/j.ijbiomac.

Abidharini, J. D., Souparnika, B. J., Elizabeth, J., Vishalini, G., Nihala, S., Velayathaprabhu S., Rengarajan, R. L., Senthilkumar, N., Arthi, B., & Vijaya Anand, A., (2023). Herbal formulations in fighting against the SARS-CoV-2 infection. In: Chen, JT. (eds) *Ethnopharmacology and Drug Discovery for COVID-19: Anti-SARS-CoV-2 Agents from Herbal Medicines and Natural Products.* Springer, Singapore. https://doi.org/10.1007/978-981-99-3664-9_4.

Adem, S., Eyupoglu, V., Ibrahim, I. M., Sarfraz, I., Rasul, A., Ali, M., & Elfiky, A. A., (2022). Multidimensional *in silico* strategy for identification of natural polyphenols-based SARS-CoV-2 main protease (Mpro) inhibitors to unveil a hope against COVID-19. *Computers in Biology and Medicine. 145*, 105452, https://doi.org/10.1016/j.compbiomed.2022.105452.

Agrawal, P. K, Agrawal, C., & Blunden, G., (2020). Quercetin: Antiviral significance and possible COVID-19 integrative considerations. *Natural Product Communications. 15*(12), 266–278. https://doi.org/10.1177/1934578X20976293.

Ajith, S. L., Poornima P. A., Nihala, S., Elizabeth, J., Kaviya, M., Bharathi, K., Balamuralikrishna, B., & Vijaya Anand, A., (2023). Phytonutrients and secondary metabolites to cease SARS-CoV-2 loop. In: Chen, JT. (eds). *Bioactive Compound Against SARS-CoV-2.* CRC Press, Boca Raton, FL.

Anand, A.V., Balamuralikrishnan, B., Kaviya, M., Bharathi, K., Parithathvi, A., Arun, M., Senthilkumar, N., Velayuthaprabhu, S., Saradhadevi, M., Al-Dhabi, N. A., Arasu, M. V., Yatoo, M. I., Tiwari, R., & Dhama, K., (2021). Medicinal plants, phytochemicals, and herbs to combat viral pathogens including SARS-CoV-2. *Molecules. 26*(6), 1775. https://doi.org/10.3390/molecules26061775.

Arthi, B., Souparnika, B., Atchaya, S., Ramya, S., Senthilkumar, N., Velayuthaprabhu, S., Rengarajan, R. L., & Vijaya Anand, A., (2023). Antiviral activities of the flavonoids compound against COVID-19 and other viruses causing ailments. In: Chen, JT. (eds). *Bioactive Compound Against SARS-CoV-2.* CRC Press, Boca Raton, FL.

Arumugam, V. A., Sangeetha, T., Zareena, F., Pavithra, R., Ann, M., Sanjeev, A., Sunantha, B., Meyyazhagan, A., Yatoo, M. I., Sharun, K., Tiwari, R., & Pandey M. K., (2020). COVID-19 and the world with co-morbidities of heart disease, hypertension and diabetes. *Journal of Pure and Applied Microbiology. 14*(3), 1623–1638 https://doi.org/10.22207/JPAM.14.3.01.

Aslam, S., Jahan, N., Rahman, K.-U., Zafar, F., & Ashraf, M. Y., (2017). Synergistic interactions of polyphenols and their effect on antiradical potential. *Pakistan Journal of Pharmaceutical Sciences. 30*, 1297–1304.

Aucoin, M., Cooley, K., Saunders, P. R., Cardozo, V., Remy, D., Cramer, H., Neyre Abad, C., & Hannan, N., (2020). The effect of quercetin on the prevention or treatment of COVID-19 and other respiratory tract infections in humans: A rapid review. *Advances in Integrative Medicine. 7*(4), 247–251. https://doi.org/10.1016/j.aimed.2020.07.007.

Bahun, M., Jukic, M., Oblak, D., Kranjc, L., Bajc, G., Butala, M., Bozovicar, K., Bratkovic, T., Podlipnik, C., & Poklar Ulrih, N., (2022). Inhibition of the SARS-CoV-2 3CL(pro) main protease by plant polyphenols. *Food Chemistry. 373*, 131594. https://doi.org/10.1016/j.foodchem.2021.131594.

Bastaminejad, S., & Bakhtiyari, S., (2020). Quercetin and its relative therapeutic potential against COVID-19: A retrospective review and prospective overview. *Current Molecular Medicine. 21*, 385–391. https://doi.org/10.2174/1566524020999200918150630.

Bharathi, K., Ajith, S. L., Jananisri, A., Bavyataa, V., Rajan, B., Balamuralikrishnan, B., Valan Arasu, M., Naif Abdullah, A. D., Beulah, C., & Vijaya Anand, A., (2023). Antiviral properties of south indian plants against SARS-CoV-2. In: Chen, JT. (eds) *Ethnopharmacology and Drug Discovery for COVID-19: Anti-SARS-CoV-2 Agents from Herbal Medicines and Natural Products.* Springer, Singapore. https://doi.org/10.1007/978-981-99-3664-9_17.

Bhat, K. K. P., (1995). Medicinal plant information databases. In: *Non-Wood Forest Products. Medicinal Plants for Conservation and Health Care.* Food and Agriculture Organization, Rome.

Bhotla, H. K., Kaul, T., Balasubramanian, B., Easwaran, M., Arumugam, V. A., Pappusamy, M., & Meyyazhagan, A., (2020). Platelets to surrogate lung inflammation in COVID-19 patients. *Medical Hypotheses, 143,* 110098. https://doi.org/10.1016/j.mehy.2020.110098.

Bhotla, H. K., Balasubramanian, B., Meyyazhagan, A., Pushparaj, K., Easwaran, M., Pappusamy, M., & Di Renzo, G. C., (2021). Opportunistic mycoses in COVID-19 patients/survivors: Epidemic inside a pandemic. *Journal of Infection and Public Health. 14*(11), 1720–1726. https://doi.org/10.1016/j.jiph.2021.10.010.

Bjorklund, G., Dadar, M., Chirumbolo, S., & Lysiuk, R., (2017). Flavonoids as detoxifying and pro-survival agents: What's new? *Food and Chemical Toxicology. 110,* 240–250. https://doi.org/10.1016/j.fct.2017.10.039.

Boretti, A., (2021). Quercetin supplementation and COVID-19. *Natural Product Communications. 16*(9), 1934578X2110427. https://doi.org/10.1155/2022/3997190.

Chandra Manivannan, A., Malaisamy, A., Eswaran, M., Meyyazhagan, A., Arumugam, V. A., Rengasamy, K. R. R., Balasubramanian, B., & Liu, W. C., (2022). Evaluation of clove phytochemicals as potential antiviral drug candidates targeting SARS-CoV-2 main protease: Computational ocking, molecular dynamics simulation, and pharmacokinetic profiling. *Frontiers in Molecular Biosciences. 9,* 918101. https://doi.org/10.3389/fmolb.2022.918101.

Chen, C. N., Lin, C. P., Huang, K. K., Chen, W. C., Hsieh, H. P., Liang, P. H., & Hsu, J. T., (2005). Inhibition of SARS-CoV 3C-like protease activity by theaflavin-3,3'-digallate (TF3). *Evidence-Based Complementary and Alternative Medicine. 2,* 209–215. https://doi.org/10.1093/ecam/neh081.

Chen, L., Li, J., Luo, C., Liu, H., Xu, W., Chen, G., & Jiang, H., (2006). Binding interaction of quercetin-3-β-galactoside and its synthetic derivatives with SARS-CoV 3CLpro: Structure-activity relationship studies reveal salient pharmacophore features. *Bioorganic & Medicinal Chemistry. 14*(24), 8295–8306. https://doi.org/10.1016/j.bmc.2006.09.014.

Chirumbolo, S., Bjørklund, G., Lysiuk, R., Vella, A., Lenchyk, L., & Upyr, T., (2018). Targeting cancer with phytochemicals via their fine tuning of the cell survival sgnaling pathways. *International Journal of Molecular Sciences. 19*(11), 3568. https://doi.org/10.3390/ijms19113568.

Cho, J. K., Curtis-Long, M. J., Lee, K. H., Kim, D. W., Ryu, H. W., Yuk, H. J., & Park, K. H., (2013). Geranylated flavonoids displaying SARS-CoV papain-like protease inhibition from the fruits of Paulownia tomentosa. *Bioorganic & Medicinal Chemistry Letters. 21,* 3051–3057 https://doi.org/10.1016/j.bmc.2013.03.027.

Colunga Biancatelli, R. M. L., Berrill, M., Catravas, J. D., & Marik, P. E., (2020). Quercetin and vitamin C: An experimental, synergistic therapy for the prevention and treatment of SARS-CoV-2 related disease (COVID-19). *Frontiers in Immunology. 11,* 1451. https://doi.org/10.3389/fimmu.2020.01451.

Day, A. J., Cañada, F. J., Díaz, J. C., Kroon, P. A., Mclauchlan, R., Faulds, C. B., Plumb, G. W., Morgan, M. R., & Williamson, G., (2000). Dietary flavonoid and isoflavone glycosides are hydrolysed by the lactase site of lactase phlorizin hydrolase. *FEBS Letters. 468*(2–3), 166–170. https://doi.org/10.1016/s0014-5793(00)01211-4.

Dabeek, W. M., & Marra, M. V., (2019). Dietary quercetin and kaempferol: Bioavailability and potential cardiovascular-related bioactivity in humans. *Nutrients. 11*(10), 2288. https://doi.org/10.3390/nu11102288.

Derosa, G., Maffioli, P., D'Angelo, A., & Di Pierro, F., (2021). A role for quercetin in coronavirus disease 2019 (COVID-19). *Phytotherapy Research. 35,* 1230–1236. https://doi.org/10.1002/ptr.6887.

Di Pierro, F., Iqtadar, S., & Khan, A., (2021). Potential clinical benefts of quercetin in the early stage of COVID-19: Results of a second, pilot, randomized, controlled and open-label clinical trial. *International Journal of General Medicine. 14,* 2807–2816. https://doi.org/10.2147/IJGM.S318949.

Elkaeed, E. B., Alsfouk, B. A., Ibrahim, T. H., Arafa, R. K., Elkady, H., Ibrahim, I. M., Eissa, I. H., & Metwaly, A. M., (2023). Computer-assisted drug discovery of potential natural inhibitors of the SARS-CoV-2 RNA-dependent RNA polymerase through a multi-phase *in silico* approach. *Antiviral Therapy. 28*(5) 13596535231199838. https://doi.org/10.1177/13596535231199838.

Evans, W. C., & Trease, G. E., (2002). *Trease and Evans Pharmacognosy.* 6th edition, WB Saunders, China.

Ferrario, C. M., Jessup, J., Chappell, M. C., Averill, D. B., Brosnihan, K. B., Tallant, E. A., & Gallagher, P. E., (2005). Effect of angiotensin-converting enzyme inhibition and angiotensin II receptor blockers on cardiac angiotensin-converting enzyme 2. *Circulation. 111*(20), 2605–2610. https://doi.org/10.1161/Circulationnaha.104.510461.

Filipa Almeida, A., Grethe Iren, A. B., Mariusz, P., Adriana, T., Liliana, T., Kateřina, V., Gary, W., & Santos, C. N., (2018). Bioavailability of quercetin in humans with a focus on interindividual variation. *Comprehensive Reviews in Food Science and Food Safety. 17*:714–731.

Gasmi, A., Noor, S., Tippairote, T., Dadar, M., Menzel, A., & Bjorklund, G., (2020). Individual risk management strategy and potential therapeutic options for the COVID-19 pandemic. *Clinical Immunology. 215,* 108409. https://doi.org/10.1016/j.clim.2020.108409.

Gasmi, A., Tippairote, T., Mujawdiya, P. K., Peana, M., Menzel, A., Dadar, M., Gasmi Benahmed, A., & Bjørklund, G., (2020). Micronutrients as immunomodulatory tools for COVID-19 management. *Clinical Immunology. 220,* 108545. https://doi.org/10.1016/j.clim.2020.108545.

Gasmi, A., Mujawdiya, P. K., Lysiuk, R., Shanaida, M., Peana, M., Gasmi Benahmed, A., Beley, N., Kovalska, N., & Bjørklund, G., (2022). Quercetin in the prevention and treatment of coronavirus infections: A focus on SARS-CoV-2. *Pharmaceuticals. 15*(9), 1049. https://doi.org/10.3390/ph15091049.

Glinsky, G. V., (2020). Tripartite combination of candidate pandemic mitigationagents: Vitamin D, quercetin, and estradiol manifest properties of medicinal agents for targeted mitigation of the COVID-19 pandemic defined by genomics-guided tracing of SARS-CoV-2 targets in human cells. *Biomedicine. 8*(5), 129. https://doi.org/10.3390/biomedicines8050129.

Hecel, A., Ostrowska, M., Stokowa-Soltys, K., Watly, J., Dudek, D., Miller, A., Potocki, S., Matera-Witkiewicz, A., Dominguez-Martin, A., & Kozlowski, H., et al., (2020). Zinc(II)-The overlooked eEminence grise of chloroquine's fight against COVID-19? *Pharmaceuticals. 13*, 228. https://doi.org/10.3390/ph13090228.

Heeba, G. H., Mahmoud, M. E., & El Hanafy, A. A., (2014). Anti-inflammatory potential of curcumin and quercetin in rats: Role of oxidative stress, heme oxygenase-1 and TNF-alpha. *Toxicology & Industrial Health. 30*, 551–560. https://doi.org/10.1177/0748233712462444.

Hsu, S., (2015). Compounds erived from epigallocatechin-3-gallate (EGCG) as a novel approach to the prevention of viral infections. *Inflammation & Allergy-Drug Targets. 14*, 13–18. https://doi.org/10.1111/cbdd.13604.

Huang, F., Li, Y., Leung, E. L., Liu, X., Liu, K., Wang, Q., & Luo, L., (2020). A review of therapeutic agents and Chinese herbal medicines against SARS-COV-2 (COVID-19). *Pharmacological Research. 158*, 104929. https://doi.org/10.1016/j.phrs.2020.104929.

Hussein, R. K., Marashdeh, M., & El-Khayatt, A. M., (2023). Molecular docking analysis of novel quercetin derivatives for combating SARS-CoV-2. *Bioinformation. 19*(2), 178–183. https://doi.org/10.6026/97320630019178.

Igoli, J. O., Ogaji, O. G., Tor-Anyiin, T. A., & Igoli, N. P., (2003). Traditional medicine practice amongst the Igede people of Nigeria. Part II. *African Journal of Traditional, Complementary and Alternative Medicines. 10*(4), 1–10.

Imran, M., Thabet, H. K., Alaqel, S. I., Alzahrani, A. R., Abida, A., Alshammari, M. K., Kamal, M., Diwan, A., Asdaq, S. M. B., & Alshehri, S., (2022). The therapeutic and prophylactic potential of quercetin against COVID-19: An outlook on the clinical studies, inventive compositions, and patent literature. *Antioxidants. 11*, 876. https://doi.org/10.3390/antiox11050876.

Jaganathan, S. K., & Mandal, M., (2009). Antiproliferative effects of honey and of its polyphenols: A review. *Journal of Biomedicine and Biotechnology. 2009*, 830616. https://doi.org/10.1155/2009/830616.

Jo, S., Kim, H., Kim, S., Shin, D. H., & Kim, M. S., (2019). Characteristics of flavonoids as potent MERS-CoV 3C-like protease inhibitors. *Chemical Biology & Drug Design. 94*, 2023–2030.

Jo, S., Kim, S., Shin, D. H., & Kim, M. S., (2020). Inhibition of SARS-CoV 3CL protease by flavonoids. *Journal of Enzyme Inhibition and Medicinal Chemistry. 35*, 145–151. https://doi.org/10.1080/14756366.2019.1690480.

Joshi, B., Sah, G. P., Basnet, B. B., Bhatt, M. R., & Sharma, D., (2011). Phytochemical extraction and antimicrobial properties of different medicinal plants: Ocimum sanctum (Tulsi), Eugenia caryophyllata (Clove), Achyranthes bidentata (Datiwan) and Azadirachta indica (Neem). *Journal of Microbiology and Antimicrobials. 3*, 1–7.

Kaihatsu, K., Yamabe, M., & Ebara, Y., (2018). Antiviral mechanism of action of epigallocatechin-3-O-gallate and Its fatty acid esters. *Molecules. 23*, 2475. https://doi.org/10.3390/molecules23102475.

Kaviya, M., Peatrise Geofferina, I., Poornima, P. A., Prem Rajan, A., Balamuralikrishnan, B., Meyyazhagan, A., Naif Abdullah, A. D., Valan Arasu, M., Karthika, P., Kallidass, S., Ramya, S., & Vijaya Anand, A., (2023). Dietary plants, spices, and fruits in curbing SARS-CoV-2 virulence. In: Chen, JT. (eds) *Ethnopharmacology and Drug Discovery for COVID-19: Anti-SARS-CoV-2 Agents from Herbal Medicines and Natural Products.* Springer, Singapore. https://doi.org/10.1007/978-981-99-3664-9_4.

Khaerunnisa, S., Kurniawan, H., Awaluddin, R., Suhartati, S., & Soetjipto, S., (2020). Potential inhibitor of COVID-19 main protease (Mpro) from several medicinal plant compounds by molecular docking study. *Preprints.* DOI:10.20944/preprints202003.0226.v1.

Kim, D. W., Seo, K. H., Curtis-Long, M. J., Oh, K. Y., Oh, J. W., Cho, J. K., & Park, K. H., (2014). Phenolic phytochemical displaying SARS-CoV papain-like protease inhibition from the seeds of Psoralea corylifolia. *Journal of Enzyme Inhibition and Medicinal Chemistry. 29*(1), 59–63. https://doi.org/10.3109/14756366.2012.753591.

Kinker, B., Comstock, A. T., & Sajjan, U. S., (2014). Quercetin: A promising treatment for the common cold. *Journal of ANC Disaster Prevention Research. 2*(2), 1–3. https://doi.org/10.4172/2329-8731.1000111.

Kuchi Bhotla, H., Balasubramanian, B., Arumugam, V. A., Pushparaj, K., Murugesh, E., & Rathinasamy, B., (2021). Insinuating cocktailed components in biocompatible-nanoparticles could act as an impressive neo-adjuvant strategy to combat COVID-19. *Natural Resources for Human Health. 1*, 3–7. https://doi.org/10.53365/nrfhh/140607.

Li, Y., Yao, J., Han, C., Yang, J., Chaudhry, M. T., Wang, S., & Yin, Y., (2016). Quercetin, inflammation and immunity. *Nutrients. 8*(3), 167. https://doi.org/10.3390/nu8030167.

Lin, C. W., Tsai, F. J., Tsai, C. H., Lai, C. C., Wan, L., Ho, T. Y., & Chao, P. D. L., (2005). Anti-SARS coronavirus 3C-like protease effects of *Isatis indigotica* root and plant-derived phenolic compounds. *Antiviral Research. 68*(1), 36–42. https://doi.org/10.1016/j.antiviral.2005.07.002.

Liu, C., Zhou, Q., Li, Y., Garner, L. V., Watkins, S. P., Carter, L. J., & Albaiu, D., (2020). Research and development on therapeutic agents and vaccines for COVID-19 and related human coronavirus diseases. *ACS Central Science. 6*, 315–331. https://doi.org/10.1021/acscentsci.0c00272.

Liu, X., Raghuvanshi, R., Ceylan, F. D., & Bolling, B. W., (2020). Quercetin and its metabolites inhibit recombinant human angiotensin-converting enzyme 2 (ACE2) activity. *Journal of Agricultural and Food Chemistry. 68*(47), 13982–13989. https://doi.org/10.1021/acs.jafc.0c05064.

Mehany, T., Khalifa, I., Barakat, H., Althwab, S. A., Alharbi, Y. M., & El-Sohaimy, S., (2021). Polyphenols as promising biologically active substances for preventing SARS-CoV-2: A review with research evidence and underlying mechanisms. *Food Bioscience. 40*, 100891. https://doi.org/10.1016/j.fbio.2021.100891.

Meyyazhagan, A., Pushparaj, K., Balasubramanian, B., Kuchi Bhotla, H., Pappusamy, M., Arumugam, V. A., & Di Renzo, G. C., (2022). COVID-19 in pregnant women and children: Insights on clinical manifestations, complexities, and pathogenesis. *International Journal of Gynecology & Obstetrics. 156*(2), 216–224. https://doi.org/10.1002/ijgo.14007.

Mora, J. R., Iwata, M., & Von Andrian, U. H., (2008). Vitamin effects on the immune system: Vitamins A and D take centre stage. *Nature Reviews Immunology. 8*(9), 685–698. https://doi.org/10.1038/nri2378.

Morán-Santibañez, K., Peña-Hernández, M. A., Cruz-Suárez, L. E., Ricque-Marie, D., Skouta, R., Vasquez, A. H., & Trejo-Avila, L. M., (2018). Virucidal and synergistic activity of polyphenol-rich extracts of seaweeds against measles virus. *Viruses. 10*(9), 465. https://doi.org/10.3390/v10090465 .

Mostafavi-Pour, Z., Zal, F., Monabati, A., & Vessal, M., (2008). Protective effects of a combination of quercetin and vitamin E against cyclosporine A-induced oxidative stress and hepatotoxicity in rats. *Hepatology Research*. *38*(4), 385–392. https://doi.org/10.1111/j.1872-034X.2007.00273.x.

Munafò, F., Donati, E., Brindani, N., Ottonello, G., Armirotti, A., & De Vivo, M., (2022). Quercetin and luteolin are single-digit micromolar inhibitors of the SARS-CoV-2 RNA-dependent RNA polymerase. *Scientific Reports*. *12*, 10571. https://doi.org/10.1038/s41598-022-146.

Munita, J. M., & Arias, C. A., (2016). Mechanisms of antibiotic resistance. *Microbiology Spectrum*. *4*, 481–511. https://doi.org/10.1128/microbiolspec.

Nguyen, T. T. H., Woo, H. J., Kang, H. K., Nguyen, V. D., Kim, Y. M., Kim, D. W., & Kim, D., (2012). Flavonoid-mediated inhibition of SARS coronavirus 3C-like protease expressed in *Pichia pastoris*. *Biotechnology Letters*. *34*, 831–838. https://doi.org/10.1007/s10529-011-0845-8.

Ożarowski, M., & Karpiński, T. M., (2021). Extracts and flavonoids of *Passiflora* species as promising anti-inflammatory and antioxidant substances. *Current Pharmaceutical Design*. *27*(22), 2582–2604. https://doi.org/10.2174/13816 12826666200526150113.

Pan, B., Fang, S., Wang, L., Pan, Z., Li, M., & Liu, L., (2023). Quercetin: A promising drug candidate against the potential SARS-CoV-2-Spike mutants with high viral infectivity. *Computational and Structural Biotechnology Journal*. *21*, 5092–5098. https://doi.org/10.1016/j.csbj.2023.10.029.

Park, J. Y., Yuk, H. J., Ryu, H. W., Lim, S. H., Kim, K. S., Park, K. H., & Lee, W. S., (2017). Evaluation of polyphenols from *Broussonetia papyrifera* as coronavirus protease inhibitors. *Journal of Enzyme Inhibition and Medicinal Chemistry*. *32*(1), 504–512. https://doi.org/10.1080/14756366.2016.12 65519.

Poochi, S. P., Easwaran, M., Balasubramanian, B., Anbuselvam, M., Meyyazhagan, A., Park, S., & Kaul, T., (2020). Employing bioactive compounds derived from *Ipomoea obscura* (L.) to evaluate potential inhibitor for SARS-CoV-2 main protease and ACE2 protein. *Food Frontiers*. *1*(2), 168–179. https://doi.org/10.1002/fft2.29.

Prabu, S. M., Shagirtha, K., & Renugadevi, J., (2010). Quercetin in combination with vitamins (C and E) improves oxidative stress and renal injury in cadmium intoxicated rats. *European Review for Medical & Pharmacological Sciences*. *14*(11), 903–914.

Pushparaj, K., Kuchi Bhotla, H., Arumugam, V. A., Pappusamy, M., Easwaran, M., Liu, W. C., Issara, U., Rengasamy, K. R..R., Meyyazhagan, A., & Balasubramanian, B. (2022). Mucormycosis (black fungus) ensuing COVID-19 and comorbidity meets-magnifying global pandemic grieve and catastrophe begins. *Science of The Total Environment*. *805*, 150355 https://doi.org/10.1016/j.scitotenv.2021.150355.

Rakotoarivelo, N. H., Rakotoarivony, F., Ramarosandratana, A. V., Jeannoda, V. H., Kuhlman, A. R., Randrianasolo, A., & Bussmann, R. W., (2015). Medicinal plants used to treat the most frequent diseases encountered in Ambalabe rural community, Eastern Madagascar. *Journal of Ethnobiology and Ethnomedicine*. *11*, 1–16. https://doi.org/10.1186/s13002-015-0050-2.

Ramya, S., Boro, A., Geofferina, I. P., Jananisri, A., Vishalini, G., Balamuralikrishna, B., & Anand, A. V., (2023). The potential contribution of vitamin K as a nutraceutical to scale down the mortality rate of COVID-19. In *Bioactive Compounds Against SARS-CoV-2*, pp. 140–159. CRC Press, Boca Raton, FL.

Rasool Hassan, B. A., (2012). Medicinal plants (importance and uses). *Pharmaceutica Analytica Acta*. *3*(10), 2153–2435.

Ryu, Y. B., Jeong, H. J., Kim, J. H., Kim, Y. M., Park, J. Y., Kim, D., & Lee, W. S., (2010). Biflavonoids from *Torreya nucifera* displaying SARS-CoV 3CLpro inhibition. *Bioorganic & Medicinal Chemistry*. *18*(22), 7940–7947. https://doi.org/10.1016/j.bmc.2010.09.035.

Saeedi-Boroujeni, A., & Mahmoudian-Sani, M. R., (2021). Anti-inflammatory potential of Quercetin in COVID-19 treatment. *Journal of Inflammation*. *18*, 1–9. https://doi.org/10.1186/s12950-021-00268-6.

Sagar, S., Rathinavel, A. K., Lutz, W. E., Struble, L. R., Khurana, S., Schnaubelt, A. T., & Radhakrishnan, P., (2020). Bromelain inhibits SARS-CoV-2 infection in VeroE6 cells. *BioRxiv*. https://doi.org/10.1101/2020.09.16.297366.

Schwarz, S., Sauter, D., Wang, K., Zhang, R., Sun, B., Karioti, A., & Schwarz, W., (2014). Kaempferol derivatives as antiviral drugs against the 3a channel protein of coronavirus. *Planta Medica*. *80*(02/03), 177–182. https://doi.org/10.1055/s-0033-1360277.

Shanmugam, R., Thangavelu, S., Fathah, Z., Yatoo, M. I., Tiwari, R., Pandey, M. K., Dhama, J., Chandra, R., Malik, Y., Dhama, K., Sha, R., & Chaicumpa, W. (2020). Shanmugam V Arumugam AV, SARS-CoV-2/COVID-19 pandemic-an update. *Journal of Experimental Biology and Agricultural Sciences*. *8*(1), S219–S245.

Shin, D., Mukherjee, R., Grewe, D., Bojkova, D., Baek, K., Bhattacharya, A., & Dikic, I., (2020). Papain-like protease regulates SARS-CoV-2 viral spread and innate immunity. *Nature*. *587*(7835), 657–662. https://doi.org/10.1038/s41586-020-2601-5.

Stanek, N., Kafarski, P., & Jasicka-Misiak, I., (2019). Development of a high performance thin layer chromatography method for the rapid qualification and quantification of phenolic compounds and abscisic acid in honeys. *Journal of Chromatography A*. *1598*, 209–215. https://doi.org/10.1016/j.chroma.

Sule, W. F., Okonko, I. O., Joseph, T. A., Ojezele, M. O., Nwanze, J. C., Alli, J. A., & Adewale, O. G., (2010). In vitro antifungal activity of *Senna alata* Linn. crude leaf extract. *Research Journal of Biological Sciences*. *5*(3), 275–284.

Timothy, S. Y., Wazis, C. H., Adati, R. G., & Maspalma, I. D., (2012). Antifungal activity of aqueous and ethanolic leaf extracts of Cassia alata Linn. *Journal of Applied Pharmaceutical Science*. *2*(7), 182–185. https://doi.org/10.7324/JAPS.2012.2728.

Tsanova-Savova, S., Ribarova, F., & Petkov, V., (2018). Quercetin content and ratios to total flavonols and total flavonoids in Bulgarian fruits and vegetables. *Bulgarian Chemical Communications*. *50*(1), 69–73.

Ubani, A., Agwom, F., Morenikeji, O. R., Shehu, N. Y., Umera, E. A., Umar, U., & Luka, P. D., (2020). Molecular docking analysis of selected phytochemicals on two SARS-CoV-2 targets. *F1000 Research*. *9*, 1157. DOI:10.12688/f1000research.25076.1.

Ul Qamar, M. T., Alqahtani, S. M., Alamri, M. A., & Chen, L. L., (2020). Structural basis of SARS-CoV-2. *Journal of Pharmaceutical Analysis*. *10*(4), 313–319. https://doi.org/10.1016/j.jpha.2020.03.009.

Yamshchikov, A. V., Desai, N. S., Blumberg, H. M., Ziegler, T. R., & Tangpricha, V., (2009). Vitamin D for treatment and prevention of infectious diseases; A systematic review of randomized controlled trials. *Endocrine Practice. 15*(5), 438–449. https://doi.org/10.4158/EP09101.ORR.

Yan, H., Zhang, R., Liu, X., Wang, Y., & Chen, Y., (2023). Reframing quercetin as a promiscuous inhibitor against SARS-CoV-2 main protease. *Proceedings of the National Academy of Sciences. 120*(37), e2309289120. https://doi.org/10.1073/pnas.2309289120.

Yang, Y., Islam, M. S., Wang, J., Li, Y., & Chen, X., (2020). Traditional Chinese medicine in the treatment of patients infected with 2019-new coronavirus (SARS-CoV-2): A review and perspective. *International Journal of Biological Sciences. 16*(10), 1708. https://doi.org/10.7150/ijbs.45538.

Yi, L., Li, Z., Yuan, K., Qu, X., Chen, J., Wang, G., & Xu, X., (2004). Small molecules blocking the entry of severe acute respiratory syndrome coronavirus into host cells. *Journal of Virology. 78*(20), 11334–11339. https://doi.org/10.1128/JVI.78.20.11334-11339.2004.

Yu, M. S., Lee, J., Lee, J. M., Kim, Y., Chin, Y. W., Jee, J. G., & Jeong, Y. J., (2012). Identification of myricetin and scutellarein as novel chemical inhibitors of the SARS coronavirus helicase, nsP13. *Bioorganic & Medicinal Chemistry Letters. 22*(12), 4049–4054. https://doi.org/10.1016/j.bmcl.2012.04.081.

Yuan, H., Ma, Q., Ye, L., & Piao, G., (2016). The traditional medicine and modern medicine from natural products. *Molecules. 21*(5), 559. https://doi.org/10.3390/molecules21050559.

Zhang, L., Lin, D., Sun, X., Curth, U., Drosten, C., Sauerhering, L., & Hilgenfeld, R., (2020). Crystal structure of SARS-CoV-2 main protease provides a basis for design of improved α-ketoamide inhibitors. *Science. 368*(6489), 409–412. https://doi.org/10.1126/science.abb3405.

Zhang, Y. N., Zhu, G. H., Liu, W., Chen, X. X., Xie, Y. Y., Xu, J. R., & Ge, G. B., (2023). Discovery of the covalent SARS-CoV-2 Mpro inhibitors from antiviral herbs via integrating target-based high-throughput screening and chemoproteomic approaches. *Journal of Medical Virology. 95*(11), e29208. https://doi.org/10.1002/jmv.29208.

Zheng, Y. Z., Deng, G., Liang, Q., Chen, D. F., Guo, R., & Lai, R. C., (2017). Antioxidant activity of quercetin and its glucosides from propolis: A theoretical study. *Scientific Reports. 7*(1), 7543.

Zhuang, M., Jiang, H., Suzuki, Y., Li, X., Xiao, P., Tanaka, T., & Hattori, T., (2009). Procyanidins and butanol extract of *Cinnamomi cortex* inhibit SARS-CoV infection. *Antiviral Research. 82*(1), 73–81. DOI:10.1016/j.antiviral.2009.02.001.

Zoroddu, M. A., Aaseth, J., Crisponi, G., Medici, S., Peana, M., & Nurchi, V. M., (2019). The essential metals for humans: A brief overview. *Journal of Inorganic Biochemistry. 195*, 120–129. https://doi.org/10.1016/j.jinorgbio.2019.03.013.

15 Preventive and Therapeutic Potentials of Epigallocatechin Gallate for the Management of COVID-19

A Mechanistic Insight

Shivani Rana, Chetna Jhagta, Arun Parashar, Minaxi Sharma, Baskaran Stephen Inbaraj, Vineet Mehta, and Kandi Sridhar

15.1 INTRODUCTION

Severe Acute Respiratory Syndrome Coronavirus 2 (SARS-CoV-2) is an extremely pathogenic virus accountable for the global pandemic known as coronavirus disease (COVID-19). The SARS-CoV outbreak has affected approximately 672,215,054 people worldwide, with 6,736,076 fatalities and 643,749,929 recovered patients till January 2023, and this number is increasing with time. Many of those who have recovered still suffer from post-COVID complications (Worldometers.info, 2022). The pandemic occurred in waves, with the first wave reported from March 2020 to February 2021, followed by the second wave from March 2021 to the first week of April 2021, and the third wave from December 2021 to February 2022.

Coronaviruses are known to cause upper respiratory tract illnesses in birds and mammals, including humans. In a group of seven coronaviruses known to affect humans, SARS-CoV, Middle East Respiratory Syndrome Coronavirus (MERS-CoV), and SARS-CoV-2 have caused previous epidemics (Singhal, 2020). SARS-CoV-2 is extremely infectious and spreads via direct or indirect exposure to infected individuals. The disease presents symptoms in three phases. Initially, the asymptomatic stage lasts 1–2 days after infection when the virus enters cells and replicates, with partial effectiveness of innate immunity. In the second stage, the virus migrates from the upper to the lower respiratory tract, leading to an initial immune response, mainly confined to the upper respiratory tract in most cases (Parasher, 2021). The third stage is described by the development of acute respiratory distress syndrome (ARDS) and hypoxia due to alveolar damage caused by the body's overactive immune response, known as cytokine storm syndrome (Singhal, 2020). Injured cells release damage-associated molecular patterns, protein-associated molecular patterns, and cytokines, which trigger the innate immune system, leading to inflammation and fluid accumulation between the capillary and alveolus, impairing gas exchange (Yuki et al., 2020). Inflammatory mediators like interleukin-2 (IL-2), IL-6, IL-10, tumor necrosis factor-alpha (TNF-α), pegylated granulocyte colony-stimulating factor (G-CSF), and monocyte chemoattractant protein-1 (MCP1) are released by phagocytic cells, leading to chronic inflammation and reduced surfactant in the alveolus (Hosseini et al., 2020; Khalil et al., 2021). The observed fever in COVID-19 is caused by these inflammatory mediators acting in the hypothalamus. In severe cases, protein-rich fluid may enter the circulation, triggering systemic inflammatory response syndrome, which can result in multi-organ failure (Catanzaro et al., 2020). The cytokines can also increase procoagulant levels, potentially leading to pulmonary embolism (Khalil et al., 2021). Additionally, the virus stimulates cranial nerve receptors, causing the central nervous system to produce a cough reflex, besides contributing to several neurological complications (Yuki et al., 2020).

Despite significant efforts, there are currently no specific therapies for SARS-CoV-2 infection, although therapies to counter associated symptoms are being used clinically. The virus weakens the immune system regardless of age. Various therapeutic options have been proposed for treating SARS-CoV-2 infection, including nucleoside analogs, remdesivir, lopinavir, ritonavir, anti-inflammatory medications, repurposed drugs, and medication combinations (Amanat and Krammer, 2020). Several clinical trials investigating these medications and others are being conducted; however, it remains uncertain whether these treatments will be effective in treating SARS-CoV-2 infection in humans.

Green tea has numerous physiological and pharmacological benefits due to its active ingredients, such as epigallocatechin gallate (EGCG) (Steinmann et al., 2013). EGCG, the most abundant catechin in green tea, possesses antioxidant and anti-inflammatory properties (Hong et al., 2011; Riegsecker et al., 2013) and it has demonstrated a broad-spectrum antiviral effect on various human viruses (Colpitts and Schang, 2014; Reid et al., 2014; Carneiro et al., 2016; Liu et al., 2018; Lai et al., 2018), including those causing respiratory diseases (Furushima et al., 2018). Previous studies have shown that green tea catechins can inhibit the activities of the influenza A virus and reduce

DOI: 10.1201/9781003433200-15

infection rates (Furushima et al., 2019; Onishi et al., 2020). Emerging evidence suggests that green tea and its major ingredients, particularly EGCG, may be beneficial in combating SARS-CoV-2 infection in humans (Mhatre et al., 2021). Green tea catechins have been associated with factors linked to the COVID-19 death rate, such as antioxidative, anti-hypertensive, anti-proliferative, anti-thrombogenic, and lipid/cholesterol-lowering effects (Storozhuk, 2022). EGCG's anti-inflammatory property may also help mitigate the severity of SARS-CoV-2 infection by reducing immune system overactivity and associated complications (Menegazzi et al., 2020). *In vitro* studies have demonstrated that EGCG can block the SARS-CoV-2 3CL-protease, an essential enzyme for coronavirus contagion and multiplication in the host, making it a potential therapeutic option for SARS-CoV-2 infection (Baum et al., 2020). Additionally, ecological investigations have proposed a possible correlation between higher green tea consumption and lower morbidity and mortality of COVID-19 in certain countries (Storozhuk, 2022).

Although the evidence supporting the beneficial effects of EGCG during SARS-CoV-2 infection seems to be promising, further experimental and mechanistic approaches are needed to assess the influence of the major constituents of green tea on SARS-CoV-2 infection. In this chapter, we sum up the state of knowledge regarding the nature of COVID-19 and SARS-CoV-2, with a focus on the importance of EGCG to provide positive effects during SARS-CoV-2 infection, including their limitations and future scope. We provide a deep analysis of the pathogenesis of SARS-CoV-2 infection based on recently published findings.

15.2 CORONAVIRUS

Corona in *Latin* means crown. Coronavirus is a positive single-stranded RNA virus having spike glycoproteins on its envelope that appear like crowns under the electron microscope. The coronavirus is categorized into different groups, i.e., alpha (α), beta (β), gamma (γ), and delta (δ). It comes under the Coronaviridae family and follows the Nidovirales order (Yi et al., 2020). The α-coronaviruses include Human Coronavirus-229E (HCoV229E) and Human Coronavirus NL63 (HCoV-NL63), while β-coronaviruses include Human Coronavirus OC43 (HCoV-OC43), SARS-CoV, HKU-1, MERS-CoV, and SARC-CoV-2. SARS-CoV-2 is a novel coronavirus (nCov-2019) (Figure 15.1), a novel strain of the coronavirus family which has not been identified earlier in humans (Muniyappa et al., 2020).

In China's Wuhan city in December 2019, the first SARS-CoV-2 infection case was reported. As a deadly and contagious virus, it spreads fast over the world and was recognized as a worldwide pandemic in 2020. The WHO declared the new coronavirus epidemic a public health emergency of global concern on January 30, 2020. The disease was designated as COVID-19 by the WHO, and the new coronavirus was assigned the name "SARS-CoV-2" by the International Committee on Taxonomy of Viruses on February 11, 2020 (Eurosurveillance Editorial Team, 2020; Coronaviridae Study Group, 2020). As of June 14, 2023, the global number of confirmed COVID-19 cases stands at approximately 767.98 million, with around 6.94 million reported deaths according to the WHO (https://covid19.who.int/). SARS-CoV-2-induced COVID-19 physiology varies from a superficial respiratory infection to severe pneumonia, hypoxia, multi-organ failure, and demise, with the elderly at higher risk (Yuki et al., 2020). Patients suffering from SARS-CoV-2 infection displayed viral pneumonia symptoms, such as pyrexia (fever), coughing, and chest pain, as well as dyspnea and bilateral lung infiltration in more severe cases (Zhu et al., 2019; Gralinski et al., 2020); however, long-term and post COVID complications have now started to emerge as a serious concern in patients recovered from SARS-CoV-2 infection (Nagu et al., 2021).

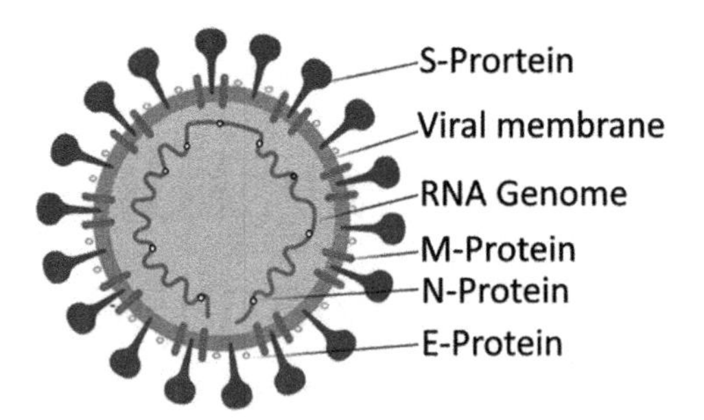

FIGURE 15.1 Structure of SARS-CoV-2 (Constructed by authors using BioRender, Toronto, Ontario, Canada).

15.3 STRUCTURE AND GENOMIC SEQUENCE OF SARS-COV-2

SARS-CoV-2 shared more than 80% of its characteristics with SARS-CoV and 50% with MERS-CoV. It has a genomic size of approximately 30 kb and is a single-stranded RNA virus (Woo et al., 2010). It also consists of 14 open reading frames (ORFs), making up 67% of the SARS-CoV-2 genome. Among them, ORF1a and ORF1a/b are major frames that encode two overlapped chains of polyproteins further divided into 16 non-structural proteins (nsp) (Chen et al., 2020). There are several cleaving enzymes for each nsp, including exoribonuclease (nsp14), CL^{pro} (chymotrypsin-like protease) (nsp5), RdRp (RNA-dependent RNA polymerase) (nsp12), and papain-like protease (PL^{pro}) (nsp1–3) (Ziebuhr et al., 2000). These enzymes are crucial for the viral transcription and replication processes that result in the formation of complete viral assembly. Only 33% of the remaining genomic sequence is occupied by structural proteins (Prajapat et al., 2020). The spike (S), membrane (M), nucleocapsid (N), and envelope (E) proteins are the four structural proteins that make up the CoV. SARS-CoV-2 has a spherical shape having spike proteins (S1 and S2) on its envelope glycoproteins (Figure 15.1) (Tang et al., 2020). The spike proteins (S) are essential for binding to the host cell. Angiotensin-converting enzyme 2 (ACE-2) receptors enable molecules to enter cells through endocytosis. The membrane protein (M), which is present on the envelope, regulates the shape of the virus. The viral envelope is created by the interaction of the E glycoprotein and the M protein. E protein, which is the smallest protein, plays a key role in viral morphogenesis, particularly during assembly development. The function of the N protein is the maintenance of the RNP complex (Hoffmann et al., 2020).

15.4 LIFE CYCLE AND PATHOPHYSIOLOGY OF SARS-COV-2

SARS-COV-2 enters inside the human host through the respiratory tract and enters into the host cell through type 2 alveolar epithelial cells. ACE-2 receptors, serine protease, and transmembrane protease serine 2 (TMPRSS2) are present in epithelial cells and help in the viral invasion of the host cell. The virus's spikes protein attaches to the host receptor and starts the infection of the host cell in two phases (Hoffmann et al., 2020). Spike protein consists of S1/S2 domains and S2¢ domain. S1 is the main receptor-binding domain, whereas S2 is a fusion peptide that makes fusion between viral and cellular membranes. Initially, the virus binds to the ACE-2 receptor through the S1 domain of spikes protein, and after binding, priming of spikes protein takes place by an enzyme TMPRSS2, stimulating the fusion between the host and virus membrane through the S2 domain. This enzyme makes a cleavage between the S1/S2 and S2¢ sites of spike protein that is likely to activate it, causing conformational changes that fuse the viral and host cell membrane (Tang et al., 2020; Shamsi et al., 2021). Post membrane fusion, the virus enters pulmonary alveolar epithelial cells, releases its RNA genome, and undergoes replication by using a host cell ribosome followed by a transcription and translation process with the help of various proteolytic enzymes. Chymotrypsin-like protease (CL^{pro}) cleaves nsp4–11, PL^{pro} cleaves nsp1–3, RdRp cleaves nsp12, and exoribonuclease cleaves nsp14 and forms viral proteins (Prajapat et al., 2020). These viral proteins bind with genomic RNA and form nucleocapsids. These newly formed nucleocapsids are then transported to the cell membrane via Golgi vesicles and then to the extracellular space via exocytosis (Vallamkondu et al., 2020). After the release of the virus, it then subsequently attacks the alveolar epithelial cells (Figure 15.2).

Angiotensin II production is increased when the ACE-2 enzyme is inhibited, which causes vasoconstriction and alters the functioning of the renin–angiotensin–aldosterone system (RAAS). Angiotensin II then binds to the angiotensin receptor 1 and activates the pro-inflammatory cytokines NF-κB and metalloprotease 17 (ADAM17), resulting in increased pulmonary vascular permeability, which in turn stimulates the production of EGFR and TNF-α and STAT3 (Eguchi et al., 2018; Murakami et al., 2019). All these events will lead to a state of hyperinflammation and inflammatory reaction in the body. Some postmortem studies have revealed the over-activation of some proinflammatory cytokines like CD4+ and CD8+ T helper cells that might be responsible for chronic immune injury in patients infected with SARS-COV-2 (Xu et al., 2020; Zhou et al., 2020). NK cell reduction is more pronounced in severe cases. The enteric virus, however, also activates macrophages, which play a significant role in the cytokine storm induced by SARS-CoV-2; this in turn activates several inflammatory mediators, including cytokines (IL-2, IL-7, IL-10, GCSF, IP-10, MCP-1, MIP-1a, and TNF-α) and chemokines (CXCL10 and CCL2), which may cause a cytokine

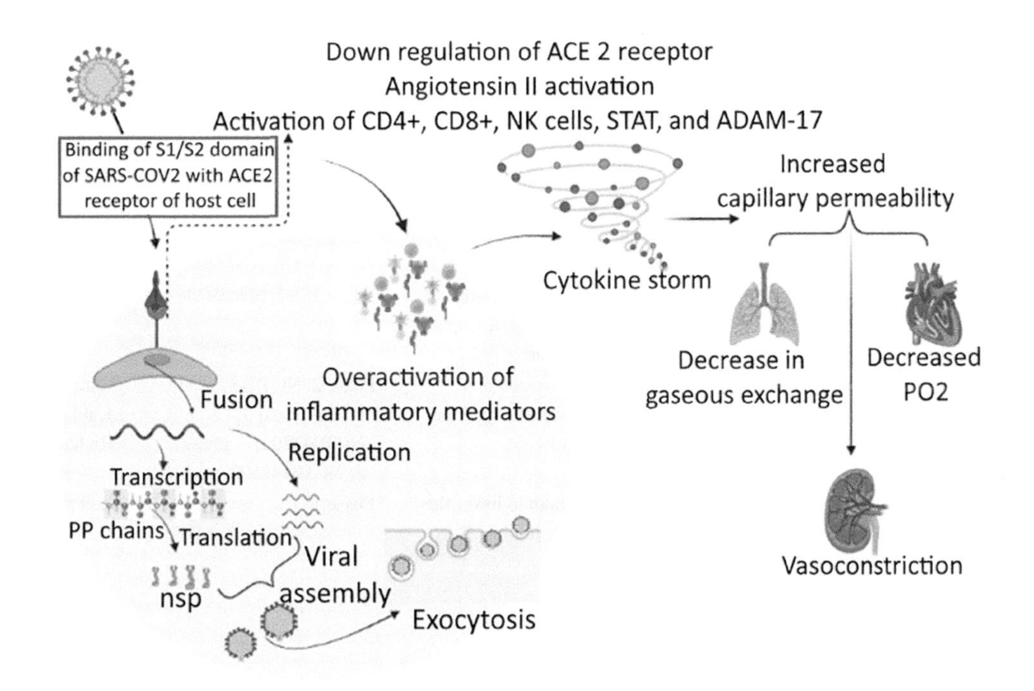

FIGURE 15.2 Life cycle and pathophysiology of SARS-COV-2. This Figure is constructed by authors using BioRender, Toronto, Ontario, Canada.

storm complication throughout the body (Huang et al., 2020; Vallamkondu et al., 2020). Additionally, these cytokines, especially IL-17 and IL-21, in conjunction with CD4 T helper cells, produce more macrophages and neutrophils, which may lead to alveolar collapse, various organ failures, and eventually leading to death (Gao et al., 2019).

15.5 DRUG TARGETS IN SARS-COV-2

The discovery of numerous SARS-CoV-2 therapeutic targets has laid the foundation for the creation of both selective inhibitors and broad-spectrum antiviral medications (Table 15.1). However, despite extensive research, an effective antiviral drug that specifically targets SARS-CoV-2 has not been discovered or reported to date. The literature highlights various drug targets in SARS-CoV-2, which are as follows:

15.5.1 Spikes Proteins (S)

Altering the binding of spikes proteins to the receptor-binding domain can be an effective drug target to design broad-spectrum antiviral drugs for SARS-CoV-2 (Prajapat et al., 2020). There are presently no medications specifically designed to treat SARS-CoV-2, but medications already in use to treat other viral illnesses have shown promise in treating COVID-19. Arbidol (umifenovir), which is known to prevent influenza viruses from fusing with host cells, has been recommended as a promising treatment approach for SARS-CoV-2 infection. It was discovered to effectively inhibit both the pre-entry and post-entry phases of SARS-CoV-2 infection, possibly by concentrating on the S protein or ACE-2 (Wang et al., 2020). According to simulations of molecular dynamics, arbidol prevents the S protein from forming trimers (Vankadari, 2020). Arbidol

TABLE 15.1
List of Drugs Found to be Useful Against SARS-CoV-2 Infection

Drug	Target	Mechanism of Action/ Activity	Clinical Trial	Reference
Arbidol (Umifenovir)	S protein	Antiviral	–	Wang et al. (2020)
Danoprevir	M^{pro}	Antiviral	–	Xu et al. (2010)
Dipyridamole	M^{pro}	Antiviral	–	Aljoundi et al. (2020)
Favipiravir	RdRp	Antiviral	–	Jena et al. (2020)
Ribavirin	RdRp	Antiviral	–	Jena et al. (2020)
EIDD-2901/ EIDD-1931	RdRp	Antiviral	–	Sheahan et al. (2020); Sticher et al. (2020)
Dexamethasone	Cytokines	Immune modulator/ steroid/ anti-inflammatory	–	Lythgoe et al. (2020)
Anakinra	Cytokines	Immune modulator	–	
Hydrocortisone	Cytokines	Immune modulator/ steroid/ anti-inflammatory	–	Lythgoe et al. (2020)
Interferon β-1a	Cytokines	Immune modulator	–	Davis et al. (1998)
Azithromycin (in combination with other antivirals)	Cytokines	Antibiotic	–	Rosa et al. (2020)
Ulinastatin	Cytokines	Immune modulator	–	Karnad et al. (2014)
Ruxolitinib	Janus-associated kinase	Immune modulator	–	Cao et al. (2020)
Atazanavir	SARS-CoV-2 M^{pro}	Adjusts in M^{pro} active site, inhibits its activity, disrupts viral replication	Phase 2: NCT04459286	Beck et al. (2020)
Baricitinib	Human AP2-linked protein kinase 1 (AAK1); Janus kinase (JAK) 1 & 2	Blocks viral invasion and contagion, impairs cytokine signaling pathways	Phase 2: NCT04321993; Phases 2 and 3: NCT04320277; Phase 3: NCT04421027	Cantini et al. (2020); Kalil et al. (2020); Richardson et al. (2020)
Mefuparib hydrochloride (CVL218)	N protein; poly-ADP-ribose polymerase 1(PARP1)	Targets N protein to lower the RNA binding, inhibits IL-6 production	Phase 1	Ge et al. (2021)
Pemirolast, nitrofurantoin isoniazid pyruvate, eriodictyol	ACE-2 receptor	Interacts with ACE-2 receptor; inhibits S protein to ACE-2 interaction	–	Spinner et al. (2020)

(Continued)

TABLE 15.1 ***(Continued)***
List of Drugs Found to be Useful Against SARS-CoV-2 Infection

Drug	Target	Mechanism of Action/ Activity	Clinical Trial	Reference
Cepharanthine, ergoloid, hypericin	S protein	Causes suitable ring–protein interaction, inhibits host identification	–	Spinner et al. (2020)
Remdesivir	RdRp	Nucleoside (adenosine) analog RdRP inhibitor, inhibits RNA synthesis	Phase 3: NCT04292899	Wang et al. (2020); Elfiky et al. (2020)
Chloroquine/ hydroxychloroquine	–	Disturbs ACE-2 glycosylation, inhibits viral protein production, endosomal acidification	Phase 2 and 3: NCT04353336	Wang et al. (2020); Hung et al. (2020); Borba et al. (2020); Yao et al. (2021); Fantini et al. (2020)
Lopinavir/ritonavir combination	$3CL^{pro}$	Disrupts viral replication and release from the cell	Phase 2: NCT0427668	Ko et al. (2020); Hoffmann et al. (2020)
Nafamostat or camostat	Serine protease TMPRSS2	Acts as TMPRSS2 opponent, prevents membrane fusion	Phase 2 and 3: NCT04418128; Phase 2: NCT04625114	Gil et al. (2020)
Famotidine	PL^{pro}	Possibly binds PL^{pro}, essential for SARS-CoV-2 entry	Phase 3: NCT04504240	Pepperrell et al. (2020)
Umifenovir	Viral lipid membrane	Binds to the viral lipoidal membrane, affects cellular trafficking of the virus	Phase 4: NCT04350684	Caly et al. (2020)
Nitazoxanide	Not known	Suppresses viral hemagglutinin maturation, activates translation INF2α	Phase 2: NCT04552483	Ceccarelli et al. (2020)
Ivermectin	Not known	Inhibits viral N protein and IL-6 expression, reduces viral replication, destabilizes cell-transport proteins	Phase 1: NCT04343092	Chakraborty et al. (2020)
Teicoplanin	Not known	Inhibits the activity of cathepsin L, potentially blocking viral entry in cells	–	Robinson et al. (2020)
Tocilizumab/ sarilumab (mAb)	IL-6 receptor antagonists	Suppression of IL-6 may attenuate pulmonary inflammation and fibrosis	Phase 3 and Phase 2/3: NCT04315298	Robinson et al. (2020)
Anti-TNF-α agents	TNF-α	Blockage of TNF-α leads to downregulation of pro-inflammatory mediators, i.e., IL-1, IL-6, and cytokines	–	Robinson et al. (2020)

has been shown in studies on other viruses to interfere with viral protein conformational changes (Kadam, 2017), obstruct virus attachment, and interfere with intracellular trafficking (Peucheur et al., 2007). Antimalarial medication chloroquine has shown activity in lowering viral load in SARS-CoV-2 infected patients (Wang et al., 2020). Even better effects have been seen with derivatives like hydroxychloroquine and hydroxychloroquine sulfate (Singh et al., 2020; Liu et al., 2020). These medications are thought to affect the virus in several ways, including affecting ACE-2 glycosylation, obstructing viral assembly, release and replication, and raising the pH of acidic vesicles to prevent viral uncoating (Singh et al., 2020). When combined with the antibiotic azithromycin, hydroxychloroquine has demonstrated positive results in clinical trials, but its efficacy as a standalone treatment has varied across studies (Rosa et al., 2020). Nafamostat and its derivative Nafamostat mesylate have been found to inhibit S protein-initiated fusion, potentially by inhibiting TMPRSS2 (Hoffman et al., 2020). Peptide inhibitors, such as lipopeptides resulting from EK1, have also shown promise in inhibiting S protein-mediated membrane fusion and pseudovirus infection. Among these peptides, EK1C4 has demonstrated the highest effectiveness (Xia et al., 2020).

15.5.2 Membrane protein (M)

It is a transmembrane glycoprotein consisting of three domains, also called M proteins. This structural protein, which consists of 222 amino acids, works with other structural proteins and leads to RNA packaging and is responsible for the framework or shape of the virus. Recent studies suggest that targeting M protein could be effective in reducing viral load and countering the symptoms and complications associated with SARS-CoV-2 infection (Shamsi et al., 2021).

15.5.3 Envelope Protein (E)

E proteins are one of the smallest and key structural proteins in the virus that plays an important role in virus morphogenesis and viral assembly. This pentameric E channel protein functions as an ion channel (Zhang et al., 2020). This feature of E protein is very important for protein trafficking and activation of inflammatory mediators. This protein also resembles a highly structural similarity with another coronavirus-like BAT-CoV and SARS-CoV. Due to the highly conserved feature, this protein can also be a potential drug target for drug design (Ullrich et al., 2020). Hexamethylene amiloride (HMA) has been investigated as a potential medication to prevent viral replication and ion channel activity in synthetic coronavirus E proteins (Pervushin et al., 2009). Additionally, modeling studies have suggested that several phytochemicals and other molecules may be able to block the E-protein's ion channel activities (Gupta et al., 2020; Chernyshev et al., 2020).

15.5.4 Nucleocapsid Protein (N)

N protein consists of three domains, viz., RNA-binding domain (NTD) also termed as N terminal binding domain, Ser/Arg (SR)-rich central linker region, and a C-terminal dimerization domain (CTD). After the formation of viral proteins, they bind with genomic RNA with the help of this protein through the N terminal domain and form viral nucleocapsids (Yin et al., 2020). These freshly formed nucleocapsids are then transported to the cell membrane via Golgi vesicles. These proteins are responsible for the packaging of the viral RNA genome, RNA transcription, and replication as well. All these functions of the N protein, especially the binding of viral protein at the RNA-binding domain (N-terminal), make it a possible drug target. The latest studies suggest that targeting the N protein could be effective in reducing viral load and countering complications associated with SARS-CoV-2 infection (Yang et al., 2003).

15.5.5 Chymotrypsin-Like Protease (CL^{pro})

CL^{pro} is also termed the main protease (M^{pro}). It is an eye-catching target in COVID-19 therapeutics and is receiving major attention as a promising drug target to counter SARS-CoV-2 infection. It indicates ~99% sequence identity with BatCoV RaTG13 M^{pro} and ~96% with the preceding SARS-CoV M^{pro}. It consists of 3 domains: Domains I, II, and III comprise 8–101, 102–184, and 201–306 amino acid residues, respectively. Domain II and domain III are linked by a loop region comprising 185–200 amino acid residues (Hsu et al., 2005; Xue et al., 2007). This enzyme plays a significant role in the cleavage of overlapped polyproteins (pp1a and pp1b), especially nsp5, and forms various viral proteins. M^{pro} involves in protein replication and it is a key enzyme in the replication cycle. Based on its specific functions and highly conserved active site region, this enzyme is the most attractive viral target for designing novel selective broad-spectrum antiviral drugs (Xue et al., 2007). The active site of M^{pro}, located between domains I and II, contains a typical Cys–His dyad (Cys145–His41), which is essential for substrate catalysis (Zhang et al, 2020). Virtual screening studies, followed by molecular dynamics (MD) simulations, have identified several potential inhibitors of M^{pro} (Balakrishnan et al., 2020; Bharadwaj et al., 2020; Balaramnavar et al., 2020; Parashar et. al., 2021). Compounds such as cobicistat, ritonavir, lopinavir, and darunavir, along with other FDA-approved drugs and molecules under clinical trials, have shown promising inhibitory effects against M^{pro} (Pant, 2020).

15.5.6 Papain-Like Protease (PL^{pro})

PL^{pro}, part of nsp3, processes the viral polyproteins, deubiquitinates IRF3 and NF-κB, and deISGylates IFN-stimulated gene 15 (ISG15), making it an attractive and novel drug target for COVID-19 therapeutics (Chou et al., 2004). Inhibiting PL^{pro} is expected to hinder the replication and translation of the virus while preserving the integrity of the host's immune system. Recent studies have demonstrated the inhibitory potential of naphthalene-based inhibitors against the PL^{pro} of SARS-CoV-2 (Freitas et al., 2020). Through virtual screening, several drugs from different categories have been identified as potential inhibitors of PL^{pro} activity, including antivirals (ribavirin, valganciclovir), antibacterials (chloramphenicol, cefamandole), antitussives (levodropropizine), and muscle relaxants (chlorphenesin carbamate) (Wu et al., 2020).

15.5.7 RNA-Dependent RNA Polymerases (RdRp)

Catalytic unit of RdRp, i.e., nsp12 in association with nsp7 and nsp8, makes a strong core component for the transcription and replication process. In complex with these two non-structural proteins, RNA is synthesized, enabling replication and transcription of the viral genome. Residues present in the active binding site of RdRp consisting of Lys545, Arg555, Asp623, Ser682, Thr687, Asn691, Ser759, Asp760, and Asp761 play an important role in designing various drugs that can make an effective, selective, and powerful inhibition of RdRp (Jin et al., 2020). Favipiravir (Baranovich et al., 2013; Lumby et al., 2020; Furuta 2017), triazavirin (Wu et al., 2020; Jena, 2020), ribavirin (Graci, 2006; Khalili et al., 2020), and galidesivir (Taylor et al., 2016; Lim et al., 2020; Julander et al., 2014) are being explored for their potential to inhibit the RdRp enzyme of SARS-CoV-2 through various mechanisms, such as interfering with RNA elongation, inducing high mutation rates (Yanai, 2020), or directly blocking viral RNA replication.

15.6 EPIGALLOCATECHIN GALLATE (EGCG) AND SARS-COV-2 INFECTION

Several secondary plant metabolites like alkaloids, glycosides, tannins, polyphenols, resins, etc. are reported to occur in plants. Among all these, polyphenols represent a major group of phytoconstituents, and based on the number of phenolic rings present in the structure, they are further subclassified into four classes, i.e., phenolic acids, flavonoids, lignans, and stilbenes. EGCG is a polyphenol that is most abundantly rich in tea (*Camellia sinensis*, Theaceae family) (Pandey et al., 2009). Catechins are the primary flavonoids consisting of 2-phenylchromane as a basic nucleus with substitutions of the hydroxyl group at certain positions. In the EGCG structure (Figure 15.3), the basic nucleus of catechin is substituted with the OH group at the C-5 and C7 position of ring A; C-3¢, C-4¢, and C5¢ position of ring B; and C-3″, C-4″, and C5″ position of ring D. As the ring B is derived from gallic acid, it is named as "*gallo*" catechin. At the C-2 position of ring C, the esterified gallate moiety attached giving rise to "*gallate*" and the term "*epi*" is designated to the compound due to its levorotatory activity (Higdon et al., 2003). Earlier studies reported in the literature indicate that EGCG exhibits stronger antioxidant potential as well as free radical scavenging activity in contrast to other catechins and therefore combat many life-threatening or infectious diseases. The gallocatechol ring of EGCG is accountable for its antioxidant and chelating properties (Granja et al., 2017).

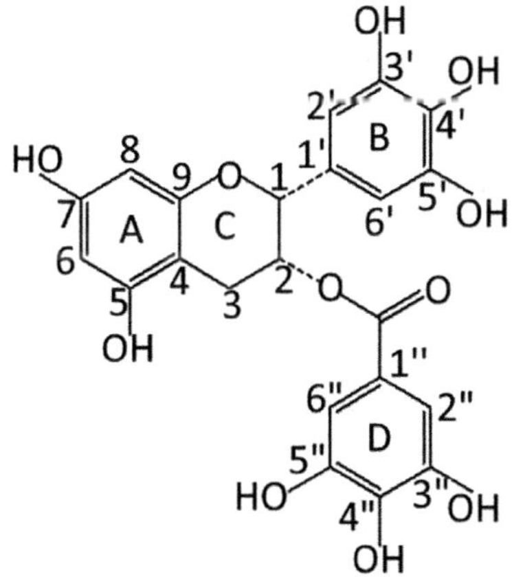

FIGURE 15.3 Structure of (–) epigallocatechin gallate (EGCG) This Figure is constructed by authors using BioRender, Toronto, Ontario, Canada.

Most of the polyphenols, specifically EGCG, regulate the functions of viruses and deactivate their activity via two processes, by sticking themselves to the viral surface and by adhering to the receptors of the host cell. In both cases, the binding site is not free, and therefore, any kind of interaction between the virus surface and host cells is not possible. This process will ultimately lead to the restraint of the multiplication of the virus and its activity (Steinmann et al., 2013). Reported preclinical studies on EGCG against SARS-CoV-2 infection suggest a promising potential in the management of SARS-CoV-2 infection. EGCG acts by interfering with the invasion of the virus to the host cell, their replication, or transcription via diverse molecular mechanisms. EGCG inhibits the replication of the virus by acting on several receptors, such as ACE-2, M^{pro}, PL^{pro}, 3CL^{pro}, RdRp, and also tends to block the active sites existing on the structure of SARS-CoV-2 virus (Mhatre et al., 2021; Singh et al., 2021). The proteins play a significant role in the life cycle of coronavirus (Menegazzi et al., 2020). *In vitro* and *in silico* studies show the inhibitory action of EGCG against SARS-CoV-2 M^{pro} and 3CL^{pro} target because of the strongest interaction with His41 and Cys145 of the catalytic pocket of SARS-CoV-2-3CL^{pro}. It is also reported that the natural polyphenol EGCG is more active than other approved anti-SARS-CoV-2 medications and safer to use because of its safety profile (Jang et al., 2021). *In silico* studies also indicate that EGCG binds to the S1 protein binding site, thus inhibiting the PL^{pro} enzyme and retarding SARS-CoV-2 infection (Jang et al., 2021). Oxidative stress of host cells is elevated in patients suffering from coronavirus as the respiratory viruses augment the production of reactive oxygen species. Oxidative stress endorses tissue damage and hastens the process of virus replication. Since EGCG is a polyphenolic compound and is well-known for its antioxidant and chelating properties, its quenching action on reactive radicals prevents the formation of reactive oxygen species (Zhang et al., 2021). The entry point of the coronavirus to the targeted host cell is viral spike glycoprotein that contains a specific receptor-binding domain where the host cell receptors interact. ACE-2 receptor of the host cell binds with the spike protein of the virus and promotes viral invasion in the host cell. *In silico* molecular docking studies suggest the inhibitory role of EGCG in the interaction of ACE-2 receptors to viral binding site (Ohishi et al., 2022).

Accumulated evidence in recent times indicates the protective role of EGCG in the inflammation damages caused by SARS-CoV-2 infection. EGCG regulates the overproduction of cellular inflammatory factors such as cytokines (IL-1, IL-6, and TNF-α), thereby preventing severe damage associated with cytokine storm syndrome (Maiti et al., 2021). Excessive release of pro-inflammatory cytokines, specifically interferon-α, interferon-γ, IL-6, IL-1β, and CXCL10, is associated with fatality of SARS-CoV-2 infection (Zhang et al., 2020). A postmortem pathological study of the lung of a patient who died from SARS-CoV-2 infection illustrated the occurrence of ARDS and uncontrolled excessive production of T-cells. This event is demonstrated by the activation of the innate and adaptive immune systems in response to SARS-CoV-2 infection. The cytokine storm results in necrosis of epithelial cells, endothelial cells, vascular leakage, and the advent of syndromes like ARDS and mortality (Tang et al., 2020). In a recent study, it was demonstrated that the anti-inflammatory action of EGCG is mediated through its inhibitory effect on JAK2-STAT-1 phosphorylation, which causes blockage of the expression target genes such as CXCL9, CXCL10, ICAM1, chemokine ligand, and nitric oxide synthase (Menegazzi et al., 2020). The widespread anti-inflammatory potential of EGCG is supported by many documented studies that suggest that its action is associated with the inhibition of noxious oxidants and STAT-1 activity. As the majority of cytokine receptors specifically interleukins such as IL-6, IL-10, IL-21, and IL-23 trigger STAT-3 activation, they are considered a crucial nuclear factor for the regulation of immune responses (Wang Y. et al., 2013; Menegazzi et al., 2020). *In silico* docking studies and surface plasmon resonance (SPR) assay demonstrated that EGCG plays a major role in the inhibition of STAT-3 activation. The molecular mechanism involved in SPR assay states that in micro-molar concentration, EGCG is effective to disrupt the STAT-3 peptide binding; however, docking studies reveal that EGCG selectively binds to STAT-3 SH2 domain, thereby inhibiting STAT-3 phosphorylation and signaling (Wang et al., 2013). During SARS-CoV-2 infection, the hyperinflammatory condition is aggravated due to the elevation in the level of IL-6 and STAT-3 activation. As EGCG has the potential to block IL-6 signaling and is a powerful blocker of STAT-3, it prevents the increased risk of inflammation. The NF-κB acts as a key regulator for the expression of pro-inflammatory cytokines (IL-16, IL-1β, TNF-α, etc.) that are involved in the progression of cytokine storm syndrome. As EGCG inhibits the activation of NF-κB and stimulates the NRF2 nuclear translocation, it may alter the hyperinflammatory status in patients suffering from SARS-CoV-2 infection. From the above discussion, it can be concluded that EGCG works through multiple targeting mechanistic approaches against SARS-CoV-2 infection (Vanoni et al., 2017). Previfenon® drug contains EGCG (250mg) as the main chemical constituent and is in the second phase of clinical trials for the determination of its efficacy in managing SARS-CoV-2 infection. The proposed mechanism of action of the drug is antiviral chemoprophylaxis of SARS-CoV-2 infection (Sapra et al., 2021).

15.7 CONCLUSION

While the potential of EGCG in combating SARS-CoV-2 infection seems to be very promising, it is important to note that most of the evidence is based on *in vitro* studies or animal models. Clinical trials are needed to assess the effectiveness and safety of EGCG in humans. Additionally, the optimal dosage and treatment duration of EGCG for SARS-CoV-2 infection are yet to be determined. Furthermore, the bioavailability of EGCG is relatively low, and its efficacy may be limited by factors such as poor absorption, rapid metabolism, and short half-life. Strategies to enhance the bioavailability and stability of EGCG need to be explored. In conclusion, EGCG shows potential as a natural compound with antiviral and immunomodulatory properties against SARS-CoV-2 infection, especially through its potential to attenuate inflammatory damage. Moreover, further research is required to fully understand or predict its mechanisms of action and to develop effective therapeutic approaches using EGCG or its derivatives.

15.8 DECLARATION OF COMPETING INTEREST

The authors declare that they have no known competing financial interests or personal relationships.

REFERENCES

Aljoundi A, Bjij I, Rashedy AE, Soliman MES. Covalent versus non-covalent enzyme inhibition: which route should we take? a justification of the good and bad from molecular modelling perspective. *Protein J* (2020); 39: 97–105.

Balakrishnan V, Lakshminarayan K. Screening of FDA approved drugs againstSARS-CoV-2 main protease: coronavirus disease. *Int J Pept Res Ther* (2020); 27: 3741–3751. https://doi.org/10.1007/s10989-020-10115-6.

Balaramnavar VM, Ahmad K, Saeed M, Ahmad I, Kamal M, Jawed T. Pharmacophore-based approaches in the rational repurposing technique for FDA approved drugs targeting SARS-CoV-2 Mpro. *RSC Adv* (2020); 10: 40264–40275.

Baranovich T, et al. T-705 (Favipiravir) induces lethal mutagenesis in influenza A H1N1 viruses, vitro. *J Virol* (2013); 87: 3741–3751.

Baum A, Fulton BO, Wloga E, Copin R, Pascal KE, Russo V, et al., Antibody cocktail to SARS-CoV-2 spike protein prevents rapid mutational escape seen with individual antibodies. *Science* (2020); 369: 1014–1018.

Beck BR, Shin B, Choi Y, Park S, Kang K. Predicting commercially available antiviral drugs that may act on the novel coronavirus (SARS-CoV-2) through a drug-target interaction deep learning model. *Comput Struct Biotechnol J* (2020); 18: 784–790.

Bharadwaj S, et al. SARS-CoV-2 Mpro inhibitors: identification of anti-SARS-CoV2 Mpro compounds from FDA approved drugs. *J Biomol Struct Dyn* (2020); 2020: 1–16. https://doi.org/10.1080/07391102.2020.1842807.

Borba MGS, Val FFA, Sampaio VS, Alexandre MAA, Melo GC, Brito M, Mourão MPG, Brito-Sousa JD, Baía-da-Silva D, Guerra MVF. Effect of high vs low doses of chloroquine diphosphate as adjunctive therapy for patients hospitalized with severe acute respiratory syndrome coronavirus 2 (SARS-CoV-2) infection: a randomized clinical trial. *JAMA Netw Open* (2020); 3(4): e208857.

Caly L, Druce JD, Catton MG, Jans DA, Wagstaff KM. The FDA-approved drug ivermectin inhibits the replication of SARS-CoV-2 in vitro. *Antivir Res* (2020); 178: 104787.

Cantini F, Niccoli L, Matarrese D, Nicastri E, Stobbione P, Goletti D. Baricitinib therapy in COVID-19: a pilot study on safety and clinical impact. *J Infect* (2020); 2: 318–356.

Cao Y, et al. Ruxolitinib in treatment of severe coronavirus disease 2019 (COVID-19): a multicenter, single-blind, randomized controlled trial. *J Allergy Clin Immunol* (2020); 146: 137–146.

Carneiro BM, Batista MN, Braga ACS, Nogueira ML, Rahal P. The green tea molecule EGCG inhibits Zika virus entry. *Virology* (2016); 496: 215–218.

Ceccarelli G, Alessandri F, d'Ettorre G, Borrazzo C, Spagnolello O, Oliva A, Ruberto F, Mastroianni CM, Pugliese F, Venditti M. Is teicoplanin a complementary treatment option for COVID-19? The question remains. *Int J Antimicrob Agents* (2020); 56(2): 106029.

Chakraborty C, Sharma AR, Bhattacharya M, Sharma G, Lee SS, Agoramoorthy G. COVID-19: consider IL6 receptor antagonist for the therapy of cytokine storm syndrome in SARS-CoV-2 infected patients. *J Med Virol* (2020); 92(11): 2260–2262.

Chen Y, Liu Q, Guo D. Emerging coronaviruses: genome structure, replication, and pathogenesis. *J Med Virol* (2020); 92(4): 418–423. https://doi.org/10.1002/jmv.25681.

Chernyshev A. Pharmaceutical targeting the envelope protein of SARS-CoV-2: the screening for inhibitors in approved drugs. *ChemBioRixv* (2020). https://doi.org/10.26434/chemrxiv.12286421.

Chou CY, Chang HC, Hsu WC, Lin TZ, Lin CH, Chang GG. Quaternary structure of the severe acute respiratory syndrome (SARS) coronavirus main protease. *Biochemistry* (2004); 43(47): 14958–14970. PubMed PMID: 15554704.

Colpitts CC, Schang LM. A small molecule inhibits virion attachment to heparan sulfate- or sialic acid-containing glycans. *J Virol* (2014); 88: 7806–7817.

Coronaviridae Study Group of the International Committee on Taxonomy of Viruses. The species severe acute respiratory syndrome-related coronavirus: classifying 2019-nCoV and naming it SARS-CoV-2. *Nat Microbiol* (2020); 5(4): 536–544. PubMed PMID: 32123347; PubMed Central PMCID: PMC7095448.

Cui J, Li F, Shi ZL. Origin and evolution of pathogenic coronaviruses. *Nat Rev Microbiol* (2019); 17: 181–192.

Dong E, Du H, Gardner L. An interactive web-based dashboard to track COVID-19 in real-time. *Lancet Infect Dis* (2020); 20(5): 533–534. PubMed PMID: 32087114; PubMed Central PMCID: PMC7159018.

Eguchi S, Kawai T, Scalia R, Rizzo V. Understanding angiotensin II Type 1 receptor signaling in vascular pathophysiology. *Hypertension* (2018); 71: 804–810.

ElfikyAA. SARS-CoV-2 RNA dependent RNA polymerase (RdRp) targeting: an in silico perspective. *J Biomol Struct Dyn* (2020); 39: 1–9.

Eurosurveillance Editorial Team. Note from the editors: World Health Organization declares novel coronavirus (2019-nCoV) sixth public health emergency of international concern. *Euro Surveill* (2020); 25(5): 200131e. PubMed PMID: 32046819.

Fantini J, Di Scala C, Chahinian H, Yahi N. Structural and molecular modeling studies reveal a new mechanism of action of chloroquine and hydroxychloroquine against SARS-CoV-2 infection. *Int J Antimicrob Agents* (2020); 5: 105960.

Freitas BT, et al., Characterization and noncovalent inhibition of the deubiquitinase and deISGylase activity of SARS-CoV-2 papain-like protease. *ACS Infect Dis* (2020); 6(8): 2099–2109.

Furushima D, Ide K, Yamada H. Effect of tea catechins on influenza infection and the common cold with a focus on epidemiological/clinical studies. *Molecules* (2018); 23: 1795.

Furushima D, Nishimura T, Takuma N, Iketani R, Mizuno T, Matsui Y, et al. Prevention of acute upper respiratory infections by consumption of catechins in healthcare workers: a randomized, placebo-controlled trial. *Nutrients* (2019); 12: 4.

Furuta Y, Komeno T, Nakamura T. Favipiravir (T-705), a broad spectrum inhibitor of viral RNA polymerase. *Proc Jpn Acad Ser B Phys Biol Sci* (2017); 93: 449–463.

Gao YM, Xu G, Wang B, Liu BC. Cytokine storm syndrome in coronavirus disease 2019: a narrative review. *J Intern Med* (2021); 289(2): 147–161. https://doi.org/10.1111/joim.13144. Epub 2020 Jul 22. PMID: 32696489; PMCID: PMC7404514.

Ge Y, Tian T, Huang S, Wan F, Li J, Li S, Wang X, Yang H, Hong L, Wu N, Yuan E. An integrative drug repositioning framework discovered a potential therapeutic agent targeting COVID-19. *Signal Transduct Target Ther.* (2021); 6(1): 165.

Gil C, Ginex T, Maestro I, Nozal V, Barrado-Gil L, Cuesta-Geijo MA, Urquiza, J Ramírez, D, Alonso C, Campillo NE. COVID-19: drug targets and potential treatments. *J Med Chem* (2020); 63(21): 12359–12386.

Graci JD, Cameron CE. Mechanisms of action of ribavirin against distinct viruses, *Rev Med Virol* (2006); 16: 37–48.

Gralinski LE, Menachery VD. Return of the coronavirus: 2019-nCoV. *Viruses* (2020); 12(2): 135. PubMed PMID: 32050646; PubMed Central PMCID: PMC7077242.

Granja A, Frias I, Neves AR, Pinheiro M, Reis S. Therapeutic potential of epigallocatechin gallate nanodelivery systems. *BioMed Res Int* (2017); 2017: 5813793.

Gupta MK, et al. In-silico approaches to detect inhibitors of the human severe acute respiratory syndrome coronavirus envelope protein ion channel. *J Biomol Struct Dyn* (2020); 39(7):2617–2627. https://doi.org/10.1080/07391102.2020.1751300.

Higdon J, Frei B. Tea catechins and polyphenols: health effects, metabolism, and antioxidant functions. In: *Oxidative Stress and Disease*. CRC Press, Boca Raton, FL; 2003. pp. 89–143.

Hoffman M, et al. Nafamostat mesylate blocks activation of SARS-CoV-2: a new treatment option for COVID-19. *Antimicrob Agents Chemother* (2020); 64(6): e00754-20.

Hoffmann M, Kleine-Weber H, Schroeder S, et al. SARS-CoV-2 cell entry depends on ACE2 and TMPRSS2 and is blocked by a clinically proven protease inhibitor. *Cell* (2020); 181(2): 271–280

Hong Y, Shahidi F. Lipophilized epigallocatechin gallate (EGCG) derivatives as novel antioxidants. *J Agric Food Chem* (2011); 59: 6526–6533.

Hsu MF, Kuo CJ, Chang KT, et al. Mechanism of the maturation process of SARSCoV 3CL protease. *J Biol Chem* (2005); 280(35): 31257–31266. PubMed PMID: 15983031.

Hu B, Guo H, Zhou P, Shi ZL. Characteristics of SARS-CoV-2 and COVID-19. *Nat Rev Microbiol* (2021); 19(3): 141–154. PubMed PMID: 33208903.

Huang C, Wang Y, Li X, et al. Clinical features of patients infected with 2019 novel coronavirus in Wuhan, China. *Lancet* (2020); 395: 497–506.

Hung IFN, Lung KC, Tso EYK, Liu R, Chung TWH, Chu MY, Ng YY, Lo J, Chan J, Tam AR. Triple combination of interferon beta-1b, lopinavir-ritonavir, and ribavirin in the treatment of patients admitted to hospital with COVID-19: an open label, randomized phase 2 trial. *Lancet* (2020); 395(10238): 1695–1704.

Jang M, Park R, Park YI, Cha YE, Yamamoto A, Lee JI, Park J. EGCG, a green tea polyphenol, inhibits human coronavirus replication in vitro. *Biochem Biophys Res Commun* (2021); 547: 23–28.

Jena N. Identification of potent drugs and antiviral agents for the treatment of the SARS-CoV-2 infection. *ChemRxiv, Preprint* (2020). https://doi.org/10.26434/chem rxiv.12330599.v1.

Jena NR. Role of different tautomers on the base-pairing abilities of some of the vital antiviral drugs used against COVID-19. *Phys Chem Phys* (2020); 22: 28115–28122. https://doi.org/10.1039/D0CP05297C.

Jin Z, Du X, Xu Y, et al. Structure of Mpro from COVID-19 virus and discovery of its inhibitors. *Nature* (2020); 582: 289–293.

Julander JG, et al. BCX4430, a novel nucleoside analog, effectively treats yellow fever in a hamster model. *Antimicrob Agents Chemother* (2014); 58: 6607–6614.

Kadam RU, Wilson IA. Structural basis of influenza virus fusion inhibition by the antiviral drug Arbidol. *Proc Natl Acad Sci USA* (2017); 114: 206–214.

Kalil AC, Patterson TF, Mehta AK, Tomashek KM, Wolfe CR, Ghazaryan V, Marconi VC, Ruiz-Palacios GM, Hsieh L, Kline S. Baricitinib plus remdesivir for hospitalized adults with Covid-19. *N Engl J Med* (2020); 384: 795–807.

Karnad DK, et al., Intravenous administration of ulinastatin (human urinary trypsin inhibitor) in severe sepsis: a multicenter randomized controlled study. *Intensive Care Med* (2014); 40: 830–838.

Khalili JS, et al., Novel coronavirus treatment with ribavirin: groundwork for an evaluation concerning COVID-19. *J Med Virol* (2020); 92: 740–746. https://doi.org/10.1002/jmv.25798.

Ko M, Jeon S, Ryu WS, Kim S. Comparative analysis of antiviral efficacy of FDA-approved drugs against SARS-CoV-2 in human lung cells. *J Med Virol* (2021); 93(3): 1403–1408.

Lai YH, Sun CP, Huang HC, Chen JC, Liu HK, Huang C. Epigallocatechin gallate inhibits hepatitis B virus infection in human liver chimeric mice. *BMC Complement Altern Med* (2018); 18: 248.

Lim SY, et al., A direct-acting antiviral drug abrogates viremia in Zika virus-infected rhesus macaques. *Sci Transl Med* (2020); 12(547): eaau9135.

Liu J, et al. Hydroxychloroquine, a less toxic derivative of chloroquine, is effective in inhibiting SARS-CoV-2 infection in vitro. *Cell Discov* (2020); 6: 16.

Liu JB, Li JL, Zhuang K, Liu H, Wang X, Xiao QH, et al. Epigallocatechin-3-gallate local pre-exposure application prevents SHIV rectal infection of macaques. *Mucosal Immunol* (2018); 11: 1230–1238.

Lumby CK, et al., Favipiravir and zanamivir cleared infection with influenza B in a severely immune compromised child. *Clin Infect Dis* (2020); 71: e191–e194. https://doi.org/10.1093/cid/ciaa023.

Lythgoe MK, Middleton P. Ongoing clinical trials for the management of the COVID-19 pandemic. *Trends Pharmacol Sci* (2020); 41: 363–382.

Maiti S, Banerjee A. Epigallocatechin gallate and theaflavin gallate interaction in SARS-CoV-2 spike-protein central channel with reference to the hydroxychloroquine interaction: bioinformatics and molecular docking study. *Drug Dev Res* (2021); 82(1): 86–96.

Menegazzi M, Campagnari R, Bertoldi M, Crupi R, Di Paola R, Cuzzocrea S. Protective effect of epigallocatechin-3-gallate (EGCG) in diseases with uncontrolled immune activation: could such a scenario be helpful to counteract COVID-19?. *Int J Mol Sci* (2020); 21(14): 5171.

Mhatre S, Srivastava T, Naik S, Patravale V. Antiviral activity of green tea and black tea polyphenols in prophylaxis and treatment of COVID-19: a review. *Phytomedicine* (2021); 85: 153286.

Muniyappa R, Gubbi S. COVID-19 pandemic, coronaviruses, and diabetes mellitus. *Am J Physiol Endocrinol Metab* (2020); 318(5): E736–E741. PubMed PMID: 32216618; PubMed Central PMCID: PMC7195309.

Murakami M, Kamimura D, Hirano T. Pleiotropy and specificity: insights from the interleukin 6 family of cytokines. *Immunity* (2019); 50: 812–831

Nagu P, Parashar A, Behl T, Mehta V. CNS implications of COVID-19: a comprehensive review. *Rev Neurosci* (2021); 32(2): 219–234.

National Human Genome Research Institute. Virus. Retrieved from https://www.genome.gov/genetics-glossary/Virus.

Ohishi T, Hishiki T, Baig MS, Rajpoot S, Saqib U, Takasaki T, Hara Y. Epigallocatechin gallate (EGCG) attenuates severe acute respiratory coronavirus disease 2 (SARS-CoV-2) infection by blocking the interaction of SARS-CoV-2 spike protein receptor-binding domain to human angiotensin-converting enzyme 2. *PLoS One* (2022); 17(7): e0271112.

Onishi S, Mori T, Kanbara H, Habe T, Ota N, Kurebayashi Y, et al. Green tea catechins adsorbed on the murine pharyngeal mucosa reduce influenza A virus infection. *J Funct Foods* (2020); 68: 103894.

Pandey KB, Rizvi SI. Plant polyphenols as dietary antioxidants in human health and disease. *Oxid Med Cell Longev* (2009); 2: 270–278.

Pant S. Peptide-like and small-molecule inhibitors against Covid-19. *J Biomol Struct Dyn* (2020); 39: 1–10. https://doi.org/10.1080/07391102.2020.1757510.

Parashar A, Shukla A, Sharma A, Behl T, Goswami D, Mehta V. Reckoning γ-Glutamyl-S-allylcysteine as a potential main protease (mpro) inhibitor of novel SARS-CoV-2 virus identified using docking and molecular dynamics simulation. *Drug Dev Ind Pharm* (2021); 47(5): 699–710.

Pepperrell T, Pilkington V, Owen A, Wang J, Hill AM. Review of safety and minimum pricing of nitazoxanide for potential treatment of COVID-19. *J Virus Erad* (2020); 6(2): 52.

Pervushin K, et al., Structure and inhibition of the SARS coronavirus envelope protein ion channel. *PLoS Pathog* (2009); 5: e1000511.

Peucheur EI, et al., Biochemical mechanism of hepatitis C virus inhibition by the broad-spectrum antiviral arbidol. *Biochemistry* (2007); 46: 6050–6059.

Prajapat M, Sarma P, Shekhar N, et al. Drug targets for coronavirus: a systematic review. *Indian J Pharmacol* (2020); 52(1): 56–65. https://doi.org/10.4103/ijp.IJP_115_20. PubMed PMID: 32201449; PubMed Central PMCID: PMC7074424.

Reid SP, Shurtleff AC, Costantino JA, Tritsch SR, Retterer C, Spurgers KB, et al. HSPA5 is an essential host factor for Ebola virus infection. *Antiviral Res* (2014); 109: 171–174.

Richardson P, Griffin I, Tucker C, Smith D, Oechsle O, Phelan A, Stebbing J. Baricitinib as potential treatment for 2019-nCoV acute respiratory disease. *Lancet (London, England)* (2020); 395(10223): e30.

Riegsecker S, Wiczynski D, Kaplan MJ, Ahmed S. Potential benefits of green tea polyphenol EGCG in the prevention and treatment of vascular inflammation in rheumatoid arthritis. *Life Sci* (2013); 93: 307–312.

Robinson PC, Richards D, Tanner HL, Feldmann M. Accumulating evidence suggests anti-TNF therapy needs to be given trial priority in COVID-19 treatment. *Lancet Rheumatol* (2020); 2(11): e653–e655.

Rosa SG, Santos WC. Clinical trials on drug repositioning for COVID-19 treatment. *Revista Panamericana de Salud Pública* (2020); 44: e40.

Sapra L, Bhardwaj A, Azam Z, Madhry D, Verma B, Rathore S, Srivastava RK. Phytotherapy for treatment of cytokine storm in COVID-19. *Front Biosci Landmark* (2021); 26(5): 51–75.

Shamsi A, Mohammad T, Anwar S, et al. Potential drug targets of SARS-CoV-2: from genomics to therapeutics. *Int J Biol Macromol* (2021); 177: 1–9.

Sheahan TP, et al. An orally bioavailable broad-spectrum antiviral inhibits SARS-CoV-2 in human airway epithelial cell cultures and multiple coronaviruses in mice. *Sci Transl Med* (2020); 12: eabb5883.

Singh AK, Singh A, Shaikh A, Singh R, Mishra A. Chloroquine and hydroxychloroquine in the treatment of COVID-19 with or without diabetes: a systematic search and a narrative review with a special reference to India and other developing countries. *Diabetes Metab Syndr* (2020); 14: 241–246.

Singh S, Sk MF, Sonawane A, Kar P, Sadhukhan S. Plant-derived natural polyphenols as potential antiviral drugs against SARS-CoV-2 via RNA-dependent RNA polymerase (RdRp) inhibition: an in-silico analysis. *J Biomol Struct Dyn* (2021); 39(16): 6249–6264.

Spinner CD, Gottlieb RL, Criner GJ, López JRA, Cattelan AM, Viladomiu AS, Ogbuagu O, Malhotra P, Mullane KM, Castagna A. Effect of remdesivir vs standard care on clinical status at 11 days in patients with moderate COVID-19: a randomized clinical trial. *JAMA* (2020); 324(11): 1048–1057.

Steinmann J, Buer J, Pietschmann T, Steinmann E. Anti-infective properties of epigallocatechin-3-gallate (EGCG), a component of green tea. *Br J Pharmacol* (2013); 168(5): 1059–1073.

Sticher ZM, et al. Analysis of the potential for N4-hydroxycytidine to inhibit mitochondrial replication and function. *Antimicrob Agents Chemother* (2020); 64: e01719–e01719.

Storozhuk M. Green tea catechins against COVID-19: Lower COVID-19 morbidity and mortality in countries with higher per capita green tea consumption. *Coronaviruses* (2022); 3(3):57–64.

Tang T, Bidon M, Jaimes JA, Whittaker GR, Daniel S. Coronavirus membrane fusion mechanism offers a potential target for antiviral development. *Antiviral Res* (2020); 178: 104792. PubMed PMID: 32871255; PubMed Central PMCID: PMC7447109.

Tang Y, Liu J, Zhang D, Xu Z, Ji J, Wen C. Cytokine storm in COVID-19: the current evidence and treatment strategies. *Front Immunol* (2020); 11: 1708.

Taylor R, et al., BCX4430-a broad-spectrum antiviral adenosine nucleoside analog under development for the treatment of Ebola virus disease. *J Infect Public Health* (2016); 9: 220–226.

Ullrich S, Nitsche C. The SARS-CoV-2 main protease as drug target. *Bioorg Med Chem Lett* (2020); 30(17): 127377. https://doi.org/10.1016/j.bmcl.2020.127377. Epub 2020 Jul 2. PMID: 32738988; PMCID: PMC7331567.

Vallamkondu J, John A, Wani WY, Ramadevi SP, Jella KK, Reddy PH, Kandimalla R. SARS-CoV-2 pathophysiology and assessment of coronaviruses in CNS diseases with a focus on therapeutic targets. *Biochim Biophys Acta Mol Basis Dis* (2020); 1866(10): 165889. doi: 10.1016/j.bbadis.2020.165889. Epub 2020 Jun 27. PMID: 32603829; PMCID: PMC7320676.

Vankadari N. Arbidol: a potential antiviral drug for the treatment of SARS-CoV 2 by blocking trimerization of the spike glycoprotein. *Int J Antimicrob Agents* (2020); 56: 105998. https://doi.org/10.1016/j.ijantimicag.2020.105998.

Vanoni S, Tsai YT, Waddell A, Waggoner L, Klarquist J, Divanovic S, Hoebe K, Steinbrecher KA, Hogan SP. Myeloid-derived NF-κB negative regulation of PU. 1 and c/EBP-β-driven pro-inflammatory cytokine production restrains LPS-induced shock. *Innate Immunity* (2017); 23(2): 175–187.

Wang M, Cao R, Zhang L, Yang X, Liu J, Xu M, Shi Z, Hu Z, Zhong W, Xiao G. Remdesivir and chloroquine effectively inhibit the recently emerged novel coronavirus (2019-nCoV) in vitro. *Cell Res* (2020); 30(3): 269–271.

Wang X, et al., The anti-influenza virus drug, arbidol is an efficient inhibitor ofSARS-CoV-2 in vitro. *Cell Discov* (2020); 6: 28.

Wang Y, Ren X, Deng C, Yang L, Yan E, Guo T, Li Y, Xu MX. Mechanism of the inhibition of the STAT3 signaling pathway by EGCG. *Oncol Rep* (2013); 30(6): 2691–2696.

Woo PCY, Huang Y, Lau SKP, Yuen K-Y. Coronavirus genomics and bioinformatics analysis. *Viruses* (2010); 2(8): 1804–1820. https://doi.org/10.3390/v2081803.

World Health Organization. WHO Coronavirus (COVID-19) Dashboard. Retrieved June 14, 2023, from https://covid19.who.int/

Wu C, et al., Analysis of therapeutic targets for SARS-CoV-2 and discovery of potential drugs by computational methods. *Acta Pharm Sin B* (2020); 10: 766–788.

Wu X, et al., Efficacy and safety of triazavirin therapy for coronavirus disease 2019: a pilot randomized controlled trial. *Engineering* (2020); 6: 1185–1191.

Xia S, et al. Inhibition of SARS-CoV-2 (previously 2019-nCoV) infection by a highly potent pan-coronavirus fusion inhibitor targeting its spike protein that harbors a high capacity to mediate membrane fusion. *Cell Res* (2020); 30: 343–355.

Xu L, et al., Cobicistat (GS-9350): a potent and selective inhibitor of human CYP3Aas a novel pharmaco enhancer. *ACS Med Chem Lett* (2010); 1: 209–213.

Xu Z, Shi L, Wang Y, et al. Pathological findings of COVID-19 associated with acute respiratory distress syndrome. *Lancet Respir Med* (2020); 8: 420–422.

Xue X, et al., Production of authentic SARS-CoVM(pro) with enhanced activity: application as a novel tag-cleavage endopeptidase for protein overproduction. *J Mol Biol* (2007); 366: 965–975.

Yanai H. Favipiravir: a possible pharmaceutical treatment for COVID-19. *J Endocrinol Metab* (2020); 10: 33–34.

Yang H, Yang M, Ding Y, et al. The crystal structures of severe acute respiratory syndrome virus main protease and its complex with an inhibitor. *Proc Natl Acad Sci USA* (2003); 100(23): 13190–13195. PubMed PMID: 14595026; PubMed Central PMCID: PMC263858.

Yao X, Ye F, Zhang M, Cui C, Huang B, Niu P, Liu X, Zhao L, Dong E, Song C. Invitro antiviral activity and projection of optimized dosing design of hydroxychloroquine for the treatment of severe acute respiratory syndrome coronavirus2 (SARS-CoV-2). *Clin Infect Dis* (2020); 71(15): 732–739.

Yi Y, Lagniton PNP, Ye S, Li E, Xu RH. COVID-19: what has been learned and to be learned about the novel coronavirus disease. *Int J Biol Sci* (2020); 16(10): 1753–1766. https://doi.org/10.7150/ijbs.45134. PubMed PMID: 32226292; PubMed Central PMCID: PMC7102625.

Yin W, Mao C, Luan X, Shen DD, Shen Q, Su H, Wang X, Zhou F, Zhao W, Gao M. Structural basis for inhibition of the RNA-dependent RNA polymerase from SARS-CoV-2 by remdesivir. *Science* (2020); 369(6509): 1499–1504. PubMed PMID: 32661059.

Zhang L, et al., Crystal structure of SARS-CoV-2 main protease provides a basis for the design of improved α-ketoamide inhibitors. *Science* (2020); 368: 409–412.

Zhang W, Zhao Y, Zhang F, Wang Q, Li T, Liu Z, Wang J, Qin Y, Zhang X, Yan X, Zeng X. The use of anti-inflammatory drugs in the treatment of people with severe coronavirus disease 2019 (COVID-19): the Perspectives of clinical immunologists from China. *Clin Immunol* (2020); 214: 108393.

Zhang Z, Zhang X, Bi K, He Y, Yan W, Yang CS, Zhang J. Potential protective mechanisms of green tea polyphenol EGCG against COVID-19. *Trends Food Sci Technol* (2021); 114: 11–24.

Zhou Y, Fu B, Zheng X et al. Pathogenic T cells and inflammatory monocytes incite inflammatory storm in severeCOVID-19 patients. *National Sci Rev* (2020); 7: 998–1002.

Zhu N, Zhang D, Wang W, et al., China novel coronavirus investigating and research team. A novel coronavirus from patients with pneumonia in China, 2019. *N Engl J Med* (2020); 382(8): 727–733. PubMed PMID: 31978945; PubMed Central PMCID: PMC7092803.

Ziebuhr J, Snijder EJ, Gorbalenya AE. Virus-encoded proteinases and proteolytic processing in the Nidovirales. *J Gen Virol* (2000); 81(4): 853–879.

16 Anti-SARS-CoV-2 Catechins and Their Roles in COVID-19 Management

Puja Gupta, Deepak Nandi, Sonu Ram, Eswar Rao Tatta, and Ranjith Kumavath

16.1 INTRODUCTION

COVID-19 has prompted the most pandemics in recent history. It impacted nearly every nation in the world and is acknowledged as an overpowering hazard to humankind. The virus SARS-CoV-2 is the root source of the devastating worldwide disease. It originated in China and then continued worldwide to expand nations. The World Health Organization declared it a pandemic on March 11, 2020 (Cucinotta and Vanelli, 2020). It existed in 229 countries as of 24:00 on September 2023. The virus was reported to have killed about 6.87 million individuals and infected ~ 687 million people worldwide in the year 2023 (John Elflein, 2023). Besides endangering people's health and lives, the pandemic significantly negatively influenced the economy and public health infrastructure. Management of SARS-CoV-2 primarily involves a combination of public health measures, medical care, and vaccination. It's crucial to remember that the management has changed over time, and it's vital to remember that COVID-19 management has changed throughout time.

SARS-CoV-2 has a unique spike protein on its surface, resembling MERS-CoV and SARS-CoV. It has a complex life cycle, including attachment, entrance, replication, assembly, and expulsion (Hatmal et al., 2020; Ravi et al., 2022). The virus produces its structural components, i.e., envelope (E), spike (S), membrane (M), and nucleocapsid (N) proteins, by synthesizing viral mRNA. These proteins are vital for developing new virus moieties, eventually released from the host cell to continue the infection cycle (Breitinger et al., 2022). Indulgently, the process of infection is crucial for developing strategies to combat SARS-CoV-2. Similar to previous viruses, SARS-CoV-2 originated from animals. It has a stretched incubation period and is contagious among humans (Wu et al., 2020). The lung's angiotensin-converting enzyme 2 (ACE2) receptor has been identified as a co-receptor for SARS-CoV (Kuba et al., 2021).

The five recognized mutants were alpha, beta, delta, gamma, and omicron. The omicron created much anxiety, spreading swiftly and affecting the immune system (Magazine et al., 2022). The typical signs of COVID-19 included fever, fatigue, headache, nausea, respiratory issues, vomiting, cough, diarrhea, and stuffy nose. Individuals with diabetes, heart or lung disease, or advanced age showed a higher chance of developing a major illness; many even passed away (Andrews et al., 2021).

There were not many COVID-19 treatments available at the beginning of the epidemic and pandemic. Due to the absence of a dedicated antiviral medication for SARS-CoV-2, the existing antiviral drugs were explored, reconnoitered, and studied for the remedy of COVID-19. The drug exhibited an inhibitory consequence on the viral RNA polymerase, a crucial enzyme in virus replication (Jiang et al., 2021). Favipiravir showed inhibitory effects on influenza through a similar mechanism (Agrawal et al., 2020). Lopinavir worked as a viral protease inhibitor. It was initially established for HIV therapy, but demonstrated an inhibitory effect on coronavirus (CoV) infected cells *in vitro* studies (Choy et al., 2020). However, the systematic reviews did not reveal any beneficial effects of lopinavir against SARS-CoV-2 (Martínez et al., 2021). Besides the vaccination, three additional oral medications, e.g., Paxlovid, Molnupiravir, and VV116, were also used globally to treat COVID-19 (Rahmah et al., 2022).

Besides antiviral drugs, the natural compounds products and flavonoids played a vital role during the COVID-19 infections (Kaul et al., 2021). Fruits and vegetables contain natural compounds called flavonoids. These flavonoids possess various valuable pharmacological activities (Table 16.1). These offer essential health advantages, e.g., anti-inflammatory, anticancer, and antiviral capabilities. Flavonoid has hindered viral pathogenesis by targeting crucial phases in the life cycle (Sadati et al., 2018). Flavonoids can target viruses and impede their ability to infect and replicate within host cells. They inhibit viruses in different ways as shown in Figure 16.1, including (i) binding to specific parts of the virus located outside of host cells, such as viral proteins on the capsids; (ii) preventing the virus from attaching and entering the host cells; (iii) inhibiting the early stage of viral replication mechanism; (iv) blocking viral transcription and translation processes; (v) inhibiting late stages of virus maturation, including assembly, packaging, and release; (vi) interfering with host factors necessary for viral infection and thereby reducing the infection rate; and (vii) modulating the immune system to decrease the amount of virus present in the body (Jo et al., 2020). The studies have further unveiled the potential antiviral attributes of herbal components against SARS-CoV-2 (Gurung et al., 2020). Among the crucial flavonoids recognized for

DOI: 10.1201/9781003433200-16

TABLE 16.1
Various Flavonoid Derivative Compounds with Anti-SARS-CoV-2 Activities and Target of Inhibition

Flavonoid Compounds	Source	Target Site	References
Epicatechin-3-gallate	Green tea, Black tea, Gooseberries,	3CLpro	Liu et al. (2022)
Gallocatechin-3- gallate	Grape seeds, Red wine, Chocolate,	3CLpro and Mpro	Ungarala et al. (2022)
Gallocatechin	Apple, Cacao liquor, Berry fruits, and Kiwi fruits	Mpro and 3CLpro	Selvaraj et al. (2021) 0
Epicatechin		Mpro and 3CLpro	Liu et al. (2022)
Catechin gallate		Mpro and 3CLpro	Khaerunnisa et al. (2022)
Epigallocatechin		3CLpro	Du et al. (2021)
Catechin		3CLpro and Mpro	Liu et al. (2022)
Epigallocatechin gallate		3CLpro/Mpro	Khaerunnisa et al. (2020)
Epigallocatechin-3,5-digallate		Mpro and PLpro	Sharma et al. (2022)
Malvidin	Red wine, Red	Mpro and RdRp	Pendyala and Patras (2020)
Pelargonidin	grapes, Strawberry, and Blueberry	S–ACE interaction	Toigo et al. (2023)
Cyanidin		PLpro	Pitsillou et al. (2021)
Delphinidin		3C-like protease	Mahmood et al. (2022)
Naringenin	Lemon, Grapes, Orange, Tomatoes, Citrus fruits, and Bergamot	3CLpro, Mpro, ACE2, and RBD-S	Adem et al. (2022); Utomo et al. (2020); Tallei et al. (2020);
Neohesperidin		TMPRSS2	Chikhale et al. (2020)
Eriodictyol		3CLpro, Mpro, and ACE2	Khaerunnisa et al. (2020); W. Liu et al. (2022)
Naringin		3CLpro and TMPRSS2	Chikhale et al. (2020)
Hesperidin		3CLpro	Peterson (2020)
Glycitein	Legumes, Tofu, Miso Soybeans,	3CL protease	Abdul-Jabar et al. (2020)
Daidzein	Soya foods, Tempeh, and	3CLpro, Mpro, and RdRp	Pendyala and Patras (2020)
Genistein	soy protein isolates	3CLpro	Kaul et al. (2021)
Tangeritin	Parsley, Red pepper, Peel of citrus,	Mpro, ACE2, and S	Khaerunnisa et al. (2020); Leal et al. (2021)
Apigenin	fruits, Celery, and Mint	3CLpro and Mpro	Khaerunnisa et al. (2020)
Kaempferol		3CLpro	Khan et al. (2021)
Silymarin		3CLpro	Saraswat et al. (2021)
Nobiletin		3CLpro	Tallei et al. (2020)
Tangeretin		3CLpro	Jo et al. (2020)
Luteolin		3CLpro and PLpro	Mouffouk et al. (2021)
Sinensetin	Apple, Tomato, Onion, Green tea,	Mpro and TMPRSS2	Khaerunnisa et al. (2020)
Fisetin	Lettuce, Berry, Black tea, and Red wine	Mpro, TMPRSS2, ACE2, and S	Chikhale et al. (2020); Khaerunnisa et al. (2020); Omar et al. (2020)
Quercetin		3CLpro, ACE2, and TMPRSS2	Chikhale et al. (2020)
Myricetin		3CLpro	Mouffouk et al. (2021)
Pectolinarin	Green tea, Fruits, wine, and	3C-like protease	Bhattacharya et al. (2022)
Rutin	vegetables	3CLpro	Rizzuti et al. (2021)
Astragalin		Spike protein and RdRp	Mouffouk et al. (2021)
Prunin		3CL protease	Masand et al. (2020)
Nicotiflorin		3CL protease	Akbaba and Karataş (2023)
Biorobin		Chymotrypsin-like protease	Mouffouk et al. (2021)

their antiviral potential, notable examples include catechins, kaempferol, quercetin, baicalein, hesperidin, and isorhamnetin (Ahmadian et al., 2020; Khazeei Tabari et al., 2021). Quercetin is commonly present in various plants, primarily glycosides (Andrea, 2015). It has potent antiviral, immunomodulatory, and anti-inflammatory actions (Gasmi et al., 2022). Quercetin, when given in combination with other antiviral drugs for the treatment of COVID-19, decreased the death rate and the lifespan of the virus (Bartleson et al., 2021). When used in higher concentrations, the findings from RT-qPCR analysis for quercetin had a notable inhibitory effect on SARS-CoV-2 infection. Alternatively, quercetin derivatives exhibited their inhibitory action at lower concentrations against the viral infection (Mangiavacchi et al., 2021). Baicalin and Baicalein are flavonoid-derived compounds from the dried roots of *Scutellaria baicalensis*. These possess pharmacological benefits, including antiviral, anti-inflammatory, antibacterial, hepatoprotective, and

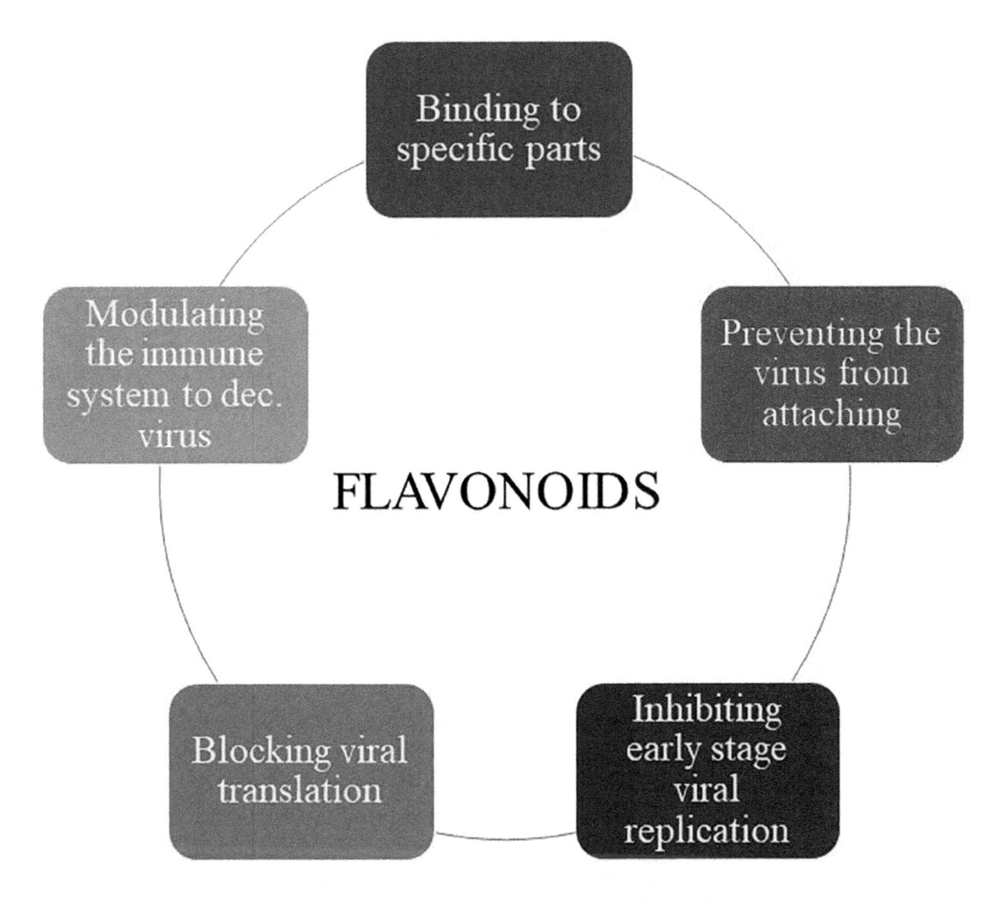

FIGURE 16.1 Antiviral activity of flavonoid group of compounds against SARS-CoV-2.

choleretic properties (Ekiert et al., 2022). Hesperidin was evaluated to block the AKT pathway, decrease collagen expression caused by Ang II, and suppress cardiac fibroblast development (Zalpoor et al., 2022). Clinical studies have shown that hesperidin can effectively alleviate specific symptoms of COVID-19, including cough, shortness of breath, diminished or loss of taste, and fever (Dupuis et al., 2022). Additionally, findings from an *in vitro* examination have shown that isorhamnetin engages with ACE2, displaying anti-SARS-CoV-2 efficacy (Zhan et al., 2021). It was discovered that naringenin demonstrates anti-SARS-CoV-2 capabilities *in vitro* by inhibiting M^{pro} (Clementi et al., 2021; Devi et al., 2020). Luteolin is predominantly found in glycosidic form in numerous plant species. It boasts a plethora of pharmacological effects, including anti-inflammatory, antiallergic, antiviral, anti-tumor, and antibacterial properties. Clinically, it is frequently employed for its capacity to alleviate inflammation, address coughs, and act as an expectorant as such complications arise in COVID positive patients (Lopez-Lazaro and López-Lázaro, 2009; Guo et al., 2017; Yao et al., 2018). In laboratory-based investigations, luteolin exhibits inhibitory effects on $3CL^{pro}$ at 20 µM concentration. Furthermore, luteolin has demonstrated the ability to inhibit RdRp activity (Munafò et al., 2022). In a study using a pseudo virus mimicking SARS-CoV-2, researchers discovered that luteolin could ascribe the S protein and ability to enter cells (Mishra et al., 2020).

Camellia sinensis (green tea) is a source of numerous flavonoids. Various studies have recorded a broad spectrum of health advantages associated with green tea, which may encompass its ability to potentially counteract cancer or lower the likelihood of its occurrence, enhance brain function, provide neuroprotection, reduce anxiety, prevent cardiovascular disease, lower cholesterol levels, offer anti-arthritic effects, inhibit angiogenesis, and exhibit antiviral, antibacterial, and anti-inflammatory activities. Catechins are natural polyphenolic compounds in the flavonoid family (Liu et al., 2014). A significant advantage of catechins is their ability to interact with multiple proteins, some of which may play a role in modulating the immune system. In the present book chapter, we will focus only on polyphenols and the role of catechins against SARS-CoV-2.

16.2 POLYPHENOL AGAINST SARS-COV-2

Numerous naturally occurring compounds called polyphenols are present in plants and have been the focus of research due to their possible health advantages, which include anti-inflammatory and antioxidant capabilities. Studies have looked into the possibility of polyphenols having antiviral characteristics, including the ability to combat SARS-CoV-2 (Mehany et al., 2021). Polyphenols have the advantage of being less toxic compared to other drugs, making them promising candidates for antiviral purposes. Polyphenols, such as theaflavin 3,30-gallate (TF3), theaflavin-30-O-gallate (TF2a), theaflavin (TF1), theaflavin-30-gallate (TF2b), and myricetin, have been identified. Thus, extensive research is to be conducted to identify potent M^{pro} inhibitors. Moreover, polyphenols have been evaluated for their potential inhibitors of SARS-CoV-2 RdRp (Munafò et al., 2022; Hendaus, 2021). Furthermore, eight polyphenols isolated from green tea leaves have exhibited potent antiviral properties (Ghosh et al., 2021). These findings highlight the diverse and promising potential of polyphenols in the field of antiviral research.

16.2.1 Green Tea as a Source of Catechins

Green tea leaves contain many polyphenols with potent antiviral properties. Green tea contains catechins that inhibit neuraminidase, disrupting the influenza virus membrane (Song et al., 2005). Epigallocatechin gallate (EGCG) is among the most abundant catechins present in green tea, roughly 60% of the overall catechin concentrations, i.e., EGC (~20%), ECG (~14%), and EC (~6%) (Chourasia et al., 2021). The plasma concentration of catechins typically reaches its highest level within 1–4 hours after consuming oral green tea or supplements of catechins. Subsequently, it returns to its baseline within 24 hours (Henning et al., 2004). EGCG and ECG are robust green tea catechins that encompass the galloyl component, which could be accountable for these compounds' more robust physiological impacts (Kim et al., 2014; Zwolak, 2021).

EGCG has received attention and found to possess an antiviral-like biological antimicrobial (Takabayashi et al., 2004), anti-inflammatory, immunomodulatory (Singh et al., 2016), and anticancer (Gan et al., 2018) activities. The EGCG has demonstrated to abrogate porcine reproductive and respiratory syndrome viruses (Ge et al., 2018), Zika virus (Mou et al., 2020), Chikungunya virus (Lu et al., 2017), Dengue virus (Raekiansyah et al., 2018), influenza A virus H_1N_1 (Mou et al., 2020), human immunodeficiency virus (Yamaguchi et al., 2002), and Ebola virus (Reid et al., 2014). Powdered green tea (matcha) had 137 times the amount of EGCG found in loose-leaf green tea (sencha) (Kaihatsu et al., 2018). A different investigation revealed that when exposed to green tea, oolong tea, or toasted green tea, SARS-CoV-2 experienced a notable reduction in its capacity to infect. SARS-CoV-2 gains entry into host cells through its interaction with the ACE2 receptor facilitated by the surface S protein. EGCG can hinder the virus from attaching to the ACE2 receptor (Liu et al., 2021). EGCG acts by inhibiting the RBD (Receptor-Binding Domain) of SARS-CoV-2 from adhering to the ACE2 receptor, effectively preventing the virus from entering cells and has low cytotoxicity (Day et al., 2021).

16.2.1.1 Therapeutic Intervention for SARS-CoV-2:

EGCG exhibits a higher binding affinity to viral proteins than the benchmark drugs, chloroquine and remdesivir, resulting in more potent antiviral effectiveness (Khan et al., 2021). EGCG effectively rendered SARS-CoV-2 inactive (Ohgitani et al., 2021). Another study suggested that EGCG could hinder viral replication by regulating the cellular oxidation–reduction environment (Ho et al., 2009). Natural flavonoids can counteract the growth of SARS-CoV-2. EGCG also alters the viral envelope, obstructing the virus's ability to infect other cells. A laboratory study conducted by Hong and colleagues (2010) demonstrated that EGCG notably suppressed the action of SARS-CoV-2's Nsp15 enzyme (Hong et al., 2021). Nsp15 substantially replicates the virus (Hackbart et al., 2020; Pillon et al., 2021). The EGCG at lower concentrations halted the Nsp15 activity, consequently inhibiting virus replication within cells.

16.2.1.2 EGCG Mode of Action

The exact mechanism of action of EGCG against SARS-CoV-2 is not fully elucidated. Still, several potential mechanisms have been proposed (Figure 16.2). EGCG may hinder the virus by inhibiting its entry into host cells (Xiao et al., 2023), potentially interfering with the binding of the viral spike protein to ACE2 receptors. It may also disrupt viral replication by targeting the viral RdRp (Mouffouk et al., 2021), which is vital for replicating the virus's genetic material. EGCG exerts various effects by engaging with diverse cell surface receptors, signaling pathways, and transcription factors within the nucleus (Kim et al., 2014). Through a 67LR-dependent mechanism, EGCG effectively inhibits toll-like receptor 4 (TLR4) signaling, demonstrating its anti-inflammatory properties (Xu et al., 2017). TLR4 activation triggers pivotal pathways involved in regulating inflammation and apoptosis (Byun et al., 2014) including nuclear factor (NF)-κB, activator protein (AP)-1 (Wang et al., 2019), which is activated via mitogen-activated protein kinases (MAPK), and interferon regulatory factor (IRF)3 (Neill et al., 2013). As a result, EGCG has the potential to interact with these mentioned pathways, yielding various beneficial effects, including anti-tumor, antioxidant, anti-inflammatory (Chen et al., 2018), and neuroprotective, as substantiated by numerous frequent studies (Zhong et al., 2019).

The MAPK pathway, profoundly affected by EGCG, is pivotal in orchestrating cellular responses to diverse stimuli (Liu et al., 2021). It responds to growth factors and mitogens by stimulating the extracellular signal-regulated kinases (ERK) 1/2 module, regulating cell proliferation and differentiation. Besides, oxidative stress and proinflammatory cytokines activate the c-Jun N-terminal kinase (JNK)/p38 module (Liu et al., 2014), which governs cell cycle arrest, cell differentiation, inflammation, and apoptosis. Lastly, the ERK5 module is triggered by morphogenic signals, leading to the formation of endothelial lumens. This intricate interplay within the MAPK pathway underscores the multifaceted effects of EGCG (Mou et al., 2020), encompassing anti-tumor, antioxidative, anti-inflammatory, and other beneficial outcomes, as evidenced by the extensive research. The use of EGCG or any other dietary supplement for the prevention or treatment of COVID-19 (Ge et al., 2018) should be done under the guidance of healthcare

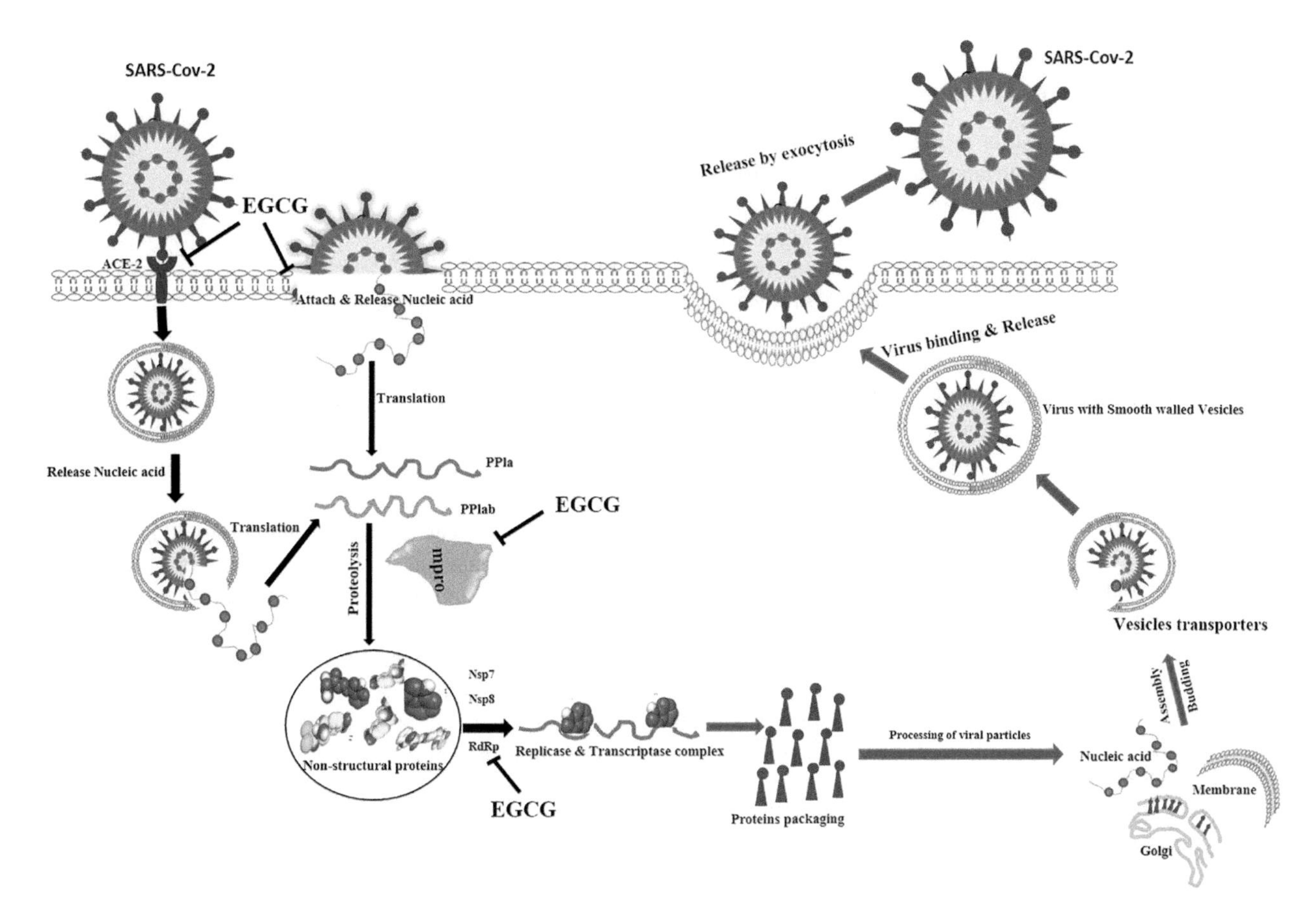

FIGURE 16.2 Schematic representation of SARS-CoV-2 virus in host cells through ACE2 receptor (I) and EGCGs are shown to inhibit the modes of action.

professionals and by the latest recommendations from public health authorities (Bimonte et al., 2021). Research on this topic is ongoing, and understanding EGCG's potential role in managing SARS-CoV-2 may continue to evolve.

The influence of EGCG on MAPK has been extensively substantiated in various studies. Notably, EGCG efficiently inhibited aflatoxin B1 biosynthesis while reducing oxidative stress by suppressing the MAPK signaling pathway (Xu et al., 2021). EGCG stimulated expression of β-defensin3, an antiviral peptide synthesized by epithelial cells, by upregulating the p38 MAPK, ERK, and JNK signaling pathways. This led to suppressing influenza A virus H_1N_1 replication (Mou et al., 2020). EGCG demonstrated the ability to mitigate *in vitro* hemolysis induced by α-hemolysin, a *Staphylococcus aureus* product. Furthermore, EGCG reduced the excessive production of ROS caused by α-hemolysin, decreased the expression of NLRP3 inflammasome (Liu et al., 2021), and generated caspase-1, IL-1β, and IL-18 in mice. These effects were linked to a diminished activation of the MAPK signaling pathway, underscoring ROS and MAPK as significant instigators of NLRP3 inflammasome activation (Liu et al., 2021). Additionally, anti-inflammatory and antioxidant properties may help mitigate excessive inflammation and oxidative stress associated with COVID-19.

16.3 GREEN TEA POLYPHENOL AND MEDICINAL USES

The polyphenols from green tea, particularly the catechins of EGCG and other polyphenols, have various medicinal uses and health benefits (Henning et al., 2004). They serve as potent antioxidants, aiding in preventing cell damage and reducing the risk of persistent diseases like cardiovascular and cancer conditions. Green tea polyphenols possess anti-inflammatory properties (Chourasia et al., 2021), potentially alleviating conditions such as arthritis (Min et al., 2015), supporting weight management, improving blood sugar control, and promoting cardiovascular and liver health (Chen et al., 2016). Additionally, they have shown promise in neuroprotection, possibly lowering the risk of neurodegenerative diseases, and they are commonly incorporated into skincare products to protect the skin from UV radiation damage (Roh et al., 2017). However, while green tea polyphenols offer numerous potential health advantages, they are not standalone treatments and should be part of a balanced approach to health and wellness. Consulting a healthcare professional for personalized guidance is advisable for their optimal use. There were many complications post-COVID-19 (Tallei et al., 2021). Green tea polyphenols were primary in healing due to their medicinal properties. Some of the medical complications are listed below.

16.3.1 Anti-inflammatory Activities of Green Tea Polyphenols

COVID-19 infection results in the interference of 18 key cytokines significantly elevated in individuals displaying symptoms (Long, 2020). Prior research has indicated that EGCG could hinder inflammatory signals, signifying its potential utility as a comprehensive treatment option for COVID-19 patients (Chourasia et al., 2021). Moreover, green tea and its primary component EGCG have been reported to display anti-inflammatory properties in cellular, animal, and human studies (Ohishi et al., 2016). This is achieved through the inhibition of the activation of nuclear factor-kappa B (NF-κB) (Syed et al., 2007), directing to reduce expression of inflammatory cytokines and enzymes associated with inflammation, such as TNF-α (Moe), MMP-9 (Guarneri et al., 2017), and COX-2 (Ke et al., 2007). EGCG inhibits the manifestation of Toll-like receptor 4 (TLR4) and TLR2 by blocking MAPK and NF-κB signaling pathways, which leads to the suppression of proinflammatory cytokine production (Byun et al., 2010). The administration of a green tea petiole extract led to the inhibition of TNF-α, IL-1, and IL-8 production (Park et al., 2019). EGCG also reduces IL-8 production in the respiratory tract epithelium of humans, thereby mitigating the intensity of the inflammatory response (Kim et al., 2014). Furthermore, EGCG has demonstrated the ability to alleviate skin inflammation and asthma in rats exposed to airborne delicate particulate matter (Wang et al., 2019). EGCG considerably decreased histamine discharge in rat cell cultures by 90% (Tallei et al., 2021).

16.3.2 Green Tea Antioxidant Activity

The green tea contains catechins as shown in Figure 16.3: epicatechin (EC), gallocatechin (GC), epigallocatechin (EGC), epigallocatechin-3-gallate (EGCG), catechin (C), epicatechin-3-gallate (ECG), and epigallocatechin 3,5-digallate (EGCD) which have potential antiviral properties. Free radicals are implicated in developing respiratory virus infections, and there is a strong association between pro-inflammatory and reactive oxygen species in numerous lung infections, including coronavirus infections, which involve mutually oxidative stress and inflammation (Ungarala et al., 2022). In severe medical conditions, oxidative stress amplifies, leading to organ dysfunction. COVID-19, for instance, provokes a vigorous inflammatory reaction called a cytokine storm, which may be driven by oxidative stress (Soto et al., 2020).

Research indicates that polyphenols in green tea function as straight antioxidants in laboratory settings by neutralizing ROS and binding to transition metals. Various reports suggested that catechins might expansively enhance

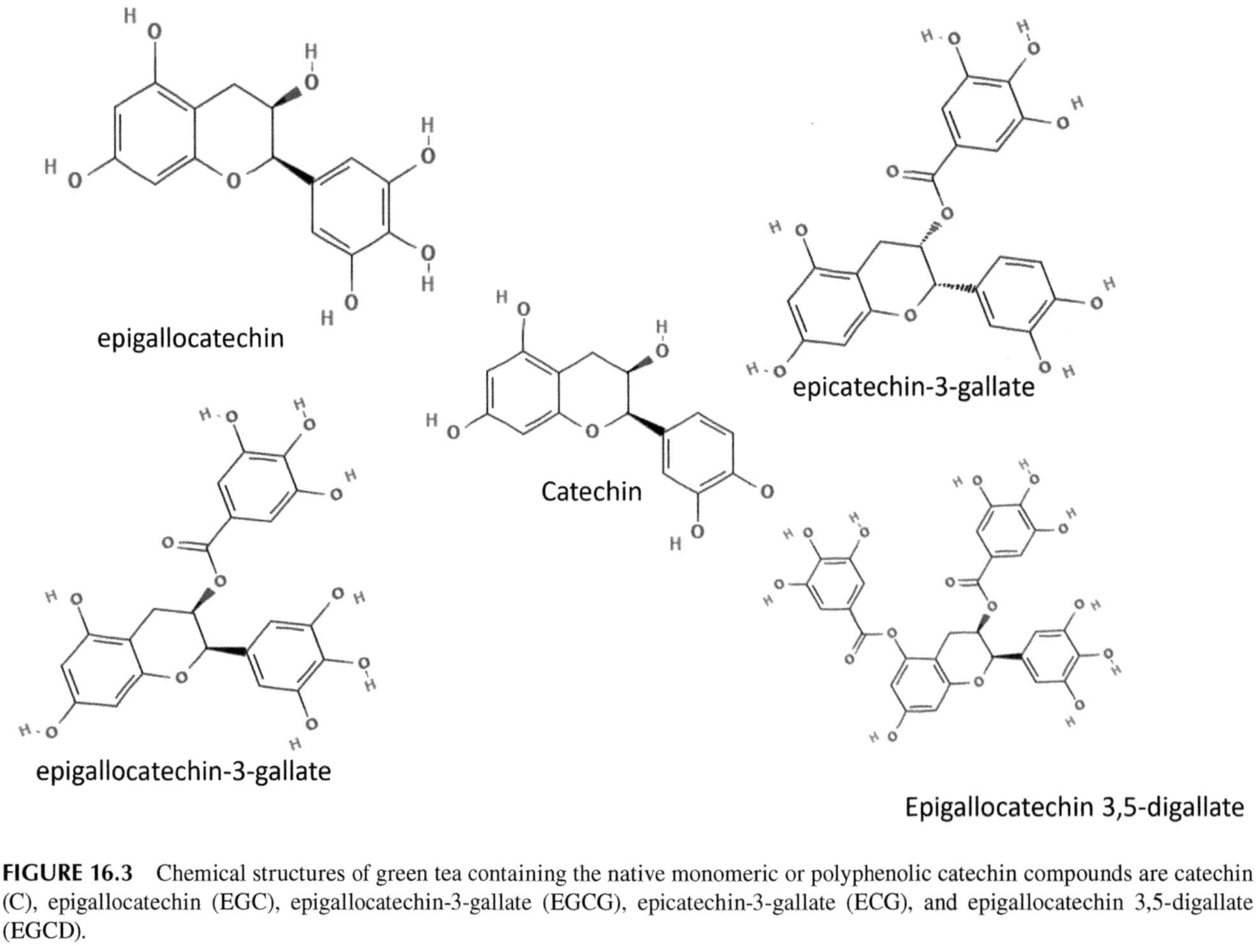

FIGURE 16.3 Chemical structures of green tea containing the native monomeric or polyphenolic catechin compounds are catechin (C), epigallocatechin (EGC), epigallocatechin-3-gallate (EGCG), epicatechin-3-gallate (ECG), and epigallocatechin 3,5-digallate (EGCD).

the activity of phase II antioxidant enzymes like glutathione peroxidase, catalase, and glutathione reductase (Forester and Lambert, 2011). COVID-19 seriousness has been correlated with heightened oxidative stress and inflammation, especially in individuals with pre-existing chronic conditions that weaken the antioxidant defense system (Lammi and Arnoldi, 2021). Consequently, using antioxidant treatment effectively reduces the inflammation associated with the elevated oxidative state observed in COVID-19 (Soto et al., 2020).

16.3.3 Immunomodulatory

EGCG and EGC, found abundantly in green tea, exhibit immunomodulatory properties by affecting the proliferation of T lymphocytes and cytokines production. While EGCG has been proposed as a potential remedy for immune disorders, it is essential to note that comprehensive human research data are still pending. Most research has reported that green tea extract could boost lymphoblast proliferation and increase lymphocyte generation. Additionally, EGCG could promote the production of IL-1α, IL-1β, monocytes, and lymphocytes, further highlighting its potential immunomodulatory effects (Zhu et al., 2020). In immune-compromised patients infected with *Candida albicans*, green tea extract has demonstrated its immunomodulatory potential. It achieves this by enhancing the expression of IL-8, IL-17A, and human β-defensin (Rahayu et al., 2018; Zhu et al., 2020). A study indicates that green tea exerts an immunostimulatory impact in a murine model of immunosuppression induced by cyclophosphamide. This effect is likely due to stimulation of both cellular and humoral immune functions, resulting in elevated total leukocyte counts (Haque and Ansari, 2014). Models suggest that green tea extract might regulate allergic responses (Kuo et al., 2014). Additionally, in animal models of arthritis, EGCG has exhibited the ability to alleviate arthritis symptoms, incorporating pathological arthritic characteristics and levels of serum CII-specific IgG2a antibodies. Moreover, EGCG treatment has been shown to notably decrease the production of cytokines associated with inflammation, such as IL-6, IFN-c, and TNF-α, while simultaneously increasing the production of IL-10 (Min et al., 2015). Furthermore, EGCG has demonstrated the capability to suppress the propagation of autoreactive T cells, reduce the production of proinflammatory cytokines, diminish the populations of T helper type (+) 1 and +17 cells, and increase the presence of TR cells in lymphoid tissue and central nervous system.

16.3.4 Effects of Polyphenols on Reduced Mucin Hypersecretion

Excessive inflammation in conditions like COVID-19 could lead to an overproduction of mucus, obstructing the respiratory tract. The mucus build-up can also lead to persistent respiratory tract infections, further complicating the respiratory tract blockage. Liang et al. (2017) discovered that EGCG could most likely alleviate excessive respiratory tract mucus production in rats by inhibiting EGFR signaling pathways. This implies that EGCG could hold promise as a therapeutic option for preventing or treating chronic inflammation of the respiratory tract and the abnormal production of mucus (Liang et al., 2017). During viral infections, such as those caused by COVID-19, there is an increase in the concentration of neutrophils in the respiratory system (Camp and Jonsson, 2017). EGCG can potentially diminish the stimulation of the TGF-β signaling pathway, which initiates inflammation (Li et al., 2019) and fibrosis (Menegazzi et al., 2020). The utilization of green tea polyphenols in treating COVID-19 is under investigation due to these beneficial properties.

16.4 THE CHALLENGES IN THE MANAGEMENT OF SARS-COV-2 INFECTIONS

While corona viruses continuously regulate their binding patterns as they evolve, the target location in the alveolus may change. Still, it remains consistent in the small intestine. This makes intestinal mucosal cells a potential reservoir for coronaviruses (Gasmi et al., 2022). During the acute phase of infection, only 10% of COVID-19 patients have virus cDNA in their blood, but nearly 50% of excrete it in their feces, suggesting the possibility of contamination through the oral–fecal route (Islam et al., 2020). Interestingly, the presence of virus in the gut may explain the fluctuations in viral load observed in repeated testing of the same individual. The Chinese researchers have studied changes in the microbiota of COVID-19 patients who succumbed to the infection. They found a significant decrease in beneficial bacteria like *Lactobacilli* and *Bifidobacteria* and an increase in opportunistic bacteria such as *Ruthenibacterium* and *Corynebacterium* (Mirzaei et al., 2021). This imbalance in gut flora could have considerable implications for the respiratory immune system, potentially increasing the COVID-19 risk-induced respiratory distress.

The emergence of SARS-CoV-2 variants continues to pose tasks for treating, preventing, and diagnosing COVID-19. One pressing question is whether previous variants will resurface over time, either from the human population or an animal reservoir or if they have become extinct (Dupuis et al., 2022). Additionally, the methods for predicting new mutations and combinations of change in mutations that could give rise to new variants may aid in preparedness. For example, recent reports outlined an approach to identify significant mutations based on the genomic and epidemiological surveillance reconnaissance data, which has successfully predicted variants of concern. More research to enhance the prediction and detection of SARS-CoV-2 variants and swift measures to control their spread remains crucial for global public health.

16.5 CONCLUSION

The COVID-19 outbreak had far-reaching implications in all aspects of society, driving an intense search for new remedies against SARS-CoV-2. Furthermore, due to restricted vaccine production and distribution, total herd immunity may take several years, and the prospect of new coronavirus infections cannot be ruled out. As a result, there is a pressing need to develop multiple antiviral drugs to treat and mitigate current and future coronavirus diseases. Flavonoids, a class of safe and widely available phytoconstituents, have garnered significant attention for their potential benefits in combating COVID-19. Flavonoids have shown possible antiviral effects *in vitro* and *in vivo* research on viruses with pathogenic processes similar to SARS-CoV-2, including HIV, influenza, ebolavirus, SARS, and MERS. Herbal sources of flavonoids are *Camellia sinensis* (Tea), mainly green tea, which is rich in catechins and EGCG, a crucial catechin in green tea, is increasingly considered a potential treatment option for SARS-CoV-2 infections. EGCG and other catechins may have the potential for repurposed use as medicines in this context. A substantial body of scientific literature on EGCG's role in inflammation suggests its potential in targeting inflammatory pathways associated with COVID-19, including genes like papain-like protease protein (PL^{pro}). Tea is a widely consumed beverage in human diets. Tea is also a promising candidate for future clinical trials due to its established safety profile. Future research will strongly emphasize green tea polyphenols and perform extensive studies to confirm that green tea has therapeutic effects in the context of coronavirus illnesses.

ACKNOWLEDGMENTS

PG acknowledges the Department of Science and Technology (DST) for the Women Scientist-A (SR/WOS-A/LS-136/2017) award and RIMT University, Mandi Govindgarh, Punjab; Department of Biotechnology, Pondicherry University; and Central University of Kerala, Kasaragod for facilities.

REFERENCES

Abdul-Jabar, R. A., & Al-Fadal, S. A. M. (2020). In-Silico study of the inhibitory effect of some flavonoids compounds and their derivatives on SARS-COV-2. *International Journal of Pharmaceutical Research*, *12*(2), 09752366.

Agrawal, U., Raju, R., & Udwadia, Z. F. (2020). Favipiravir: A new and emerging antiviral option in COVID-19. *Medical Journal Armed Forces India*, *76*(4), 866–872. https://doi.org/10.1016/j.mjafi.2020.08.004

Ahmadian, R., Rahimi, R., & Bahramsoltani, R. (2020). Kaempferol: An encouraging flavonoid for COVID-19. *Boletin Latinoamericano y del Caribe de Plantas Medicinales y Aromaticas*, *19*(5), 492–494. https://doi.org/10.37360/blacpma.20.19.5.33

Akbaba, E., & Karataş, D. (2023). Phytochemicals of Hibiscus sabdariffa with therapeutic potential against SARS-CoV-2: A molecular docking study. *Journal of the Institute of Science and Technology*, *13*(2), 872–888.

Andrews, P. L. R., Cai, W., Rudd, J. A., & Sanger, G. J. (2021). COVID-19, nausea, and vomiting. *Journal of Gastroenterology and Hepatology (Australia)*, *36*(3), 646–656. https://doi.org/10.1111/jgh.15261

Bartleson, J. M., Radenkovic, D., Covarrubias, A. J., Furman, D., Winer, D. A., & Verdin, E. (2021). SARS-CoV-2, COVID-19, and the aging immune system. *Nature Aging*, *1*(9), 769–782. https://doi.org/10.1038/s43587-021-00114-7

Bhattacharya, K., Bordoloi, R., Chanu, N. R., Kalita, R., Sahariah, B. J., & Bhattacharjee, A. (2022). In silico discovery of 3 novel quercetin derivatives against papain-like protease, spike protein, and 3C-like protease of SARS-CoV-2. *Journal of Genetic Engineering and Biotechnology*, *20*(1), 1–20.

Breitinger, U., Farag, N. S., Sticht, H., & Breitinger, H. G. (2022). Viroporins: Structure, function, and their role in the life cycle of SARS-CoV-2. *The International Journal of Biochemistry & Cell Biology*, *145*, 106185. https://doi.org/10.1016/j.biocel.2022.106185

Byun, E.-B., Yang, M.-S., Kim, J.-H., Song, D.-S., Lee, B.-S., Park, J.-N., Park, S.-H., Park, C., Jung, P.-M., Sung, N.-Y., & Byun, E.-H. (2014). Epigallocatechin-3-gallate-mediated Tollip induction through the 67-kDa laminin receptor negatively regulates TLR4 signaling in endothelial cells. *Immunobiology*, *219*(11), 866–872. https://doi.org/10.1016/j.imbio.2014.07.010

Camp, J. V., & Jonsson, C. B. (2017). A role for neutrophils in viral respiratory disease. *Frontiers in Immunology*, *8*, 550. https://doi.org/10.3389/fimmu.2017.00550

Chen, C.-Y., Kao, C.-L., & Liu, C.-M. (2018). The cancer prevention, anti-inflammatory and anti-oxidation of bioactive phytochemicals targeting the TLR4 signaling pathway. *International Journal of Molecular Sciences*, *19*(9), 2729. https://doi.org/10.3390/ijms19092729

Chen, I. J., Liu, C. Y., Chiu, J. P., & Hsu, C. H. (2016). Therapeutic effect of high-dose green tea extract on weight reduction: A randomized, double-blind, placebo-controlled clinical trial. *Clinical Nutrition*, *35*(3), 592–599.

Chourasia, M., Koppula, P. R., Battu, A., Ouseph, M. M., & Singh, A. K. (2021). EGCG, a green tea catechin, as a potential therapeutic agent for symptomatic and asymptomatic SARS-CoV-2 infection. *Molecules*, *26*(5), 1200. https://doi.org/10.3390/molecules26051200

Choy, K.-T., Wong, A. Y.-L., Kaewpreedee, P., Sia, S. F., Chen, D., Hui, K. P. Y., Chu, D. K. W., Chan, M. C. W., Cheung, P. P.-H., Huang, X., Peiris, M., & Yen, H.-L. (2020). Remdesivir, lopinavir, emetine, and homoharringtonine inhibit SARS-CoV-2 replication in vitro. *Antiviral Research*, *178*, 104786. https://doi.org/10.1016/j.antiviral.2020.104786

Clementi, N., Scagnolari, C., D'Amore, A., Palombi, F., Criscuolo, E., Frasca, F., Pierangeli, A., Mancini, N., Antonelli, G., Clementi, M., Carpaneto, A., & Filippini, A. (2021). Naringenin is a powerful inhibitor of SARS-CoV-2 infection in vitro. *Pharmacological Research*, *163*, 105255. https://doi.org/10.1016/j.phrs.2020.105255

Cucinotta, D., & Vanelli, M. (2020). WHO declares COVID-19 a pandemic. *Acta Biomedica*, *91*(1), 157–160. https://doi.org/10.23750/abm.v91i1.9397

Day, C. J., Bailly, B., Guillon, P., Dirr, L., Jen, F. E. C., Spillings, B. L., Mak, J., von Itzstein, M., Haselhorst, T., & Jennings, M. P. (2021). Multidisciplinary approaches identify compounds that bind to human ACE2 or SARS-CoV-2 spike protein as candidates to block SARS-CoV-2-ACE2 receptor interactions. *MBio*, *12*(2), 1–14. https://doi.org/10.1128/mBio.03681-20

Devi, S. K., & Girija, A. S. S. (2020). Structural basis of SARS COV 2 3CL PRO drug and antiCOVID-19 discovery from medicinal plants - A review. *International Journal of Current Research and Review*, *12*(21 Special Issue), 101–105. https://doi.org/10.31782/IJCRR.2020.SP46

Du, A., Zheng, R., Disoma, C., Li, S., Chen, Z., Li, S., & Xia, Z. (2021). Epigallocatechin-3-gallate, an active ingredient of traditional Chinese medicines, inhibits the 3CLpro activity of SARS-CoV-2. *International Journal of Biological Macromolecules*, *176*, 1–12.

Dupuis, J., Laurin, P., Tardif, J. C., Hausermann, L., Rosa, C., Guertin, M. C., Thibaudeau, K., Gagnon, L., Cesari, F., Robitaille, M., & Moran, J. E. (2022). Fourteen-day evolution of COVID-19 symptoms during the third wave in nonvaccinated subjects and effects of hesperidin therapy: A randomized, double-blinded, placebo-controlled study. *Evidence-Based Complementary and Alternative Medicine*, *2022*, 3125662. https://doi.org/10.1155/2022/3125662

Ekiert, H. M., Kubica, P., Kwiecień, I., Jafernik, K., Klimek-Szczykutowicz, M., & Szopa, A. (2022). Cultures of medicinal plants in vitro as a potential rich source of antioxidants. In *Reference Series in Phytochemistry*. Chem: Springer. https://doi.org/10.1007/978-3-030-78160-6_37

Forester, S. C., & Lambert, J. D. (2011). The role of antioxidant versus pro-oxidant effects of green tea polyphenols in cancer prevention. *Molecular Nutrition & Food Research*, *55*(6), 844–854. https://doi.org/10.1002/mnfr.201000641

Gan, R.-Y., Li, H.-B., Sui, Z.-Q., & Corke, H. (2018). Absorption, metabolism, anti-cancer effect and molecular targets of epigallocatechin gallate (EGCG): An updated review. *Critical Reviews in Food Science and Nutrition*, *58*(6), 924–941. https://doi.org/10.1080/10408398.2016.1231168

Gasmi, A., Mujawdiya, P. K., Lysiuk, R., Shanaida, M., Peana, M., Gasmi Benahmed, A., Beley, N., Kovalska, N., & Bjørklund, G. (2022). Quercetin in the prevention and treatment of coronavirus infections: A focus on SARS-CoV-2. *Pharmaceuticals*, *15*(9), 1049. https://doi.org/10.3390/ph15091049

Ge, M., Xiao, Y., Chen, H., Luo, F., Du, G., & Zeng, F. (2018). Multiple antiviral approaches of (−)-epigallocatechin-3-gallate (EGCG) against porcine reproductive and respiratory syndrome virus infection in vitro. *Antiviral Research*, *158*, 52–62. https://doi.org/10.1016/j.antiviral.2018.07.012

Ghosh, R., Chakraborty, A., Biswas, A., & Chowdhuri, S. (2021). Identification of polyphenols from Broussonetia papyrifera as SARS CoV-2 main protease inhibitors using in silico docking and molecular dynamics simulation approaches. *Journal of Biomolecular Structure and Dynamics*, *39*(17), 6747–6760. https://doi.org/10.1080/07391102.2020.1802347

Guarneri, C., Bevelacqua, V., Polesel, J., Falzone, L., Cannavò, P. S., Spandidos, D. A., Malaponte, G., & Libra, M. (2017). NF-κB inhibition is associated with OPN/MMP-9 downregulation in cutaneous melanoma. *Oncology Reports*, *37*(2), 737–746. https://doi.org/10.3892/or.2017.5362

Guo, Y. F., Xu, N. N., Sun, W., Zhao, Y., Li, C. Y., & Guo, M. Y. (2017). Luteolin reduces inflammation in Staphylococcus aureus-induced mastitis by inhibiting NF-kB activation and MMP expression. *Oncotarget*, *8*(17), 28481–28493. https://doi.org/10.18632/oncotarget.16092

Gurung, A. B., Ali, M. A., Lee, J., Farah, M. A., & Al-Anazi, K. M. (2020). Unravelling lead antiviral phytochemicals for the inhibition of SARS-CoV-2 Mpro enzyme through in silico approach. *Life Sciences*, *255*, 117831. https://doi.org/10.1016/j.lfs.2020.117831

Hackbart, M., Deng, X., & Baker, S. C. (2020). Coronavirus endoribonuclease targets viral polyuridine sequences to evade activating host sensors. *Proceedings of the National Academy of Sciences*, *117*(14), 8094–8103. https://doi.org/10.1073/pnas.1921485117

Haque, M., & Ansari, S. (2014). Immunostimulatory effect of standardised alcoholic extract of green TeaTea (Camellia sinensis L.) against cyclophosphamide-induced immunosuppression in murine model. *International Journal of Green Pharmacy*, *8*(1), 52. https://doi.org/10.4103/0973-8258.126824

Hatmal, M. M., Alshaer, W., Al-Hatamleh, M. A. I., Hatmal, M., Smadi, O., Taha, M. O., Oweida, A. J., Boer, J. C., Mohamud, R., & Plebanski, M. (2020). Comprehensive structural and molecular comparison of spike proteins of SARS-CoV-2, SARS-CoV and MERS-CoV, and their interactions with ACE2. *Cells*, *9*(12), 2638. https://doi.org/10.3390/cells9122638

Hendaus, M. A. (2021). Remdesivir in the treatment of coronavirus disease 2019 (COVID-19): A simplified summary. *Journal of Biomolecular Structure and Dynamics*, *39*(10), 3787–3792. https://doi.org/10.1080/07391102.2020.1767691

Henning, S. M., Niu, Y., Lee, N. H., Thames, G. D., Minutti, R. R., Wang, H., Go, V. L. W., & Heber, D. (2004). Bioavailability and antioxidant activity of tea flavanols after consumption of green TeaTea, black TeaTea, or a green tea extract supplement. *American Journal of Clinical Nutrition*, *80*(6), 1558–1564. https://doi.org/10.1093/ajcn/80.6.1558

Ho, H.-Y., Cheng, M.-L., Weng, S.-F., Leu, Y.-L., & Chiu, D. T.-Y. (2009). Antiviral effect of epigallocatechin gallate on enterovirus 71. *Journal of Agricultural and Food Chemistry*, *57*(14), 6140–6147. https://doi.org/10.1021/jf901128u

Hong, S., Seo, S. H., Woo, S.-J., Kwon, Y., Song, M., & Ha, N.-C. (2021). Epigallocatechin gallate inhibits the uridylate-specific endoribonuclease Nsp15 and efficiently neutralizes the SARS-CoV-2 strain. *Journal of Agricultural and Food Chemistry*, *69*(21), 5948–5954. https://doi.org/10.1021/acs.jafc.1c02050

Hong Byun, E., Fujimura, Y., Yamada, K., & Tachibana, H. (2010). TLR4 signaling inhibitory pathway induced by green tea polyphenol epigallocatechin-3-gallate through 67-kDa laminin receptor. *The Journal of Immunology*, *185*(1), 33–45. https://doi.org/10.4049/jimmunol.0903742

Jiang, Y., Yin, W., & Xu, H. E. (2021). RNA-dependent RNA polymerase: Structure, mechanism, and drug discovery for COVID-19. *Biochemical and Biophysical Research Communications*, *538*, 47–53. https://doi.org/10.1016/j.bbrc.2020.08.116

Jo, S., Kim, S., Kim, D. Y., Kim, M. S., & Shin, D. H. (2020). Flavonoids with inhibitory activity against SARS-CoV-2 3CLpro. *Journal of Enzyme Inhibition and Medicinal Chemistry*, *35*(1), 1539–1544.

Kaihatsu, K., Yamabe, M., & Ebara, Y. (2018). Antiviral mechanism of action of epigallocatechin-3-O-gallate and its fatty acid esters. *Molecules*, *23*(10), 2475. https://doi.org/10.3390/molecules23102475

Kaul, R., Paul, P., Kumar, S., Büsselberg, D., Dwivedi, V. D., & Chaari, A. (2021). Promising antiviral activities of natural flavonoids against SARS-CoV-2 targets: Systematic review. *International Journal of Molecular Sciences*, *22*(20), 11069.

Ke, J., Long, X., Liu, Y., Zhang, Y. F., Li, J., Fang, W., & Meng, Q. G. (2007). Role of NF-κB in TNF-α-induced COX-2 expression in synovial fibroblasts from human TMJ. *Journal of Dental Research*, *86*(4), 363–367. https://doi.org/10.1177/154405910708600412

Khaerunnisa, S., Kurniawan, H., Awaluddin, R., Suhartati, S., & Soetjipto, S. (2020). Potential inhibitor of COVID-19 main protease (Mpro) from several medicinal plant compounds by molecular docking study. *Preprints*, *2020*, 2020030226.

Khan, A., Heng, W., Wang, Y., Qiu, J., Wei, X., Peng, S., ... & Wei, D. Q. (2021). In silico and in vitro evaluation of kaempferol as a potential inhibitor of the SARS-CoV-2 main protease (3CLpro). *Phytotherapy Research*, *35*(6), 2841.

Khan, R. J., Jha, R. K., Amera, G. M., Jain, M., Singh, E., Pathak, A., Singh, R. P., Muthukumaran, J., & Singh, A. K. (2021). Targeting SARS-CoV-2: A systematic drug repurposing approach to identify promising inhibitors against 3C-like proteinase and 2′-O-ribose methyltransferase. *Journal of Biomolecular Structure and Dynamics*, *39*(8), 2679–2692. https://doi.org/10.1080/07391102.2020.1753577

Khazeei Tabari, M. A., Iranpanah, A., Bahramsoltani, R., & Rahimi, R. (2021). Flavonoids as promising antiviral agents against SARS-CoV-2 infection: A mechanistic review. *Molecules*, *26*(13), 3900. https://doi.org/10.3390/molecules26133900

Kim, H.-S., Quon, M. J., & Kim, J. (2014). New insights into the mechanisms of polyphenols beyond antioxidant properties; lessons from the green tea polyphenol, epigallocatechin 3-gallate. *Redox Biology*, *2*(1), 187–195. https://doi.org/10.1016/j.redox.2013.12.022

Kuba, K., Yamaguchi, T., & Penninger, J. M. (2021). Angiotensin-converting enzyme 2 (ACE2) in the pathogenesis of ARDS in COVID-19. *Frontiers in Immunology*, *12*, 732690. https://doi.org/10.3389/fimmu.2021.732690

Kuo, C.-L., Chen, T.-S., Liou, S.-Y., & Hsieh, C.-C. (2014). Immunomodulatory effects of EGCG fraction of green tea extract in innate and adaptive immunity via T regulatory cells in murine model. *Immunopharmacology and Immunotoxicology*, *36*(5), 364–370. https://doi.org/10.3109/08923973.2014.953637

Lammi, C., & Arnoldi, A. (2021). Food-derived antioxidants and COVID-19. *Journal of Food Biochemistry*, *45*(1), e13557. https://doi.org/10.1111/jfbc.13557

Li, T., Zhao, N., Lu, J., Zhu, Q., Liu, X., Hao, F., & Jiao, X. (2019). Epigallocatechin gallate (EGCG) suppresses epithelial-mesenchymal transition (EMT) and invasion in anaplastic thyroid carcinoma cells through blocking of TGF-β1/Smad signaling pathways. *Bioengineered*, *10*(1), 282–291. https://doi.org/10.1080/21655979.2019.1632669

Liang, Y., Liu, K. W. K., Yeung, S. C., Li, X., Ip, M. S. M., & Mak, J. C. W. (2017). (−)-epigallocatechin-3-gallate reduces cigarette smoke-induced airway neutrophilic inflammation and mucin hypersecretion in rats. *Frontiers in Pharmacology*, *8*, 618. https://doi.org/10.3389/fphar.2017.00618

Liu, C., Hao, K., Liu, Z., Liu, Z., & Guo, N. (2021). Epigallocatechin gallate (EGCG) attenuates staphylococcal alpha-hemolysin (Hla)-induced NLRP3 inflammasome activation via ROS-MAPK pathways and EGCG-Hla interactions. *International Immunopharmacology*, *100*, 108170. https://doi.org/10.1016/j.intimp.2021.108170

Liu, J., Bodnar, B. H., Meng, F., Khan, A. I., Wang, X., Saribas, S., Wang, T., Lohani, S. C., Wang, P., Wei, Z., Luo, J., Zhou, L., Wu, J., Luo, G., Li, Q., Hu, W., & Ho, W. (2021). Epigallocatechin gallate from green tea effectively blocks infection of SARS-CoV-2 and new variants by inhibiting spike binding to ACE2 receptor. *Cell & Bioscience*, *11*(1), 168. https://doi.org/10.1186/s13578-021-00680-8

Liu, S. Y., Wang, W., Ke, J. P., Zhang, P., Chu, G. X., & Bao, G. H. (2022). Discovery of Camellia sinensis catechins as SARS-CoV-2 3CL protease inhibitors through molecular docking, intra and extra cellular assays. *Phytomedicine*, *96*, 153853.

Liu, W., Dong, M., Bo, L., Li, C., Liu, Q., Li, Y., Ma, L., Xie, Y., Fu, E., Mu, D., Pan, L., Jin, F., & Li, Z. (2014). Epigallocatechin-3-gallate ameliorates seawater aspiration-induced acute lung injury via regulating inflammatory cytokines and inhibiting JAK/STAT1 pathway in rats. *Mediators of Inflammation*, 2014, 1–12. https://doi.org/10.1155/2014/612593

Lopez-Lazaro, M., & López-Lázaro, M. (2009). Distribution and biological activities of the flavonoid luteolin. *Mini Reviews in Medicinal Chemistry*, *9*(1), 31–59.

Lu, J.-W., Hsieh, P.-S., Lin, C.-C., Hu, M.-K., Huang, S.-M., Wang, Y.-M., Liang, C.-Y., Gong, Z., & Ho, Y.-J. (2017). Synergistic effects of combination treatment using EGCG and suramin against the chikungunya virus. *Biochemical and Biophysical Research Communications*, *491*(3), 595–602. https://doi.org/10.1016/j.bbrc.2017.07.157

Magazine, N., Zhang, T., Wu, Y., McGee, M. C., Veggiani, G., & Huang, W. (2022). Mutations and evolution of the SARS-CoV-2 spike protein. *Viruses*, *14*(3), 640. https://doi.org/10.3390/v14030640

Mahmood, R. A., Hasan, A., Rahmatullah, M., Paul, A. K., Jahan, R., Jannat, K., ... & Wilairatana, P. (2022). Solanaceae family phytochemicals as inhibitors of 3C-like protease of SARS-CoV-2: An in silico analysis. *Molecules*, *27*(15), 4739.

Mangiavacchi, F., Botwina, P., Menichetti, E., Bagnoli, L., Rosati, O., Marini, F., Fonseca, S. F., Abenante, L., Alves, D., Dabrowska, A., Kula-Pacurar, A., Ortega-Alarcon, D., Jimenez-Alesanco, A., Ceballos-Laita, L., Vega, S., Rizzuti, B., Abian, O., Lenardão, E. J., Velazquez-Campoy, A., Santi, C. (2021). Seleno-functionalization of quercetin improves the non-covalent inhibition of Mpro and its antiviral activity in cells against SARS-CoV-2. *International Journal of Molecular Sciences*, *22*(13), 7048. https://doi.org/10.3390/ijms22137048

Masand, V. H., Akasapu, S., Gandhi, A., Rastija, V., & Patil, M. K. (2020). Structure features of peptide-type SARS-CoV main protease inhibitors: Quantitative structure activity relationship study. *Chemometrics and Intelligent Laboratory Systems*, *206*, 104172.

Mehany, T., Khalifa, I., Barakat, H., Althwab, S. A., Alharbi, Y. M., & El-Sohaimy, S. (2021). Polyphenols as promising biologically active substances for preventing SARS-CoV-2: A review with research evidence and underlying mechanisms. *Food Bioscience*, *40*, 100891.

Menegazzi, M., Campagnari, R., Bertoldi, M., Crupi, R., Di Paola, R., & Cuzzocrea, S. (2020). Protective effect of epigallocatechin-3-gallate (EGCG) in diseases with uncontrolled immune activation: Could such a scenario be helpful to counteract COVID-19? *International Journal of Molecular Sciences*, *21*(14), 5171. https://doi.org/10.3390/ijms21145171

Min, S.-Y., Yan, M., Kim, S. B., Ravikumar, S., Kwon, S.-R., Vanarsa, K., Kim, H.-Y., Davis, L. S., & Mohan, C. (2015). Green Tea epigallocatechin-3-gallate suppresses autoimmune arthritis through indoleamine-2,3-dioxygenase expressing dendritic cells and the nuclear factor, erythroid 2-like 2 antioxidant pathway. *Journal of Inflammation*, *12*(1), 53. https://doi.org/10.1186/s12950-015-0097-9

Mirzaei, R., Attar, A., Papizadeh, S., Jeda, A. S., Hosseini-Fard, S. R., Jamasbi, E., & Karampoor, S. (2021). The emerging role of probiotics as a mitigation strategy against coronavirus disease 2019 (COVID-19). *Archives of Virology*, *166*, 1819–1840.

Mishra, S., Yang, X., & Singh, H. B. (2020). Evidence for positive response of soil bacterial community structure and functions to biosynthesized silver nanoparticles: An approach to conquer nanotoxicity? *Journal of Environmental Management*, *253*, 109584. https://doi.org/10.1016/j.jenvman.2019.109584

Mou, Q., Jiang, Y., Zhu, L., Zhu, Z., & Ren, T. (2020). EGCG induces β-defensin 3 against influenza A virus H1N1 by the MAPK signaling pathway. *Experimental and Therapeutic Medicine*, *20*, 3017–3024. https://doi.org/10.3892/etm.2020.9047

Mouffouk, C., Mouffouk, S., Mouffouk, S., Hambaba, L., & Haba, H. (2021). Flavonols as potential antiviral drugs targeting SARS-CoV-2 proteases (3CLpro and PLpro), spike protein, RNA-dependent RNA polymerase (RdRp) and angiotensin-converting enzyme II receptor (ACE2). *European Journal of Pharmacology*, *891*, 173759.

Munafò, F., Donati, E., Brindani, N., Ottonello, G., Armirotti, A., & De Vivo, M. (2022). Quercetin and luteolin are single-digit micromolar inhibitors of the SARS-CoV-2 RNA-dependent RNA polymerase. *Scientific Reports*, *12*(1), 10571. https://doi.org/10.1038/s41598-022-14664-2

Neill, L. A. J., Golenbock, D., & Bowie, A. G. (2013). The history of toll-like receptors - redefining innate immunity. *Nature Reviews Immunology*, *13*(6), 453–460. https://doi.org/10.1038/nri3446

Ohgitani, E., Shin-Ya, M., Ichitani, M., Kobayashi, M., Takihara, T., Kawamoto, M., Kinugasa, H., & Mazda, O. (2021). Significant inactivation of SARS-CoV-2 in vitro by a green tea catechin, a catechin-derivative, and black tea galloylated theaflavins. *Molecules*, *26*(12), 3572. https://doi.org/10.3390/molecules26123572

Ohishi, T., Goto, S., Monira, P., Isemura, M., & Nakamura, Y. (2016). Anti-inflammatory action of green tea. *Anti-Inflammatory & Anti-Allergy Agents in Medicinal Chemistry*, *15*(2), 74–90. https://doi.org/10.2174/1871523015666160915154443

Park, N. H., Bae, I.-H., Han, S., Kim, M., Lee, S. H., Park, W. S., Lee, C. S., & Hwang, J. S. (2019). Anti-inflammatory effect of green tea petiole extracts in poly(I:C)-stimulated human epidermal keratinocytes and a human 3D skin equivalent. *European Journal of Inflammation*, *17*, 205873921985757. https://doi.org/10.1177/2058739219857576

Peterson, L. (2020). COVID-19 and flavonoids: In silico molecular dynamics docking to the active catalytic site of SARS-CoV and SARS-CoV-2 main protease. Available at SSRN 3599426.

Pillon, M. C., Frazier, M. N., Dillard, L. B., Williams, J. G., Kocaman, S., Krahn, J. M., Perera, L., Hayne, C. K., Gordon, J., Stewart, Z. D., Sobhany, M., Deterding, L. J., Hsu, A. L., Dandey, V. P., Borgnia, M. J., & Stanley, R. E. (2021). Cryo-EM structures of the SARS-CoV-2 endoribonuclease Nsp15 reveal insight into nuclease specificity and dynamics. *Nature Communications*, *12*(1), 636. https://doi.org/10.1038/s41467-020-20608-z

Pitsillou, E., Liang, J., Ververis, K., Hung, A., & Karagiannis, T. C. (2021). Interaction of small molecules with the SARS-CoV-2 papain-like protease: In silico studies and in vitro validation of protease activity inhibition using an enzymatic inhibition assay. *Journal of Molecular Graphics and Modelling*, *104*, 107851.

Raekiansyah, M., Buerano, C. C., Luz, M. A. D., & Morita, K. (2018). Inhibitory effect of the green tea molecule EGCG against dengue virus infection. *Archives of Virology*, *163*(6), 1649–1655. https://doi.org/10.1007/s00705-018-3769-y

Rahayu, R. P., Prasetyo, R. A., Purwanto, D. A., Kresnoadi, U., Iskandar, R. P. D., & Rubianto, M. (2018). The immunomodulatory effect of green TeaTea (Camellia sinensis) leaves extract on immunocompromised Wistar rats infected by Candida albicans. *Veterinary World*, *11*(6), 765–770. https://doi.org/10.14202/vetworld.2018.765-770

Rahmah, L., Abarikwu, S. O., Arero, A. G., Essouma, M., Jibril, A. T., Fal, A., Flisiak, R., Makuku, R., Marquez, L., Mohamed, K., Ndow, L., Zarębska-Michaluk, D., Rezaei, N., & Rzymski, P. (2022). Oral antiviral treatments for COVID-19: Opportunities and challenges. *Pharmacological Reports*, *74*(6), 1255–1278. https://doi.org/10.1007/s43440-022-00388-7

Ravi, V., Saxena, S., & Panda, P. S. (2022). Basic virology of SARS-CoV 2. *Indian Journal of Medical Microbiology*, *40*(2), 182–186. https://doi.org/10.1016/j.ijmmb.2022.02.005

Rizzuti, B., Grande, F., Conforti, F., Jimenez-Alesanco, A., Ceballos-Laita, L., Ortega-Alarcon, D., & Velazquez-Campoy, A. (2021). Rutin is a low micromolar inhibitor of SARS-CoV-2 main protease 3CLpro: Implications for drug design of quercetin analogs. *Biomedicines*, *9*(4), 375.

Roh, E., Kim, J. E., Kwon, J. Y., Park, J. S., Bode, A. M., Dong, Z., & Lee, K. W. (2017). Molecular mechanisms of green tea polyphenols with protective effects against skin photoaging. *Critical Reviews in Food Science and Nutrition*, *57*(8), 1631–1637.

Sadati, S., Gheibi, N., Ranjbar, S., & Hashemzadeh, M. (2018). Docking study of flavonoid derivatives as potent inhibitors of influenza H1N1 virus neuraminidase. *Biomedical Reports*, *10*(1), 33–38. https://doi.org/10.3892/br.2018.1173

Saraswat, J., Singh, P., & Patel, R. (2021). A computational approach for the screening of potential antiviral compounds against SARS-CoV-2 protease: Ionic liquid vs herbal and natural compounds. *Journal of Molecular Liquids*, *326*, 115298.

Selvaraj, J., Rajan, L., Selvaraj, D., Palanisamy, D., Pk, K. N., & Mohankumar, S. K. (2021). Identification of (2 R, 3 R)-2-(3, 4-dihydroxyphenyl) chroman-3-yl-3, 4, 5–trihydroxy benzoate as multiple inhibitors of SARS-CoV-2 targets; a systematic molecular modelling approach. *RSC Advances*, *11*(22), 13051–13060.

Sharma, M., Mahto, J. K., Dhaka, P., Neetu, N., Tomar, S., & Kumar, P. (2022). MD simulation and MM/PBSA identifies phytochemicals as bifunctional inhibitors of SARS-CoV-2. *Journal of Biomolecular Structure and Dynamics*, *40*(22), 12048–12061.

Singh, A. K., Umar, S., Riegsecker, S., Chourasia, M., Ahmed, S., & Egcg, B. (2016). Regulation of TAK1 activation by epigallocatechin-3-gallate in RA synovial fibroblasts: Suppression of K63-linked autoubiquitination of TRAF6. *Arthritis Rheumatol*, *68*(2), 347–358.

Song, J. M., Lee, K. H., & Seong, B. L. (2005). Antiviral effect of catechins in green TeaTea on influenza virus. *Antiviral Research*, *68*(2), 66–74. https://doi.org/10.1016/j.antiviral.2005.06.010

Soto, M. E., Guarner-Lans, V., Soria-Castro, E., Manzano Pech, L., & Pérez-Torres, I. (2020). Is antioxidant therapy a useful complementary measure for Covid-19 treatment? An algorithm for its application. *Medicina*, *56*(8), 386. https://doi.org/10.3390/medicina56080386

Syed, D. N., Afaq, F., Kweon, M.-H., Hadi, N., Bhatia, N., Spiegelman, V. S., & Mukhtar, H. (2007). Green tea polyphenol EGCG suppresses cigarette smoke condensate-induced NF-κB activation in normal human bronchial epithelial cells. *Oncogene*, *26*(5), 673–682. https://doi.org/10.1038/sj.onc.1209829

Takabayashi, F., Harada, N., Yamada, M., Murohisa, B., & Oguni, I. (2004). Inhibitory effect of green tea catechins in combination with sucralfate on Helicobacter pylori infection in Mongolian gerbils. *Journal of Gastroenterology*, *39*(1), 61–63. https://doi.org/10.1007/s00535-003-1246-0

Tallei, T. E., Fatimawali, N. N. J., Idroes, R., Zidan, B. M. R. M., Mitra, S., Celik, I., Nainu, F., Ağagündüz, D., Emran, T. B., & Capasso, R. (2021). A comprehensive review of the potential use of green tea polyphenols in the management of COVID-19. *Evidence-Based Complementary and Alternative Medicine, 2021*, 1–13. https://doi.org/10.1155/2021/7170736

Tallei, T. E., Tumilaar, S. G., Niode, N. J., Kepel, B. J., Idroes, R., Effendi, Y., ... & Emran, T. B. (2020). Potential of plant bioactive compounds as SARS-CoV-2 main protease (M pro) and spike (S) glycoprotein inhibitors: A molecular docking study. *Scientifica*, *2020*, 6307457.

Toigo, L., dos Santos Teodoro, E. I., Guidi, A. C., Gancedo, N. C., Petruco, M. V., Melo, E. B., & Sanches, A. C. C. (2023). Flavonoid as possible therapeutic targets against COVID-19: A scoping review of in silico studies. *DARU Journal of Pharmaceutical Sciences*, *31*, 1–18.

Ungarala, R., Munikumar, M., Sinha, S. N., Kumar, D., Sunder, R. S., & Challa, S. (2022). Assessment of antioxidant, immunomodulatory activity of oxidised epigallocatechin-3-Gallate (green tea polyphenol) and its action on the main protease of SARS-CoV-2-an in vitro and in silico approach. *Antioxidants*, *11*(2), 294.

Wang, L., Lee, W., Cui, Y. R., Ahn, G., & Jeon, Y.-J. (2019). Protective effect of green tea catechin against urban fine dust particle-induced skin aging by regulation of NF-κB, AP-1, and MAPKs signaling pathways. *Environmental Pollution*, *252*, 1318–1324. https://doi.org/10.1016/j.envpol.2019.06.029

Wu, D., Wu, T., Liu, Q., & Yang, Z. (2020). The SARS-CoV-2 outbreak: What we know. *International Journal of Infectious Diseases*, *94*, 44–48. https://doi.org/10.1016/j.ijid.2020.03.004

Xiao, Z., Xu, H., Qu, Z. Y., Ma, X. Y., Huang, B. X., Sun, M. S., Wang, B.Q, & Wang, G. Y. (2023). Active ingredients of reduning injection maintain high potency against SARS-CoV-2 variants. *Chinese Journal of Integrative Medicine*, *29*(3), 205–212. https://doi.org/10.1007/s11655-022-3686-5

Xu, D., Peng, S., Guo, R., Yao, L., Mo, H., Li, H., Song, H., & Hu, L. (2021). EGCG alleviates oxidative stress and inhibits aflatoxin B1 biosynthesis via MAPK signaling pathway. *Toxins*, *13*(10), 693. https://doi.org/10.3390/toxins13100693

Xu, M., Liu, B., Wang, C., Wang, G., Tian, Y., Wang, S., Li, J., Li, P., Zhang, R., Wei, D., Tian, S., & Xu, T. (2017). Epigallocatechin-3-gallate inhibits TLR4 signaling through the 67-kDa laminin receptor and effectively alleviates acute lung injury induced by H9N2 swine influenza virus. *International Immunopharmacology*, *52*, 24–33. https://doi.org/10.1016/j.intimp.2017.08.023

Yamaguchi, K., Honda, M., Ikigai, H., Hara, Y., and Shimamura, T. (2002). Inhibitory effects of (−)-epigallocatechin gallate on the life cycle of human immunodeficiency virus type 1 (HIV-1). *Antiviral Research*, *53*(1), 19–34. https://doi.org/10.1016/S0166-3542(01)00189-9

Yao, Z. H., Yao, X. l., Zhang, Y., Zhang, S. F., & Hu, J. C. (2018). Luteolin could improve cognitive dysfunction by inhibiting neuroinflammation. *Neurochemical Research*, *43*(4), 840–849. https://doi.org/10.1007/s11064-018-2482-2

Zalpoor, H., Bakhtiyari, M., Shapourian, H., Rostampour, P., Tavakol, C., & Nabi-Afjadi, M. (2022). Hesperetin as an anti-SARS-CoV-2 agent can inhibit COVID-19-associated cancer progression by suppressing intracellular signaling pathways. *Inflammopharmacology*, *30*(5), 1533–1539. https://doi.org/10.1007/s10787-022-01054-3

Zhan, Y., Ta, W., Tang, W., Hua, R., Wang, J., Wang, C., & Lu, W. (2021). Potential antiviral activity of isorhamnetin against SARS-COV -2 spike pseudotyped virus in vitro. *Drug Development Research*, *82*(8), 1124–1130. https://doi.org/10.1002/ddr.21815

Zhong, X., Liu, M., Yao, W., Du, K., He, M., Jin, X., Jiao, L., Ma, G., Wei, B., & Wei, M. (2019). Epigallocatechin-3-gallate attenuates microglial inflammation and neurotoxicity by suppressing the activation of canonical and noncanonical inflammasome via TLR4/NF-κB pathway. *Molecular Nutrition & Food Research*, *63*(21), 1801230. https://doi.org/10.1002/mnfr.201801230

Zhu, W., Li, M. C., Wang, F. R., Mackenzie, G. G., & Oteiza, P. I. (2020). The inhibitory effect of ECG and EGCG dimeric procyanidins on colorectal cancer cell growth is associated with their actions at lipid rafts and the inhibition of the epidermal growth factor receptor signaling. *Biochemical Pharmacology*, *175*, 113923. https://doi.org/10.1016/j.bcp.2020.113923

Zwolak, I. (2021). Epigallocatechin gallate for management of heavy metal-induced oxidative stress: Mechanisms of action, efficacy, and concerns. *International Journal of Molecular Sciences*, *22*(8), 4027. https://doi.org/10.3390/ijms22084027

17 Green Tea Catechins against SARS-CoV-2

Niusha Esmaealzadeh, Roodabeh Bahramsoltani, and Roja Rahimi

17.1 INTRODUCTION

Green tea is the unfermented leaves of *Camellia sinensis* (L.) Kuntze accounts for more than 20% of tea consumption globally [1]. Green tea contains a considerable amount of polyphenols and is often consumed as an antioxidant infusion and energy booster. The history of green tea export dates back to the 17th century when the first green tea shipping from India to Japan was reported [2]. However, its consumption was highly recorded in ancient manuscripts of China and was the beverage of preference for Chinese emperors [3].

Since then, tons of green tea leaves have been processed and sent to different parts of the world, with China playing a part as a prominent producer in the Far East [4]. The unfermented leaves, however, are hugely imported by certain regions such as East Asia, some European countries, e.g., England, northern countries of Africa, and the United States [5]. To maintain its green color, the fresh leaves are steamed to prevent decomposition. This will help with preserving the content of therapeutic polyphenols. The leaves would be rolled and dried to be packed in the further steps [6]. Based on the undertaken process, as well as the cultivation conditions and the origin of the plants, green tea can be classified into different types [7]. The most common type of Japanese green tea, for example, is called Sencha which will be classified as Matcha, Bancha, and Gyokuro, with Matcha having the highest content of caffeine and L-theanine, and Bancha having the lowest [8]. Japanese green teas are not limited to these three kinds, though [9]. Similar to Japanese green teas, Chinese green teas are also of various kinds, Lung Ching, Mao Feng, and Chun Mee, to name a few. It should be noted that the concentration of caffeine and L-theanine is crucial for anticipating the therapeutic properties of the tea products. L-theanine can balance the stimulatory effects of caffeine and helps with maintaining the nervous and cardiovascular systems [10].

Tea leaves can be prepared by brewing or infusing; many factors determine the therapeutic effects and organoleptic features of the obtained beverage, though. Brewing the leaves, especially milled ones, for 5–10 minutes at 80°C–90°C, releases a considerable content of tannins which gives the beverage a rather dark color and bitter astringent taste with a low antioxidant capacity [11]. Higher-quality unfermented leaves are usually infused in warm water and produce a light-colored beverage with high antioxidant capacity [12].

Green tea is made up of different groups of phytonutrients and phytochemicals, namely, proteins, amino acids, fatty acids, as well as phenolic acids, polyphenols, and trace elements. It also contains special flavanols called catechins, which are intermediate phytochemicals in the synthesis of herbal polymers and make up almost 90% of the total content of the flavonoids in green tea [13]. The percentage of catechins is higher in unfermented green tea, compared to black tea. During the fermentation, the green tea leaves turn into Oolong and then black tea. During this process, catechins undergo conversion to become specific phytochemicals termed theaflavins, which alters their therapeutic properties [9].

Having phenolic structure, catechins exhibit significant free-radical scavenging and metal-ion chelating properties, which makes them a considerable candidate for being used as an adjuvant treatment for relieving inflammatory diseases, preventing tissue degeneration, restricting carcinogenesis and tumorigenesis, and maintaining normal functions of the cardiovascular system. Besides, they are able to induce antioxidant enzymes such as glutathione peroxidase (GPx) and superoxide dismutase (SOD). Catechins have also demonstrated notable antibacterial and anti-allergic properties [14].

17.2 CHARACTERISTICS OF GREEN TEA CATECHINS

Catechins include five main substances, i.e., catechin, epicatechin (EC), epicatechin-3-gallate (ECG), epigallocatechin (EGC), and epigallocatechin-3-gallate (EGCG) (Figure 17.1). The stability and effectiveness of catechins are mostly determined by the presence of galloyl moiety at the C-ring position and the suitable pH for catechins is 4–6 [15]. Catechins' antioxidant activity, and thus their therapeutic activity, is highly dependent on this factor, as well as the place and number of hydroxyl groups and their attached substitutions. The presence of the catechol group, however, lowers the radical scavenging activity.

In this regard, EGCG and EC have been recorded to be the most and the least effective ones, respectively. EGCG is among the most important and well-studied catechins, as it has shown many therapeutic properties. After drinking a cup of green tea, an average amount of 250 mg EGCG would be taken [16]. The structure of the EGCG is characterized by a benzenediol ring attached to a tetrahydropyran group, which is accompanied by a galloyl and a pyrogallol

DOI: 10.1201/9781003433200-17

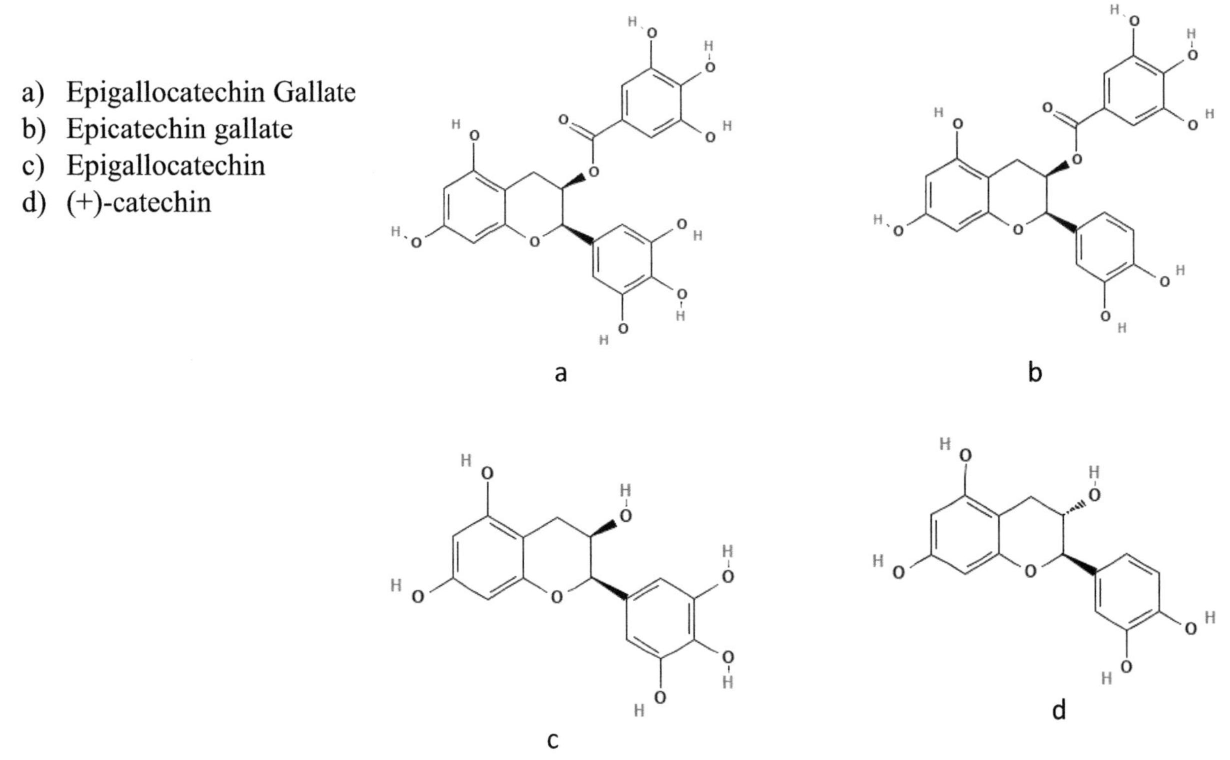

a) Epigallocatechin Gallate
b) Epicatechin gallate
c) Epigallocatechin
d) (+)-catechin

FIGURE 17.1 Molecular structures of green tea catechins.

component and 8 phenol groups [17]. This specific structure allows electron dislocation and amplifies antioxidant activity, leading to effective attachments to different enzymes and regulation of protein expression [18].

During the COVID-19 pandemic, catechins, especially EGCG, gained a lot of attention, as they had previously shown significant antiviral activities against various RNA viruses such as influenza H1N1, hepatitis C virus (HCV), human immunodeficiency virus (HIV), and different types of coronaviruses [19,20]. Therefore, scientists started to perform diverse studies to determine the possible mechanisms by which EGCG and other catechins could attack SARS-CoV-2 and limit its reproduction [21]. Common drug targets are SARS-COV-2's 3CL-protease ($3CL^{pro}$), RNA-dependent RNA polymerase (RdRp), non-structural proteins endoribonuclease (nsp)-16, nsp15, nsp12, Papain-like protease (PL^{pro}), and angiotensin-converting enzyme 2 (ACE2), which will be discussed in detail.

In this regard, this chapter reviews the current *in vitro*, *in vivo*, and *in silico* evidence for examining the effects of green tea catechins on SARS-COV-2 with a focus on their underlying pharmacological actions.

17.3 *IN VIVO* AND *IN VITRO* STUDIES

In an *in vitro* study, EGCG was studied for its inhibitory effects against $3CL^{pro}$. This protein is one of the replication proteins of the SARS-COV-2 virus and is an important drug target. Jang et al. used fluorescence resonance energy transfer (FRET) protease assay to show the inhibitory effect of EGCG on $3CL^{pro}$. One hour before the addition of the substrate to $3CL^{pro}$, EGCG was also added. After 5 hours, the activity of the enzyme was measured at 528 nm. The difference between the activities of EGCG-free incubation and EGCG-mixed one showed the inhibitory effects of EGCG against $3CL^{pro}$, dose-dependently. In this cellular study, the half-maximal inhibitory concentration (IC_{50}) for EGCG was tested using HEK293T cells and found to be 16.5 μM. The raised controversy was that the maximum concentration of EGCG in the blood circulation would not exceed 1 μM; EGCG would not last long in the blood and eventually oxidize to EGCG auto-oxidation products (EAOPs). Despite this fact, EAOPs preserve their inhibitory activity, and together with the remained EGCG, limit the activities of $3CL^{pro}$ [22]. Jang et al. also suggest that $3CL^{pro}$ is an essential protein for SARS-COV-2 proteolytic cleavage; so, blockage of this protein could significantly limit the virus and increase host cell viability. They chose less invasive SARS-COV-2, alpha, and beta coronavirus, and observed that EGCG restricted the viral protein and RNA production, thereby preventing viral replication in RD cells [23]. The research on $3CL^{pro}$ or SARS-COV-2 main protease (M^{pro}) continued with another study conducted by Kato et al. in which 37 phytonutrients were incubated with recombinant enzymes to evaluate their inhibitory effect. The product was measured with chromatographic separation. In this study, the importance of quinone moiety for the inhibitory effect of EGCG was highlighted [24].

In a study by Henss et al., different mechanisms by which EGCG inhibits SARS-COV-2 were examined and the results were compared with EC. First, EGCG was observed to limit entry of the spike protein pseudo-typed vectors to the HEK293T-ACE2 cells more efficiently than EC. The IC_{50} for EGCG was 2.47 vs. 20 μg/mL of EC. Second, EGCG could inhibit the infection of Vero cells in the virus replication test by 50% at the concentration of 1.73 μg/mL. Finally, this experiment showed that the EGCG is the most efficient one when added before infection. Although this result was not statistically significant, it could suggest that EGCG is the most effective one when it is added to the culture at the onset of the viral infection [25]. These results were supported by the results of another study by Liu et al. who observed new variants of SARS-COV-2 had a mutation in their spike protein (S^{pro}) and are therefore resistant to the early SARS-COV-2 vaccines. In this research, green tea catechins, especially EGCG, showed an inhibitory effect against SARS-COV-2. EGCG was even more effective when administered before viral infection, as it connects to the receptor-binding domain (RBD) of S^{pro} and prevents it from attaching to the ACE2 receptor of the HEK293T-hACE2 cells. Pre-treatment with EGCG at 100 μM also restricted nucleocapsid (N) protein and the produced plaques after viral introduction [26]. S^{pro} consists of two units, $S1^{pro}$, which connects SARS-COV-2 to the ACE2 receptors, and $S2^{pro}$, which plays a part in fusing the viral envelope to the membrane of the targeted cell. In a study by Takeda et al., a phenol-rich concentrated extract of green tea with a high content of catechins and theaflavins was prepared. Although no purification was undertaken, this extract only affected $S2^{pro}$ unit, not $S1^{pro}$. It also denatured the viral genome and restricted further infection of the host cells, VeroE6/TMPRSS2, in a concentration-dependent and time-dependent manner [27].

Hurst et al. compared the SARS-COV-2 inhibitory effect of EGCG in Vero 76 cells, epithelial cells of monkeys, and Caco-2 cells (colorectal adenocarcinoma cell line of humans). In the cytopathic effect assay, which was used for Vero 76, EGCG could suppress SARS-COV-2 by 50% in a low concentration of 0.59 μM. The cytotoxicity appeared in protracted exposure with EGCG at the concentration of 5.03 μM. However, this concentration was much higher with Caco-2 cells, >217 μM, which was aligned with the earlier published studies [28].

PL^{pro} of the SARS-COV-2 is another target for the antiviral drug substances. Montone et al. provided different fractions of green tea to understand which one is most active against PL^{pro}. The result showed that the fraction mostly containing ECG is the most effective one in terms of inhibiting PL^{pro}. Interestingly, the inhibitory effect of this fraction was ten times higher than pure ECG, showing the synergistic effects of other compounds in this fraction with ECG [29].

The catechin types and their amount are closely connected to the time of tea leaf processing. Ohgitani et al. tested freeze-dried extracts of different types of tea, i.e., green, black, and oolong tea, and their main constituent for their antiviral properties. In the antiviral assay in VeroE6/TMPRSS2 cells, EGCG (1 mM) could notably restrict RNA replication and ACE2–RBD interaction. The virus production and infectivity were therefore suppressed; however, theasinensin A and theaflavin 3,3′-di-O-gallate suppressed the SCV virus in much lower concentrations (40 μM and 60 μM). Hopefully, these concentrations are almost the same as the content in tea beverages [30].

Another target for anti-SARS-COV-2 agents is the RNA assembly stage in the replication process. As SARS-COV-2 bears the largest genome in RNA viruses, the virus needs to form protein complexes with the correct spatial structure to keep normal function. The nucleocapsid protein, N protein, of the SARS-COV-2 packages the viral genome and has a specific stage called liquid–liquid phase separation (LLPS). This phase is important as it forms finally the RNA–protein complex. It is demonstrated that gallocatechin gallate (GCG) can interfere with this stage at 10 μM and reduce infection of the H1299 cells [31].

LeBlanc et al. performed more profound research to discover the underlying mechanism by which EGCG suppresses coronavirus attachment to the cell surface. An essential factor in this process is Heparan sulfate proteoglycans (HSP). Heparin can attach to this glycan and interfere with the cellular attachment phase of various viruses with different spike protein targets. This is also observed in most variants of coronavirus. Similar to heparin, EGCG also connects to the HSP and reduces further cell fusion. LeBlanc's study showed that this mechanism is responsible for the inhibitory effect of EGCG against all types of coronaviruses [32].

In 2021, a Korean research group published the results of their animal studies for examining the effects of green tea catechins and EGCG on a safer version of beta-SARS-COV-2, which was HCoV-OC43. The virus was cultured in human HCT8 cells and administered intranasally to infect C57BL male mice. As a treatment, 10 mg/kg of EGCG or 30 mg/kg of Polyphenon 60, containing 60% green catechins, were orally administered in a 2-week study. Both EGCG and Polyphenon 60 restricted viral replication, protein expression, and surface projection. After dose conversions, the estimated dose for human consumption of EGCG was found to be 600 mg/kg. Although EGCG could ameliorate the health status of the mice, it should be taken into account that the effects of HCoV infection on mice were mild and did not cause weight loss or death. Thus, the effect of EGCG in more severe infections needs to be further evaluated [33].

17.4 *IN SILICO* STUDIES

During the COVID-19 pandemic, the attention of scientists redirected to reprofiling the existing drug agents, especially phytochemical agents to examine whether their spatial structures are able to efficiently interact with binding sites of SARS-COV-2, and therefore inhibit them. *In silico* studies are preferable for the primary investigations of possible drug candidates, as they are a low-budget safer way of studying new agents compared to other ones. With massive online data regarding proteins and their structure, *in silico* studies are now favorable in preliminary examinations. The information regarding plant phytochemicals is also widely available [34].

Green tea polyphenols have been screened for their ligand structures and their behavior toward SARS-COV-2 target proteins in various studies. In a study by Mhatre et al., 3CL^{pro}, PL^{pro}, RdRp, and RBD were selected as viral targets to be virtually interacted with green tea polyphenols. Promising docking scores were observed in the virtual interaction of EGCG with the viral targets via the formation of hydrogen bonds (H-bond), van der Waals, and protein residue interactions. A high docking score was observed in the interaction of EGCG and RBD via the formation of five H-bonds with Tyrosine-91 amino acid residue of the L chain, Pi-cation bonds with Lysine-417 and Arginine-403 of the E chain, and Pi-Sigma bonds with valine-50 amino acid residue of RBD [35]. Also, in the docking analysis of RdRp, EGCG interestingly showed a better docking score than remdesivir and favipiravir [36]. In another molecular docking study regarding the interaction of several phytochemicals against S^{pro} and M^{pro}, it was shown that EGCG is a better S^{pro} inhibitor, compared to nelfinavir and hydroxychloroquine. Also, EGCG showed a relatively stable position in interaction with M^{pro} and was introduced as a suitable lead compound for further antiviral analysis against SARS-COV-2 [37].

Another docking was further carried out to deeply understand which 3CL^{pro} subunits interact with EGCG more efficiently. From the four subunits of 3CL^{pro} (S1, S1¢, S2, and S4), EGCG engages with the S1, S1¢, and S2 subsites and creates H-bond with Histidine-41 and Cysteine-145 residues [38]. These results were further supported by Zhu et al. [21,39], investigating eight polyphenols, of which three were selected as the strong interactors with Histidine-41 and Cysteine-145. EGCG, ECG, and GCG exhibited significantly stable complexes with durable conformation and compactness in simulation analysis. The authors suggest that in phytochemicals with a flavan-3-ol structure, including catechins, a higher number of hydroxyl functional groups on the B-ring increases the affinity to M^{pro}. Also, a comparison of the results obtained on catechin, EC, GC, and EGC, as well as EC dimer, showed that oligomerization and galloylation on the C3-OH of the catechins increase their tendency to interact with M^{pro}. The M^{pro} inhibitory activity of ECG, GCG, and EGCG was also confirmed in the *in vitro* assay; however, catechin, EC, GC, and EGC did not show any inhibitory activity in this assay [40]. In another docking analysis reported by Gogoi et al., ECG was evaluated regarding its binding affinity with M^{pro} and PL^{pro} which showed a higher affinity for M^{pro}, in comparison with PL^{pro}. Also, the conformational stability of SARS-CoV-2 M^{pro}-ECG was almost similar to the complex formed in interaction with the control agent N3, while ECG formed a higher number of H-bonds with M^{pro} and was better fixed inside the pocket [41]. Further studies estimated the bonding affinity of EGCG for matrix metalloproteinase (MMP) of M^{pro} which also affirmed the H-bond formation and acceptable drug-likeness of EGCG [42].

Nsp15 is a specific target of the coronavirus family which plays multiple roles in the replication and assembly of SARS-COV-2; it also helps with host cell evasion. Therefore, targeting nsp15 is a novel strategy for drug discovery. In a study by Khan et al. on the interaction of 17 phytochemicals with SARS-COV-2 targets, EGCG showed the highest molecular interaction with nsp15 endoribonuclease and M^{pro} of the virus, as well as human ACE. The overall interaction of EGCG with the viral proteins was approximately two times higher than the standard antiviral drugs, i.e., remdesivir and nafamostat [43]. Additionally, a recently discovered nsp10–nsp16 system exhibited a critical role in staying under the radar while facing the immune system of the host. This system changes the genetic package of the virus so that it becomes undetectable, as it mirrors the RNA of the host cells, and provides enough time for virus mass replication. Targeting this system can help the immune system detect and restrict further infection. Catechin was docked to see whether it could interrupt the nsp10–nsp16 complex. The energy released from the interaction of catechin, and this complex was promising as H-bonds were formed during the process with the involvement of Glutamine-127, Glutamic acid-288, Arginine-4, and Lysine-5 residues [44].

RdRp or nsp12 is the main effective site for nucleotide-analog antivirals such as remdesivir, favipiravir, and sofosbuvir, and thus, is another target considered in screening natural antivirals. Bhardwaj et al. docked three green tea catechins, epigallocatechin-3,4-di-O-gallate, epicatechin-3,5-di-O-gallate, and epigallocatechin-3,5-di-O-gallate, for understanding their interactions with RdRp–RNA complex. All three catechins could interact more efficiently than remdesivir, with the catalytic part of RdRp, and form H-bonds rather than temporary van der Waals bonds, which was formed in the case of remdesivir [45]. In another study, EGC virtually interacted with nsp12 and its cofactors, nsp7 and nsp8, which showed acceptable results regarding hydrophobic interaction, S-score, and drug-likeness. Also, the most engaging residues in RdRp were found to be Glutamic acid-665, Phenylalanine-326, and Serine-664 [46].

Furthermore, the interruption of RBD–ACE2 receptor interaction by EGCG was examined *in silico*. The promising docking score was –7.8 kcal/mol with a hydrogen binding length of 2.95 Å between EGCG and R393 residue strengthened the hypothesis that EGCG binding to ACE2 receptors blocks further connections of RBD [47]. However, EGCG can directly bind to RBD efficiently, with free energy of –16.7 kcal/mol for site 1 and –11.8 kcal/mol for site 2 [48]. Maiti et al. showed that EGCG binds to sites 1, 2, and 3 of S^{pro}, while a standard drug such as hydroxychloroquine only binds to site 3. The affinity for the ACE2 receptor was calculated as an ACE value which was stronger for EGCG (–265.13) in comparison to hydroxychloroquine (–154.06) [49].

A more comprehensive approach is running multitarget studies for a single phytochemical; as Mishra et al. evaluated the interaction between catechin and all the above-mentioned targets, S^{pro}, ACE2, $3CL^{pro}$, cathepsin L (CTSL), nucleocapsid protein, RdRp, and nsp_s, in the same computational study. Catechin formed an effective bind with $3CL^{pro}$, CTSL (–5.09 kcal/mol), S^{pro}, and nsp6 (–26.09 kcal/mol) which were stabilized mostly with hydrophobic interactions [50]. This approach helps with tackling fast-developing mutation on SARS-COV-2, resulting in resistant variants. Gupta et al. also examined the effect of some natural spices against 17 proteins of SARS-COV-2 from which EC demonstrated a better inhibitory effect against multiple targets such as PL^{pro}, $3CL^{pro}$, and RdRp, compared with remdesivir, ribavirin, and hydroxychloroquine [51].

Catechin was also investigated for radical scavenging properties *in silico*. The affinity of catechin for NADPH oxidase was evaluated and GKT136901 was the control molecule. Catechin was also docked for its affinity for M^{pro} and the obtained results were compared with remdesivir. The released energy for the formation of NADPH oxidase–catechin system and M^{pro}–catechin was –8.30 and –7.89 kcal/mol, respectively, whereas the free energy was –8.72 and –7.50 kcal/mol for NADPH oxidase–GKT136901 and M^{pro}–remdesivir, respectively [52].

17.5 DISCUSSION

Green tea catechins are among the top candidates for preliminary research on viral infections. Each catechin bears specific pharmacokinetics which is different from the others. For example, EGC is quick to rise in the blood, unlike EGCG which is the last to elevate but remains in the blood for about 8 hours.

EGCG is among the best-docked plant-derived substances against SARS-COV-2, *in silico*, and has demonstrated antiviral properties against multiple RNA viruses such as hepatitis C, influenza, *Herpes simplex*, and SARS in various studies. Therefore, it has also been selected as the phytochemical of choice for many preclinical studies to find effective catechins that can ameliorate COVID-19 symptoms. Research in this field is mostly limited to preclinical studies such as *in vitro*, *in silico*, and in rare cases, animal studies. This can be due to a lack of information in the animal phase or a preference of the healthcare system to focus more on fast-acting chemical drugs rather than herbal supplements. This is also noteworthy to mention some serious side effects of EGCG supplements such as liver and kidney failure [53,54]. Liver and kidney damage is also frequently recorded in patients with COVID-19 virus.

Although the mentioned issues made EGCG rather unpreferable for the clinical phase, it is still worth further examination, especially for managing long-term COVID-19 symptoms. EGCG, with its phenol-rich structure, has a strong radical-scavenging property which makes it a considerable anti-inflammatory and immunomodulatory agent. The available supplements in the market suggest daily doses of 400 and 200 mg for whole green tea catechin and pure EGCG intake, respectively. These dosages need to be adjusted according to the patient's physiological condition, weight, and drug regimen. The source of extracted catechins is also important as it may contain toxic elements such as lead, cadmium, and aluminum [55].

Another controversy about catechins is their theoretically low stability. According to Lipinski's rule of five, catechins with a molecular weight of 300–450 kDa and more than five groups of hydroxyls in their main structure bear poor gastrointestinal bioavailability. Preclinical studies suggest that the concentration of catechins can drop down to 50 times while taken orally, thereby dropping it out of the therapeutic range. It is suggested that catechins are mostly absorbed by passive diffusion. They also undergo various phases I and II metabolization, in which they convert to numerous biologically active phenolic acids that are easily absorbable [56]. Future studies may focus on the biological activity of catechin metabolites to clarify their antiviral potential. Also, novel pharmaceutical formulations of catechins are needed in order to improve the oral bioavailability of these compounds.

Catechins have shown promising effects on viral inflammatory diseases such as COVID-19; however, similar to any other phytochemicals, more preclinical and clinical studies are needed to determine the exact therapeutic dose range, possible adverse effects, drug–herb interactions, and the impact of long-term catechin consumption on vital organs.

REFERENCES

1. Chacko, S.M., et al., Beneficial effects of green tea: a literature review. *Chinese Medicine*, 2010. **5**(1): 13.
2. Shrivastava, R.R.S.R.R., P.P.P. Pateriya, and M.S.M. Singh, Green tea-a short review. *International Journal of Indigenous Herbs and Drugs*, 2018. **30**: 12–21.
3. Benn, J.A., *Tea in China: A religious and cultural history*. 2015: University of Hawaii Press, Honolulu, Hawaii.
4. Ma, L.-l., et al., A comparative analysis of the volatile components of green tea produced from various tea cultivars in China. *Turkish Journal of Agriculture and Forestry*, 2019. **43**(5): 451–463.
5. Hayat, K., et al., Tea and its consumption: benefits and risks. *Critical Reviews in Food Science and Nutrition*, 2015. **55**(7): 939–954.
6. Lin, Y.-S., et al., Factors affecting the levels of tea polyphenols and caffeine in tea leaves. *Journal of Agricultural and Food Chemistry*, 2003. **51**(7): 1864–1873.
7. Cardoso, R.R., et al., Kombuchas from green and black teas have different phenolic profile, which impacts their antioxidant capacities, antibacterial and antiproliferative activities. *Food Research International*, 2020. **128**: 108782.
8. Rusak, G., I. Šola, and V.V. Bok, Matcha and Sencha green tea extracts with regard to their phenolics pattern and antioxidant and antidiabetic activity during in vitro digestion. *Journal of Food Science and Technology*, 2021. **58**: 3568–3578.

9. Musial, C., A. Kuban-Jankowska, and M. Gorska-Ponikowska, Beneficial properties of green tea catechins. *International Journal of Molecular Sciences*, 2020. **21**(5): 1744.
10. Yanagimoto, K., et al., Antioxidative activities of volatile extracts from green tea, oolong tea, and black tea. *Journal of Agricultural and Food Chemistry*, 2003. **51**(25): 7396–7401.
11. Pastoriza, S., S. Pérez-Burillo, and J.Á. Rufián-Henares, How brewing parameters affect the healthy profile of tea. *Current Opinion in Food Science*, 2017. **14**: 7–12.
12. Pérez-Burillo, S., et al., Effect of brewing time and temperature on antioxidant capacity and phenols of white tea: relationship with sensory properties. *Food Chemistry*, 2018. **248**: 111–118.
13. Babu, P.V. and D. Liu, Green tea catechins and cardiovascular health: an update. *Current Medicinal Chemistry*, 2008. **15**(18): 1840–1850.
14. Isemura, M., Catechin in human health and disease. *Molecules*, 2019. **24**(3): 528.
15. Graham, H.N., Green tea composition, consumption, and polyphenol chemistry. *Preventive Medicine*, 1992. **21**(3): 334–350.
16. Chu, K.O. and C.C. Pang, Pharmacokinetics and disposition of green tea catechins. In *Pharmacokinetics and Adverse Effects of Drugs: Mechanisms and Risks Factors*, edited by Ntambwe Malangu, 2018. Vol. 17. IntechOpen, London, UK
17. Balentine, D.A., S.A. Wiseman, and L.C.M. Bouwens, The chemistry of tea flavonoids. *Critical Reviews in Food Science and Nutrition*, 1997. **37**(8): 693–704.
18. Alam, M., et al., Epigallocatechin 3-gallate: from green tea to cancer therapeutics. *Food Chemistry*, 2022. **379**: 132135.
19. Sodagari, H.R., et al., Tea polyphenols as natural products for potential future management of HIV infection-an overview. *Journal of Natural Remedies*, 2016. **16**(2): 60–72.
20. Bahramsoltani, R., et al., The preventive and therapeutic potential of natural polyphenols on influenza. *Expert Review of Anti-Infective Therapy*, 2016. **14**(1): 57–80.
21. Diniz, L.R.L., et al., Catechins: therapeutic perspectives in COVID-19-associated acute kidney injury. *Molecules*, 2021. **26**(19): 5951.
22. Jang, M., et al., Tea polyphenols EGCG and theaflavin inhibit the activity of SARS-CoV-2 3CL-protease in vitro. *Evidence-Based Complementary and Alternative Medicine*, 2020. **2020**: 5630838.
23. Jang, M., et al., EGCG, a green tea polyphenol, inhibits human coronavirus replication in vitro. *Biochemical and Biophysical Research Communications*, 2021. **547**: 23–28.
24. Kato, Y., et al., Food phytochemicals, epigallocatechin gallate and myricetin, covalently bind to the active site of the coronavirus main protease in vitro. *Advances in Redox Research*, 2021. **3**: 100021.
25. Henss, L., et al., The green tea catechin epigallocatechin gallate inhibits SARS-CoV-2 infection. *Journal of General Virology*, 2021. **102**(4): 001574.
26. Liu, J., et al., Epigallocatechin gallate from green tea effectively blocks infection of SARS-CoV-2 and new variants by inhibiting spike binding to ACE2 receptor. *Cell & Bioscience*, 2021. **11**(1): 168.
27. Takeda, Y., et al., Severe acute respiratory syndrome coronavirus-2 inactivation activity of the polyphenol-rich tea leaf extract with concentrated theaflavins and other virucidal catechins. *Molecules*, 2021. **26**(16): 4803.
28. Hurst, B.L., D. Dickinson, and S. Hsu, Epigallocatechin-3-gallate (EGCG) inhibits SARS-CoV-2 infection in primate epithelial cells: (a short communication). *Microbiology & Infectious Diseases*, 2021. **5**(2): 1116.
29. Montone, C.M., et al., Characterization of the trans-epithelial transport of green tea (C. sinensis) catechin extracts with in vitro inhibitory effect against the SARS-CoV-2 papain-like protease activity. *Molecules*, 2021. **26**(21): 6744.
30. Ohgitani, E., et al., Significant inactivation of SARS-CoV-2 in vitro by a green tea catechin, a catechin-derivative, and black tea galloylated theaflavins. *Molecules*, 2021. **26**(12): 3572.
31. Zhao, M., et al., GCG inhibits SARS-CoV-2 replication by disrupting the liquid phase condensation of its nucleocapsid protein. *Nature Communications*, 2021. **12**(1): 2114.
32. LeBlanc, E.V. and C.C. Colpitts, The green tea catechin EGCG provides proof-of-concept for a pan-coronavirus attachment inhibitor. *Scientific Reports*, 2022. **12**(1): 12899.
33. Park, R., et al., Epigallocatechin gallate (EGCG), a green tea polyphenol, reduces coronavirus replication in a mouse model. *Viruses*, 2021. 13(12): 2533.
34. Onyango, O.H., In silico models for anti-COVID-19 drug discovery: a systematic review. *Advances in Pharmacological and Pharmaceutical Sciences,* 2023. **2023**: 4562974.
35. Podstawczyk, D., et al., Reactivity of (+)-catechin with copper(II) ions: the green synthesis of size-controlled sub-10nm copper nanoparticles. *ACS Sustainable Chemistry & Engineering*, 2019. **7**(20): 17535–17543.
36. Mhatre, S., S. Naik, and V. Patravale, A molecular docking study of EGCG and theaflavin digallate with the druggable targets of SARS-CoV-2. *Computers in Biology and Medicine*, 2021. **129**: 104137.
37. Tallei, T.E., et al., Potential of plant bioactive compounds as SARS-CoV-2 main protease (M(pro)) and spike (S) glycoprotein inhibitors: a molecular docking study. *Scientifica (Cairo)*, 2020. **2020**: 6307457.
38. Chiou, W.C., et al., The inhibitory effects of PGG and EGCG against the SARS-CoV-2 3C-like protease. *Biochemical and Biophysical Research Communications*, 2022. **591**: 130–136.
39. Zhu, Y. and D.Y. Xie, Docking characterization and in vitro inhibitory activity of flavan-3-ols and dimeric proanthocyanidins against the main protease activity of SARS-Cov-2. *Frontiers in Plant Science*, 2020. **11**: 601316.
40. Ghosh, R., et al., Evaluation of green tea polyphenols as novel corona virus (SARS CoV-2) main protease (Mpro) inhibitors: an in silico docking and molecular dynamics simulation study. *Journal of Biomolecular Structure and Dynamics*, 2021. **39**(12): 4362–4374.
41. Gogoi, B., et al., Identification of potential plant-based inhibitor against viral proteases of SARS-CoV-2 through molecular docking, MM-PBSA binding energy calculations and molecular dynamics simulation. *Molecular Diversity*, 2021. **25**(3): 1963–1977.
42. Kanbarkar, N. and S. Mishra, Matrix metalloproteinase inhibitors identified from Camellia sinensis for COVID-19 prophylaxis: an in silico approach. *Advances in Traditional Medicine*, 2021. **21**(1): 173–188.
43. Khan, M.F., et al., In-silico study to identify dietary molecules as potential sars-cov-2 agents. *Letters in Drug Design and Discovery*, 2021. **18**(6): 562–573.

44. Bhardwaj, A., S. Sharma, and S.K. Singh, Molecular docking studies to identify promising natural inhibitors targeting SARS-CoV-2 Nsp10-Nsp16 protein complex. *Turkish Journal of Pharmaceutical Sciences*, 2022. **19**(1): 93–100.
45. Bhardwaj, V.K., et al., Bioactive molecules of tea as potential inhibitors for RNA-dependent RNA polymerase of SARS-CoV-2. *Frontiers in Medicine (Lausanne)*, 2021. **8**: 684020.
46. Mahrosh, H.S. and G. Mustafa, An in silico approach to target RNA-dependent RNA polymerase of COVID-19 with naturally occurring phytochemicals. *Environment, Develo pment and Sustainability*, 2021. **23**(11): 16674–16687.
47. Ohishi, T., et al., Epigallocatechin gallate (EGCG) attenuates severe acute respiratory coronavirus disease 2 (SARS-CoV-2) infection by blocking the interaction of SARS-CoV-2 spike protein receptor-binding domain to human angiotensin-converting enzyme 2. *PLoS One*, 2022. **17**(7): e0271112.
48. Tsvetkov, V., et al., EGCG as an anti-SARS-CoV-2 agent: preventive versus therapeutic potential against original and mutant virus. *Biochimie*, 2021. **191**: 27–32.
49. Maiti, S. and A. Banerjee, Epigallocatechin gallate and theaflavin gallate interaction in SARS-CoV-2 spike-protein central channel with reference to the hydroxychloroquine interaction: bioinformatics and molecular docking study. *Drug Development Research*, 2021. **82**(1): 86–96.
50. Mishra, C.B., et al., Identifying the natural polyphenol catechin as a multi-targeted agent against SARS-CoV-2 for the plausible therapy of COVID-19: an integrated computational approach. *Briefings in Bioinformatics*, 2021. **22**(2): 1346–1360.
51. Gupta, S., et al., Secondary metabolites from spice and herbs as potential multitarget inhibitors of SARS-CoV-2 proteins. *Journal of Biomolecular Structure and Dynamics*, 2022. **40**(5): 2264–2283.
52. Zainuddin, A., et al., Prediction of the mechanism of action of catechin as superoxide anion antioxidants and natural antivirals for COVID-19 infection with in silico study. *Journal of Advanced Pharmaceutical Technology & Research*, 2022. **13**(3): 191–196.
53. Patel, S.S., et al., Green tea extract: a potential cause of acute liver failure. *World Journal of Gastroenterology*, 2013. **19**(31): 5174–5177.
54. Inoue, H., et al., High-dose green tea polyphenols induce nephrotoxicity in dextran sulfate sodium-induced colitis mice by down-regulation of antioxidant enzymes and heat-shock protein expressions. *Cell Stress and Chaperones*, 2011. **16**: 653–662.
55. Costa, L.M., S.T. Gouveia, and J.A. Nóbrega, Comparison of heating extraction procedures for Al, Ca, Mg, and Mn in tea samples. *Analytical Science*, 2002. **18**(3): 313–318.
56. Ferenczyová, K., et al., Pharmacology of catechins in ischemia-reperfusion injury of the heart. *Antioxidants (Basel)*, 2021. **10**(9): 1390.

18 Antiviral Activities of Naringenin and Its Derivatives as Adjuvant Treatment against SARS-CoV-2 Infections

Pallab Chakraborty, Krishnendu Acharya, Joy Sarkar, Solomon Habtemariam, Javad Sharifi-Rad, and William C. Cho

18.1 INTRODUCTION

The World Health Organization (WHO) announced the discovery of a new coronavirus in late 2019, which spread rapidly from person to person. This coronavirus was named Severe Acute Respiratory Syndrome Coronavirus 2 or SARS-CoV-2, which was called coronavirus disease 2019 (COVID-19). Shortly after, COVID-19 was declared a pandemic disease of paramount global public health importance. Coronaviruses (CoVs) can cause severe infections in both humans and animals' respiratory, gastrointestinal, and other organs/systems. Within the Coronaviridae family, they belong to the subfamily Orthocoronavirinae, which also includes beta, gamma, delta, and alpha CoVs (Sehn, 2020; Shereen et al., 2020). In the absence of a specific drug against COVID-19, drug repurposing received significant attention due to its potential to reduce time, cost, and the knowledge gap in novel drug discovery. In this context, natural products such as flavonoids discovered in fruits, flowers, seeds, tea, red wine, and spices can be considered potential sources of antiviral drugs. Furthermore, an *in silico* investigation has identified associations between different flavonoids and the critical structural proteins of CoV-2 that are necessary for viral endocytosis and replication (Mutha et al., 2021).

Phenolic compounds are one of the many natural products that plants produce and are responsible for the primary organoleptic features of foods and beverages derived from plants. They play a crucial role in determining fruits' and vegetables' colour and flavour characteristics (Horvath et al., 2005; Tapas et al., 2008). Fruits, vegetables, tea, wine, seeds, herbs, spices, and whole grains are among the foods with a high content of phenolic compounds, with flavonoids being a significant class of bioactive constituents (Hughes et al., 2008).

To date, several thousands of natural flavonoids have been identified and grouped into structural classes such as flavonols, flavones, flavanones, isoflavones, catechins, and anthocyanidins (Felgines et al., 2000). The relevance of flavonoids to human health is evident, as several epidemiological studies have suggested an inverse relationship between flavonoid consumption and oxidative damage or disease (Choi et al., 1991; Ameer et al., 1996; Cavia-Saiz et al., 2010; Dou et al., 2013). One such flavonoid is naringenin, which is abundantly found in fruits such as grapefruit. In this chapter, we aim to summarise the antiviral activities of naringenin and its derivatives against SARS-CoV-2 and other viruses.

18.2 NARINGENIN: STRUCTURE AND OCCURRENCE

Naringenin (Figure 18.1) has a molecular formula of $C_{15}H_{12}O_5$. In the scientific literature, it is known by the following names: [(2*S*)-5,7-Dihydroxy-2-(4-hydroxyphenyl)-2,3-dihydro-4H-chromen-4-one and (2*S*)-5,7-dihydroxy-2-(4-hydroxyphenyl)-3,4-dihydro-2H-1-benzopyran-4-one] (ChemSpider ID: 388383). Naringenin can be obtained from a wide variety of herbs and fruits, including grapefruit (Ho et al., 2000), sour orange (Gel-Moreto et al., 2003), bergamot (Gattuso et al., 2007), tart cherries (Wang et al., 1999), cocoa (Sánchez-Rabaneda et al., 2003), tomatoes (Minoggio et al., 2003; Vallverdú-Queralt et al., 2012), Greek oregano (Exarchou et al., 2003), beans (Hungria, 1992), and water mint (Olsen et al., 2008). Naringenin dissolves well in organic solvents like alcohol but is nearly insoluble in water. The hydrolysis of naringenin's glycone forms, such as naringin or narirutin, yields naringenin. In the grapefruit (*Citrus paradisi*), naringin (naringenin-7-rhamnoglucoside) is considered the bitter component and is present in the fruit's juice, blossom, and rind, making up to 10% of its dry weight. Naringin and

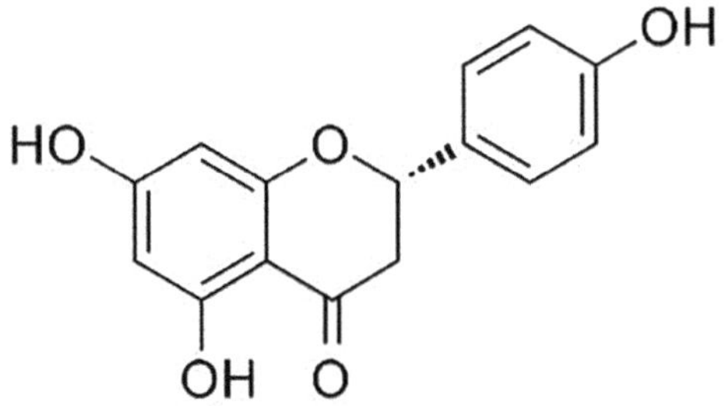

HO OH OH O

FIGURE 18.1 Chemical structure of naringenin [ChemSpider ID: 388383].

DOI: 10.1201/9781003433200-18

other naringenin glycosides can also be sourced through a synthesis route (Fuhr and Kummert, 1995).

For instance, naringenin chalcone with a diphenylpropane unit can be produced by condensation of *p*-coumaroyl-CoA with three malonyl-CoA molecules. This chalcone is then transformed into naringenin with the flavone (2-phenylchromen-4-one) backbone by conjugate ring closure. A range of structural forms, such as dihydroflavonols, chalcones, flavanones, flavans, anthocyanins, flavones and flavonols, and isoflavonoids, are produced through this and other synthesis approaches (https://www.kegg.jp/dbget-bin/www_bget?ko00941).

18.3 NARINGENIN: A FLAVONOID WITH ANTIVIRAL POTENTIAL

Citrus fruits, such as grapefruit and oranges, contain natural flavonoids, including naringenin (Den Hartogh and Tsiani, 2019), which has potent analgesic, antioxidant, anti-inflammatory, anti-tumoural, and antiviral properties (Alberca et al., 2020).

Numerous *in vitro* studies have demonstrated the antiviral efficacy of naringenin both before and after infection (Cataneo et al., 2019). While naringenin has been extensively studied *in vitro,* like other natural products, limited information is available on its *in vivo* effects (Gonçalves et al., 2017). However, the anti-inflammatory properties of naringenin, both *in vitro* and *in vivo,* have been demonstrated in several animal models, including respiratory disorders (Salehi et al., 2019). Naringenin exhibits significant antiviral activity against multiple viruses, which are listed in Table 18.1.

18.4 ANTIVIRAL TECHNIQUES OF NARINGENIN AGAINST SARS-COV-2

According to recent research, CoV-encoded proteins, namely 3-chymotrypsin-like protease (3CL^{pro}) and papain-like

TABLE 18.1
Potential Antiviral Activity of Naringenin against a Range of Viruses

Source of Naringenin	Effective Against	Mode of Action	Reference
Citrus	Dengue virus	Naringenin blocked the infection of Huh7.5 cells by four different dengue virus serotypes. Studies using the sub-genomic RepDV-1 and RepDV-3 replicon systems demonstrated the effectiveness of naringenin in preventing dengue virus reproduction. Naringenin's antiviral effects were still noticeable 24 hours after dengue virus infection in Huh7.5 cells. The potential application of naringenin to limit dengue virus replication was further supported by its anti-dengue virus activity in primary human monocytes infected with dengue virus sertoype-4	Frabasile et al. (2017)
Grapefruit	Hepatitis C virus	Infected cells actively release hepatitis C virus (HCV) via a Golgi-dependent process, with the virus attached to very low density lipoprotein (vLDL). Silencing the apolipoprotein B (ApoB) messenger RNA in infected cells reduced the discharge of ApoB-100 and HCV was reduced by 70%. Naringenin stimulation resulted in an 80% reduction of HCV formation in infected cells by suppressing microsomal triglyceride transfer protein activity, as well as the transcription of 3-hydroxy-3-methylglutaryl-coenzyme A reductase and acyl-coenzyme A: cholesterol acyltransferase 2, leading to decreased vLDL production	Nahmias et al. (2008)
Citrus	Chikungunya	Naringenin exhibited strong antiviral action against the chikungunya virus, decreasing the virus's ability to replicate and reducing the number of viral proteins produced during replication	Ahmadi et al. (2016)
Citrus and tea	Herpes simplex 1 and 2	Naringenin showed anti-replicative effects or direct deactivation against herpes simplex viruses 1 and 2	Lyu et al. (2005)
Orange and grapefruit	Yellow fever	Naringenin has the capacity to obstruct viral replication beyond the entry threshold in yellow fever virus	Castrillo et al. (2015)
Propolis (bee glue) is a product of *Apis mellifera* L.	Human immunodeficiency virus	Naringenin moderately inhibits the HIV-1 reverse transcriptase enzyme. It also exhibits strong antiviral action against CHIKV, decreasing the virus's ability to replicate and reducing the number of viral proteins produced during replication	Silva et al. (2019)

protease (PL^{pro}), play an essential role in suppressing the host's innate immune responses and reproducing virus replication (Báez-Santos et al., 2015). By inhibiting the action of these proteases, we can strengthen our defence mechanisms and establish a formidable barrier against infection (Tahir ul Qamar et al., 2020). Recent studies have suggested that certain metabolites, such as flavonoids, have the ability to hinder the function of these proteins by binding to their active sites. Naringenin and its derivatives, which are flavones, have been extensively examined for their antiviral efficacy against various viruses, including Chikungunya virus, Dengue virus, Zika virus, and the unprecedented challenge posed by COVID-19 (Sawikowska, 2020; Tutunchi et al., 2020).

In silico studies are crucial for finding substances that precisely target viral protein active sites and obstruct the interactions needed for the entrance and reproduction of the virus. These investigations concentrate on targets such as proteases, the ACE2 receptor, and the spike protein's RBD domain (Figure 18.2). In CoV-2 pathogenesis, the spike glycoprotein (type 1) serves as the initial point of contact with the host and enters the cell by interacting with the ACE2 receptor. The ACE2 receptor initiates this connection by identifying the spike protein's RBD domain. The spike protein is then broken down into its S1 subunit and S2 subunits by the serine proteases TMPRSS2, which promotes endocytosis (Yan et al., 2020; Kundu et al., 2022). An investigation suggested that naringenin has a moderate affinity for interacting with the RBD/ACE2 protein complex (around −6.44 Kilocalorie/mol) (Júnior et al., 2022). The binding sites' study reveals a preference for amino acid residues from ACE2 and RBD, including Asn15, Glu19, and Arg375 (Júnior et al., 2022). A docking study conducted by Utomo and Meiyanto (2020) reveals that naringenin exhibits low energy-binding and significant antiviral potential. Another experiment revealed that Naringenin has a lower binding affinity to the spike protein compared to remdesivir (Ubani et al., 2020). However, out of ten docked flavonoids, naringin (naringenin-7-*O*-neohesperidoside), a derivative of naringenin, exhibited the most significant binding affinity to the spike glycoprotein protein, making it the leading contender (Vijayakumar et al., 2020).

Naringenin also shows a significant affinity for TMPRSS2 (−7.3 kcal/mol), thereby preventing viral entry (Alzaabi et al., 2022). ACE2 is abundantly present in cardiorespiratory neurons within the brainstem, vasculature, lungs, heart, and kidney (Limanaqi et al., 2020). Therefore, various flavonoids, including naringenin, have undergone molecular docking studies to evaluate their binding affinity to ACE2. According to a present-day study, naringenin binds to ACE2 at specific locations including Pro146, Leu143, and Lys131 with a docking energy of roughly −6.05 kcal/mol (Liu et al., 2022). Another interaction site for 7-*O*-rutinoside of naringenin, or naringin, with the highest binding affinity towards ACE2 (docking energy of −6.85 kcal/mol) involves Asn394, Glu398, Glu402, and Tyr515 (Liu et al., 2022). Due to its strong binding affinity for the ACE2 receptor, naringenin can block SARS-CoV-2 attachment through the ACE2 pathway.

Furthermore, a deliberate focus is on blocking viral activity by targeting the protease enzymes to explore pathways beyond ACE2. One of the crucial coronavirus enzymes is papain-like protease or PL^{pro}, which is essential to convert viral proteins into active replicase protein complexes. Thus, PL^{pro} becomes a justified target for antiviral drugs, highlighting its significance in the search for practical therapeutic approaches (Shin et al., 2020). The study by Cho et al. (2013) confirms that flavonoids derived from *Paulownia tomentosa*, such as naringenin, geranylated flavonoids, and

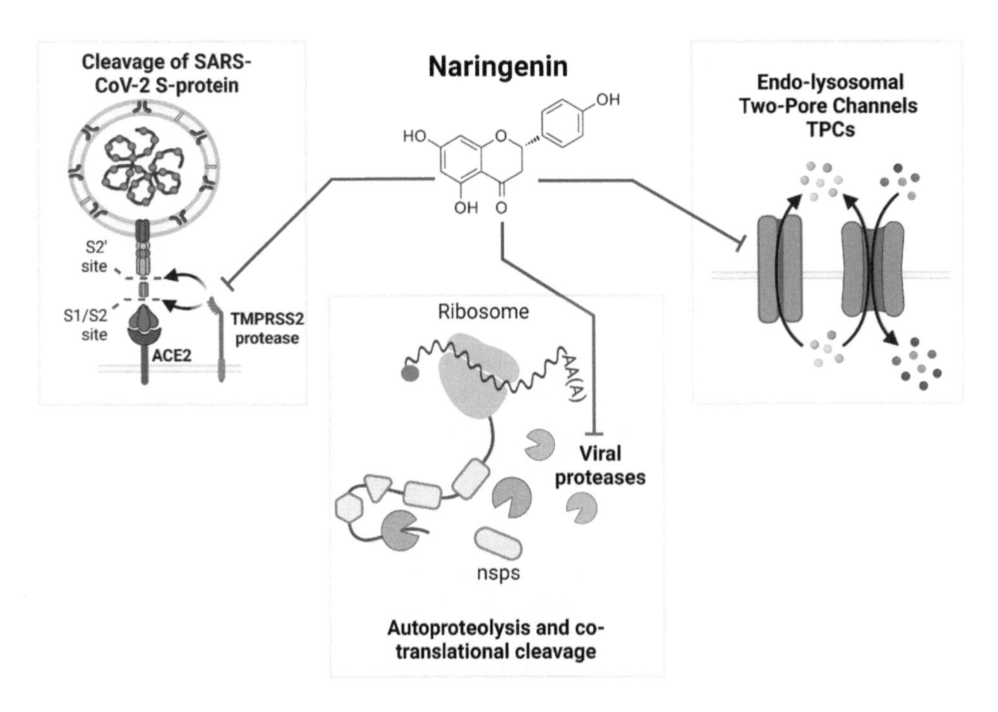

FIGURE 18.2 Inhibitory effect of naringenin in various stages of the SARS-CoV-2 life cycle (created with BioRender.com).

quercetin, effectively prevent PL^{pro}. Notably, these compounds demonstrated dual activity by not only inhibiting the viral protease but also reducing cytokinin concentration, including interleukin-1β and tumour necrosis factor-α, suggesting their potential to modulate the immune response during viral infection (Cho et al., 2013).

Another set of main proteases, such as M^{pro}, $3CL^{pro}$, and nsp5, are also targeted as inhibitors due to their capability to facilitate the viral polypeptide's proteolytic cleavage in 11 non-structural proteins, which is essential for the virus's reproduction (Cui et al., 2020). When naringenin was applied to the viral protein, it exhibited a high activity level, resulting in about 98% reduction in enzyme activity at a concentration of 100 μM (Allam et al., 2020). A docking investigation showed minimal energy binding (12.44 kcal/mol) for naringenin to the M^{pro} molecule (Yudi Utomo and Meiyanto, 2020). In another site, naringenin interacts with the substrate binding pocket situated between cleft I and cleft II domains of protease $3CL^{pro}$ to inhibit the activity of the protein (Eko Prasetyo et al., 2020). Using an *in silico* study, naringenin from grapefruit seed extract, along with other components like citric acid, narirutin, ascorbic acid, and naringin, demonstrated a binding attraction of –8.2 kcal/mol against the main protease of CoV-2 (Saric et al., 2021).

Expanding on antiviral strategies, there have been concerted efforts to target the transfer of the virus to lysosomes (Khan et al., 2020; Nile et al., 2021). Endolysosomal ion channels like Two-Pore Channels (TPCs) are crucial for the CoV life cycle (Pafumi et al., 2017). Interestingly, the study by Benkerrou et al. (2019) showcases that TPC1 and TPC2 can be hindered by naringenin in humans and plants. Significantly, naringenin demonstrated a gradual hang-up of CoV replication, observed initially at 24 hpi in cells with TPC2 silencing. This inhibitory effect intensified 48 and 72 hours after infection, highlighting naringenin's time-dependent efficiency in impeding viral replication (Clementi et al., 2021). Thus, TPC inhibition by naringenin presents a potential avenue for antiviral intervention that merits thorough investigation along with other pathways.

18.5 CONCLUSION AND FUTURE PERSPECTIVE

In conclusion, naringenin has emerged as a capable candidate to fight against SARS-CoV-2. *In silico* analysis has shown that naringenin effectively prevents viral infection at almost every stage. However, its therapeutic potential needs to be verified through specific biochemical tests as it is a new compound of interest in this field. Nutritional approaches targeting the ACE2 entry receptor of SARSCoV-2 should be cautiously assessed as lowering ACE2 levels could increase inflammation and lung damage. Additionally, the attachment of naringenin to the S protein still needs to be verified through experiments, and further validation is required through *in vivo* studies. Future research should focus on three main areas: identifying the most effective naringenin derivatives, conducting large-scale *in vivo* studies, and carrying out clinical trials to thoroughly assess the safety and effectiveness of these compounds in treating COVID-19. With advancements in genetic modifications, large-scale production of flavonoids is now possible. Ongoing advancements will provide new insights of using flavonoids in medications to treat various infectious and diseases.

ACKNOWLEDGEMENT

The authors would like to express their gratitude to the academics whose works are referenced and acknowledged in this study for their invaluable assistance. They also extend their thanks to the publishers, editors, and writers of all the books, journals, and articles that served as source material for the reviews and discussions included in this work.

REFERENCES

Ahmadi, A. et al. (2016) Inhibition of chikungunya virus replication by hesperetin and naringenin, *RSC Advances*, 6(73), 69421–69430. Available at: https://doi.org/10.1039/C6RA16640G.

Alberca, R.W. et al. (2020) Perspective: The potential effects of naringenin in COVID-19, *Frontiers in Immunology*, 11, 570919. Available at: https://doi.org/10.3389/fimmu.2020.570919.

Allam, A.E. et al. (2020) An: In silico perception for newly isolated flavonoids from peach fruit as privileged avenue for a countermeasure outbreak of COVID-19, *RSC Advances*, 10(50), 29983–29998. Available at: https://doi.org/10.1039/d0ra05265e.

Alzaabi, M.M. et al. (2022) Flavonoids are promising safe therapy against COVID-19, *Phytochemistry Reviews*, 21(1), 291–312. Available at: https://doi.org/10.1007/s11101-021-09759-z.

Ameer, B. et al. (1996) Flavanone absorption after naringin, hesperidin, and citrus administration, *Clinical Pharmacology & Therapeutics*, 60(1), 34–40. Available at: https://doi.org/10.1016/S0009-9236(96)90164-2.

Báez-Santos, Y.M., St. John, S.E. and Mesecar, A.D. (2015) The SARS-coronavirus papain-like protease: Structure, function and inhibition by designed antiviral compounds, *Antiviral Research*, 115, 21–38. Available at: https://doi.org/10.1016/j.antiviral.2014.12.015.

Benkerrou, D. et al. (2019)A perspective on the modulation of plant and animal two pore channels (TPCs) by the flavonoid naringenin, *Biophysical Chemistry*, 254, 106246. Available at: https://doi.org/10.1016/j.bpc.2019.106246.

Cataneo, A.H.D. et al. (2019) The citrus flavonoid naringenin impairs the in vitro infection of human cells by Zika virus, *Scientific Reports*, 9(1), 16348. Available at: https://doi.org/10.1038/s41598-019-52626-3.

Cavia-Saiz, M. et al. (2010) Antioxidant properties, radical scavenging activity and biomolecule protection capacity of flavonoid naringenin and its glycoside naringin: A comparative study, *Journal of the Science of Food and Agriculture*, 90(7), 1238–1244. Available at: https://doi.org/10.1002/jsfa.3959.

Cho, J.K. et al. (2013) Geranylated flavonoids displaying SARS-CoV papain-like protease inhibition from the fruits of Paulownia tomentosa, *Bioorganic & Medicinal Chemistry*, 21(11), 3051–3057. Available at: https://doi.org/10.1016/j.bmc.2013.03.027.

Choi, J.S., Yokozawa, T. and Oura, H. (1991) Antihyperlipidemic effect of flavonoids from prunus davidiana, *Journal of Natural Products*, 54(1), 218–224. Available at: https://doi.org/10.1021/np50073a022.

Clementi, N. et al. (2021) Naringenin is a powerful inhibitor of SARS-CoV-2 infection in vitro, *Pharmacological Research*, 163, 105255. Available at: https://doi.org/10.1016/j.phrs.2020.105255.

Cui, W., Yang, K. and Yang, H. (2020) Recent progress in the drug development targeting SARS-CoV-2 main protease as treatment for COVID-19, *Frontiers in Molecular Biosciences*, 7, 616341. Available at: https://doi.org/10.3389/fmolb.2020.616341.

Den Hartogh, D.J. and Tsiani, E. (2019) Antidiabetic properties of naringenin: A citrus fruit polyphenol, *Biomolecules*, 9(3), 99. Available at: https://doi.org/10.3390/biom9030099.

Dou, W. et al. (2013) Protective effect of naringenin against experimental colitis via suppression of Toll-like receptor 4/NF-κB signalling, *British Journal of Nutrition*, 110(4), 599–608. Available at: https://doi.org/10.1017/S0007114512005594.

Eko Prasetyo, W., Kusumaningsih, T. and Firdaus, M. (2020) Nature as a treasure trove for anti-COVID-19: Luteolin and naringenin from Indonesian traditional herbal medicine reveal potential SARS-CoV-2 Mpro inhibitors insight from in silico studies, *ChemRxiv*, 2020, 1–6. Available at: https://doi.org/10.26434/chemrxiv.13356842.v1.

Exarchou, V. et al. (2003) LC-UV-solid-phase extraction-NMR-MS combined with a cryogenic flow probe and its application to the identification of compounds present in greek oregano, *Analytical Chemistry*, 75(22), 6288–6294. Available at: https://doi.org/10.1021/ac0347819.

Felgines, C. et al. (2000) Bioavailability of the flavanone naringenin and its glycosides in rats, *American Journal of Physiology-Gastrointestinal and Liver Physiology*, 279(6), G1148–G1154. Available at: https://doi.org/10.1152/ajpgi.2000.279.6.G1148.

Frabasile, S. et al. (2017) The citrus flavanone naringenin impairs dengue virus replication in human cells, *Scientific Reports*, 7(1), 41864. Available at: https://doi.org/10.1038/srep41864.

Fuhr, U. and Kummert, A.L. (1995) The fate of naringin in humans: A key to grapefruit juice-drug interactions?, *Clinical Pharmacology & Therapeutics*, 58(4), 365–373. Available at: https://doi.org/10.1016/0009-9236(95)90048-9.

Gattuso, G. et al. (2007) Flavonoid composition of citrus juices, *Molecules*, 12(8), 1641–1673. Available at: https://doi.org/10.3390/12081641.

Gel-Moreto, N., Streich, R. and Galensa, R. (2003) Chiral separation of diastereomeric flavanone-7- O -glycosides in citrus by capillary electrophoresis, *Electrophoresis*, 24(15), 2716–2722. Available at: https://doi.org/10.1002/elps.200305486.

Gonçalves, D. et al. (2017) Orange juice as dietary source of antioxidants for patients with hepatitis C under antiviral therapy, *Food & Nutrition Research*, 61(1), 1296675. Available at: https://doi.org/10.1080/16546628.2017.1296675.

Ho, P.C. et al. (2000) Content of CYP3A4 inhibitors, naringin, naringenin and bergapten in grapefruit and grapefruit juice products, *Pharmaceutica Acta Helvetiae*, 74(4), 379–385. Available at: https://doi.org/10.1016/S0031-6865(99)00062-X.

Horvath, C.R., Martos, P.A. and Saxena, P.K. (2005) Identification and quantification of eight flavones in root and shoot tissues of the medicinal plant Huang-qin (Scutellaria baicalensis Georgi) using high-performance liquid chromatography with diode array and mass spectrometric detection, *Journal of Chromatography A*, 1062(2), 199–207. Available at: https://doi.org/10.1016/j.chroma.2004.11.030.

Hughes, L.A.E. et al. (2008) Higher dietary flavone, flavonol, and catechin intakes are associated with less of an increase in BMI over time in women: A longitudinal analysis from the Netherlands Cohort Study, *The American Journal of Clinical Nutrition*, 88(5), 1341–1352. Available at: https://doi.org/10.3945/AJCN.2008.26058.

Hungria, M. (1992) Effects of flavonoids released naturally from bean (Phaseolus vulgaris) on nodD -regulated gene transcription in Rhizobium leguminosarum bv. phaseoli, *Molecular Plant-Microbe Interactions*, 5(3), p. 199. Available at: https://doi.org/10.1094/MPMI-5-199.

Júnior, M.L.P. et al. (2022) Evaluation of peppermint leaf flavonoids as SARS-CoV-2 spike receptor-binding domain attachment inhibitors to the human ACE2 receptor: A molecular docking study, *Open Journal of Biophysics*, 12(2), 132–152. Available at: https://doi.org/10.4236/ojbiphy.2022.122005.

Khan, N. et al. (2020) Two-pore channels regulate Tat endolysosome escape and Tat-mediated HIV-1 LTR transactivation, *The FASEB Journal*, 34(3), 4147–4162. Available at: https://doi.org/10.1096/fj.201902534R.

Kundu, A. et al. (2022) Clinical aspects and presumed etiology of multisystem inflammatory syndrome in children (MIS-C): A review, *Clinical Epidemiology and Global Health*, 14, 100966. Available at: https://doi.org/10.1016/j.cegh.2022.100966.

Limanaqi, F. et al. (2020) Cell clearing systems as targets of polyphenols in viral infections: Potential implications for covid-19 pathogenesis, *Antioxidants*, 9(11), 1105. Available at: https://doi.org/10.3390/antiox9111105.

Liu, W. et al. (2022) Citrus fruits are rich in flavonoids for immunoregulation and potential targeting ACE2, *Natural Products and Bioprospecting*, 12(1), 4. Available at: https://doi.org/10.1007/s13659-022-00325-4.

Lyu, S.-Y., Rhim, J.-Y. and Park, W.-B. (2005) Antiherpetic activities of flavonoids against herpes simplex virus type 1 (HSV-1) and type 2 (HSV-2)in vitro, *Archives of Pharmacal Research*, 28(11), 1293–1301. Available at: https://doi.org/10.1007/BF02978215.

Minoggio, M. et al. (2003) Polyphenol pattern and antioxidant activity of different tomato lines and cultivars, *Annals of Nutrition and Metabolism*, 47(2), 64–69. Available at: https://doi.org/10.1159/000069277.

Mutha, R.E., Tatiya, A.U. and Surana, S.J. (2021) Flavonoids as natural phenolic compounds and their role in therapeutics: An overview, *Future Journal of Pharmaceutical Sciences*, 7(1), 25. Available at: https://doi.org/10.1186/s43094-020-00161-8.

Nahmias, Y. et al. (2008) Apolipoprotein B-dependent hepatitis C virus secretion is inhibited by the grapefruit flavonoid naringenin, *Hepatology*, 47(5), 1437–1445. Available at: https://doi.org/10.1002/hep.22197.

Nile, S.H. et al. (2021) Recent advances in potential drug therapies combating COVID-19 and related coronaviruses-A perspective, *Food and Chemical Toxicology*, 154, 112333. Available at: https://doi.org/10.1016/j.fct.2021.112333.

Olsen, H.T. et al. (2008) Isolation of the MAO-inhibitor naringenin from Mentha aquatica L., *Journal of Ethnopharmacology*, 117(3), 500–502. Available at: https://doi.org/10.1016/j.jep.2008.02.015.

Pafumi, I. et al. (2017) Naringenin impairs two-pore channel 2 activity and inhibits VEGF-induced angiogenesis, *Scientific Reports*, 7(1), 5121. Available at: https://doi.org/10.1038/s41598-017-04974-1.

Salehi, B. et al. (2019) The Therapeutic potential of naringenin: A review of clinical trials, *Pharmaceuticals*, 12(1), 11. Available at: https://doi.org/10.3390/ph12010011.

Sánchez-Rabaneda, F. et al. (2003) Liquid chromatographic/electrospray ionization tandem mass spectrometric study of the phenolic composition of cocoa (Theobroma cacao), *Journal of Mass Spectrometry*, 38(1), 35–42. Available at: https://doi.org/10.1002/jms.395.

Saric, B. et al. (2021) In silico analysis of selected components of grapefruit seed extract against SARS-CoV-2 main protease, *The EuroBiotech Journal*, 5(s1), 5–12. Available at: https://doi.org/10.2478/ebtj-2021-0015.

Sawikowska, A. (2020) Meta-analysis of flavonoids with antiviral potential against coronavirus, *Biometrical Letters*, 57(1), 13–22. Available at: https://doi.org/10.2478/bile-2020-0002.

Sehn, L.H. (2020) Balancing risk and benefit during coronavirus, *Blood*, 135(21), 1817–1817. Available at: https://doi.org/10.1182/blood.2020006279.

Shereen, M.A. et al. (2020) COVID-19 infection: Emergence, transmission, and characteristics of human coronaviruses, *Journal of Advanced Research*, 24, 91–98. Available at: https://doi.org/10.1016/j.jare.2020.03.005.

Shin, D. et al. (2020) Papain-like protease regulates SARS-CoV-2 viral spread and innate immunity, *Nature*, 587(7835), 657–662. Available at: https://doi.org/10.1038/s41586-020-2601-5.

Silva, C.C.F. da et al. (2019) Chemical characterization, antioxidant and anti-HIV activities of a Brazilian propolis from Ceará state, *Revista Brasileira de Farmacognosia*, 29(3), 309–318. Available at: https://doi.org/10.1016/j.bjp.2019.04.001.

Tahir ul Qamar, M. et al. (2020) Structural basis of SARS-CoV-2 3CLpro and anti-COVID-19 drug discovery from medicinal plants, *Journal of Pharmaceutical Analysis*, 10(4), 313–319. Available at: https://doi.org/10.1016/j.jpha.2020.03.009.

Tapas, A., Sakarkar, D. and Kakde, R. (2008) Flavonoids as nutraceuticals: A review, *Tropical Journal of Pharmaceutical Research*, 7(3), 14693. Available at: https://doi.org/10.4314/tjpr.v7i3.14693.

Tutunchi, H. et al. (2020) Naringenin, a flavanone with antiviral and anti-inflammatory effects: A promising treatment strategy against COVID -19, *Phytotherapy Research*, 34(12), 3137–3147. Available at: https://doi.org/10.1002/ptr.6781.

Ubani, A. et al. (2020) Molecular docking analysis of selected phytochemicals on two SARS-CoV-2 targets, *F1000Research*, 9, 1157. Available at: https://doi.org/10.12688/f1000research.25076.1.

Vallverdú-Queralt, A. et al. (2012) Changes in the polyphenol profile of tomato juices processed by pulsed electric fields, *Journal of Agricultural and Food Chemistry*, 60(38), 9667–9672. Available at: https://doi.org/10.1021/jf302791k.

Vijayakumar, B.G. et al. (2020) In silico pharmacokinetic and molecular docking studies of natural flavonoids and synthetic indole chalcones against essential proteins of SARS-CoV-2, *European Journal of Pharmacology*, 886, 173448. Available at: https://doi.org/10.1016/j.ejphar.2020.173448.

Wang, H. et al. (1999) Antioxidant polyphenols from tart cherries (Prunus cerasus), *Journal of Agricultural and Food Chemistry*, 47(3), 840–844. Available at: https://doi.org/10.1021/jf980936f.

Yan, R. et al. (2020) Structural basis for the recognition of SARS-CoV-2 by full-length human ACE2, *Science*, 367(6485), 1444–1448. Available at: https://doi.org/10.1126/science.abb2762.

Yudi Utomo, R. and Meiyanto, E. (2020) Revealing the potency of citrus and galangal constituents to halt SARS-CoV-2 infection, *Preprints* 2020, 2020030214. Available at: https://doi.org/10.20944/preprints202003.0214.v1.

19 Luteolin and Chrysin as Inhibitors of SARS-CoV-2 Infection

Keenau Pearce and Burtram Fielding

LIST OF ABBREVIATIONS

ACE2	Angiotensin-converting enzyme 2
ARDs	Acquired respiratory distress syndrome
BTK	Bruton's tyrosine kinase
CFB	Compliment factor B
IFN	Interferon
IL	Interleukin
MAPK	Mitogen-activated protein kinases
MPO	Myeloperoxidase
NF-κB	Nuclear factor-κB
NO	Nitric oxide
OH	Hydroxyl
RBD	Receptor-binding domain
SARS-CoV-2	Severe Acute Respiratory Syndrome Coronavirus type 2
TMPRSS2	Transmembrane serine protease 2

19.1 INTRODUCTION

The Severe Acute Respiratory Syndrome Coronavirus 2 (SARS-CoV-2) resulted in a global pandemic, not seen since the influenza outbreak in 1918. Despite the availability of coronavirus disease 2019 (COVID-19) vaccines and, later, FDA-approved drugs, there continues to be a steady increase in the number of infections, complicated by the constant emergence of new variants, which could lead to loss of the efficacy of current vaccines and drugs. Thus, drug development to manage SARS-CoV-2 infections remains a necessity (Robinson et al., 2020; Aleem et al., 2023).

For anti-COVID-19 drug development, various methodologies have been followed. Overall, these include the direct targeting of SARS-CoV-2 proteins, interference of host enzymes and proteins, and the blocking of specific immunoregulatory pathways. Using this as the principle, computer-aided lead compound screening and design, natural product discovery, drug repurposing, and combination therapy have all been used in the search for effective anti-COVID-19 therapies (Murrell et al., 2020; Lei et al., 2022).

Flavonoids and their subclasses have long been an essential resource for drug discovery and are increasingly evaluated as antiviral therapeutics (Malla et al., 2023). These compounds typically offer several advantages over conventional drugs, including excellent safety profiles, bioavailability, and fewer unwanted effects (Cazarolli et al., 2008). Within the context of SARS-CoV-2, recent studies have identified several flavonoids with the potential to either prevent infection or manage the symptoms of COVID-19; the latter is often through the regulation of the associated immune and inflammatory dysfunction linked to moderate to severe COVID-19 (Liskova et al., 2021). This chapter aims to provide an overview of the flavonoids luteolin and chrysin as COVID-19 treatment options.

19.2 SARS-COV-2 GENOME

SARS-CoV-2 is a positive-strand RNA virus, and like all coronaviruses (CoVs), the genome is organized in a conserved 5′-replicase-S-E-M-N-3′ order (Figure 19.1). The genome has a 5′-cap and is 3′-polyadenylated (Woo et al., 2012). The 5′ two-thirds of the genome encodes for a large frameshifted polyprotein (ORF1a/ORF1ab). The open reading frames (ORFs) produced by this polyprotein encode for 16 non-structural proteins (nsps) (Cao et al., 2021), which are essential for viral RNA replication (van der Hoek et al., 2006); these nsps include the main protease (M^{pro}), papain-like protease (PL^{pro}), and RNA-dependent RNA polymerase (RdRp).

While M^{pro} and PL^{pro} are indispensable for the synthesis and functioning of the nsps, the RdRp drives the synthesis of viral RNA, an essential process for RNA transcription and viral replication (Astuti and Ysrafil, 2020; Jiang et al., 2021). More specifically, M^{pro} (also referred to as $3C^{pro}$) is a critical enzyme for viral replication and transcription, and its primary function involves the cleavage of polyproteins PP1a and PP1ab, derived from ORF 1a/1b, which results in the release of nsps. Simultaneously, PL^{pro} (alternatively known as nsp3) is another essential enzyme for the virus replication cycle. This enzyme specifically digests the polyproteins, recognizing the conserved sequence LXGG. Beyond its direct involvement in viral replication, PL^{pro} is thought to contribute to the ubiquitination and inhibition of ISGylation on host proteins. This multifaceted role serves as an evasion mechanism against host antiviral immune responses (Shi and Lai, 2005; V'Kovski et al., 2021; Malone et al., 2022).

The remaining 3′ one-third of the genome contains the genes encoding for the major structural proteins, viz., spike protein (S), membrane glycoprotein (M), envelope protein (E), and nucleocapsid (N) (Cao et al., 2021). The S protein consists of two functional subunits, viz., the N-terminal S1

DOI: 10.1201/9781003433200-19

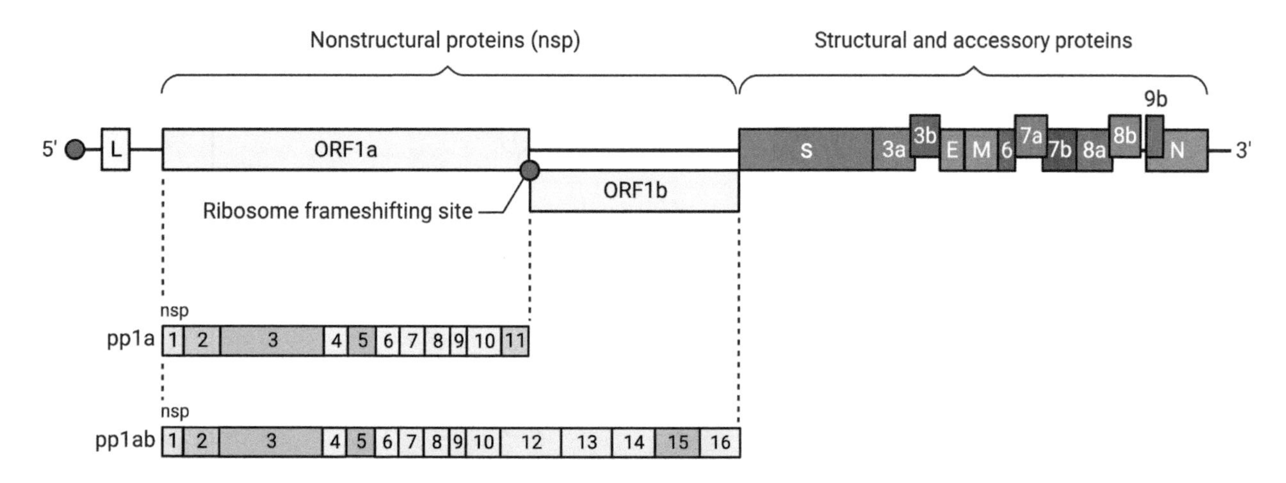

FIGURE 19.1 Genomic organization of SARS-CoV-2. The 5′ open reading frame (ORF) comprises approximately one-third of the genome and encodes for 16 non-structural proteins (nsps). The remaining 3′ ORFs encode for accessory and structural proteins (created with Biorender).

subunit and the C-terminal S2 region, and is responsible for virus binding and entry into host cells. This is a complex process, but in brief, during interaction with a host cell, the S1 subunit recognizes and binds to ACE2 on the host cell surface – we now know that the S protein's receptor-binding domain (RBD) recognizes the peptidase domain (PD) of the host cell's ACE2 receptor. The S2 subunit then facilitates the fusion of the virus's envelope with the host cell membrane, allowing viral entry into the cell (Li, 2016; Huang et al., 2020).

Interspersed among, or even overlapping, these structural ORFs are a number of accessory genes. These subgroup-specific ORFs vary in number, location, and size among human coronaviruses (hCoVs) (Liu et al., 2014), but the majority have been shown to play a role in virus infectivity and/or pathogenicity (Fang et al., 2021). Knowing the central role of the various viral proteins in the SARS-CoV-2 life cycle therefore makes them ideal targets for inhibiting SARS-CoV-2 infection and replication (Lei et al., 2022).

19.3 SARS-COV-2 INFECTION, INFLAMMATION, AND CYTOKINE STORM

COVID-19 symptoms typically develop 5–6 days after infection, and while disease severity may vary, symptoms frequently include fatigue, fever, headache, cough, diarrhea, lymphopenia, and dyspnea. Symptoms are usually mild or absent for the vast majority of cases (Alimohamadi et al., 2020). However, a small percentage of cases experience severe respiratory failure, multi-organ failure, or acute respiratory distress syndrome (ARDS) (Attaway et al., 2021). A propensity toward severe symptoms is reported for those of advanced age and with one or more co-morbidities, such as hypertension, diabetes mellitus, kidney disease, obstructive pulmonary disease, or cardiovascular disease (Jiang et al., 2022).

In the more severe cases, complications may include coagulation dysfunction, septic shock, kidney dysfunction, heart failure, and other secondary infections. Systemic hyperinflammation, leading to multiple organ failure and lung damage, has also been reported. Moreover, SARS-CoV-2 infection may result in dispersed intravascular coagulation and thrombosis through increased serum ferritin, fibrinogen, D-dimer, interleukin-6, C-reactive protein, and procalcitonin (Hadjadj et al., 2020). Several studies document inflammatory changes in olfactory bulbs with COVID-19, which is consistent with an explanation involving neuroinflammation. In fact, SARS-CoV-2 has been shown to induce pro-inflammatory microglia in the olfactory bulb, neuroinflammation, and spreads to other brain parts, contributing to long-term neurological symptoms (Morbini et al., 2020; Cocco et al., 2021). Further complications may also arise through the hyperactivation of pro-inflammatory mediator synthesis, known as cytokine storm (Figure 19.2) (Jiang et al., 2022).

19.3.1 Inflammatory Pathways and Cytokine Storm

The severity of COVID-19 is often closely linked to the immune-inflammatory response. Following SARS-CoV-2 infection, the immune response can become dysregulated and overactivated, leading to reduced lymphocytes, altered INF-I response, and the eventual cytokine storm (Ye et al., 2020). These responses can cause severe systemic damage, leading to multi-organ failure and even death. The inflammatory markers that are activated include IL-6, IL-1β, IL-2R, IL-8, TNF-α, and IFN-γ, as well as the chemokines CCL-2, CCL-3, and CCL-10 (Jiang et al., 2022). The exaggerated cytokine/chemokine response recruits more innate immune cells, leading to hyperactivation of the immune response, ultimately resulting in massive inflammation and cytokine storm. Upregulation of inflammatory factors exacerbates ARDs, which is the leading cause of death in COVID-19. Several inflammatory pathways are activated by SARS-CoV-2 infection, including Jak/STAT signaling

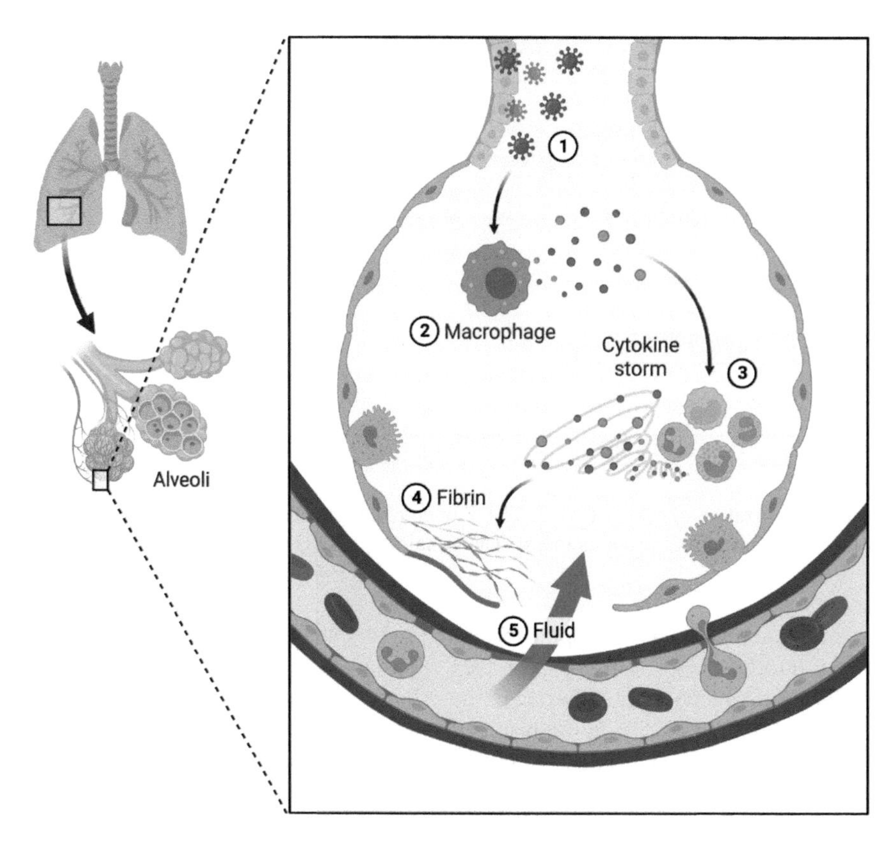

FIGURE 19.2 SARS-CoV-2 infection can lead to the development of cytokine storm in high-risk individuals. (1) SARS-CoV-2 infects cells; (2) immune cells, including macrophages, identify the virus and produce cytokines; (3) cytokines attract more immune cells, such as white blood cells, which in turn produce more cytokines, creating a cycle of inflammation that damages the lung cells; (4) damage can occur through the formation of fibrin; and (5) weakened blood vessels allow fluid to seep in and fill the lung cavities leading to respiratory failure (created with Biorender).

(Goker Bagca and Biray Avci, 2020), NF-κB signaling, BTK signaling, MAPK signaling, and the growth factor pathway (Jiang et al., 2022). Therefore, understanding and targeting these inflammatory pathways and cytokines is essential in the battle against SARS-CoV-2 (Catanzaro et al., 2020; Madden and Diamond, 2022).

19.4 FLAVONOIDS

Natural products serve as targets for novel drug discovery and may therefore provide an excellent therapeutic starting point in the fight against SAR-CoV-2 (Liu et al., 2022). These compounds exert several health-promoting effects, including antioxidant support, anti-inflammatory effects, and antiviral properties (Cazarolli et al., 2008; Liskova et al., 2021).

Flavonoids are hydroxylated phenolic phytomolecules typically found in the secondary metabolites of plants. They are characterized by a diphenylpropane structure, and the A- and B-rings of flavonoids are usually functionalized by –OH, isoprenyl, and glycosyl groups (Hostetler et al., 2017; Wang et al., 2018). Flavonoids are further divided into several subgroups based on their structural differences. The subgroups include flavones, flavan-3-ols, isoflavones, flavanols, flavanones, and anthocyanidins (Figure 19.3).

As antiviral agents, flavonoids have been reported to (i) act as a direct viricide, (ii) block viral binding, (iii) block transcription, and (iv) block co-option of intracellular enzymes required for replication (Li et al., 2000; Seong et al., 2018). -Flavonoids have also been studied as possible anti-coronavirus agents and several studies report activity against SARS-CoV, MERS-CoV, and HCoV-229E (Jo et al., 2019; Jo et al., 2020b; Mori et al., 2023). Within this context, flavonoids were explored for their anti-SARS-CoV-2 activity during the COVID-19 pandemic, and several flavonoids with SARS-CoV-2 disruptive activity were identified (Jo et al., 2020a), many belonging to the flavanol and flavone subclass (Kaul et al., 2021). The flavones, the focus of this chapter, have a 3-hydroxyflavone backbone, and although they have minor differences in their hydroxyl groups, these dramatically affect their biological activities (Butun et al., 2018).

19.4.1 Flavones

Flavones have recently seen increased scientific study due to their noteworthy biological activities both *in vitro* and *in vivo* (Hostetler et al., 2017). In plants, the flavones play various important roles, such as defending against pathogens, insects, and environmental stressors. These compounds

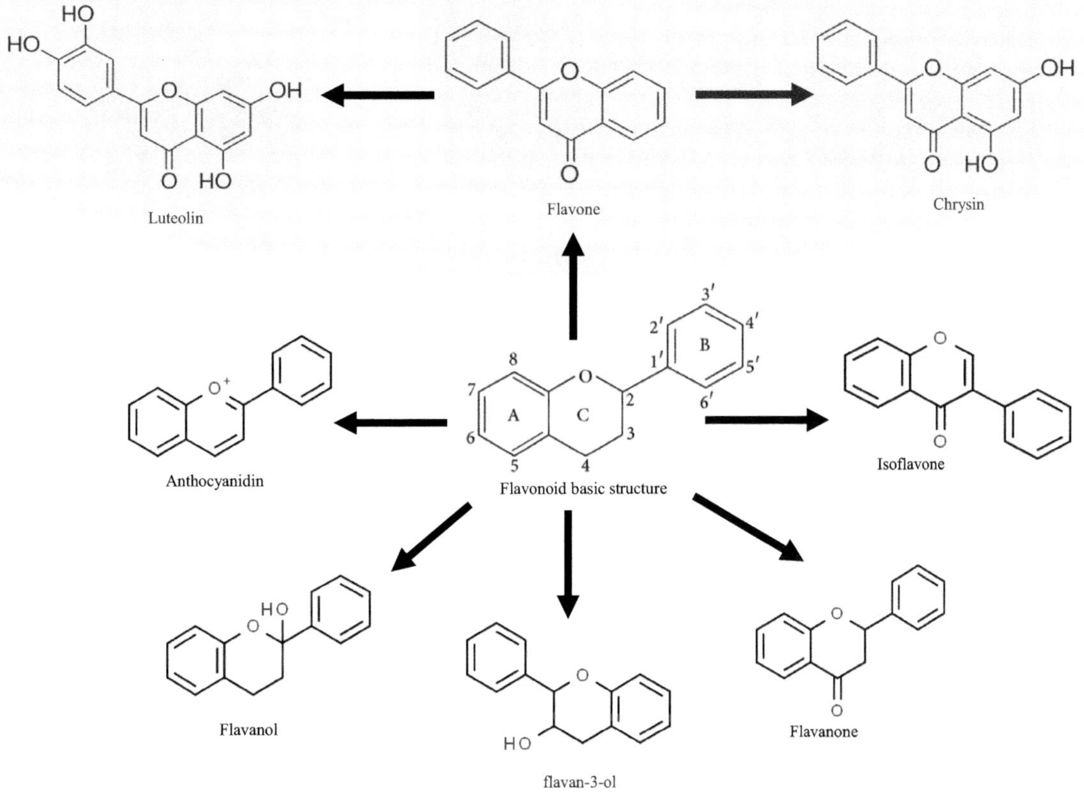

FIGURE 19.3 Chemical structures of flavonoids and subclasses of flavonoids, with emphasis on the flavones.

generally have excellent safety profiles and bioavailability in humans, and numerous health-promoting benefits have been reported (Wang et al., 2018). Health benefits include antimicrobial, antioxidant, anti-inflammatory, anti-mutagenic, anticancer, and antiviral effects (Jiang et al., 2016).

19.4.1.1 Luteolin

Luteolin (also known as 3,4,5,7-tetrahydroxyflavone) is naturally occurring in over 300 plant species, including *Asteraceae*, *Lamiaceae*, *Poaceae*, *Leguminosae*, and *Scrophulariaceae*. It's often found glycosylated in plants, with the glycoside being hydrolyzed to free luteolin during absorption (Muruganathan et al., 2022). This flavone is a cornerstone of numerous traditional and phytomedicine modalities, stimulating widespread scientific interest and investigation (Ntalouka and Tsirivakou, 2023). To date, luteolin boasts an impressive collection of recorded pharmacological properties, including antioxidant support, immune regulation, anti-parasitic, antimicrobial, anti-inflammation, neuroprotection, analgesic, anticancer, and antiviral effects (Lehane and Saliba, 2008; Hussain et al., 2023; Rakoczy et al., 2023).

As early as 2021, luteolin was predicted to be a potent blocker of SARS-Cov-2 entry into cells (Shadrack et al., 2021). This was confirmed by Zhu and colleagues (2023) who used a competitive *in vitro* binding assay and an *in vivo* SARS-CoV-2 pseudovirus study to show that luteolin indeed blocks the binding of the S RBD to the cell receptor ACE2 (Figure 19.4). Interestingly, in the same study, the authors reported that luteolin inhibits SARS-CoV-2 S-induced platelet spreading, suggesting a potential application for treating thrombosis in SARS-CoV-2 infection by lowering spike-induced platelet adhesion to fibrinogen (Zhu et al., 2023).

Even though time-of-addition assays identified that luteolin acts as an inhibitor of HCoV-229E replication, no obvious antiviral activity was detected against SARS-CoV-2 and MERS-CoV in this particular study (Hakem et al., 2023). In bioinformatic studies, however, luteolin was predicted to exhibit a stable docking structure in the binding pocket of PL^{pro} and M^{pro} (Wang et al., 2023a), potentially inhibiting SARS-CoV-2 replication. Further experimental evidence supports the flavone inhibiting the M^{pro} and RdRp (Munafo et al., 2022) (Figure 19.5), ultimately leading to the inhibition of viral replication (Munafò et al., 2022). The contradictory findings on the effect of luteolin on SARS-CoV-2 replication could indicate that it requires specific conditions to optimally interfere with virus replication, at least in *in vitro* assays. To date, studies indicate that luteolin is able to prevent SARS-CoV-2 entry into the cell, in addition to inhibiting virus replication and transcription by acting as an allosteric modulator when binding to M^{pro}, PL^{pro}, and/or RdRp (Munafo et al., 2022).

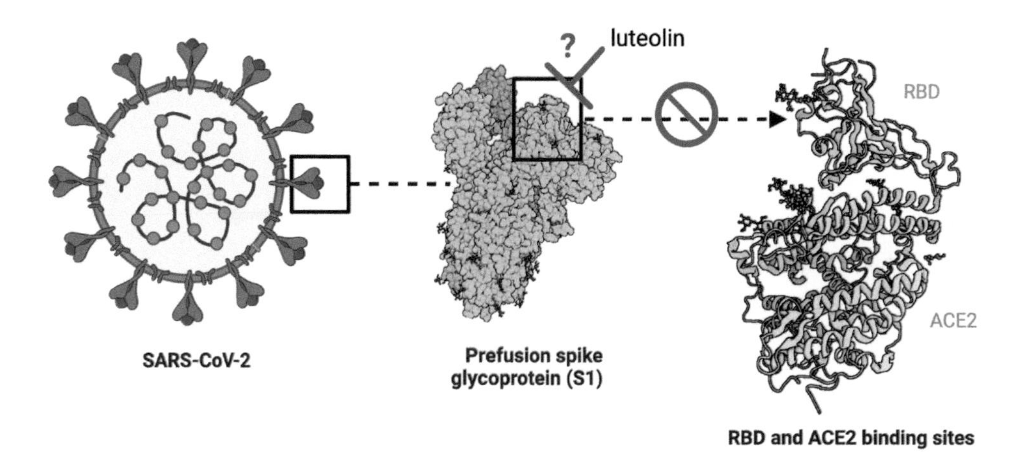

FIGURE 19.4 Luteolin interferes with the stable interaction between the SARS-CoV-2 S protein RBD and ACE2. Luteolin is predicted to interfere with residues LYS353, ASP30, and TYR83 in the cellular ACE2 receptor and GLY496, GLN498, TYR505, LEU455, GLN493, and GLU484 in the S protein RBD to prevent the entry into the susceptible cell (created with Biorender).

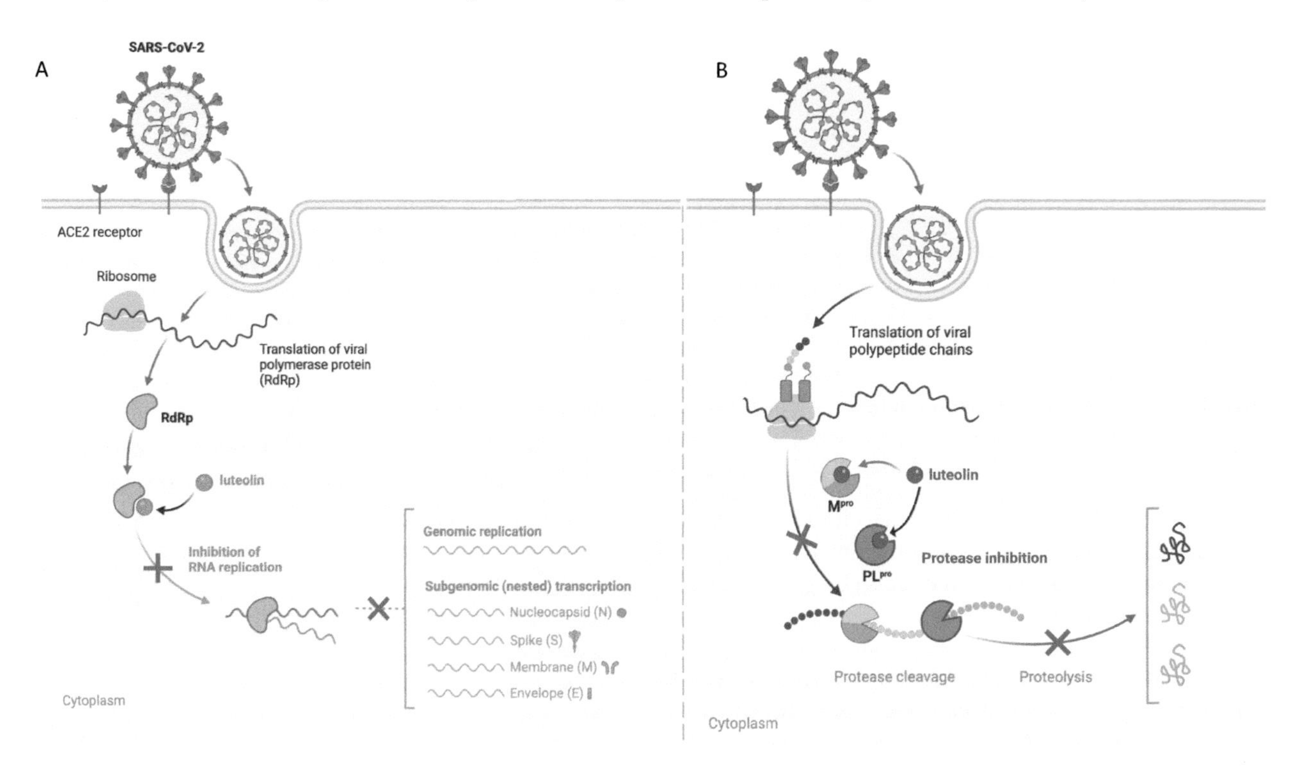

FIGURE 19.5 Luteolin interacts with SARS-CoV-2 PL^{pro}, M^{pro}, and RdRp. Various bioinformatics studies predict that luteolin binds to PL^{pro}, M^{pro}, and RdRp. (a) Luteolin binds to the binding pocket of RdRp which inhibits RNA replication, resulting in the interruption of RNA transcription and viral replication. (b) Luteolin binds to the binding pocket of PL^{pro} and/or M^{pro}. This inhibits the cleavage and processing of the viral polyprotein, which results in the cessation of viral replication and transcription (created with Biorender).

Luteolin has been reported to lower the number of CD4+ and CD25+ regulatory T-cells, which play a crucial role in immune balance, along with decreasing CD19+B, CD4+T, CD3-CCR3+, and NF-κB in murine models, indicating a potential role in the modulation of the body's inflammatory response and signaling. Specifically, the flavone functions by lowering the concentration of pro-inflammatory cytokines TNF-α, IL-6, and IL-1β and reducing myeloperoxidase activity (Kim et al., 2018; Liu et al., 2018). Luteolin has also been reported to have a strong binding affinity for several immune and stress response regulators, including CFB, EIF2AK2, OAS1, MAPK11, OAS3, and STAT1. This indicates the potential for luteolin to combat SARS-CoV-2-related symptoms by modulating the often exaggerated immune and inflammatory response (Xie et al., 2021). In fact, recent clinical trials have demonstrated the regulatory effects of luteolin, combined with olfactory training and palmitoylethanolamide supplementation, longstanding olfactory inflammation and dysfunction common to SARS-CoV-2 infection (D'Ascanio et al., 2021; De Luca et al., 2022).

19.4.1.2 Chrysin

Chrysin (also known as 5,7-hidroxyflavone or 5,7-dihydroxy-2-phenyl-4H-chromen-4-one) is a dietary phytochemical, present in different species of passion flowers, such as *Passiflora caerulea*, *Passiflora incarnata*, and *Oroxylum indicum*, as well as chamomile, mushrooms, *Pleurotus ostreatus*, propolis, and honey (Garg and Chaturvedi, 2022).

Pharmacologically, chrysin has been reported to have anti-diabetic, anticonvulsant, anticancer, anxiolytic, antidepressant, neuroprotective, anticancer, and anti-arthritis activity (Stompor-Goracy et al., 2021). It has also been shown to exert protective effects in various tissues, such as the heart, brain, kidney, and lungs. Moreover, chrysin has been shown to possess antiviral activity (Du et al., 2016; Bhat et al., 2023), but clinical relevance has not yet reliably been achieved due to poor aqueous solubility, pre-systemic metabolism in the intestines, and first-pass metabolism in the liver (Garg and Chaturvedi, 2022).

As mentioned earlier, chrysin has antiviral activity against many medically relevant viruses, including enterovirus 71 (EV71) (Wang et al., 2014), coxsackievirus B3 (CBV3) (Song et al., 2015), influenza virus (Kim et al., 2021), hepatitis B virus (HBV) (Bhat et al., 2023), and dengue and Zika virus (Suroengrit et al., 2017). Chrysin also has antiviral activity against the veterinary important coronavirus, porcine epidemic diarrhea virus (PEDV), by interacting with the 3CL^{pro} or PLP2 proteins; this inhibits the formation of PEDV nsps, and thus, interferes with virus replication (Kim et al., 2021).

Recent studies show that chrysin inhibits SARS-CoV-2 entry into susceptible cells by interfering with the interaction between ACE2 and S (Alzaabi et al., 2022). Whereas some studies conclude that chrysin binds to the cellular receptor ACE2, inducing changes to the compactness of the ACE2 structure which abrogates the ACE2–S interaction (Shahbazi et al., 2022), others report that chrysin in fact targets both ACE2, as well as the RBD region of SARS-CoV-2 (Guler et al., 2021). Going one step further,

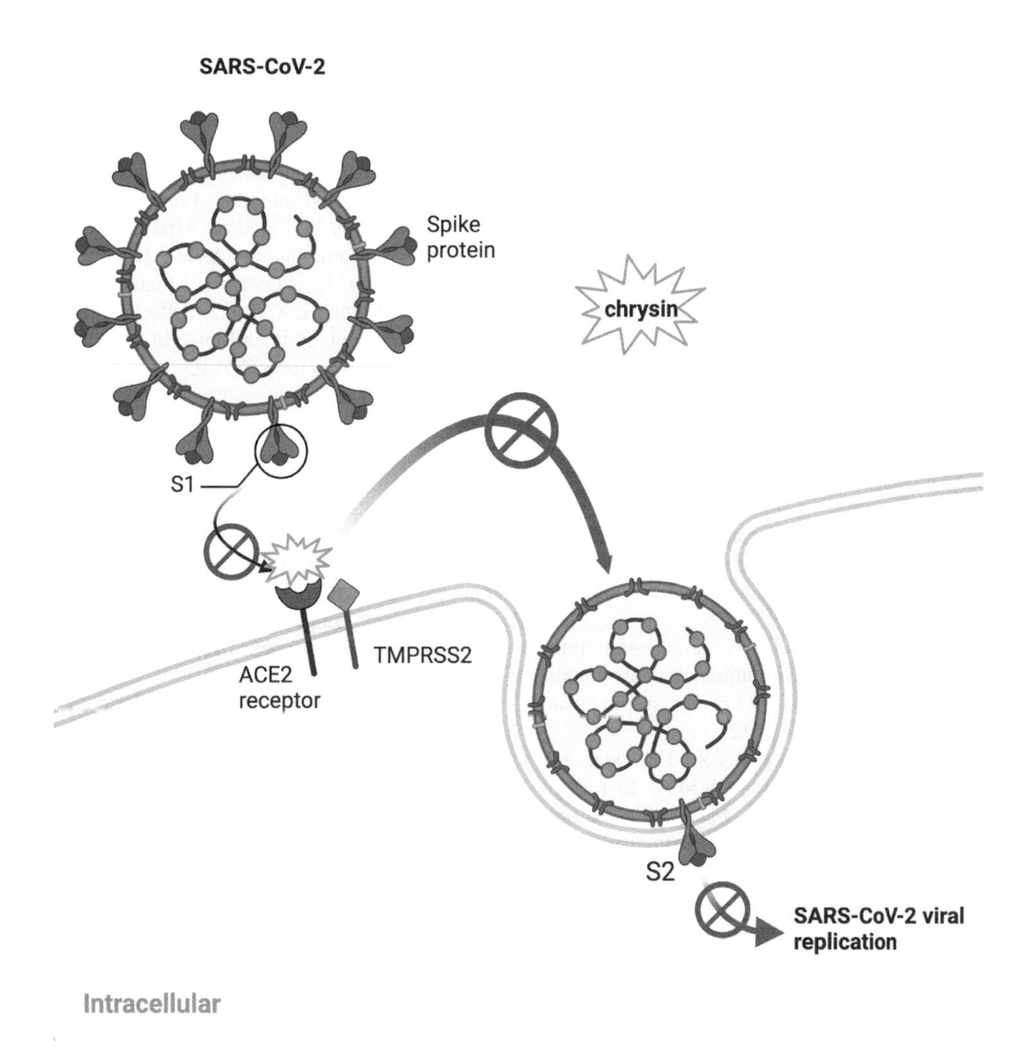

FIGURE 19.6 Chrysin interferes with the stable interaction between the SARS-CoV-2 S protein RBD and ACE2. Chrysin binds to the ACE2 and induces changes to the compactness of the ACE2 structure, causing positional changes in the ACE2 surface residues; this likely results in lower availability of the critical amino acids on the ACE2 receptor for the attachment to SARS-CoV-2 spike protein (created with Biorender).

Shahbazi and colleagues (2022) using bioinformatic analysis predict that chrysin interacts with ACE2 in such a way that it specifically blocks the S-binding pocket present in ACE2 (Figure 19.5), potentially inhibiting viral entry into the cell (Shahbazi et al., 2022).

As with the other viruses mentioned, chrysin (or at least the chrysin derivative chrysin-7-O-β-D-glucuronide) also appears to interfere with SARS-CoV-2 replication by binding specifically to the SARS-CoV-2 M^{pro} and PL^{pro}. This translates to inhibition of productive infection of SARS-CoV-2 in Vero E6 cells in antiviral *in vivo* studies. As a bonus, the authors of this study also reported anti-inflammatory activity as a result of chrysin-7-O-β-D-glucuronide decreasing the level of cytokines (Yi et al., 2023). This is evidence that research into the anti-SARS-CoV-2 potential of semisynthetic chrysin derivatives could prove useful (Figure 19.6).

19.5 CONCLUSION

The study of the anti-SARS-CoV-2 potential of flavonoids, informed by prior research on other viruses (Badshah et al., 2021), including other coronaviruses (Mori et al., 2023), and their potential in mitigating post-COVID-19 syndrome through their anti-inflammatory properties (Liskova et al., 2021; Wang et al., 2023b), highlights the need to explore their anti-coronavirus – in particular its anti-SARS-CoV-2 – activity. In summary, luteolin and chrysin show promising results in anti-SARS-CoV-2 activity (Wang et al., 2023a; Yi et al., 2023), indicating their potential as leads for further drug development in COVID-19 treatments. Nevertheless, further research is required to determine their clinical relevance, either in combination with other anti-COVID-19 therapies or as standalone treatments.

REFERENCES

Aleem, A., Akbar Samad, A.B., and Vaqar, S. (2023). *Emerging Variants of SARS-CoV-2 and Novel Therapeutics Against Coronavirus (COVID-19)*. Tampa, FL: StatPearls Publishing.

Alimohamadi, Y., Sepandi, M., Taghdir, M., and Hosamirudsari, H. (2020). Determine the most common clinical symptoms in COVID-19 patients: A systematic review and meta-analysis. *J Prev Med Hyg* 61, E304–E312.

Alzaabi, M.M., Hamdy, R., Ashmawy, N.S., Hamoda, A.M., Alkhayat, F., Khademi, N.N., Al Joud, S.M.A., El-Keblawy, A.A., and Soliman, S.S.M. (2022). Flavonoids are promising safe therapy against COVID-19. *Phytochem Rev* 21, 291–312.

Astuti, I., and Ysrafil, Y. (2020). Severe acute respiratory syndrome coronavirus 2 (SARS-CoV-2): An overview of viral structure and host response. *Diabetes Metab Syndr* 14, 407–412.

Attaway, A.H., Scheraga, R.G., Bhimraj, A., Biehl, M., and Hatipoglu, U. (2021). Severe covid-19 pneumonia: Pathogenesis and clinical management. *BMJ* 372, n436.

Badshah, S.L., Faisal, S., Muhammad, A., Poulson, B.G., Emwas, A.H., and Jaremko, M. (2021). Antiviral activities of flavonoids. *Biomed Pharmacother* 140, 111596.

Bhat, S.A., Hasan, S.K., Parray, Z.A., Siddiqui, Z.I., Ansari, S., Anwer, A., Khan, S., Amir, F., Mehmankhah, M., Islam, A., Minuchehr, Z., and Kazim, S.N. (2023). Potential antiviral activities of chrysin against hepatitis B virus. *Gut Pathog* 15, 11.

Butun, B., Topcu, G., and Ozturk, T. (2018). Recent advances on 3-hydroxyflavone derivatives: Structures and properties. *Mini Rev Med Chem* 18, 98–103.

Cao, C., Cai, Z., Xiao, X., Rao, J., Chen, J., Hu, N., Yang, M., Xing, X., Wang, Y., Li, M., Zhou, B., Wang, X., Wang, J., and Xue, Y. (2021). The architecture of the SARS-CoV-2 RNA genome inside virion. *Nat Commun* 12, 3917.

Catanzaro, M., Fagiani, F., Racchi, M., Corsini, E., Govoni, S., and Lanni, C. (2020). Immune response in COVID-19: Addressing a pharmacological challenge by targeting pathways triggered by SARS-CoV-2. *Signal Transduct Target Ther* 5, 84.

Cazarolli, L.H., Zanatta, L., Alberton, E.H., Figueiredo, M.S., Folador, P., Damazio, R.G., Pizzolatti, M.G., and Silva, F.R. (2008). Flavonoids: Prospective drug candidates. *Mini Rev Med Chem* 8, 1429–1440.

Cocco, A., Amami, P., Desai, A., Voza, A., Ferreli, F., and Albanese, A. (2021). Neurological features in SARS-CoV-2-infected patients with smell and taste disorder. *J Neurol* 268, 1570–1572.

D'ascanio, L., Vitelli, F., Cingolani, C., Maranzano, M., Brenner, M.J., and Di Stadio, A. (2021). Randomized clinical trial "olfactory dysfunction after COVID-19: Olfactory rehabilitation therapy vs. intervention treatment with palmitoylethanolamide and luteolin": Preliminary results. *Eur Rev Med Pharmacol Sci* 25, 4156–4162.

De Luca, P., Camaioni, A., Marra, P., Salzano, G., Carriere, G., Ricciardi, L., Pucci, R., Montemurro, N., Brenner, M.J., and Di Stadio, A. (2022). Effect of ultra-micronized palmitoylethanolamide and luteolin on olfaction and memory in patients with long COVID: Results of a longitudinal study. *Cells* 11, 2552.

Du, J., Chen, Z., Zhang, T., Wang, J., and Jin, Q. (2016). Inhibition of dengue virus replication by diisopropyl chrysin-7-yl phosphate. *Sci China Life Sci* 59, 832–838.

Fang, P., Fang, L., Zhang, H., Xia, S., and Xiao, S. (2021). Functions of coronavirus accessory proteins: Overview of the state of the art. *Viruses* 13, 1139.

Garg, A., and Chaturvedi, S. (2022). A comprehensive review on chrysin: Emphasis on molecular targets, pharmacological actions and bio-pharmaceutical aspects. *Curr Drug Targets* 23, 420–436.

Goker Bagca, B., and Biray Avci, C. (2020). The potential of JAK/STAT pathway inhibition by ruxolitinib in the treatment of COVID-19. *Cytokine Growth Factor Rev* 54, 51–62.

Guler, H.I., Ay Sal, F., Can, Z., Kara, Y., Yildiz, O., Belduz, A.O., Canakci, S., and Kolayli, S. (2021). Targeting CoV-2 spike RBD and ACE-2 interaction with flavonoids of Anatolian propolis by in silico and in vitro studies in terms of possible COVID-19 therapeutics. *Turk J Biol* 45, 530–548.

Hadjadj, J., Yatim, N., Barnabei, L., Corneau, A., Boussier, J., Smith, N., Pere, H., Charbit, B., Bondet, V., Chenevier-Gobeaux, C., Breillat, P., Carlier, N., Gauzit, R., Morbieu, C., Pene, F., Marin, N., Roche, N., Szwebel, T.A., Merkling, S.H., Treluyer, J.M., Veyer, D., Mouthon, L., Blanc, C., Tharaux, P.L., Rozenberg, F., Fischer, A., Duffy, D., Rieux-Laucat, F., Kerneis, S., and Terrier, B. (2020). Impaired type I interferon activity and inflammatory responses in severe COVID-19 patients. *Science* 369, 718–724.

Hakem, A., Desmarets, L., Sahli, R., Malek, R.B., Camuzet, C., Francois, N., Lefevre, G., Samaillie, J., Moureu, S., Sahpaz, S., Belouzard, S., Ksouri, R., Seron, K., and Riviere, C. (2023). Luteolin isolated from *Juncus acutus* L., a potential remedy for human coronavirus 229E. *Molecules* 28, 4263.

Hostetler, G.L., Ralston, R.A., and Schwartz, S.J. (2017). Flavones: Food sources, bioavailability, metabolism, and bioactivity. *Adv Nutr* 8, 423–435.

Huang, Y., Yang, C., Xu, X.F., Xu, W., and Liu, S.W. (2020). Structural and functional properties of SARS-CoV-2 spike protein: Potential antivirus drug development for COVID-19. *Acta Pharmacol Sin* 41, 1141–1149.

Hussain, M.S., Gupta, G., Goyal, A., Thapa, R., Almalki, W.H., Kazmi, I., Alzarea, S.I., Fuloria, S., Meenakshi, D.U., Jakhmola, V., Pandey, M., Singh, S.K., and Dua, K. (2023). From nature to therapy: Luteolin's potential as an immune system modulator in inflammatory disorders. *J Biochem Mol Toxicol* 37, e23482.

Jiang, N., Doseff, A.I., and Grotewold, E. (2016). Flavones: From biosynthesis to health benefits. *Plants (Basel)* 5, 27.

Jiang, Y., Tong, K., Yao, R., Zhou, Y., Lin, H., Du, L., Jin, Y., Cao, L., Tan, J., Zhang, X.D., Guo, D., Pan, J.A., and Peng, X. (2021). Genome-wide analysis of protein-protein interactions and involvement of viral proteins in SARS-CoV-2 replication. *Cell Biosci* 11, 140.

Jiang, Y., Zhao, T., Zhou, X., Xiang, Y., Gutierrez-Castrellon, P., and Ma, X. (2022). Inflammatory pathways in COVID-19: Mechanism and therapeutic interventions. *MedComm (2020)* 3, e154.

Jo, S., Kim, H., Kim, S., Shin, D.H., and Kim, M.S. (2019). Characteristics of flavonoids as potent MERS-CoV 3C-like protease inhibitors. *Chem Biol Drug Des* 94, 2023–2030.

Jo, S., Kim, S., Kim, D.Y., Kim, M.S., and Shin, D.H. (2020a). Flavonoids with inhibitory activity against SARS-CoV-2 3CLpro. *J Enzyme Inhib Med Chem* 35, 1539–1544.

Jo, S., Kim, S., Shin, D.H., and Kim, M.S. (2020b). Inhibition of SARS-CoV 3CL protease by flavonoids. *J Enzyme Inhib Med Chem* 35, 145–151.

Kaul, R., Paul, P., Kumar, S., Busselberg, D., Dwivedi, V.D., and Chaari, A. (2021). Promising antiviral activities of natural flavonoids against SARS-CoV-2 targets: Systematic review. *Int J Mol Sci* 22, 11069.

Kim, S.H., Saba, E., Kim, B.K., Yang, W.K., Park, Y.C., Shin, H.J., Han, C.K., Lee, Y.C., and Rhee, M.H. (2018). Luteolin attenuates airway inflammation by inducing the transition of CD4(+)CD25(-) to CD4(+)CD25(+) regulatory T cells. *Eur J Pharmacol* 820, 53–64.

Kim, S.R., Jeong, M.S., Mun, S.H., Cho, J., Seo, M.D., Kim, H., Lee, J., Song, J.H., and Ko, H.J. (2021). Antiviral activity of chrysin against influenza virus replication via inhibition of autophagy. *Viruses* 13, 1350.

Lehane, A.M., and Saliba, K.J. (2008). Common dietary flavonoids inhibit the growth of the intraerythrocytic malaria parasite. *BMC Res Notes* 1, 26.

Lei, S., Chen, X., Wu, J., Duan, X., and Men, K. (2022). Small molecules in the treatment of COVID-19. *Signal Transduct Target Ther* 7, 387.

Li, F. (2016). Structure, function, and evolution of coronavirus spike proteins. *Annu Rev Virol* 3, 237–261.

Li, B.Q., Fu, T., Dongyan, Y., Mikovits, J.A., Ruscetti, F.W., and Wang, J.M. (2000). Flavonoid baicalin inhibits HIV-1 infection at the level of viral entry. *Biochem Biophys Res Commun* 276, 534–538.

Liskova, A., Samec, M., Koklesova, L., Samuel, S.M., Zhai, K., Al-Ishaq, R.K., Abotaleb, M., Nosal, V., Kajo, K., Ashrafizadeh, M., Zarrabi, A., Brockmueller, A., Shakibaei, M., Sabaka, P., Mozos, I., Ullrich, D., Prosecky, R., La Rocca, G., Caprnda, M., Busselberg, D., Rodrigo, L., Kruzliak, P., and Kubatka, P. (2021). Flavonoids against the SARS-CoV-2 induced inflammatory storm. *Biomed Pharmacother* 138, 111430.

Liu, D.X., Fung, T.S., Chong, K.K., Shukla, A., and Hilgenfeld, R. (2014). Accessory proteins of SARS-CoV and other coronaviruses. *Antiviral Res* 109, 97–109.

Liu, B., Yu, H., Baiyun, R., Lu, J., Li, S., Bing, Q., Zhang, X., and Zhang, Z. (2018). Protective effects of dietary luteolin against mercuric chloride-induced lung injury in mice: Involvement of AKT/Nrf2 and NF-kappaB pathways. *Food Chem Toxicol* 113, 296–302.

Liu, X.H., Cheng, T., Liu, B.Y., Chi, J., Shu, T., and Wang, T. (2022). Structures of the SARS-CoV-2 spike glycoprotein and applications for novel drug development. *Front Pharmacol* 13, 955648.

Madden, E.A., and Diamond, M.S. (2022). Host cell-intrinsic innate immune recognition of SARS-CoV-2. *Curr Opin Virol* 52, 30–38.

Malla, A.M., Dar, B.A., Isaev, A.B., Lone, Y., and Banday, M.R. (2023). Flavonoids: A reservoir of drugs from nature. *Mini Rev Med Chem* 23, 772–786.

Malone, B., Urakova, N., Snijder, E.J., and Campbell, E.A. (2022). Structures and functions of coronavirus replication-transcription complexes and their relevance for SARS-CoV-2 drug design. *Nat Rev Mol Cell Biol* 23, 21–39.

Morbini, P., Benazzo, M., Verga, L., Pagella, F.G., Mojoli, F., Bruno, R., and Marena, C. (2020). Ultrastructural evidence of direct viral damage to the olfactory complex in patients testing positive for SARS-CoV-2. *JAMA Otolaryngol Head Neck Surg* 146, 972–973.

Mori, M., Quaglio, D., Calcaterra, A., Ghirga, F., Sorrentino, L., Cammarone, S., Fracella, M., D'auria, A., Frasca, F., Criscuolo, E., Clementi, N., Mancini, N., Botta, B., Antonelli, G., Pierangeli, A., and Scagnolari, C. (2023). Natural flavonoid derivatives have pan-coronavirus antiviral activity. *Microorganisms* 11, 314.

Munafo, F., Donati, E., Brindani, N., Ottonello, G., Armirotti, A., and De Vivo, M. (2022). Quercetin and luteolin are single-digit micromolar inhibitors of the SARS-CoV-2 RNA-dependent RNA polymerase. *Sci Rep* 12, 10571.

Murrell, D.F., Rudnicka, L., Shivakumar, S., Kassir, M., Jafferany, M., Galadari, H., Lotti, T., Sadoughifar, R., Sitkowska, Z., and Goldust, M. (2020). Biologics and small molecules in the treatment of COVID-19. *J Drugs Dermatol* 19, 673–675.

Muruganathan, N., Dhanapal, A.R., Baskar, V., Muthuramalingam, P., Selvaraj, D., Aara, H., Shiek Abdullah, M.Z., and Sivanesan, I. (2022). Recent updates on source, biosynthesis, and therapeutic potential of natural flavonoid luteolin: A review. *Metabolites* 12, 1145.

Ntalouka, F., and Tsirivakou, A. (2023). Luteolin: A promising natural agent in management of pain in chronic conditions. *Front Pain Res (Lausanne)* 4, 1114428.

Rakoczy, K., Kaczor, J., Soltyk, A., Szymanska, N., Stecko, J., Sleziak, J., Kulbacka, J., and Baczynska, D. (2023). Application of luteolin in neoplasms and nonneoplastic diseases. *Int J Mol Sci* 24, 15995.

Robinson, B.W.S., Tai, A., and Springer, K. (2020). Why we still need drugs for COVID-19 and can't just rely on vaccines. *Respirology* 27, 3.

Seong, R.K., Kim, J.A., and Shin, O.S. (2018). Wogonin, a flavonoid isolated from Scutellaria baicalensis, has anti-viral activities against influenza infection via modulation of AMPK pathways. *Acta Virol* 62, 78–85.

Shadrack, D.M., Deogratias, G., Kiruri, L.W., Onoka, I., Vianney, J.M., Swai, H., and Nyandoro, S.S. (2021). Luteolin: A blocker of SARS-CoV-2 cell entry based on relaxed complex scheme, molecular dynamics simulation, and metadynamics. *J Mol Model* 27, 221.

Shahbazi, B., Mafakher, L., and Teimoori-Toolabi, L. (2022). Different compounds against angiotensin-converting enzyme 2 (ACE2) receptor potentially containing the infectivity of SARS-CoV-2: An in silico study. *J Mol Model* 28, 82.

Shi, S.T., and Lai, M.M. (2005). Viral and cellular proteins involved in coronavirus replication. *Curr Top Microbiol Immunol* 287, 95–131.

Song, J.H., Kwon, B.E., Jang, H., Kang, H., Cho, S., Park, K., Ko, H.J., and Kim, H. (2015). Antiviral activity of chrysin derivatives against coxsackievirus B3 in vitro and in vivo. *Biomol Ther (Seoul)* 23, 465–470.

Stompor-Goracy, M., Bajek-Bil, A., and Machaczka, M. (2021). Chrysin: Perspectives on contemporary status and future possibilities as pro-health agent. *Nutrients* 13, 2038.

Suroengrit, A., Yuttithamnon, W., Srivarangkul, P., Pankaew, S., Kingkaew, K., Chavasiri, W., and Boonyasuppayakorn, S. (2017). Halogenated chrysins inhibit dengue and zika virus infectivity. *Sci Rep* 7, 13696.

V'kovski, P., Kratzel, A., Steiner, S., Stalder, H., and Thiel, V. (2021). Coronavirus biology and replication: Implications for SARS-CoV-2. *Nat Rev Microbiol* 19, 155–170.

Van Der Hoek, L., Pyrc, K., and Berkhout, B. (2006). Human coronavirus NL63, a new respiratory virus. *FEMS Microbiol Rev* 30, 760–773.

Wang, J., Zhang, T., Du, J., Cui, S., Yang, F., and Jin, Q. (2014). Anti-enterovirus 71 effects of chrysin and its phosphate ester. *PLoS One* 9, e89668.

Wang, T.Y., Li, Q., and Bi, K.S. (2018). Bioactive flavonoids in medicinal plants: Structure, activity and biological fate. *Asian J Pharm Sci* 13, 12–23.

Wang, W., Yang, C., Xia, J., Li, N., and Xiong, W. (2023a). Luteolin is a potential inhibitor of COVID-19: An in silico analysis. *Medicine (Baltimore)* 102, e35029.

Wang, X., Yao, Y., Li, Y., Guo, S., Li, Y., and Zhang, G. (2023b). Experimental study on the effect of luteolin on the proliferation, apoptosis and expression of inflammation-related mediators in lipopolysaccharide-induced keratinocytes. *Int J Immunopathol Pharmacol* 37, 3946320231169175.

Woo, P.C., Lau, S.K., Lam, C.S., Lau, C.C., Tsang, A.K., Lau, J.H., Bai, R., Teng, J.L., Tsang, C.C., Wang, M., Zheng, B.J., Chan, K.H., and Yuen, K.Y. (2012). Discovery of seven novel Mammalian and avian coronaviruses in the genus deltacoronavirus supports bat coronaviruses as the gene source of alphacoronavirus and betacoronavirus and avian coronaviruses as the gene source of gammacoronavirus and deltacoronavirus. *J Virol* 86, 3995–4008.

Xie, Y.Z., Peng, C.W., Su, Z.Q., Huang, H.T., Liu, X.H., Zhan, S.F., and Huang, X.F. (2021). A practical strategy for exploring the pharmacological mechanism of luteolin against COVID-19/asthma comorbidity: Findings of system pharmacology and bioinformatics analysis. *Front Immunol* 12, 769011.

Ye, Q., Wang, B., and Mao, J. (2020). The pathogenesis and treatment of the 'Cytokine Storm' in COVID-19. *J Infect* 80, 607–613.

Yi, Y., Yu, R., Xue, H., Jin, Z., Zhang, M., Bao, Y.O., Wang, Z., Wei, H., Qiao, X., and Yang, H. (2023). Chrysin 7-O-beta-D-glucuronide, a dual inhibitor of SARS-CoV-2 3CL(pro) and PL(pro), for the prevention and treatment of COVID-19. *Int J Antimicrob Agents* 63, 107039.

Zhu, J., Yan, H., Shi, M., Zhang, M., Lu, J., Wang, J., Chen, L., Wang, Y., Li, L., Miao, L., and Zhang, H. (2023). Luteolin inhibits spike protein of severe acute respiratory syndrome coronavirus-2 (SARS-CoV-2) binding to angiotensin-converting enzyme 2. *Phytother Res* 37, 3508–3521.

20 Flavonoids

Inhibitors at Stages of Infection by SARS-CoV-2

Jothi Dheivasikamani Abidharini, Ayyadurai Pavithra, Arthi Boro, Palanisamy Sampathkumar, Ramalingam Sivakumar, Saravanan Renuka, Gunna Sureshbabu Suruthi, and Arumugam Vijaya Anand

LIST OF ABBREVIATIONS

ACE2 Angiotensin-Converting Enzyme 2
ARDS Acute Respiratory Distress Syndrome
E Envelope Proteins
ETM Transmembrane Domain
EV71 Enterovirus 71
HCMV Human Cytomegalovirus
HIV Human Immunodeficiency Virus
HR Heptad Repeat
M Membrane Protein
NF-κB Nuclear Factor
ORFs Open Reading Frames
RBD Receptor-Binding Domain
RBD-S RBD of S Protein
S Spike Glycoprotein
SARS Severe Acute Respiratory Syndrome
ssRNA Single-Stranded Positive-Sense RNA
TNF-α Tumor Necrosis Factor-α
VLPs Virus-Like Particles
WHO World Health Organization

20.1 INTRODUCTION

In southern China, Guangdong Province, outbreaks of peculiar pneumonia were caused by an infectious pathogen in November 2002 (Peng et al., 2003; Shanmugam et al., 2020). The sickness typically developed with a high temperature and minor respiratory symptoms but promptly grew into pneumonia within a short period (Zhong et al., 2003). The epidemic had spread to nearby countries and territories by the end of February 2003 (Tsang et al., 2003; Lee et al., 2003). The illness was severe and was highly contagious from person to person and indiscriminately affected all individuals; hence, it appeared to pile up the concern of healthcare professionals worldwide. The sickness was called Severe Acute Respiratory Syndrome (SARS), and on March 13, 2003, WHO issued a global alert regarding it (Zhong et al., 2004; Meyyazhagan et al., 2022). SARS explosion happened in North America, Europe, and Southeast Asia, and they ended up creating the first pandemic of the twenty-first century. Single-stranded RNA viruses called coronaviruse pose health risks to people. Among the four coronavirus genera, alpha, beta, gamma, and delta, betacoronavirus has the most increased medical significance. When an infected individual sneezes or coughs, respiratory droplets are the main way that SARS-CoV spreads. The virus affects the respiratory system and causes symptoms like coughing, hemoptysis, sputum production, drowsiness, myalgia, and fever that damage the immune system. It may culminate organ dysfunction, respiratory infections, and acute respiratory distress syndrome (ARDS) in potential deadly cases, all with a high mortality rate (Chen et al., 2020; Yang and Shen, 2020; Gorbalenya et al., 2020; Wong et al., 2020; Zumla et al., 2020). The epidemiology data reports that there were a total of 8,098 reported infections and 774 mortalities as a result of the SARS outbreak in 2002–2003. Millions of confirmed illnesses and fatalities have been caused worldwide by the COVID-19 pandemic during September 2021. As more cases and data are recorded, the actual numbers are subject to change. At that time, the virus had already claimed over four million lives and infected many more individuals in varying degrees. There have been 3,029,811 deaths worldwide and 142,097,799 verified cases as of April 20, 2021. The number of deaths is increased quickly, by about 10,000 per day. There's a substantial likelihood that a few major SARS-CoV-2 strains gained over in many nations and spread uncontrollably over the world (MacLean et al., 2020; Ke et al., 2020; Plante et al., 2021).

The possible entry of the SARS virus is exhibited by surface receptors that bind with receptors in the host cells that permit the virus entry, and some studies have shown that the TLR4 receptor in the platelets interacts with angiotensin II which is released intracellularly through angiotensin-converting enzyme 2 (ACE2) due to virus entry, impairing the function and causing inflammation and platelet stacking in the lung and liver (Bhotla et al., 2020). The infections caused by the virus lead to other morbidities like hypertension, diabetes, heart disease, and kidney and liver disease that are suspected to decrease the survival of the patient (Arumugam et al., 2020).

DOI: 10.1201/9781003433200-20

20.2 SARS-COV

20.2.1 The Birth and Rise of SARS-CoV

The SARS-CoV-2 exhibited traits common to the 2B lineage of the betacoronaviruses family. SARS-CoV-2 samples collected from Wuhan showed genome sequences identical of 79.5% to SARS-CoV. From all this evidence, we can conclude that SARS-CoV-2 is a unique betacoronavirus that contaminates humans which has a divergent evolution from SARS-CoV. The whole genome investigation of the SARS-CoV-2 sequence revealed 96% identity to the strain BatCov RaTG13 obtained from bats. This study suggests that RaTG13, a bat coronavirus, might have developed into SARS-CoV-2 (Lai et al., 2020; Zhou et al., 2020). 13 mutation types in the spike protein have been recognized and studied to date. Out of which the mutation type Spike-D614G showed 8 times more efficacy at infecting cells than wild-type spike proteins (Zhang et al., 2020).

20.2.2 The Propagation of SARS-CoV

The SARS-CoV-2 virus is a natural reservoir in bats. They can infect human organs and intestinal epithelium as well as intestinal cells of bats. Although it does not cause a serious illness in bats, it can spread to other animals or people and cause disease epidemics. The closely contacted domestic animal species like dogs, chickens, ducks, and pigs are not excessively affected, but some evidence shows that this virus can infect cats, ferrets, and golden hamsters (Sia et al., 2020; Shi et al., 2020; Zhou et al., 2020). When the infected and uninfected people are in close proximity to one another, the infection is spread via fomites and droplets. SARS-CoV-2 aerosol transmission was also seen during the spread of COVID-19 (Chan et al., 2020; Meselson et al., 2020; Sommerstein et al., 2020). Until this, there is no proper proof of the communication of disease to babies from the affected pregnant women (Chen et al., 2020).

20.2.3 The Structure of SARS-CoV

The SARS-CoV has four well-known genera, all falling into the same family Coronaviridae. The SARS-CoV contains ssRNA (single-stranded positive-sense RNA) of about 27–32 kb long, which exceeds the size of all other known RNA viruses. The SARS-CoV-2 genome is as long as ~29.9 Kb in size which consists of 13–15 (12 functional) open reading frames (ORFs) with a total of around 30,000 nucleotides. The GC content of the genome is 38% with 11 genes coding for the expression of 12 proteins (Rota et al., 2002). The ORFs of SARS-CoV-2 compare favorably to those of SARS-CoV and MERS-CoV with an increased rate of resemblance (Lu et al., 2020). The genome is directly covered and protected by the nucleoproteins (N). Apart from this cover, there are three other proteins called envelope protein (E), membrane protein (M), and spike proteins (S) which make up the viral structure (Brian et al., 2005).

20.2.3.1 Nucleocapsid Protein (N)

An important component of the replica process is the SARS-CoV-2N, which encases the genome of the virus, guards it from the host cell surroundings, and controls the transcription of the gene. Nucleoprotein is a crucial indicator of infection since it is produced in large quantities by infected cells and has also been related to the disruption of various host functions (Almazán et al., 2004; McBride et al., 2014; Wu et al., 2014).

20.2.3.2 Envelope Proteins (E)

All coronaviruses has the envelope (E) protein, a short polypeptide with a minimum of one helical transmembrane domain (ETM) and 76 amino acids in SARS-CoV. The transmembrane region of SARS-CoV E extends over about 25 residues, roughly from position 10 to 35. The lipidic coat of the virion incorporates the M, S, and coronavirus E proteins. The E protein is not crucial to coronavirus replication *in vitro* or *in vivo*, even though the M protein is essential for envelope synthesis and budding. The S protein is crucial in merging with host membranes during entrance inside the cells. Viral morphogenesis, or the production and release of virus-like particles (VLPs), is influenced by the E protein. The assembly and maturation of the MHV virus were hampered by mutations in the extramembrane domain of the E protein. When E protein was absent in TGEV, virus trafficking via the secretory channel was blocked, and virus maturation was halted (Torres et al., 2006; Ortego et al., 2002; Curtis et al., 2002).

20.2.3.3 Membrane Protein (M)

The coronavirus M acts as a viral protein, i.e., structural, and is required for the assemblage and morphogenesis of the SARS-CoV. The most prevalent structural viral protein, M protein, is thought to recreate a key role in driving membrane budding and virus assembly. M protein is synthesized in the host's endoplasmic reticulum–golgi intermediate compartment, which delivers a medium for taking in other structural viral proteins. Three helices are transmembrane and a domain, i.e., infravision, which is found in the M protein, in which two transmembrane domains replaced the two intravirion domains and the three-helix bundles to generate a dimer that has a mushroom structure. Higher-order oligomers of the M protein continue to develop from this, and for any conformational changes, a highly conserved hinge region is required. Also, the viral ion channel SARS-CoV-2 ORF3a and the M protein dimer are found to be very similar in a recent study (Fehr and Perlman, 2015; Zhang et al., 2022).

20.2.3.4 Spike Glycoprotein (S)

The glycoprotein known as viral spike (S) is used to identify coronavirus infection. The 180-kDa oligomeric S protein of the murine coronavirus strain A59 that causes mouse hepatitis is cleaved post-translationally to produce the membrane fusion unit S2 and the binding receptor S1. The latter has two numbers of 4,3 hydrophobic (heptad) repeat sections, HR1 and HR2, and an internal fusion peptide. The membrane anchor is not far from HR2, and HR1 is about 170 amino acids upstream of it. Many distinct viruse fusion proteins contain heptad repeat (HR) regions which are a crucial component of class I viral fusion proteins. The coronavirus fusion protein S and class I virus fusion proteins have several similarities. After being produced in the endoplasmic reticulum, the type I M is translocated to the plasma membrane. It consists of two HR sequences: one near the transmembrane region and one downstream of the fusion peptide. The coronavirus S protein differs from class I fusion proteins in several ways, despite some similarities. The membrane-anchored component's lack of an N-terminal or even N-proximal fusion peptide is the first. The HR zones' relatively cnormous size (between 100 and 40 amino acids) is another feature. Third, membrane fusion does not need cleavage of the S protein; in fact, it does not occur in class 1 coronavirus (Bosch et al., 2003; Figure 20.1).

20.2.4 Mechanism of Infection of SARS-CoV and SARS-CoV-2

Infection mechanisms of SARS-CoV and SARS-CoV-2 are almost the same. This mechanism follows various steps in which the infection progresses. These steps are as follows:

Step 1: Contamination and Entry of the virus.

When an infected person coughs, sneezes, and breathes, respiratory droplets are released, which are then transmitted to another healthy person. The SARS virus primarily accesses the body through these droplets from individuals who are infected. The virus can also spread by touching the face, especially the mouth, nose, or eyes, after getting into contact with contaminated surfaces (Horve et al., 2022).

Step 2: Binding of the virus to the target receptor.

After the virus enters the respiratory tract through inhalation, it binds to particular receptor sites on the surface of human cells. The ACE2 receptor, which is found in cells in the lungs, heart, kidneys, intestines, and other organs, is the principal target for both SARS-CoV-2 and SARS-CoV (Ge et al., 2013).

Step 3: Membrane Fusion and incorporation of viral genetic material.

After it interacts with the receptors of ACE2, the envelope of the virus fuses with the cell membrane of the host, allowing the genetic material from the virus to enter the host cell. This is how viral genetic material enters the host cell and is incorporated into the host cell genome (Türk et al., 2021).

Step 4: Replication and transcription of viral RNA.

The viral RNA has entered the host cell and the machinery is used to release the viral genome for replication and to create viral proteins. The host's original transcription is manipulated by the virus to favor its replication (Malone et al., 2022).

Step 5: Spread of infection throughout the host body.

New viral particles are assembled inside the infected cell before being released out of the infected host cell. These particles then exit the cell, typically by budding through the cell membrane, and spread the infection to other healthy cells throughout the host (Baggen et al., 2021).

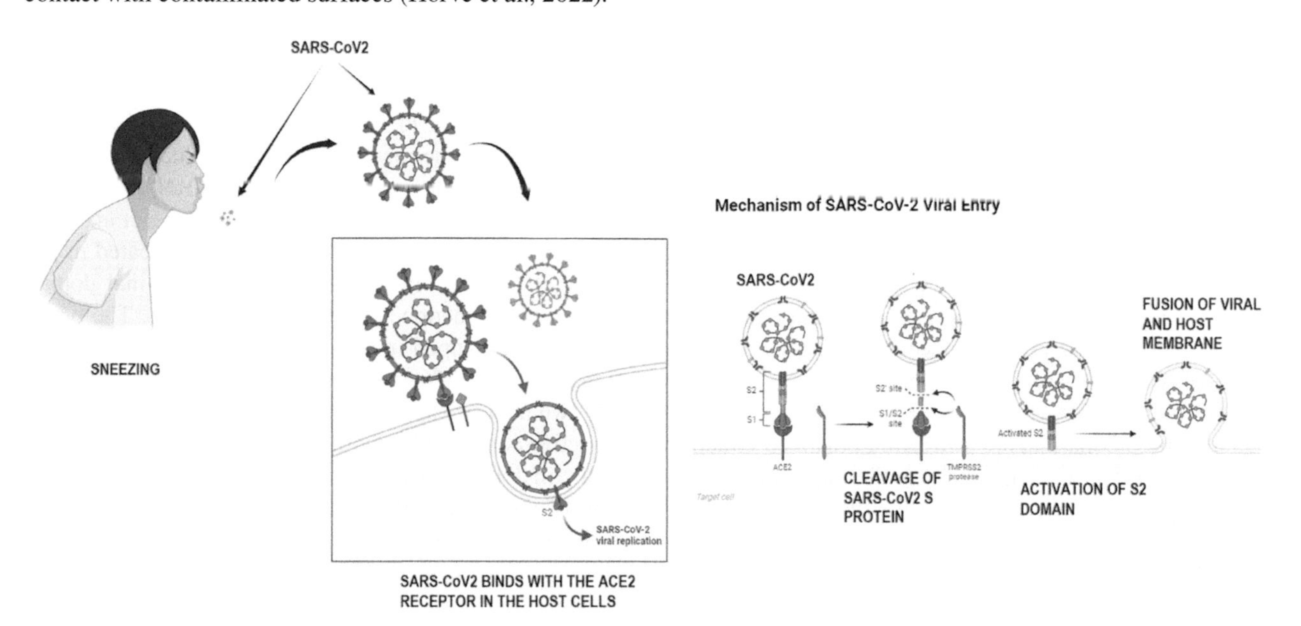

FIGURE 20.1 Structure of SARS-CoV-2 virus and its associated proteins.

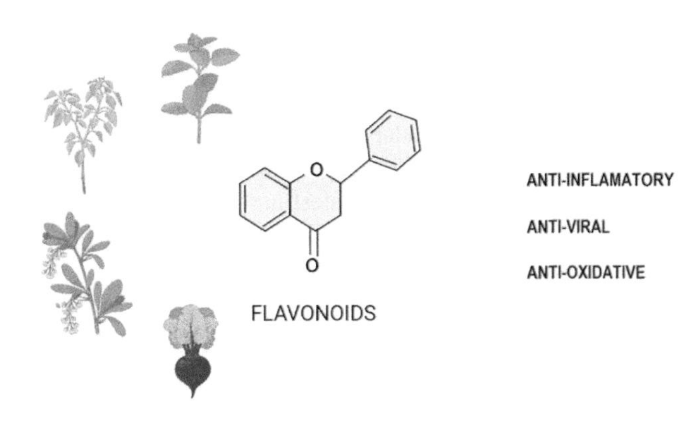

FIGURE 20.2 Mechanism of protection of the SARS-CoV-2 by plants.

Step 6: Immune reaction of the host.

As the infection grows more severe, the host's immune system will start tracking the virus and mount an immune defense. One of the variables that contribute to the stringency of COVID-19 and SARS is the body's immunological ability to become dysregulated, which can result in a cytokine storm and serious inflammation (Kumar et al., 2020; Figure 20.2).

20.3 FLAVONOIDS IN SARS

For several disorders, secondary metabolites obtained from plants have demonstrated as potential effective alternative to traditional treatments. Since SARS-CoV-2 is the cause of COVID-19, an acute medical issue with efficient antiviral treatments, the compounds' strong antiviral qualities offer hope in the form of secondary metabolites (Anand et al., 2021; Latha et al., 2024). Flavonoids are secondary metabolites and include many combinations that are found in vegetables, fruits, and several other parts of the plants. These are hydroxylated phenolic compounds which are characterized by the structural class of these compounds and their degree of hydroxylation. These are biologically active substances and have the potential to interrupt the life cycle of the coronavirus life cycle because they are non-toxic when administered and can synergically work with different medicines; additionally, the active groups present in these compounds are able to interfere with other cellular targets in the signaling pathways (Kaul et al., 2021). These bioactive compounds in plants can be preferred over synthetically produced drugs as they are found to effective, and these are the compounds that are present in the traditional Indian medicine system like Ayurvedic, Unani, and Siddha that are still being practiced (Bharathi et al., 2023). These compounds are the large class of phytochemicals that are required for the diets of humans and have numerous pharmacological properties which also include antiviral properties (Jo et al., 2020). The primarily used medicine in combating COVID-19 is corticosteroids (Bhotla et al., 2021). For effective treatment of infections, diagnosing the symptoms caused in the earlier stage is much more effective (Pushparaj et al., 2022). Flavonoids can help modulating the COVID-19-related inflammatory strategies and immune response. These compounds could modulate the synthesis of inflammatory mediators, such as the induction of luteolin, apigenin, and catechin, a glycosylated derivative of quercetin, which inhibited the inflammatory response (Liskova et al., 2021). One of the main target proteins of SARS-COVID-19 is a main protease (M^{pro}); other than this target, papain-like protease, spike protein, and RNA-dependent RNA polymerase (RdRp) are important; in an *in silico* docking analysis, the flavonoid compounds herbacetin, pectolinarin, and baicalin are found to have proteolytic activity against the protein M^{pro}. Baicalin showed effective inhibitory activity against the protein than the two other flavonoids and the docking mode is different (Jo et al., 2020; Kaviya et al., 2023; Boro et al., 2024). The potential therapeutic target of COVID-19 is an ACE2, hesperidin compound found to inhibit the exchange between the receptor-binding domain (RBD) and the ACE2 receptor in humans (Muchtaridi et al., 2020). Flavonoid compounds derived from Anatolian propolis are found to have good binding energy against receptors like S protein and ACE2 receptor; the compound pinocembrin is found to have the highest binding energy observed than chrysin, hesperetin, and caffeic acid phenethyl ester (Güler et al., 2021). Naringin, a phytochemical of the citrus fruit, can prevent the cytokine storm, and the phytochemicals found in the citrus fruit may have potential anti-coronavirus and anti-inflammatory properties (Cheng et al., 2020).

20.3.1 Flavonoid Compound against Nucleocapsid Protein

The structural protein of SARS-CoV-2 includes the nucleocapsid protein, which is crucial for the virus's entrance and survival in host cells. Its main function is to be involved in interactions with the viral genome to encapsulate it within a ribonucleoprotein structure. It is a multifunctional protein that has been identified to play an array of roles in many stages of the virus's life cycle, including regulating the host cell cycle, modulating viral mRNA replication, and viral budding and assemblage (Kwarteng et al., 2020). The nucleocapsid protein has seven motifs located in its C terminal domain, linker region, and N terminal domain; these motifs function as hotspots for epitopes. The protein sequence is mapped in three different nuclear localization signals: 36–41, 256–260, and 363–389; there are additionally two nuclear export signals (Kumar et al., 2020). Three bioactive compounds, anolide D, hypericin, and silymarin, result in long-lasting binding free energy in three other sites of the protein nucleocapsid in protein docking along with oxyacanthine, with aferin A, acetyl aleuritolic acid, and rhein having druggable ADME properties and toxicity (Kashyap et al., 2023). The plant-derived compounds epicatechin, embelin, carbazole, hydroxychloroquine, and

1-phenylethanethiol are potential inhibitors of the nucleocapsid protein (Gupta et al., 2022).

20.3.2 Flavonoid Compound against Spike (S) Protein

S protein is the structural protein that is encoded by the viral genome of coronavirus. It is comprised of 3 domains: the outermost domain is the ectodomain where the RBD domain binds with the receptor of ACE2 in the cells of the host; the other two domains are a brief intracellular tail and a single-pass transmembrane. It is one of the targets that is focused more on the development of drugs and vaccines (Jain et al., 2021). The flavonol compounds myricetin and linebacker are revealed to have possible binding efficiency with the spike protein; other flavonoids like herbacetin, kaempferol, and morin are also found to have good binding ability with the S protein (Mouffouk et al., 2021). Six flavonoid compounds found in the water extract of *Salvadora persica* docked with S protein are found to have acceptable binding affinities against the viral protein (Owis et al., 2021). The host cell receptor known as ACE2 serves as the site where the viral S protein binds. The flavonoids, such as theaflavin gallate and epigallocatechin in tea, have a higher atomic contact energy with ACE2 protein. This indicates the compounds' potential, but further study must be conducted to reach any definitive inferences (Maiti and Banerjee, 2021). Cyanidin through molecular dynamics and molecular docking is found to bind with S protein. Red grapes extract containing the cyanidin in *in vitro* assay is found to have achieved a 50% reduction in binding of ACE2 with S protein; it can be an advantageous adjuvant medication against the SARS-CoV-2 spike protein, but further research is needed to be carried out (Shrestha et al., 2022).

20.3.3 Flavonoids against SARS-CoV Proteins

SARS-CoV-2 has several proteins that have important roles in different functions in interacting with the host cells or have roles in other functions. $3CL^{pro}$ comprehended as the M^{pro} has an essential role in the replication of the viral genome (Ramya et al., 2023). The necessary protein for the life cycle of SARS-CoV is called RdRp. It can be referred to as nsp12 because it is essential for both transcription and virus replication. The phytochemical substances that are prevalent in *Dysphania ambrosioides*, including rutin, nicotiflorin, glucuronide derivative, and sulfate derivative, have been demonstrated through docking research to be potential inhibitors of both M^{pro} and RdRp (daSilva et al., 2020). *Anastatica hierochuntica*, *Citrus reticulata*, and *Kickxia aegyptiaca* are a source of five flavonoid aglycone compounds that have been identified and isolated: taxifolin, pectolinarigenin, tangeretin, gardenin, and hispidulin. Pectolinarigenin, tangeretin, and gardenin have been documented to have a potential effect against SARS-CoV-2 in *in vitro* and *in vivo* experiments. In addition, these compounds also exhibited good binding to M^{pro} in molecular docking research (Al-Karmalawy et al., 2021). Rutin also inhibits M^{pro} in SARS-CoV-2 and is found to possess a docking score of –9.2 kcal/mol. Eleven potent compounds are identified in the ethanolic extract of *Ipomoea obscura* leaves; five of these potent compounds – ursodeoxycholic acid, demeclocycline, tetracycline, chlorotetracycline, and ethyl iso-allocholate – are found to have antiviral activities and also exhibited good binding capacity against ACE2 and M^{pro} proteins (Poochi et al., 2020). The polyherbal formulations sitopaladi and kaempferol in zein-chitosan nanoparticles can be good options for the medicine of COVID-19, as these formulations can modulate the immune response. These formulations are the traditional medicines that are used in place of conventional medicines in many regions worldwide and have better potential (Bhotla et al., 2021; Abidharini et al., 2023). Clove has four phytochemicals: eugenie, syzyginin B, eugenol, and casuarictin. These phytochemicals have been shown to have enhanced binding abilities with M^{pro} of SARS-CoV-2 and inhibit the viral genome's replication (Chandra Manivannan et al., 2022).

20.4 SCOPE OF FLAVONOIDS

Phytochemicals such as flavonoids are now used in the evaluation for therapeutics; nowadays, researchers search for novel and effective therapies that are safe, have multiple receptor sites and don't cause any side effects. The flavonoids have already several therapeutic potentials against various diseases (Khan et al., 2018).

20.4.1 Synthetic and Semisynthetic Flavonoids

In the production of flavonoids, there are two types of synthesis: synthetic flavonoids and semisynthetic flavonoids; construction of a flavone scaffold is one of the key syntheses of flavonoids (Gaspar et al., 2014). When compared to natural flavonoids, synthetic flavonoids are developed in multiple laboratories, and also, there is a difficulty in producing natural flavonoids in drug development. Synthetic flavonoids are used to overcome the problem faced by natural flavonoids and their bioavailability and stability. Several experiments are conducted to determine the specific functional group of the natural flavonoids that exhibit the advantages of the antiplatelet effects (Vallance et al., 2019).

20.4.1.1 Antiviral Activity of the Synthetic and Semisynthetic Flavonoid Derivatives

Studies state that the excellent antiviral activity in the naturally occurring flavonoids which are quercetin and baicalein shows the potential activity against the human cytomegalovirus (HCMV) and the human immunodeficiency virus (HIV). HIV-1 protease is inhibited by chalcone butein, while herpes virus including HCMV is inhibited by the xanthohumol; enterovirus 71 (EV71) is resistant to chrysin (Zainuri et al., 2017).

20.4.1.2 Antidepressant Activity of the Synthetic and Semisynthetic Flavonoid Derivatives

The flavonoids, which have the potential to treat antidepressant activity, are isorhamnetin, eriocitrin, diosmetin, taxifolin, apigenin, and naringin; compared to the antidepressant drugs, the flavonoids play a vital role in treating conditions such as depression and anxiety; the drugs that increase the level of monoamine in the neurotransmitters of the brain are noradrenaline, dopamine, and serotonin (Uivarosi et al., 2019)

20.4.1.3 Anti-inflammatory Activity of the Synthetic and Semisynthetic Flavonoid Derivatives

Eicosanoids, histamine, chemokines, serotonin, and cytokines are the mediators of chemicals which control acute and chronic inflammation; the main transcription factor in the strategies of inflammation is tumor necrosis factor-α (TNF-α), a type of cytokine that activates the nuclear factor (NF-κB). The NF-κB process produces new chemokines, cytokines, and proinflammatory enzymes which include cyclooxygenases COX-1 and COX-2 that are encoded by the genes in transcription (Uivarosi et al., 2019).

20.4.1.4 Synthetic Flavonoids Affecting Function of Platelet

Naturally occurring flavonoids, such as chrysin, has the ability to suppress platelet activation after isolation. When this procedure is repeated, the energy level in the PRP experiment decreases (presence of plasma proteins) (Liu et al., 2016). The interference of the plasma proteins will be reduced and it is regulated even by the synthetic modification of the chrysin chemical structure through which a library of compounds is synthesized. For instance, conjugation with ruthenium (Ru) substantially enhanced the prohibitory effects of chrysin in PRP and whole blood (Page et al., 2012). Furthermore, the compound's potency in modifying platelet function is increased when a thiocarbonyl group is added to the carbonyl group (for instance, 25 μM of chrysin caused 25% approximate prohibition in PRP, whereas 25 μM of Ru-thiochrysin caused 65% approximate inhibition) (Page et al., 2012). Ru-thiochrysin is just as effective as chrysin in isolated platelets, indicating that the observed variations are not the result of greater potency or binding capacity at effector molecules, but rather from the lack of binding to plasma proteins. It has been proposed that conjugating flavonoids to metals like ruthenium enhances the stability, membrane permeability, and binding efficiency of the molecule to the target molecules (Page et al., 2012; Del Turco et al., 2015). Additionally, it has been proposed that conjugation to ruthenium might lessen the negative effects of chemicals; in fact, the mice that were given the Ru-conjugated molecule did not exhibit any harmful effects (Page et al., 2012). Several synthetics were discovered by Del Turco: the compounds having methoxy groups are 2,3-diphenyl-4H-pyrido(1,2-α) and pyrimidine-4-one which are compound 6 and compound 7; these two compounds react to produce apigenin and quercetin when PRP is stimulated by U46619 or collagen (Del Turco et al., 2015). The modification of chemical structure takes place when the synthetic flavonoids are developed; the antiplatelet ability of the compounds is examined by different functional groups to know their effect; the structure–activity relationship will be concluded by the very basic structure of the flavonoids in the synthetic flavonoids (Ravishankar et al., 2018).

20.4.1.5 The Principal Actions of Flavonoids Semisynthetic Acylation

The primary functions of flavonoids are about reports on their *in vitro* alteration through the addition of fatty acids. Lipase is the most often used acylation enzyme. The acylation of flavonoid glycosides can be catalyzed by a variety of lipases (De Araújo et al., 2017; Biely et al., 2014). In an earlier study (De Araújo et al., 2017), a range of compounds were acylated using the *Candida antarctica* lipase B enzyme (CALB) in different organic solvents. For instance, hesperidin, a flavanone from citrus, was acylated with decanoic acid using CALB as a catalyst. In a distinct piece of writing (Ardhaoui et al., 2004), rutin and esculin sugar residues were acylated with fatty acids employing immobilized lipase enzymes derived from *C. antarctica*. Using aliphatic fatty acids with more than 12 carbons resulted in a 70% increase in the yield of esterified flavonoids. Spectrophotometric analyses revealed that esterification occurred on rutin's rhamnose sugar at 4″¢ hydroxyl group and esculin glucose moiety, a primary hydroxyl group. It is highlighted that the increased acylation outcome of esculin is linked to its "existing better acyl acceptor" when compared to rutin's primary hydroxyl group in the glucoside structure; additionally, the bulkiness of the structure of rutin negatively impacted the yield of acylation reaction. The acylation yield is also influenced by the lipase enzyme's supply. For example, *Pseudomonas cepacia* lipase acylates the flavonoid aglycones and hydroxy groups, while lipase from *C. antarctica* can esterify the hydroxy groups of sugar moieties. The lipase enzymes acylated several flavonoids by utilizing decanoic acid as an acyl donor and hesperidin as an acceptor, resulting in a 55% hesperidin decanoate product (De Araújo et al., 2017; Chebil et al., 2007). Another study (De Oliveira et al., 2009) evaluated the lipase enzyme regioselectivity on the esterification of isoquercetin and rutin using molecular simulation. The aglycone portion of both flavonoids was found to be localized near the binding pocket's entrance and stabilized by hydrophobic contacts and hydrogen bonds, according to the results. While both the 6″-OH of glucose, i.e., the primary (isoquercitrin), and 4‴-OH of rhamnose, i.e., the secondary (rutin), were acetylated (Biely et al., 2014), the sugar component appeared near the bottom of the pocket. Using the esterase enzyme of carbohydrate "acetyl esterase" in a water-based solvent, short carboxylic acids were used to investigate the acylation of rutin and esculin sugars. In a non-buffered aqueous media, esculin was effectively metabolized to its propionate and acetate by-products.

As in non-buffered water media, esculin was effectively metabolized to the derivative of acetate and propionate. Because of its lower solubility in water, it was effectively acylated in vinyl acetate and 2-propanol and worked as a donor for acyl for rutin. The converted derivatives that are acetylated appear sufficiently with vinyl acetate and decrease by an increase in the number of carbons in the ester of carboxylic acid which is employed as the donor for acyl. Because the enzyme is responsible for catalyzing transacetylation and transpropionation in the 3-position of the sugar moiety, like esculin due to glucosyl residue, it was unaffected despite the bulkiness of the aglycone. When vinyl propionates were employed as the acyl donor rather than vinyl butyrate, the reaction's efficiency was significantly reduced (Biely et al., 2014).

20.4.2 Vaccine from Flavonoids

For SARS-CoV-2, flavonoids have been identified as potential inhibitors. For the SARS-CoV replication process, the M^{pro} is required (Yang et al., 2020). SARS-CoV and SARS-CoV-2 have a similar M^{pro} (Tahir Ul Qamar et al., 2020). The naringenin during the molecular docking releases the binding of M^{pro} to it by forming the hydrogen bond with the amino acids allowing M^{pro} to interfere with SARS-CoV-2 (Khaerunnisa et al., 2020). Current studies have demonstrated that hesperidin has inhibitory action against SARS-CoV-2; SARS-CoV-2 binds to the M^{pro} (PD-ACE-2) which is the peptidase domain of ACE-2 and the RBD of S protein (RBD-S) (Adem et al., 2020; Tallei et al., 2020; Utomo et al., 2020). In the case of SARS-CoV-2, quercetin has the potential inhibition activity as observed in the *in silico* study (Sekiou et al., 2020); M^{pro} has a greater binding affinity with quercetin (Sekiou et al., 2020); some food bioactive flavonoids also have the inhibitory action against SARS-CoV-2, and in another study, *in silico* analysis showed that cyanidin and genistein show the affinity to M^{pro} and RdRp when compared to nelfinavir and lopinavir (Pendyala and Patras, 2020). Several other computational studies have identified potential flavonoids that inhibit the SARS-CoV-2 M^{pro} including kaempferol, apigenin-7-glucoside, luteolin-7-glucoside, quercetin, epigallocatechin, catechin, and naringenin (Khaerunnisa et al., 2020). A potential anti-SARS-CoV-2 is the rutin (Xu et al., 2020).

20.5 CHALLENGES

Flavonoids are important compounds present in the parts of plants. The use of these compounds against several diseases is being documented and some of the compounds are also developed into potent treatment. In the studies regarding COVID-19 treatment, these compounds are found to exhibit better binding capacities with different viral proteins in molecular docking studies as well as better effects both *in vitro* and *in vivo*. These compounds present a challenge for their adoption as future therapeutics due to their limited availability as these are found in lower amounts in plants.

20.6 CONCLUSION

Flavonoids are studied for their efficiency against several diseases. In COVID-19, these compounds are found to have better binding abilities with viral proteins like M^{pro}, RdRp, PL^{pro}, spike protein, nucleocapsid protein, and others in molecular docking studies. Further research using different models is required to be carried out and further clinical trials are needed to conclude.

ACKNOWLEDGMENT

All the authors are thankful to their respective University and Institution for their valuable supports.

DATA AVAILABILITY

The authors confirm that the data supporting the findings of this study are available within the book/chapter.

REFERENCE

Abidharini, J. D., Souparnika, B. R., Elizabeth, J., Vishalini, G., Nihala, S., Shanmugam, V., & Anand, A. V. (2023). Herbal formulations in fighting against the SARS-CoV-2 infection. In *Ethnopharmacology and Drug Discovery for COVID-19: Anti-SARS-CoV-2 Agents from Herbal Medicines and Natural Products* (pp. 85–113). Singapore: Springer Nature Singapore.

Al-Karmalawy, A. A., Farid, M. M., Mostafa, A., Ragheb, A. Y., H. Mahmoud, S., Shehata, M., & Marzouk, M. M. (2021). Naturally available flavonoid aglycones as potential antiviral drug candidates against SARS-CoV-2. *Molecules*, *26*(21), 6559.

Almazán, F., Galán, C., & Enjuanes, L. (2004). The nucleoprotein is required for efficient coronavirus genome replication. *Journal of Virology*, *78*(22), 12683–12688.

Anand, A. V., Balamuralikrishnan, B., Kaviya, M., Bharathi, K., Parithathvi, A., Arun, M., & Dhama, K. (2021). Medicinal plants, phytochemicals, and herbs to combat viral pathogens including SARS-CoV-2. *Molecules*, *26*(6), 1775.

Ardhaoui, M., Falcimaigne, A., Ognier, S., Engasser, J. M., Moussou, P., Pauly, G., & Ghoul, M. (2004). Effect of acyl donor chain length and substitutions pattern on the enzymatic acylation of flavonoids. *Journal of Biotechnology*, *110*(3), 265–272.

Arumugam, V. A., Thangavelu, S., Fathah, Z., Ravindran, P., Sanjeev, A. M. A., Babu, S., & Dhama, K. (2020). COVID-19 and the world with co-morbidities of heart disease, hypertension and diabetes. *Journal of Pure and Applied Microbiology*, *14*(3), 1623–1638.

Baggen, J., Vanstreels, E., Jansen, S., & Daelemans, D. (2021). Cellular host factors for SARS-CoV-2 infection. *Nature Microbiology*, *6*(10), 1219–1232.

Bharathi, K., Sivasangar Latha, A., Jananisri, A., Bavyataa, V., Rajan, B., Balamuralikrishnan, B., & Anand, A. V. (2023). Antiviral properties of south Indian plants against SARS-CoV-2. In *Ethnopharmacology and Drug Discovery for COVID-19: Anti-SARS-CoV-2 Agents from Herbal Medicines and Natural Products* (pp. 447–478). Singapore: Springer Nature Singapore.

Bhotla, H. K., Kaul, T., Balasubramanian, B., Easwaran, M., Arumugam, V. A., Pappusamy, M., & Meyyazhagan, A. (2020). Platelets to surrogate lung inflammation in COVID-19 patients. *Medical Hypotheses*, *143*, 110098.

Bhotla, H. K., Balasubramanian, B., Arumugam, V. A., Pushparaj, K., Easwaran, M., Baskaran, R., & Meyyazhagan, A. (2021). Insinuating cocktailed components in biocompatible-nanoparticles could act as an impressive neo-adjuvant strategy to combat COVID-19. *Natural Resources for Human Health*, *1*(1), 3–7.

Bhotla, H. K., Balasubramanian, B., Meyyazhagan, A., Pushparaj, K., Easwaran, M., Pappusamy, M., & Di Renzo, G. C. (2021). Opportunistic mycoses in COVID-19 patients/survivors: Epidemic inside a pandemic. *Journal of Infection and Public Health*, *14*(11), 1720–1726.

Biely, P., Cziszárová, M., Wong, K. K., & Fernyhough, A. (2014). Enzymatic acylation of flavonoid glycosides by a carbohydrate esterase of family 16. *Biotechnology Letters*, *36*, 2249–2255.

Boro, A., Souparnika, B. R., Atchaya, S., Ramya, S., Senthilkumar, N., Velayuthaprabhu, S., & Anand, A. V. (2024). Antiviral activities of flavonoids against COVID-19 and other virus-causing ailments. In *Bioactive Compounds Against SARS-CoV-2* (pp. 68–81). Boca Raton, FL: CRC Press.

Bosch, B. J., Van der Zee, R., De Haan, C. A., & Rottier, P. J. (2003). The coronavirus spike protein is a class I virus fusion protein: Structural and functional characterization of the fusion core complex. *Journal of Virology*, *77*(16), 8801–8811.

Brian, D. A., & Baric, R. S. (2005). Coronavirus genome structure and replication. *Current Topics in Microbiology and Immunology*, *287*(1), 1–30.

Chan, J. F. W., Yuan, S., Kok, K. H., To, K. K. W., Chu, H., Yang, J., & Yuen, K. Y. (2020). A familial cluster of pneumonia associated with the 2019 novel coronavirus indicating person-to-person transmission: A study of a family cluster. *The Lancet*, *395*(10223), 514–523.

Chandra Manivannan, A., Malaisamy, A., Eswaran, M., Meyyazhagan, A., Arumugam, V. A., Rengasamy, K. R., & Liu, W. C. (2022). Evaluation of clove phytochemicals as potential antiviral drug candidates targeting SARS-CoV-2 main protease: Computational Docking, molecular dynamics simulation, and pharmacokinetic profiling. *Frontiers in Molecular Biosciences*, *9*, 918101.

Chebil, L., Anthoni, J., Humeau, C., Gerardin, C., Engasser, J. M., & Ghoul, M. (2007). Enzymatic acylation of flavonoids: Effect of the nature of the substrate, origin of lipase, and operating conditions on conversion yield and regioselectivity. *Journal of Agricultural and Food Chemistry*, *55*(23), 9496–9502.

Chen, B., Tian, E. K., He, B., Tian, L., Han, R., Wang, S., & Cheng, W. (2020). Overview of lethal human coronaviruses. *Signal Transduction and Targeted Therapy*, *5*(1), 89.

Chen, N., Zhou, M., Dong, X., Qu, J., Gong, F., Han, Y., & Zhang, L. (2020). Epidemiological and clinical characteristics of 99 cases of 2019 novel coronavirus pneumonia in Wuhan, China: A descriptive study. *The Lancet*, *395*(10223), 507–513.

Chikhale, R. V., Gupta, V. K., Eldesoky, G. E., Wabaidur, S. M., Patil, S. A., & Islam, M. A. (2021). Identification of potential anti-TMPRSS2 natural products through homology modelling, virtual screening and molecular dynamics simulation studies. *Journal of Biomolecular Structure and Dynamics*, *39*(17), 6660–6675.

Curtis, K. M., Yount, B., & Baric, R. S. (2002). Heterologous gene expression from transmissible gastroenteritis virus replicon particles. *Journal of Virology*, *76*(3), 1422–1434.

da Silva, F. M. A., da Silva, K. P. A., de Oliveira, L. P. M., Costa, E. V., Koolen, H. H., Pinheiro, M. L. B., & de Souza, A. D. L. (2020). Flavonoid glycosides and their putative human metabolites as potential inhibitors of the SARS-CoV-2 main protease (Mpro) and RNA-dependent RNA polymerase (RdRp). *Memórias do Instituto Oswaldo Cruz*, *115*, e200207.

de Araújo, M. E. M., Franco, Y. E., Messias, M. C., Longato, G. B., Pamphile, J. A., & Carvalho, P. D. O. (2017). Biocatalytic synthesis of flavonoid esters by lipases and their biological benefits. *Planta Medica*, *83*(01/02), 7–22.

De Oliveira, E. B., Humeau, C., Chebil, L., Maia, E. R., Dehez, F., Maigret, B., & Engasser, J. M. (2009). A molecular modelling study to rationalize the regioselectivity in acylation of flavonoid glycosides catalyzed by Candida antarctica lipase B. *Journal of Molecular Catalysis B: Enzymatic*, *59*(1–3), 96–105.

Del Turco, S., Sartini, S., Cigni, G., Sentieri, C., Sbrana, S., Battaglia, D., & Basta, G. (2015). Synthetic analogues of flavonoids with improved activity against platelet activation and aggregation as novel prototypes of food supplements. *Food Chemistry*, *175*, 494–499.

Fehr, A. R., & Perlman, S. (2015). Coronaviruses: An overview of their replication and pathogenesis. *Methods in Molecular Biology*, *1282*, 1–23.

Gaspar, A., Matos, M. J., Garrido, J., Uriarte, E., & Borges, F. (2014). Chromone: A valid scaffold in medicinal chemistry. *Chemical Reviews*, *114*(9), 4960–4992.

Ge, X. Y., Li, J. L., Yang, X. L., Chmura, A. A., Zhu, G., Epstein, J. H., & Shi, Z. L. (2013). Isolation and characterization of a bat SARS-like coronavirus that uses the ACE2 receptor. *Nature*, *503*(7477), 535–538.

Gorbalenya, A. E., Baker, S. C., Baric, R. S., de Groot, R. J., Drosten, C., Gulyaeva, A. A., & Ziebuhr, J. (2020). Severe acute respiratory syndrome-related coronavirus: The species and its viruses-a statement of the Coronavirus Study Group. *BioRxiv. Preprints*.

Güler, H. I., Şal, F. A., Can, Z., Kara, Y., Yildiz, O., Beldüz, A. O., & Kolayli, S. (2021). Targeting CoV-2 spike RBD and ACE-2 interaction with flavonoids of Anatolian propolis by *in silico* and *in vitro* studies in terms of possible COVID-19 therapeutics. *Turkish Journal of Biology*, *45*(7), 530–548.

Gupta, S., Singh, V., Varadwaj, P. K., Chakravartty, N., Katta, A. K. M., Lekkala, S. P., & Reddy Lachagari, V. B. (2022). Secondary metabolites from spice and herbs as potential multitarget inhibitors of SARS-CoV-2 proteins. *Journal of Biomolecular Structure and Dynamics*, *40*(5), 2264–2283.

Horve, P. F., Dietz, L. G., Bowles, G., MacCrone, G., Olsen-Martinez, A., Northcutt, D., & Van Den Wymelenberg, K. G. (2022). Longitudinal analysis of built environment and aerosol contamination associated with isolated COVID-19 positive individuals. *Scientific Reports*, *12*(1), 7395.

Hou, Q., Wang, C., Guo, H., Xia, Z., Ye, J., Liu, K., & Ding, Y. (2015). Draft genome sequence of Delftia tsuruhatensis MTQ3, a strain of plant growth-promoting rhizobacterium with antimicrobial activity. *Genome Announcements*, *3*(4), 10–1128.

Jain, A. S., Sushma, P., Dharmashekar, C., Beelagi, M. S., Prasad, S. K., Shivamallu, C., & Prasad, K. S. (2021). *In silico* evaluation of flavonoids as effective antiviral agents on the spike glycoprotein of SARS-CoV-2. *Saudi Journal of Biological Sciences*, *28*(1), 1040–1051.

Jo, S., Kim, S., Kim, D. Y., Kim, M. S., & Shin, D. H. (2020). Flavonoids with inhibitory activity against SARS-CoV-2 3CLpro. *Journal of Enzyme Inhibition and Medicinal Chemistry*, *35*(1), 1539–1544.

Kashyap, D., Roy, R., Kar, P., & Jha, H. C. (2023). Plant-derived active compounds as a potential nucleocapsid protein inhibitor of SARS-CoV-2: An in-silico study. *Journal of Biomolecular Structure and Dynamics*, *41*(10), 4770–4785.

Kaul, R., Paul, P., Kumar, S., Büsselberg, D., Dwivedi, V. D., & Chaari, A. (2021). Promising antiviral activities of natural flavonoids against SARS-CoV-2 targets: Systematic review. *International Journal of Molecular Sciences*, *22*(20), 11069.

Kaviya, M., Geofferina, I. P., Poornima, P., Rajan, A. P., Balamuralikrishnan, B., Arun, M., & Anand, A. V. (2023). Dietary plants, spices, and fruits in curbing SARS-CoV-2 virulence. In *Ethnopharmacology and Drug Discovery for COVID-19: Anti-SARS-CoV-2 Agents from Herbal Medicines and Natural Products* (pp. 265–316). Singapore: Springer Nature Singapore.

Khaerunnisa, S., Kurniawan, H., Awaluddin, R., Suhartati, S., & Soetjipto, S. (2020). Potential inhibitor of COVID-19 main protease (Mpro) from several medicinal plant compounds by molecular docking study. *Preprints*, *2020*, 2020030226.

Khan, H., Amin, S., Kamal, M. A., & Patel, S. (2018). Flavonoids as acetylcholinesterase inhibitors: Current therapeutic standing and future prospects. *Biomedicine & Pharmacotherapy*, *101*, 860–870.

Kumar, S., Nyodu, R., Maurya, V. K., & Saxena, S. K. (2020). Host immune response and immunobiology of human SARS-CoV-2 infection. *Coronavirus Disease 2019 (COVID-19) Epidemiology, Pathogenesis, Diagnosis, and Therapeutics*, *2020*, 43–53.

Kwarteng, A., Asiedu, E., Sakyi, S. A., & Asiedu, S. O. (2020). Targeting the SARS-CoV2 nucleocapsid protein for potential therapeutics using immuno-informatics and structure-based drug discovery techniques. *Biomedicine & Pharmacotherapy*, *132*, 110914.

Lai, C. C., Shih, T. P., Ko, W. C., Tang, H. J., & Hsueh, P. R. (2020). Severe acute respiratory syndrome coronavirus 2 (SARS-CoV-2) and coronavirus disease-2019 (COVID-19): The epidemic and the challenges. *International Journal of Antimicrobial Agents*, *55*(3), 105924.

Latha, A. S., Anita, P. P., Sidhic, N., James, F., Kaviya, M., Bharathi, K., & Anand, A. V. (2024). Phytonutrients and secondary metabolites to cease SARS-CoV-2 loop. In *Bioactive Compounds Against SARS-CoV-2* (pp. 111–124). Boca Raton, FL: CRC Press.

Lee, N., Hui, D., Wu, A., Chan, P., Cameron, P., Joynt, G. M., & Sung, J. J. (2003). A major outbreak of severe acute respiratory syndrome in Hong Kong. *New England Journal of Medicine*, *348*(20), 1986–1994.

Liskova, A., Samec, M., Koklesova, L., Samuel, S. M., Zhai, K., Al-Ishaq, R. K., & Kubatka, P. (2021). Flavonoids against the SARS-CoV-2 induced inflammatory storm. *Biomedicine & Pharmacotherapy*, *138*, 111430.

Liu, G., Xie, W., He, A. D., Da, X. W., Liang, M. L., Yao, G. Q., & Ming, Z. Y. (2016). Antiplatelet activity of chrysin via inhibiting platelet αIIbβ3-mediated signaling pathway. *Molecular Nutrition & Food Research*, *60*(9), 1984–1993.

Liu, W., Zheng, W., Cheng . L., Li, M., Huang, J., Bao, S., Xu, Q., & Ma, Z. (2022). Citrus fruits are rich in flavonoids for immunoregulation and potential targeting ACE2. *Natural Products and Bioprospecting*, *12*(1), 4.

Lu, R., Zhao, X., Li, J., Niu, P., Yang, B., Wu, H., & Tan, W. (2020). Genomic characterisation and epidemiology of 2019 novel coronavirus: Implications for virus origins and receptor binding. *The Lancet*, *395*(10224), 565–574.

MacLean, O. A., Orton, R. J., Singer, J. B., & Robertson, D. L. (2020). No evidence for distinct types in the evolution of SARS-CoV-2. *Virus Evolution*, *6*(1), veaa034.

Maiti, S., & Banerjee, A. (2021). Epigallocatechin gallate and theaflavin gallate interaction in SARS-CoV-2 spike-protein central channel with reference to the hydroxychloroquine interaction: Bioinformatics and molecular docking study. *Drug Development Research*, *82*(1), 86–96.

Malone, B., Urakova, N., Snijder, E. J., & Campbell, E. A. (2022). Structures and functions of coronavirus replication-transcription complexes and their relevance for SARS-CoV-2 drug design. *Nature Reviews Molecular Cell Biology*, *23*(1), 21–39.

McBride, R., Van Zyl, M., & Fielding, B. C. (2014). The coronavirus nucleocapsid is a multifunctional protein. *Viruses*, *6*(8), 2991–3018.

Meselson, M. (2020). Droplets and aerosols in the transmission of SARS-CoV-2. *New England Journal of Medicine*, *382*(21), 2063–2063.

Meyyazhagan, A., Pushparaj, K., Balasubramanian, B., Kuchi Bhotla, H., Pappusamy, M., Arumugam, V. A., & Di Renzo, G. C. (2022). COVID-19 in pregnant women and children: Insights on clinical manifestations, complexities, and pathogenesis. *International Journal of Gynecology & Obstetrics*, *156*(2), 216–224.

Mouffouk, C., Mouffouk, S., Mouffouk, S., Hambaba, L., & Haba, H. (2021). Flavonols as potential antiviral drugs targeting SARS-CoV-2 proteases (3CLpro and PLpro), spike protein, RNA-dependent RNA polymerase (RdRp) and angiotensin-converting enzyme II receptor (ACE2). *European Journal of Pharmacology*, *891*, 173759.

Muchtaridi, M., Fauzi, M., Khairul Ikram, N. K., Mohd Gazzali, A., & Wahab, H. A. (2020). Natural flavonoids as potential angiotensin-converting enzyme 2 inhibitors for anti-SARS-CoV-2. *Molecules*, *25*(17), 3980.

Ortego, J., Escors, D., Laude, H., & Enjuanes, L. (2002). Generation of a replication-competent, propagation-deficient virus vector based on the transmissible gastroenteritis coronavirus genome. *Journal of Virology*, *76*(22), 11518–11529.

Owis, A. I., El-Hawary, M. S., El Amir, D., Refaat, H., Alaaeldin, E., Aly, O. M., & Kamel, M. S. (2021). Flavonoids of Salvadora persica L. (meswak) and its liposomal formulation as a potential inhibitor of SARS-CoV-2. *RSC Advances*, *11*(22), 13537–13544.

Page, S., & Wheeler, R. (2012). Ruthenium compounds as anticancer agents. *Education in Chemistry*, *49*(1), 26.

Pendyala, B., & Patras, A. (2020). *In silico* screening of food bioactive compounds to predict potential inhibitors of COVID-19 main protease (Mpro) and RNA-dependent RNA polymerase (RdRp). *ChemRxiv*.

Peng, G. W., He, J. F., Lin, J. Y., Zhou, D. H., Yu, D. W., Liang, W. J., & Xu, R. H. (2003). Epidemiological study on severe acute respiratory syndrome in Guangdong province. *Zhonghua Liu Xing Bing Xue Za Zhi= Zhonghua Liuxingbingxue Zazhi*, *24*(5), 350–352.

Plante, J. A., Liu, Y., Liu, J., Xia, H., Johnson, B. A., Lokugamage, K. G., & Shi, P. Y. (2021). Spike mutation D614G alters SARS-CoV-2 fitness. *Nature, 592*(7852), 116–121.

Poland, G. A., Ovsyannikova, I. G., Crooke, S. N., & Kennedy, R. B. (2020). SARS-CoV-2 vaccine development: Current status. In *Mayo Clinic Proceedings* (Vol. 95, No. 10, pp. 2172–2188). Amsterdam, The Netherlands: Elsevier.

Poochi, S. P., Easwaran, M., Balasubramanian, B., Anbuselvam, M., Meyyazhagan, A., Park, S., & Kaul, T. (2020). Employing bioactive compounds derived from Ipomoea obscura (L.) to evaluate potential inhibitor for SARS-CoV-2 main protease and ACE2 protein. *Food Frontiers, 1*(2), 168–179.

Pushparaj, K., Bhotla, H. K., Arumugam, V. A., Pappusamy, M., Easwaran, M., Liu, W. C., & Balasubramanian, B. (2022). Mucormycosis (black fungus) ensuing COVID-19 and comorbidity meets-magnifying global pandemic grieve and catastrophe begins. *Science of The Total Environment, 805*, 150355.

Ramya, S., Boro, A., Geofferina, I. P., Jananisri, A., Vishalini, G., Balamuralikrishna, B., & Anand, A. V (2023). The potential contribution of vitamin K as a nutraceutical to scale down the mortality rate of COVID-19. In *Bioactive Compounds Against SARS-CoV-2* (pp. 140–159). Boca Raton, FL: CRC Press.

Ravishankar, D., Salamah, M., Akimbaev, A., Williams, H. F., Albadawi, D. A., Vaiyapuri, R., & Vaiyapuri, S. (2018). Impact of specific functional groups in flavonoids on the modulation of platelet activation. *Scientific Reports, 8*(1), 9528.

Rota, P. A., Oberste, M. S., Monroe, S. S., Nix, W. A., Campagnoli, R., Icenogle, J. P., & Bellini, W. J. (2003). Characterization of a novel coronavirus associated with severe acute respiratory syndrome. *Science, 300*(5624), 1394–1399.

Sekiou, O., Bouziane, I., Frissou, N., Bouslama, Z., Honcharova, O., Djemel, A., & Benselhoub, A. (2020). In-silico identification of potent inhibitors of COVID-19 main protease (Mpro) from natural products. *International Journal of Biochemistry & Physiology, 5*(3), 000189.

Shanmugam, R., Thangavelu, S., Fathah, Z., Yatoo, M. I., Tiwari, R., Pandey, M. K., & Arumugam, V. A. (2020). SARS-CoV-2/COVID-19 pandemic-an update. *Journal of Experimental Biology and Agricultural Sciences, 8*, S219–S245.

Shi, J., Wen, Z., Zhong, G., Yang, H., Wang, C., Huang, B., & Bu, Z. (2020). Susceptibility of ferrets, cats, dogs, and other domesticated animals to SARS-coronavirus 2. *Science, 368*(6494), 1016–1020.

Shrestha, A., Marahatha, R., Basnet, S., Regmi, B. P., Katuwal, S., Dahal, S. R., & Parajuli, N. (2022). Molecular docking and dynamics simulation of several flavonoids predict cyanidin as an effective drug candidate against SARS-CoV-2 spike protein. *Advances in Pharmacological and Pharmaceutical Sciences, 2022,* 3742318.

Sia, S. F., Yan, L. M., Chin, A. W., Fung, K., Choy, K. T., Wong, A. Y., & Yen, H. L. (2020). Pathogenesis and transmission of SARS-CoV-2 in golden hamsters. *Nature, 583*(7818), 834–838.

Sommerstein, R., Fux, C. A., Vuichard-Gysin, D., Abbas, M., Marschall, J., Balmelli, C., & Swissnoso Carlo Balmelli Marie-Christine Eisenring Stephan Harbarth Jonas Marschall Didier Pittet Hugo Sax Matthias Schlegel Alexander Schweiger Laurence Senn Nicolas Troillet Andreas F. Widmer Giorgio Zanetti. (2020). Risk of SARS-CoV-2 transmission by aerosols, the rational use of masks, and protection of healthcare workers from COVID-19. *Antimicrobial Resistance & Infection Control, 9*, 1–8.

Tallei, T. E., Tumilaar, S. G., Niode, N. J., Kepel, B. J., Idroes, R., Effendi, Y., & Emran, T. B. (2020). Potential of plant bioactive compounds as SARS-CoV-2 main protease (M pro) and spike (S) glycoprotein inhibitors: A molecular docking study. *Scientifica, 2020,* 6307457.

Torres, J., Parthasarathy, K., Lin, X., Saravanan, R., Kukol, A., & Liu, D. X. (2006). Model of a putative pore: The pentameric α-helical bundle of SARS coronavirus E protein in lipid bilayers. *Biophysical Journal, 91*(3), 938–947.

Tsang, K. W., Ho, P. L., Ooi, G. C., Yee, W. K., Wang, T., Chan-Yeung, M., & Lai, K. N. (2003). A cluster of cases of severe acute respiratory syndrome in Hong Kong. *New England Journal of Medicine, 348*(20), 1977–1985.

Türk, S., Türk, C., Malkan, Ü. Y., Temirci, E. S., PEKER, M. Ç., & Haznedaroğlu, İ. C. (2021). Current community transmission and future perspectives on the COVID-19 process. *Turkish Journal of Medical Sciences, 51*(3), 1001–1011.

Uivarosi, V., Munteanu, A. C., & Nițulescu, G. M. (2019). An overview of synthetic and semisynthetic flavonoid derivatives and analogues: Perspectives in drug discovery. *Studies in Natural Products Chemistry, 60*, 29–84.

ul Qamar, M. T., Alqahtani, S. M., Alamri, M. A., & Chen, L. L. (2020). Structural basis of SARS-CoV-2 3CLpro and anti-COVID-19 drug discovery from medicinal plants. *Journal of Pharmaceutical Analysis, 10*(4), 313–319.

Utomo, R. Y., Ikawati, M., & Meiyanto, E. (2020). Revealing the potency of citrus and galangal constituents to halt SARS-CoV-2 infection. *Preprints.*

Vallance, T. M., Ravishankar, D., Albadawi, D. A., Osborn, H. M., & Vaiyapuri, S. (2019). Synthetic flavonoids as novel modulators of platelet function and thrombosis. *International Journal of Molecular Sciences, 20*(12), 3106.

Wong, R. S. (2020). The SARS-CoV-2 outbreak: An epidemiological and clinical perspective. *SN Comprehensive Clinical Medicine, 2*(11), 1983–1991.

Wu, C. H., Chen, P. J., & Yeh, S. H. (2014). Nucleocapsid phosphorylation and RNA helicase DDX1 recruitment enables coronavirus transition from discontinuous to continuous transcription. *Cell Host & Microbe, 16*(4), 462–472.

Xu, Z., Yang, L., Zhang, X., Zhang, Q., Yang, Z., Liu, Y., & Liu, W. (2020). Discovery of potential flavonoid inhibitors against COVID-19 3CL proteinase based on virtual screening strategy. *Frontiers in Molecular Biosciences, 7*, 556481.

Yang, N., & Shen, H. M. (2020). Targeting the endocytic pathway and autophagy process as a novel therapeutic strategy in COVID-19. *International Journal of Biological Sciences, 16*(10), 1724

Yang, X., Yu, Y., Xu, J., Shu, H., Liu, H., Wu, Y., & Shang, Y. (2020). Clinical course and outcomes of critically ill patients with SARS-CoV-2 pneumonia in Wuhan, China: A single-centered, retrospective, observational study. *The Lancet Respiratory Medicine, 8*(5), 475–481.

Yurkovetskiy, L., Wang, X., Pascal, K. E., Tomkins-Tinch, C., Nyalile, T. P., Wang, Y., & Luban, J. (2020). Structural and functional analysis of the D614G SARS-CoV-2 spike protein variant. *Cell, 183*(3), 739–751.

Zainuri, D. A., Arshad, S., Khalib, N. C., Razak, I. A., Pillai, R. R., Sulaiman, S. F., & Van Alsenoy, C. (2017). Synthesis, XRD crystal structure, spectroscopic characterization (FT-IR, 1H and 13C NMR), DFT studies, chemical reactivity and bond dissociation energy studies using molecular dynamics simulations and evaluation of antimicrobial and antioxidant activities of a novel chalcone derivative,(E)-1-(4-bromophenyl)-3-(4-iodophenyl) prop-2-en-1-one. *Journal of Molecular Structure*, *1128*, 520–533.

Zhang, C., Zheng, W., Huang, X., Bell, E. W., Zhou, X., & Zhang, Y. (2020). Protein structure and sequence reanalysis of 2019-nCoV genome refutes snakes as its intermediate host and the unique similarity between its spike protein insertions and HIV-1. *Journal of Proteome Research*, *19*(4), 1351–1360.

Zhang, Z., Nomura, N., Muramoto, Y., Ekimoto, T., Uemura, T., Liu, K., & Shimizu, T. (2022). Structure of SARS-CoV-2 membrane protein essential for virus assembly. *Nature Communications*, *13*(1), 4399.

Zhong, N. S., & Wong, G. W. (2004). Epidemiology of severe acute respiratory syndrome (SARS): Adults and children. *Paediatric Respiratory Reviews*, *5*(4), 270–274.

Zhong, N. S., Zheng, B. J., Li, Y. M., Poon, L. L. M., Xie, Z. H., Chan, K. H., & Guan, Y. (2003). Epidemiology and cause of severe acute respiratory syndrome (SARS) in Guangdong, People's Republic of China, in February, 2003. *The Lancet*, *362*(9393), 1353–1358.

Zhou, J., Li, C., Liu, X., Chiu, M. C., Zhao, X., Wang, D., & Yuen, K. Y. (2020). Infection of bat and human intestinal organoids by SARS-CoV-2. *Nature Medicine*, *26*(7), 1077–1083.

Zhou, P., Yang, X. L., Wang, X. G., Hu, B., Zhang, L., Zhang, W., & Shi, Z. L. (2020). A pneumonia outbreak associated with a new coronavirus of probable bat origin. *Nature*, *579*(7798), 270–273.

Zumla, A., Hui, D. S., Azhar, E. I., Memish, Z. A., & Maeurer, M. (2020). Reducing mortality from 2019-nCoV: Host-directed therapies should be an option. *The Lancet*, *395*(10224), e35–e36.

21 Traditional Medicines and Functional Foods in Indochina for COVID-19 Management

The Impact of Flavonoids

Sora Yasri and Viroj Wiwanitkit

21.1 INTRODUCTION

This virus typically spreads via respiratory contact between individuals. This method of transmission involves the disease being passed from one person to another by infected respiratory droplets or secretions. This uncommon respiratory virus can cause a range of clinical problems, including mortality. Despite the fact that coronavirus disease 2019 (COVID-19) has not yet been contained, efforts are being made worldwide to do so [1,2]. The public health must act quickly to combat this respiratory virus. The two main types of therapy used to treat infections are supportive and symptomatic. The two main types of therapy for infection control are supportive and symptomatic. The genuine effectiveness and safety of current therapeutic modalities must be evaluated because there is currently no evidence-based therapy for COVID-19. There is currently no evidence-based treatment for COVID-19, and once appropriate multi-site clinical data are available, it will be necessary to further assess the genuine efficacy and safety of current therapeutic options. There aren't many treatment options available because this ailment is new. According to reports as of July 2020, the sickness appears to be most prevalent in the Americas and Europe. The majority of Severe Acute Respiratory Syndrome Coronavirus 2 (SARS-CoV-2) patients won't show any symptoms or will only have minor to moderate respiratory symptoms, and they will recover without the need for any specialized antiviral medication. People over 65 and those with underlying medical conditions such as cancer, cardiovascular disease, diabetes, chronic respiratory disease, and diabetes are more prone to develop serious illnesses [1,2].

COVID-19 is a recently identified viral disease that has spread globally. This condition is caused by the SARS-CoV-2. It was found in China. Fever, dry cough, and dyspnea are early COVID-19 symptoms that are comparable to those of other viral respiratory infections such as the common flu. As a result, anamnesis-based COVID-19 diagnosis remains challenging. The normal incubation period is 15 days, but it has been noted that it can range from 0 to 24 days [1,2]. This viral infection might result in a fever and respiratory issues. The COVID-19 virus, which has expanded globally, has killed millions of people. The disease appears to remain important despite the worldwide immunization against the virus. "Alternative COVID cures" are popular. The database for such treatments is often tiny, if not nonexistent, as of 2019. Strong, unsupported claims about the benefits and drawbacks of herbal therapies, on the other hand, are commonly made, providing persons with COVID-19 who are at risk or ill with both genuine and unwarranted hopes or significant fears. Other medical research studies are underway to identify plants and other natural chemicals that may be useful in treating coronavirus disease [3]. Traditional medicine performs antiviral, anti-inflammatory, and immunoregulatory roles in the treatment of COVID-19 through a variety of components that work on several targets and pathways. Clinical trials have also proved the effectiveness of traditional medicine in treating patients. The recently released findings offer important and pertinent data for the creation of future COVID-19 and other antiviral infectious illness drugs [3].

By offering patients complementary care, traditional medicine can help patients manage COVID-19. Numerous conventional drugs and treatments have demonstrated the potential to boost the immune system and ease COVID-19 symptoms. For instance, herbal treatments and dietary supplements like vitamin C, ginger, and turmeric can help raise immunity and fend off illness. Traditional techniques like acupuncture and meditation can also support stress and anxiety reduction, which can improve general health and recovery. It is crucial to remember that traditional medicine should only be used in conjunction with therapies that are supported by scientific evidence and with the help of a healthcare practitioner. The first article by Sriwijitalai and Wiwanitkit after the discovery of COVID-19 showed that using certain natural product regimens against COVID-19 may have some advantages [4]. *Hesperethusa crenulata*, *Perilla frutescens*, *Ephedra equisetina*, *Shiraia bambusicola*, and *Panax ginseng* are a few examples of these herbs [4]. In East Asia, where COVID-19 first arose, these plants are some of the most well-known and traditional ones. These results are useful for further investigation into a variety of herbal remedies and their potential utility

DOI: 10.1201/9781003433200-21

in the treatment of COVID-19. Many of these herbs have already been shown to successfully treat viral respiratory infections. For example, *E. equisetina* and *P. ginseng* have demonstrated anti-influenza properties [4]. Complementary and alternative medicine has a position in the current world. There are now several traditional medical systems available. Traditional medicine makes extensive use of historical herbs, and it's intriguing to see how these ancient botanicals are employed to cure various illnesses. Many plants are thought to be beneficial in the treatment of a variety of illnesses, and current research on the advantages has been published [5,6].

SARS-CoV-2 is crucial in the search for a novel disease cure. The use of bioactive chemicals derived from natural sources is one of many choices. The authors give a succinct overview of flavonoid-based COVID-19 agents found in Indochina's marine natural products. There may be bioactive compounds in some well-known conventional ethnopharmacological regimens that can be used to treat COVID-19.

21.2 TRADITIONAL MEDICINES FOR COVID-19 MANAGEMENT

Traditional medicine refers to the knowledge, skills, and practices based on theories, beliefs, and experiences indigenous to different cultures that are used in the maintenance of health and the prevention, diagnosis, improvement, or treatment of physical and mental illness. It can include various practices, such as herbalism, acupuncture, manual therapies, and energy healing. The evidence from the included trials suggests that the inclusion of traditional medicine in COVID-19 treatment may improve clinical results when compared to conventional Western medical therapy alone [7]. The quality of the traditional medicine evidence for COVID-19 was generally moderate to low [7]. By taking into account the experiences of clinical professionals, medical policies, and other factors, meta-analyses of the use of conventional medicine in the treatment of COVID-19 can be used for clinical decision-making [7]. Studies have shown that traditional Chinese medicine treatment can significantly improve clinical symptoms, according to Ding et al. [8]. Therefore, Ding et al. noted that more research may have a high translational value in discovering potential targeted treatments for COVID-19 [8].

There have already been numerous medications suggested for the treatment of lung disorders. A small number of them are being tested in clinical settings and could treat infectious disorders. Traditional Chinese Medicine (TCM) and Indian Ayurveda have both made substantial use of plant extracts or herbal medicines. In addition, it has been used to promote the regulation of signaling pathways by inhibiting the effects of specific genes and proteins [9]. Natural remedies are regarded as an affordable and secure treatment for lung disease because they have been scientifically demonstrated to have extraordinary bioactivities. Natural plant metabolites (such as flavonoids, alkaloids, and terpenoids) have been studied *in vitro*, *in vivo*, and computationally for their potential to treat various lung disorders [8]. To find potential treatments for the most difficult lung diseases, particularly SARS-CoV-2, natural metabolites that have been repurposed for use in different lung diseases should be evaluated more by advanced computational applications and experimental models in the biological system, and need to be validated by clinical trials [9]. Even though traditional medicine has been used for generations to treat a variety of medical disorders, including infectious diseases, there is currently no scientific data to back up its use in the treatment of COVID-19. The World Health Organization (WHO) and health professionals from all over the world highly advise using evidence-based COVID-19 treatments and therapies. It's crucial to remember that evidence-based COVID-19 treatments should not be replaced by conventional medication. The WHO states that traditional medicine can be used as a supplement to COVID-19 treatment. This implies that traditional medicine can be utilized in addition to conventional medical treatment to help treat symptoms and enhance general health. Herbal medicines, acupuncture, and breathing exercises are a few traditional medical procedures that may be beneficial.

21.3 FUNCTIONAL FOODS FOR COVID-19 MANAGEMENT

Functional foods are those that have been created to offer extra health advantages over and above basic nutrition. These foods include bioactive substances, which are said to support well-being, health, and disease prevention. Fortified cereals, beverages with additional vitamins and minerals, and probiotic yogurts are a few examples of functional foods. Functional foods are significant in medicine because they provide a natural means of enhancing health and preventing illness. They may be used to treat certain medical conditions or as an addition to conventional medical care. Some functional meals, for instance, are believed to be advantageous for lowering inflammation, enhancing digestion, increasing immunity, and controlling blood sugar levels. Functional foods are predicted to become an increasingly significant food category as more study is done to determine their health advantages. Functional foods are set to play a bigger role in modern medicine as further study reveals their health benefits [10].

Functional foods can aid in the management of infections by supplying vital nutrients that can strengthen the immune system, enhance gastrointestinal health, and lower bodily inflammation. For instance, probiotic-rich foods like yogurt and kefir can stimulate the development of good gut flora, which can defend against dangerous pathogens that can cause illnesses. Furthermore, foods high in antioxidants, including berries and leafy greens, can aid the body in minimizing inflammation, which can lower the chance of infection and speed up healing. Overall, treating infections and promoting general health can both benefit from a well-balanced diet that contains functional foods. Conceptually, the human immune system's intricate network of internal mechanisms offers defense against a variety of illnesses [11]. These defenses combine to generate innate and adaptive immunity, wherein certain immune elements cooperate to thwart pathogens [11]. In addition to inherited traits, factors including lifestyle decisions, age, and environmental determinants may also have an impact on a person's vulnerability to certain diseases. It has been demonstrated that specific dietary chemical constituents influence pathophysiology by controlling signal transduction and cell morphologies [11]. Many studies are currently being conducted on a variety of functional foods, which are frequently marketed as immune system boosters, to find evidence of their potential protective role against viral diseases like the influenza viruses (A and B), herpes simplex virus (HSV), and SARS-CoV-2 [11]. Nevertheless, at this time, there is no proof that functional foods can effectively prevent or treat COVID-19. However, a nutritious and balanced diet rich in functional foods can improve overall health and strengthen the immune system, which may assist in lowering the risk of infections like COVID-19.

21.4 FLAVONOIDS AS A NATURALLY DERIVED COMPOUND AND ITS ROLE IN MEDICINE

The discovery of compounds derived from plants that can be used as nutraceuticals and can halt the spread of new viruses is exciting and good for the environment. In general, studies have demonstrated that plants can serve as a source of antivirals that can fight harmful viruses. Natural compounds produced from medicinal herbs exhibit inhibitory effects on the viral life cycle, including COVID-19 virion entrance, replication, assembly, and release, in addition to having anti-inflammatory and antiviral characteristics. In Indochina, there is a separate local alternative healthcare system. Particularly noteworthy is the traditional knowledge of the herbal plant, and some indigenous herbal plants have been recommended for their possible anti-COVID-19 activity.

Two aromatic rings (A and B) connected by a three-carbon bridge (C) define the chemical structure of the class of naturally occurring chemicals known as flavonoids. The addition of other functional groups, such as hydroxyl, methoxy, and glycosyl groups, can alter this flavone backbone structure, resulting in a vast variety of flavonoid molecules with varying physical and biological properties. A class of naturally occurring plant substances known as flavonoids have anti-inflammatory and antioxidant effects (Table 21.1). They can be found in a wide range of foods, including tea, wine, beans, fruits, and vegetables. Flavonoids include substances like quercetin, kaempferol, and catechins. Numerous plants have higher flavonoid content. Fruits like berries, apples, citrus fruits, and grapes are a few of the more popular ones. Other excellent sources of flavonoids are spinach, kale, broccoli, and onions. Additionally, several herbs including mint, thyme, and parsley contain flavonoids. Other additional plants are abundant in flavonoids in addition to these.

Scientists have been researching viral entry and pathogenesis ever since the SARS-CoV-2 outbreak. The creation of medications using pharmacological components sourced from natural sources may be a viable COVID-19 treatment method. In the published literature, therapeutic plants have

TABLE 21.1
Possible Usefulness of Flavonoids Against Infection

Mechanisms	Explanations
Anti-inflammatory effects	Flavonoids are naturally occurring chemicals present in many plants that have anti-inflammatory actions. This means that they can assist in reducing inflammation in the body, which can be very beneficial in the treatment of infections. When a virus or bacteria infects the body, the immune system responds by generating inflammation in the affected area. While inflammation is required to aid in the battle against infection, it can also cause harm to good tissue and lead to symptoms such as pain, redness, and swelling. Flavonoids can aid in the reduction of inflammation by inhibiting the activity of specific enzymes and signaling molecules involved in the inflammatory response. This can assist in preventing inflammation's harm, lessen symptoms, and enhance overall health. Additionally, it has been demonstrated that certain flavonoids have direct antiviral and antibacterial activities, which can aid in the direct treatment of infections. For instance, it has been demonstrated that the flavonoid quercetin, which is present in many fruits and vegetables, has antibacterial properties that can fight a variety of pathogens while also preventing the replication of some viruses. Overall, flavonoids are a promising tool for treating infections and promoting general health thanks to their anti-inflammatory and antibacterial capabilities.
Antioxidant effects	Known for their antioxidant effects, flavonoids are naturally occurring substances that can be found in a wide variety of plants. Free radicals are chemicals that can destroy cells and lead to diseases like infection. Antioxidants shield cells from this damage. By scavenging free radicals and lowering inflammation, flavonoids can strengthen the immune system in the context of infection control. In turn, this may lessen the severity of illnesses and speed up healing. Additionally, some research has revealed that flavonoids might directly combat microbes, making them potentially effective in the treatment of some infections. Overall, including foods high in flavonoids in your diet can support the immune system and improve general health.

received more attention than biodiverse sources such as products generated from plants. Flavonoids, at concentrations mainly between 2 and 300 μM, are capable of regulating platelet aggregation, blood coagulation, fibrinolysis, and nitric oxide production due to their action on multiple receptors and enzymes [12]. The only flavonoid currently being clinically tested for its impact on D-dimer levels in COVID-19 patients is quercetin [12]. It is hypothesized that flavonoids could serve as a therapeutic alternative for the management of thrombotic side effects related to COVID-19 [12]. Therefore, research into the use of naturally occurring compounds based on flavonoids to control COVID-19 is worthwhile. As earlier noted, several flavonoid compounds might be useful for the management of infection including COVID-19 (Table 21.2) [13–15]. The following topics will provide examples of significant marine products that are the subject of extensive research and are utilized locally by public health systems for the control of COVID-19.

21.5 FLAVONOIDS IN TRADITIONAL MEDICINE AND THEIR USE FOR MANAGEMENT OF COVID-19

Because of their possible health benefits, flavonoids have been employed in traditional medicine for millennia. Anti-inflammatory, antioxidant, and anticancer qualities are some of their current roles in traditional medicine. Flavonoids have also been utilized to treat a variety of ailments, including allergies, asthma, gastrointestinal disorders, and cardiovascular disease. It is crucial to highlight, however, that the efficacy of flavonoids in traditional medicine is still being studied, and more research is needed to completely appreciate their advantages.

21.5.1 Chinese Medicine

A class of naturally occurring substances called flavonoids is widely present in plants, including many of the herbs used in traditional Chinese medicine. Flavonoids are said to have a variety of health advantages in Chinese traditional medicine, including anti-inflammatory, antioxidant, and antiviral properties. Flavonoids may aid in the treatment of illnesses by enhancing the immune system of the body [16]. According to certain studies, some flavonoids may have antiviral characteristics that can help prevent viral infections. Flavonoids may also possess antibacterial characteristics that can aid in the treatment of infections.

21.5.2 Ayurveda

Since ancient times, flavonoids have been employed in Ayurveda for their therapeutic benefits, including their capacity to treat infections. Flavonoids are regarded in Ayurveda as having anti-inflammatory, antibacterial, and antiviral qualities that can aid in supporting the immune system and ward against illnesses. Many plants, including fruits, vegetables, and herbs, contain flavonoids, which can be used as a nutritional supplement or incorporated into a balanced diet.

For the management of COVID-19, there are many traditional medicine regimens with flavonoid compounds that are studied for their effectiveness against the infection. The examples will be further listed.

1. Jin-Hua-Qing-Gan

   The Chinese Food and Drug Administration has also suggested Jin-Hua-Qing-Gan for the treatment of COVID-19, and it has played an important

TABLE 21.2
A List of Flavonoids and Their Biological Property Against Infections [13–15]

Mechanisms	Explanations
Quercetin	A flavonoid, or type of plant pigment, known as quercetin has anti-inflammatory and antioxidant characteristics. It has been demonstrated to have antiviral properties and could aid in the treatment of infections. According to some research, quercetin may be able to stop viruses from entering cells and lessen the inflammation brought on by infections. Quercetin may also strengthen the immune system by producing more white blood cells. To completely comprehend the potential advantages of quercetin for treating infections, more study is necessary.
Kaempferol	A flavonoid called kaempferol is present in many plant-based foods like kale, broccoli, and tea. Kaempferol has been proven in studies to possess antibacterial characteristics, which implies that it can aid in the defense against bacteria, viruses, and other infectious pathogens. Additionally, kaempferol has been demonstrated to have anti-inflammatory properties, which may lessen the severity of infections and encourage healing. Kaempferol has been proven to have immunomodulatory effects, which indicates that it can assist in regulating the immune system in addition to its antibacterial and anti-inflammatory qualities. This is particularly helpful in the treatment of infections because an overactive immune system can occasionally do more harm than good. Overall, kaempferol may be a viable natural choice for those wishing to enhance their immune system and lower the risk of infection, even if further research is required to fully grasp its potential advantages in the management of infection.
Catechins	Catechins are a kind of flavonoid present in many forms of tea, but they are most prevalent in green tea. According to studies, catechins contain antiviral and antimicrobial characteristics that can help fight off infections and stop the spread of dangerous germs and viruses. Catechins have been demonstrated to support immune system enhancement and inflammation reduction, which can improve the body's ability to combat infections. By limiting the development of bacteria and viruses inside the body, they may also aid in halting the spread of infectious diseases. Overall, as part of a larger therapeutic strategy, catechins can be a helpful tool in treating infections.

role in the prevention of a range of viral infections [17]. A traditional Chinese herbal remedy called Jin-Hua-Qing-Gan is used to cure conditions of the liver. It does contain flavonoids, plant-based substances with anti-inflammatory and antioxidant effects. The preparation and origin of the herbs employed in the formula can affect the precise flavonoid composition of Jin-Hua-Qing-Gan. *Scutellaria baicalensis* (Huang Qin) and *Gardenia jasminoides* (Zhi Zi), two often utilized herbs in the recipe, are known to contain flavonoids called baicalin and geniposide, respectively.

2. Qu Du Qiang Fei I Hao Fang

   According to relevant traditional Chinese medicine patterns of COVID-19 presentations, such as heat and cold patterns, damp and phlegm syndromes, toxicity, and deficiency patterns, its specific herbs and herbal combinations are examined for their suitability [18]. It is advised to do additional research in a randomized, double-blind, and placebo-controlled trial of QDQF1 to evaluate its therapeutic efficacy in the management of COVID-19 [18].

3. Lian-Hua-Qing-Wen

   TCM, or Lian-Hua-Qing-Wen, is used to treat COVID-19 and other respiratory tract infections. It is a complex blend of herbs that has been discovered to contain a number of bioactive substances, such as flavonoids. Due to their demonstrated anti-inflammatory and antioxidant capabilities, flavonoids may help Lian-Hua-Qing-Wen's therapeutic effects. By modulating host immunological reactions and inflammation as well as the viral life cycle, Lian-Hua-Qing-Wen is endowed with a variety of antiviral properties [17]. The China Food and Drug Administration advises Lian-Hua-Qing-Wen for the treatment of COVID-19, and it has been crucial in preventing a number of viral infections [17]. Patients with compromised immune systems or those whose symptoms have subsided following therapy with Jin-Hua-Qing-Gan may benefit more from oral administration of Lian-Hua-Qing-Wen [17]. According to Zheng et al.'s examination of a thorough network, this herbal traditional medicine treatment controls the inflammatory process, has antiviral effects, and heals lung damage [19]. Additionally, it reduces the "cytokine storm" and ameliorates symptoms brought on by ACE2-expression disease [19]. These ground-breaking discoveries provide a sound pharmacological foundation and justification for using this medication to treat COVID-19 and perhaps other disorders as well [19]. Hu et al. sought to thoroughly present the clinical evidence of LHQWG in treating COVID-19 in a different paper, which will have major implications for future study and clinical use [20]. There is yet insufficient evidence to conclude that the medicine is effective for COVID-19, according to Hu et al. [20]. In conclusion, there is some indication that Lian-Hua-Qing-Wen may help with COVID-19 management. In actuality, the substance is a traditional Chinese herbal remedy that has been used for more than 1,000 years to treat respiratory infections. According to several recent research, the regimen may have antiviral and anti-inflammatory properties that could be helpful for COVID-19 patients. To validate these results and establish the ideal dosages and treatment plans for this medication, additional study is required.

4. Triphala

   The three fruits Amalaki (*Emblica officinalis*), Bibhitaki (*Terminalia bellirica*), and Haritaki (*Terminalia chebula*) make up the traditional Ayurvedic herbal remedy known as triphala. Numerous bioactive substances, including flavonoids, tannins, and polyphenols, are present in these fruits. Triphala is usually regarded as a good source of flavonoids, though the precise quantity can vary depending on the source and preparation technique. This regimen is currently adopted in many areas of the world. For example, it is directly quoted into the Thai traditional medicine regimen. Regarding COVID-19, to test their ability to suppress SARS-CoV-2, Rudrapal et al. docked bioactive compounds from Triphala, an Ayurvedic medicine, against M^{pro} and then ran a molecular dynamics simulation [21]. Terflavin A, chebulagic acid, chebulinic acid, and corilagin from Triphala formulation were found by Rudrapal et al. as prospective inhibitors of SARS-CoV-2 M^{pro}, and experimental (*in vitro*/*in vivo*) investigations to further investigate their inhibitory mechanisms are recommended [21].

21.6 FLAVONOIDS IN FUNCTIONAL FOODS AND ITS USE FOR MANAGEMENT OF COVID-19

Flavonoids have been discovered to serve critical functions in functional meals. They are natural substances found in many fruits, plants, and herbs that have been proven to provide a variety of health benefits. Antioxidant, anticancer, anti-inflammatory, and antibacterial capabilities are among these benefits. Flavonoids can be utilized in functional meals to increase the nutritional content of the product, improve its flavor and texture, and bring health advantages to the user. Green tea, dark chocolate, and berries are examples of useful foods that include flavonoids. Even though there is no certainty that any meal can prevent or treat COVID-19, some functional foods have been demonstrated to have immune-supportive qualities that may enhance general health and wellness. Here are a few well-known Asian meals with functional advantages against COVID-19, that contain flavonoids, which will be further discussed.

- Miso

  Miso is a traditional Japanese seasoning made by fermenting soybeans with salt and a specific fungus called koji. Probiotics are found in the fermented soybean paste known as miso, and they are believed to enhance intestinal health. Miso can be used as a marinade for meats and vegetables or even to make soups. Miso has a salty and savory taste and is commonly used in soups, marinades, and sauces. As for flavonoids, miso does contain certain types of flavonoids such as isoflavones, which are also present in soybeans. However, the amount of flavonoids in miso may vary depending on the type of miso and the fermentation process used to make it. Regarding COVID-19, there are reports of biological activity and recently investigated potential of fermented soybean compounds against SARS-CoV-2 [22].
- Garlic

  Garlic includes flavonoids, a class of antioxidants that aid in defending the body against dangerous free radicals. A range of Asian meals, including stir-fries, fried rice, and soups, can benefit from the antibacterial and antiviral properties of garlic. Garlic contains a number of flavonoids, including apigenin, kaempferol, and quercetin. These substances are believed to provide a number of health advantages, including lowering the risk of some chronic diseases and reducing inflammation. According to traditional belief, in Indochina, the local abbot usually recommends the local people take a lot of garlic during the COVID-19 outbreak to promote their health against COVID-19. Due to its efficiency and safety in treating COVID-19 patients, garlic essential oil is suggested as a preventive strategy or supportive therapy during the COVID-19 pandemic, according to Wang et al. [23].
- Ginger

  Ginger is widely ingested in Asian countries. Fresh ginger can be used to season soups, stir-fries, and even to produce a calming ginger tea by steeping it in hot water. Ginger does include flavonoids, a class of plant substances with anti-inflammatory and antioxidant activities. Gingerols, quercetin, and kaempferol are a few of the flavonoids that can be found in ginger. These substances are thought to provide a variety of health advantages, such as lowering inflammation, enhancing heart health, and even assisting in the prevention of some cancers. According to Li et al.'s research, COVID-19 patients who took a ginger supplement spent less time in the hospital overall [24].
- Turmeric

  Curcumin, a flavonoid found in turmeric, has been demonstrated to have antioxidant and anti-inflammatory properties. While studies are still being conducted, some have claimed that curcumin's anti-inflammatory properties may hold promise for the treatment of COVID-19. As earlier noted, curcumin, a substance found in turmeric, has anti-inflammatory and antioxidant characteristics that may benefit the immune system. We can include turmeric in soups, curries, and even golden milk. A recent study by Askari et al. suggested that co-supplementing curcumin and piperine could considerably lessen weakness in outpatients with COVID-19 [25]. The other indices in this trial, including the biochemical and clinical indices, were not significantly impacted by the co-supplementation of curcumin and piperine [25].
- Tea

  Flavonoids, which are organic substances with antioxidant and anti-inflammatory properties, are present in tea. These characteristics might help one strengthen their immune system and lower their risk of contracting specific chronic diseases. There is no evidence, nevertheless, that flavonoids or tea can cure or stop COVID-19. In some regions of Indochina, like Myanmar, tea is a popular beverage and part of the cuisine (Figure 21.1). The

FIGURE 21.1 Local Myanmar cuisine (La Phet Thoke) contains tea that is rich in flavonoids and possibly useful against COVID-19.

locals who routinely drink tea in isolated underdeveloped areas where there is a shortage of protective equipment and vaccine are still not heavily afflicted by serious COVID-19 [26]. As part of a balanced diet, including foods strong in flavonoids and tea can be beneficial, but it's also crucial to follow public health advice and get vaccinated against COVID-19.

It should be noted that there is presently no data to show that any one food or diet will prevent or cure COVID-19; however, a balanced diet can improve overall immunological function.

21.7 CURRENT RESEARCH AND PERSPECTIVES IN SOUTHEAST ASIA

Infection is one of the most frequent diseases and a leading cause of death. The global incidence of antibiotic-resistant bacteria indicates that pharmacological treatment options for infectious diseases are dwindling. Furthermore, the emergence of drug resistance has overtaken the development of new treatments. As a result, new medications or alternative kinds of treatment are required. Herbal supplements are now more popular than ever, thanks to a new trend in alternative treatment. Indochina offers a few advantages in this area due to the country's unique biodiversity and the government's tireless attempts to discover the benefits of natural medicine. Several countries have reported on anti-COVID-19 chemical research employing flavonoid-based products found in local traditional remedies and functional foods. The amount and quality of reports vary depending on the resources available in each country.

A. Vietnam

There have been reports of anti-COVID-19 chemical experiments using natural resources derived from the ocean from a number of different nations. The number and caliber of reports specifically about Vietnam vary. Vietnam, a still-developing country in Indochina, is aggressively investigating natural goods. Due to the abundance of local herbal specimens, worldwide cooperation can help in identifying natural remedies against COVID-19. Studies on marine products and their ability to combat COVID-19 have been conducted in Vietnam. One study found that a number of compounds originating from seaweed and sea cucumber were antiviral against the coronavirus. The creation of the medication Chuan Xin Lian (*Andrographis paniculata*), derived from a traditional herb, is an illustration of Vietnamese achievement. The plant *A. paniculata*, sometimes known as the "King of Bitters," contains flavonoids, a group of plant chemicals with antioxidant properties. A few types of flavonoids present in *A. paniculata* are apigenin, luteolin, and quercetin. These flavonoids have been shown to have a wide range of potential health benefits, from reducing inflammation to boosting immune system performance [27].

B. Malaysia

Another Indochina nation is aggressively investigating the natural products Malaysia has to offer. Due to the large number of local natural product specimens, cooperation with other nations can aid in the identification of natural goods against COVID-19. An excellent example of success is the recent publication on active compounds derived from a local natural product with anti-SARS-CoV-2 action. For example, the native Malaysian population consumes a lot of pea eggplant (*Solanum torvum* Swartz.), also known as turkey berry or "terung pipit" in Malay [28]. The shrub produces turkey berry fruits (TBFs), which resemble peas and are a source of valuable phytochemicals [28]. *S. torvum* polyphenols have good therapeutic promise for COVID-19 treatment, according to Govender et al. [28].

C. Thailand

Thailand encountered problems early in the COVID-19 pandemic, and those problems are still there as of the year 2023. Thailand has a native traditional medical system that is influenced by Ayurveda. There is no doubt that there are reports on natural active components from natural products for the treatment of COVID-19. Due to a lack of qualified experts, the research that is now available is typically preliminary and not very detailed for future reference. The problem with the research methodology used by the Thai researchers is another serious concern.

Tomyam is a hot, sour soup that is frequently found in Thai and Malaysian cuisine, according to the local functional foods menu. Lemongrass, shrimp, chili peppers, lime juice, and occasionally, other vegetables are among the usual ingredients. Tomyam includes a number of components that are known to contain flavonoids. For instance, lemongrass is a good source of flavonoids, and other components like lime and chili peppers also have flavonoids in them. However, there is no proof that tomyam is effective against COVID-19.

D. Myanmar

Myanmar is still a developing country, but local academics are collaborating with countries like China and India to explore naturally occurring active chemicals.

E. Cambodia

Due to the nature of Cambodia being a resource-poor country, there is no international publication on the active ingredient from a natural source for the management of COVID-19.

F. Laos

Laos is a resource-poor nation, hence there are no worldwide publications on the active ingredient from a natural source for the management of COVID-19. Focusing on local cuisine, shredded papaya, tomatoes, carrots, green beans, peanuts, and a hot dressing consisting of fish sauce, lime juice, garlic, and chilies make up the popular Lao salad known as som tam. It is a tasty and nutritious dish that's loaded with fiber, antioxidants, vitamins, and minerals. Flavonoids, a class of plant pigment with several health advantages, are present in papaya. However, there is no proof that papaya or flavonoids can specifically guard against COVID-19.

Many herbal products contain flavonoids in the tropical region of Indochina. Due to limitations in the region's infrastructure and resources, there aren't many articles on natural anti-COVID-19 treatments. Additionally, several of the offered reports are unqualified. The management of COVID-19 in this area can, however, be accomplished by utilizing a widely used alternative medical strategy. Additional research and development are urgently needed to support the widespread adoption of non-verified management of the condition in this area.

21.8 CONCLUSIONS

Natural components, both ancient and modern, are commonly used as sources in traditional treatment. It's amazing to learn how traditional botanicals are utilized to cure medical conditions. Many natural therapies are regarded to be effective in the treatment of a wide range of ailments. Several recent studies have emphasized the benefits of standard therapies for a variety of SARS-CoV-2 infection-related symptoms. This chapter provides an overview of flavonoids active compounds in natural products for traditional medicines and functional foods for a different approach to COVID-19 management, with a focus on their use in Indochina (Vietnam, Malaysia, Thailand, Myanmar, Cambodia, and Laos). A variety of well-known conventional ethnopharmacological regimens have been proposed as sources of bioactive compounds for the treatment of COVID-19. Many naturally occurring flavonoids in Indochina include active components that may be beneficial in the treatment of COVID-19. There have been reports from many Indochina countries on the search for candidate chemicals with anti-coronavirus capabilities, and it was revealed that the quality and quantity of research on this topic varied by country.

REFERENCES

1. Hsia W. Emerging new coronavirus infection in Wuhan, China: Situation in early 2020. *Case Study Case Rep.* 2020;10:8–9.
2. Yasri S, Wiwanitkit V. Editorial: Wuhan coronavirus outbreak and imported case. *Adv Trop Med Pub Health Int.* 2019;9:1–2.
3. Ling CQ. Traditional Chinese medicine is a resource for drug discovery against 2019 novel coronavirus (SARS-CoV-2). *J Integr Med.* 2020;18:87–88.
4. Sriwijitalai W, Wiwanitkit V. Herbs that might be effective for the management of COVID-19: A bioinformatics analysis on anti-tyrosine kinase property. *J Res Med Sci.* 2020;25:44.
5. Singh PK, Rawat P. Evolving herbal formulations in management of COVID-19. *J Ayurveda Integr Med.* 2017;8:207–210.
6. Chawla P, Yadav A, Chawla V. Clinical implications and treatment of COVID-19. *Asian Pac J Trop Med.* 2014;7:169–178.
7. Wu HT, Ji CH, Dai RC, Hei PJ, Liang J, Wu XQ, Li QS, Yang JC, Mao W, Guo Q. J Traditional Chinese medicine treatment for COVID-19: An overview of systematic reviews and meta-analyses. *Integr Med.* 2022;20(5):416–426.
8. Ding X, Fan LL, Zhang SX, Ma XX, Meng PF, Li LP, Huang MY, Guo JL, Zhong PZ, Xu LR. Traditional chinese medicine in treatment of COVID-19 and viral disease: Efficacies and clinical evidence. *Int J Gen Med.* 2022;15:8353–8363.
9. Rahman MM, Bibi S, Rahaman MS, Rahman F, Islam F, Khan MS, Hasan MM, Parvez A, Hossain MA, Maeesa SK, Islam MR, Najda A, Al-Malky HS, Mohamed HRH, AlGwaiz HIM, Awaji AA, Germoush MO, Kensara OA, Abdel-Daim MM, Saeed M, Kamal MA. Natural therapeutics and nutraceuticals for lung diseases: Traditional significance, phytochemistry, and pharmacology. *Biomed Pharmacother.* 2022;150:113041.
10. Essa MM, Bishir M, Bhat A, Chidambaram SB, Al-Balushi B, Hamdan H, Govindarajan N, Freidland RP, Qoronfleh MW. Functional foods and their impact on health. *J Food Sci Technol.* 2023;60(3):820–834.
11. Ricci A, Roviello GN. Exploring the protective effect of food drugs against viral diseases: Interaction of functional food ingredients and SARS-CoV-2, influenza virus, and HSV. *Life (Basel).* 2023;13(2):402.
12. Quintal Martínez JP, Segura Campos MR. Flavonoids as a therapeutical option for the treatment of thrombotic complications associated with COVID-19. *Phytother Res.* 2023;37(3):1092–1114.
13. Wang G, Wang Y, Yao L, Gu W, Zhao S, Shen Z, Lin Z, Liu W, Yan T. Pharmacological activity of quercetin: An updated review. *Evid Based Complement Alternat Med.* 2022;2022:3997190.
14. Periferakis A, Periferakis K, Badarau IA, Petran EM, Popa DC, Caruntu A, Costache RS, Scheau C, Caruntu C, Costache DO. Kaempferol: Antimicrobial properties, sources, clinical, and traditional applications. *Int J Mol Sci.* 2022;23(23):15054.
15. Kim JM, Heo HJ. The roles of catechins in regulation of systemic inflammation. *Food Sci Biotechnol.* 2022;31(8):957–970.
16. Wei Z, Chen J, Zuo F, Guo J, Sun X, Liu D, Liu C. Traditional Chinese medicine has great potential as candidate drugs for lung cancer: A review. *J Ethnopharmacol.* 2023;300:115748.
17. Shi M, Peng B, :Li A, Li Z, Song P, Li J, Xu R, Ning broad anti-viral capacities of Lian-Hua-Qing-Wen capsule and Jin-Hua-Qing-Gan granule and rational use against COVID-19 based on literature mining. *Front Pharmacol.* 2021;12:640782.

18. Cruz J, Trombley J, Carrington L, Cheng X. Properties of the novel Chinese herbal medicine formula Qu Du Qiang Fei I Hao Fang warrant further Research to determine its clinical efficacy in COVID-19 treatment. *Med Acupunct.* 2021;33(1):71–82.
19. Zheng S, Baak JP, Li S, Xiao W, Ren H, Yang H, Gan Y, Wen C. Network pharmacology analysis of the therapeutic mechanisms of the traditional Chinese herbal formula Lian Hua Qing Wen in Corona virus disease 2019 (COVID-19), gives fundamental support to the clinical use of LHQW. *Phytomedicine.* 2020;79:153336.
20. Hu Z, Yang M, Xie C. Efficacy and safety of Lian-Hua Qing-Wen granule for COVID-2019: A protocol for systematic review and meta-analysis. *Medicine (Baltimore).* 2020;99(23):e20203.
21. Rudrapal M, Celik I, Khan J, Ansari MA, Alomary MN, Yadav R, Sharma T, Tallei TE, Pasala PK, Sahoo RK, Khairnar SJ, Bendale AR, Zothantluanga JH, Chetia D, Walode SG. Identification of bioactive molecules from Triphala (Ayurvedic herbal formulation) as potential inhibitors of SARS-CoV-2 main protease (Mpro) through computational investigations. *J King Saud Univ Sci.* 2022;34(3):101826.
22. do Prado FG, Pagnoncelli MGB, de Melo Pereira GV, Karp SG, Soccol CR. Fermented soy products and their potential health benefits: A review. *Microorganisms.* 2022;10(8):1606.
23. Wang Y, Wu Y, Fu P, Zhou H, Guo X, Zhu C, Tu Y, Wang J, Li H, Chen Z. Effect of garlic essential oil in 97 patients hospitalized with covid-19: A multi-center experience. *Pak J Pharm Sci.* 2022;35(4):1077–1082.
24. Li Y, Yang D, Gao X, Ju M, Fang H, Yan Z, Qu H, Zhang Y, Xie L, Weng H, Bai C, Song Y, Sun Z, Geng W, Gao X. Ginger supplement significantly reduced length of hospital stay in individuals with COVID-19. *Nutr Metab (Lond).* 2022;19(1):84.
25. Askari G, Sahebkar A, Soleimani D, Mahdavi A, Rafiee S, Majeed M, Khorvash F, Iraj B, Elyasi M, Rouhani MH, Bagherniya M. The efficacy of curcumin-piperine co-supplementation on clinical symptoms, duration, severity, and inflammatory factors in COVID-19 outpatients: A randomized double-blind, placebo-controlled trial. *Trials.* 2022;23(1):472.
26. Ge J, Song T, Li M, Chen W, Li J, Gong S, Zhao Y, Ma L, Yu H, Li X, Fu K. The medicinal value of tea drinking in the management of COVID-19. *Heliyon.* 2023;9(1):e12968.
27. Nguyen HT, Do VM, Phan TT, Huynh DTN. The potential of ameliorating COVID-19 and sequelae from andrographis paniculata via bioinformatics. *Bioinform Biol Insights.* 2023;17:11779322221149622.

Index

Note: **Bold** page numbers refer to tables and *italic* page numbers refer to figures.

9781032556215